ANALYSIS OF OILSEEDS, FATS AND FATTY FOODS

ANALYSIS OF OILSEEDS, FATS AND FATTY FOODS

Edited by

J. B. ROSSELL

Leatherhead Food Research Association,
Randalls Road, Leatherhead, Surrey KT22 7RY, UK

and

J. L. R. PRITCHARD

Formerly of:
J. Bibby & Sons plc
Liverpool, UK

ELSEVIER APPLIED SCIENCE
LONDON and NEW YORK

ELSEVIER SCIENCE PUBLISHERS LTD
Crown House, Linton Road, Barking, Essex IG11 8JU, England

Sole Distributor in the USA and Canada
ELSEVIER SCIENCE PUBLISHING CO., INC.
655 Avenue of the Americas, New York, NY 10010, USA

WITH 124 TABLES AND 60 ILLUSTRATIONS

© 1991 ELSEVIER SCIENCE PUBLISHERS LTD

British Library Cataloguing in Publication Data

Analysis of oilseeds, fats and fatty foods.
1. Food. Oils
I. Rossell, J. B.   II. Pritchard, J. L. R.
664.3

ISBN 1-85166-614-1

Library of Congress Cataloging-in-Publication Data

Analysis of oilseeds, fats, and fatty foods/edited by J. B. Rossell
and J. L. R. Pritchard.
   p. cm.
Includes bibliographical references and index.
ISBN 1-85166-614-1
1. Oils and fats, Edible—Analysis. 2. Oilseeds—Analysis.
I. Rossell, J. B. II. Pritchard, J. L. R.
TP680.A53 1991   91–9711
664'.3—dc20   CIP

Printed in Northern Ireland by The Universities Press (Belfast) Ltd.

Contents

Foreword

This book is concerned with analysis.

The products covered are essential parts of the human diet and provide essential nutrients. However, they must be of good quality if they are to fulfil their proper role in the human diet. Analysis is the tool to ensure that they are of good quality. As Mr. D. Pocklington says in the first chapter '. . . the use of dependable and validated methods is essential if the required quality in terms of accuracy is to be reached'. He mentions the circumstances in which it is essential to use a nationally or internationally validated method.

Oilseeds are an important source of oils and fats (together with animal and marine sources) and of animal feed. In the second chapter, Mr. J. L. R. Pritchard deals with the analysis and properties of oilseeds. He addresses the main analytical indices specified for trading purposes and reinforces the point made in Chapter 1 about the use of nationally or internationally validated methods. He also describes the properties of common commercial oilseeds and touches on new oilseeds.

In his chapter on oilseed residues, Mr. Pritchard deals not only with their analysis and properties but with government controls, nutritional value, protein denaturation and anti-nutritional factors.

Two major anti-nutritional factors—mycotoxins and glucosinolates—are treated in the next two chapters.

The chapter on mycotoxins classifies them, dealing with the type of food at risk and mentioning regulations relating to mycotoxins in food and foodstuffs designed to protect the consumer. It touches on legislation for animal feeds and feeding-stuffs and treats environmental factors affecting the production of mycotoxins. There is a section on methods for the detection and determination of mycotoxins including the important question of sampling. Finally, the author of this chapter deals with more recent developments in mycotoxin methodology.

The presence of glucosinolates in cruciferous oilseeds has hindered the use of their meals in animal feed. Seeds with low glucosinolate levels have been developed which has brought with it a great deal of interest in glucosinolate analysis. The fifth chapter presents information on the current state of glucosinolate analysis and recommends those methods best for particular purposes including methods for the analysis of individual glucosinolates and for the determination of total glucosinolate content of oilseeds crops and commodities.

Chapter 6 covers the extraction of fats from fatty foods and the determination of fat content, discussing the use of standard methods for the determination of fat content. There is a discussion of standard methods for:

meat and meat products;
milk and dairy products;
oilseeds and residues;
animal feedingstuffs, and
cereal products.

The chapter also deals with non-standard, routine procedures and physical methods for the determination of fat (such as near infra-red reflectance).

Dr. Rossell is concerned with crude vegetable oils and fats in the work which he has contributed. In particular, he discusses analyses for oil quality and oil composition, including fatty acid composition. In his chapter he deals with tests for oil quality (including the identity characteristics of vegetable oils and fats) and sterols and tocopherols.

In Chapter 8, Animal Carcass Fats and Fish Oils, Dr. Enser deals first with animal fats including their composition. He discusses extraction procedures and covers their application to sheep, beef and pig carcases. So far as fish oil is concerned, he deals with extraction procedures, the fatty acid composition of fish oil, its triacylglycerol structure and unsaponifiable matter.

Butters, margarines and other spreads are covered in Chapter 9. The author discusses the nature and properties of milk fats and associated methods of analysis and margarine, including its manufacture and routine analysis.

He finishes with a discussion on reduced and low fat spreads. He also considers their routine examination.

Quality control is an essential tool for oils and fats processors world-wide. In Chapter 10, Mr. Mcginley deals with this subject in great detail. He suggests the components of a quality control system, covering feedstock quality, refining, analysis for feedstock quality, bleaching and testing methods to control quality. He deals with modification processes such as hydrogenation and process control in these procedures. Finally, he discusses the applications of processed oils and fats.

Last, but not least, is sampling. It needs to be carried out properly because the analysis of an unsatisfactory sample is of no use. The late Dr. A. Thomas deals with sampling in the final chapter in which he covers the subject from the taking of the sample increments from a parcel of goods through to the test portion on which the analysis is performed.

The majority of the authors are actively involved in national and international standardisation of methods through the BSI, the AOCS, ISO, IUPAC and FOSFA International Committees. This, together with the international origins of the contributions, broadens the scope and usefulness of the text in support of current trade developments.

DERICK REFFOLD
Chief Executive and Secretary
FOSFA International

List of Contributors

R. K. COLWELL
Laboratory of the Government Chemist, Queens Road, Teddington, Middlesex TW11 OLY, UK

J. K. DAUN
Canadian Grain Commission, Grain Lab. Division, Room 1404–303 Main Street, Winnipeg, Manitoba, Canada R3C 3G8

M. B. ENSER
Department of Meat Animal Science, University of Bristol Verterinary School, Churchill Building, Langford, Bristol BS18 7DY, UK

I. D. LUMLEY
Laboratory of the Government Chemist, Queens Road, Teddington, Middlesex TW11 OLY, UK

L. MCGINLEY
21 Buttermere Avenue, Noctorum, Birkenhead, Merseyside L43 9RH, UK

D. I. MCGREGOR
Agriculture Canada, Research Station, 107 Science Crescent, Saskatoon, Saskatchewan, Canada S7N OX2

W. D. POCKLINGTON
Chairman, IUPAC Commission VI. 3 on Oils, Fats and Derivatives, Laboratory of the Government Chemist, Queens Road, Teddington, Middlesex TW11 OLY, UK

J. L. R. PRITCHARD
*Lingmoor, 7 Hisnams Field, Bishops Cleeve, Cheltenham, Glos.
GL52 4LQ, UK*

J. B. ROSSELL
*Leatherhead Food R.A., Randalls Road, Leatherhead, Surrey
KT22 7RY, UK*

P. M. SCOTT
*Health and Welfare Canada, Health Protection Branch, Bureau of
Chemical Safety, Ottawa, Canada K1A OL2*

THE LATE A. THOMAS
Unimills, Postfach 101509, Dammtorwall 15, 2000 Hamburg 36, FRG

R. A. WILBEY
*Department of Food Science and Technology, University of Reading,
Whiteknights, P.O. Box 226, Reading RG6 2AP, UK*

1

Precision and Accuracy of Analysis; Standardisation of Analytical Methods

W. Dennis Pocklington

Chairman, IUPAC Commission VI.3 on Oils, Fats and Derivatives Laboratory of the Government Chemist, UK

1.1. INTRODUCTION

A surgeon may have the highest of qualifications, he may have at his disposal one of the most well-equipped operating theatres in the country, but if he were to undertake major surgery using an unproved technique, the results could be quite disastrous. This is perhaps a statement of the obvious, yet in some analytical laboratories it is not always fully appreciated that the expertise of analysts, and the quality of equipment and reagents, cannot compensate for any deficiencies in the methods used for analysis.

The quality performance of laboratories is therefore directly related to the use of validated or standardised analytical methods. In this context the terms *validated* and *standardised*, when applied to an analytical method, should be understood to indicate that the performance of the method has been demonstrated to meet a required degree of precision and, where appropriate, its *accuracy* (or rather its *trueness*—see Table 1.1 in which the revised ISO (International Organisation for Standardisation) definitions of accuracy, precision and trueness are given) has been shown to be satisfactory.

The vital role that the analytical method plays in the quality performance of analytical laboratories cannot be over-emphasised. Whilst analyst expertise, quality reagents and equipment all contribute to the quality of results produced by a laboratory, the use of dependable and validated methods is essential if the required quality in terms of accuracy is to be reached.

TABLE 1.1
ISO definitions of accuracy, trueness, etc

Accuracy
   Total displacement of a result from a reference value
   This displacement is due to:
      random errors—related to precision, and
      systematic errors—related to trueness
Trueness
   Can only be measured if a reference value is available
   May be defined as 'the closeness of agreement between the average value
   obtained from a large set of observations and the accepted reference value'
Precision
   As a measure of random error is usually expressed in terms of repeatability[a]
   and reproducibility,[a] these being computed as standard deviations of the set
   of observations
   May be defined as 'the closeness of agreement between independent test
   results obtained under prescribed conditions'
   It is not related to the true value
Bias
   Is a measure of the displacement from a true or reference value; this
   displacement, which is a systematic error, may be attributed to:

      (a) bias of the test method, and
      (b) bias of the laboratory using that test method

   It may be defined as 'the difference between the expectation of the observed
   values or test results and the accepted reference value'

[a] See Table 1.14 for the ISO definitions of these terms.
Reference: ISO 5725 (draft revision) *Accuracy (trueness and precision) of measurement methods and results. Part 1: General principles and definitions.*

Laboratory in-house quality control obviously relies on validated methods. Such methods may have been developed by the laboratory itself or have been published by standards organisations. For many purposes a laboratory may find it satisfactory to use methods it has developed and validated, but when results are to be considered by other laboratories or organisations, the use of a nationally or internationally validated method will normally be essential.

The AOAC (Association of Offical Analytical Chemists) has a well-deserved excellent reputation for publishing in its Journal methods which have been validated by collaborative study, and simultaneously, a full report on the study. The final texts of the methods are subsequently reproduced in the AOAC's book of *Official Methods of Analysis.*[1] This book of methods was first published in 1920 and is constantly being revised and expanded—the last edition (15th) appearing in 1989. Supplements containing newly validated methods are issued each year and incorporated into a new edition of the methods book at about 5-yearly intervals.

One might be justified in assuming that all methods published by ISO, and national standardising bodies such as the British Standards Institution have been validated by collaborative study. However, it was in 1981 that the ISO Technical Committee 34 (Agricultural food products) first drew the attention of its sub-committees to the necessity of ensuring that all methods being considered for adoption as ISO standards were first to be thoroughly validated for their performance in terms of precision before being published as definitive standards. It was also emphasised by this ISO Committee that this necessary validation could only be accomplished by properly conducted collaborative studies, and it may therefore come as a surprise to learn that, until relatively recently, few standards organisations (the AOAC, as indicated above, is a notable exception) stipulated that a method must be validated by collaborative study *before* it could be adopted as a standard.

An ISO TC 34 committee document[2] circulated in 1986 indicated that the number of ISO methods (for the analysis of food and agricultural materials) supplied with precision data is relatively low in the case of general methods, and that very few of the reference methods have such data. The reason for this lack of precision data would appear to be that the methods had not been validated by collaborative study and hence no precision data were available for inclusion in the standard methods.

If we examine the *British Standard 684* (methods of analysis of fats and fatty oils)[3] we find that the situation is only marginally better. Although some information as regards precision will be found in 14 of the 50 or so methods in the Standard, it would appear that in many cases arbitrary repeatability values have been cited, rather than values derived from the statistical evaluation of collaborative study results.

In the Foreword to the 6th Edition (1979) of the *IUPAC Standard Methods for the Analysis of Oils, Fats and Derivatives*[4] it is stated that 'all the methods have been studied in collaborative tests, and adopted only when concordant results have been obtained'. This may well be the case, but unfortunately it has not been the policy, until relatively recently, for this organisation to publish results of the collaborative studies which led to standardisation of the methods. Such a claim does not appear in the Foreword to the 7th Edition (1987)[5] of this work and here again it will be found that the citation of precision characteristics is the exception rather than the rule.

A consequence of this lack of information regarding the precision and accuracy of standard methods is, understandably, that a government department may be inclined to recommended AOAC methods, rather than methods published by its own national standards body (or even international standard methods), to be used in connection with the enforcement of legislation such as labelling directives. This is because, as indicated above, the AOAC has adhered to its policy of publishing precision data and reports on the collaborative studies conducted for the validation of its standard methods, thereby making this data readily accessible for scrutiny by legislative bodies.

It is a cause for considerable concern that perhaps the majority of published standards for oils and fats analysis are enjoying a status to which they may not be entitled, i.e. being viewed as definitive validated methods. Furthermore, even where some reference is made in the text of a standard method to its precision, such information is often of little practical value to the analyst, and may be intelligible only to a statistician. The lack of uniformity in presenting statistical data, as it relates to the precision of the method, certainly warrants the attention of those who are directly responsible for the drafting of standard methods of analysis.

Accordingly it is proposed to consider in this chapter:

(1) what has been achieved in the way of harmonisation in protocols for the organisation of collaborative studies designed for the validation and standardisation of methods of analysis;
(2) the need to make the best possible application of data derived from statistical analysis of collaborative study results;
(3) how precision and other performance data can be effectively drafted for analytical quality control purposes; and
(4) the progress made internationally towards harmonisation in the presentation of statistical parameters in standardised methods.

1.2. HARMONISATION OF COLLABORATIVE STUDY PROTOCOLS

A standards organisation considering for 'adoption' a method that has already been published by another organisation, would normally want to satisfy itself that the method had been validated by a properly organised collaborative study. Furthermore it would require to know whether the precision of the method had been evaluated by an accepted method of statistical analysis of the results obtained during the collaborative study. The considerable variation in the approach by standards organisations to method validation has impeded the adoption process and in certain cases has resulted in the duplication of analytical work. The latter has sometimes taken the form of actually repeating a collaborative study, generally following a much more clearly defined protocol.

Fortunately the need for harmonisation in the approach to collaborative studies of methods has now been recognised by standards organisations and it is satisfying to see the successful outcome of the symposia organised by IUPAC (International Union for Pure and Applied Chemistry) to achieve this harmonisation.

The IUPAC Working Party on the Harmonisation of Collaborative Study Protocols, at its meeting in Boston (USA) during August 1987, put the finishing touches to a document which was subsequently published under the title *IUPAC-1987 Protocol for the Design, Conduct, and Interpretation of Collaborative Studies*.[6] This document provides standards organisations with specific recommendations for meeting certain minimum requirements for collaborative studies. If a collaborative study is to be indicated as complying with these requirements then it must be in conformity with the minimum rules outlined in the document. A summary of the main recommendations will be found in Table 1.2, but three of what may be viewed as the most important recommendations can be mentioned here:

(a) not less than five materials should normally be provided for analysis by participants in the collaborative study;
(b) a required minimum of eight laboratories reporting valid data for each material analysed;
(c) two analyses to be carried out on each material, this replication being produced by blind duplicates or split-levels.

1.2.1 Models for Collaborative Studies

Studies organised over the last 5 years or so have generally been based on one of the five models illustrated in Table 1.3. The first model (uniform-level without 'blind duplicates') has been extensively used but is not now regarded as being very satisfactory for accurately

TABLE 1.2
1987-IUPAC harmonised protocol for the design, conduct, and interpretation
of collaborative studies

Note: Only   the   main   recommendations   agreed   at   the   joint
ISO/IUPAC/AOAC Harmonization Workshop (held in Geneva, May 1987
and confirmed at the meeting of the IUPAC Working Party on the
Harmonization of Collaborative Analytical Studies at Boston (USA) in August
1987) have been reproduced

(1) The results of collaborative studies should be analysed by one-way
     analysis of variance ('material-by-material'), but more complex analyses
     are not precluded
(2) The absolute minimum number of materials to be used in a collaborative
     study is five. However, when a single level specification is involved this
     may be reduced to an absolute minimum of three
(3) The most important objective of a collaborative study should be attaining
     a reliable estimate of reproducibility[a] parameters. To the extent that the
     performance of known (parallel) replicate analyses detracts from this
     objective, a requirement for the use of this type of replicate analysis
     should be discouraged
(4) The best estimate of repeatability[a] parameters is obtained by the
     following procedures (listed approximately in order of desirability):

     (a) the use of a split-level design (single values from the analysis of each
          of two closely related materials);
     (b) the use of both split-levels and blind duplicate analyses in the same
          study;
     (c) the use of blind duplicate analyses;
     (d) the use of known duplicate analyses (two repeat analyses from the
          same test sample), but only when it is not practical to use one of the
          preceding designs;

(5) The minimum number of participating laboratories in a collaborative
     study is eight. Only when it is impossible to obtain this number may the
     study be conducted with less, but with an absolute minimum of five
     laboratories. (Although it is desirable to have more than eight labora-
     tories participating, studies containing more than about fifteen labora-
     tories become unwieldy)
(6) The precision estimates are to be calculated both with no outliers
     removed and with outliers removed, using the Cochran and Grubbs
     outlier tests. The Grubbs tests should be applied only to laboratory
     means, not to individual values of replicated designs. Outlier removal
     should be stopped when more than 22% (i.e. more than two out of nine
     laboratories) would be removed as a result of the sequential application
     of the outlier tests

[a] As defined in ISO 5725-1986—see Table 1.14.

TABLE 1.3
Examples of collaborative study models

Level	Sample code	Sub-level	Replicate analyses	Example of results
(1) 'Uniform-level' study (without 'blind duplicates')				
Low	A	—	2	0·4, 0·5
Medium	B	—	2	5·6, 5·4
High	C	—	2	9·7, 9·7

In this model each of the levels is represented by one sample, and each sample is required to be analysed in duplicate. For each level the analyses must be carried out within a short interval of time

Level	Sample code	Sub-level	Replicate analyses	Example of results
(2) 'Uniform-level' study (with 'blind duplicates')				
Low	A, E	—	1	0·4, 0·6
Medium	B, D	—	1	5·2, 5·6
High	C, F	—	1	9·7, 9·4

In this model two identical samples are provided ('blind coded') for each of the levels and one analysis only of each sample is required. For each level the analyses must be carried out within a short interval of time

Level	Sample code	Sub-level	Replicate analyses	Example of results
(3) 'Split-level' study (without 'blind duplicates')				
Low	M	1	1	10·8
	H	2	1	9·1
Medium	L	1	1	35·5
	G	2	1	38·8
High	J	1	1	61·4
	K	2	1	65·5

In this model the samples for the two sub-levels (batches) of each of the three levels are selected (or prepared) so that the difference between the concentration of the analyte in the two associated sublevels is about two to three times the expected standard deviation (at that concentration level) of the method under study.

Only one sample for each of the sublevels is provided and only one analysis of each test sample is required. For each level the analyses must be carried out within a short interval of time

Level	Sample code	Sub-level	Replicate analyses	Example of results
(4) 'Double split-level' study (with 'blind duplicates')				
Low	M, P	1	1	10·8, 11·1
	H, S	2	1	9·1, 8·9
Medium	L, Q	1	1	35·5, 35·9
	G, T	2	1	38·8, 38·3
High	E, N	1	1	61·4, 61·8
	K, R	2	1	65·5, 64·7

In a 'double split-level' study two identical samples for each sub-level, i.e. four samples for each level, are provided. Thus in a three concentration-level study, as outlined above, participants would receive twelve randomly coded samples in total.

TABLE 1.3—*contd.*

TABLE 1.3—contd.

The samples for the two sub-levels (batches) for each of the levels are selected (or prepared) so that the difference between the concentration of the analyte in the two associated sub-levels is about two to three times the expected standard deviation (at the concentration level) of the method under study. Only one analysis of each test sample is required and the analyses for each level must be carried out within a short interval of time

Level	Sample code	Sub-level	Replicate analyses	Example of results
(5) Multiple analyses 'uniform-level' study (without 'blind duplicates')				
Low	A	—	5	0·4, 0·5, 0·4, 0·6, 0·5
Medium	B	—	5	5·5, 5·5, 5·6, 5·2, 5·4
High	C	—	5	9·8, 9·9, 9·7 9·8, 9·7

In this model one sample is provided for each level and a specified number (>2) of analyses of each test sample is required (to be carried out within a short interval of time for each level).

In the version of this model recommended by the German Federal Office of Health (BGA) not less than five replicate analyses are required of each test sample for each of the concentration levels represented. The participant is required to apply outlier tests to each set of five results and, if any outliers are found, to carry out a further three analyses on that sample.

Note: See ISO 5725-1986 for a full explanation of the basis for these models and the statistical analysis of results obtained when using them, except in the case of the 'double split-level' model which is fully described in the NNI standard NEN 603 (April 1988).

assessing the precision of the method. The model can, however, be considered quite useful for a preliminary assessment of a method, which may be carried out by a limited number of laboratories (e.g. two, or possibly three). The results obtained from such a preliminary study may then allow a conclusion to be reached as to whether or not the method warrants a fully comprehensive study.

The main criticism of this first model arises from the fact that when analysts are required to submit duplicate results for the analysis of a single sample, analyst bias will often result in the second result submitted being influenced by the first result obtained.

The second model (uniform-level with 'blind duplicates') represents an improvement over the previous model, inasmuch as the analyst is required to carry out only one determination on each of the samples provided. Since the samples are coded the analyst should be unaware that he may be carrying out analyses on identical samples. For this model it is essential that clear instructions are given as to the order in which the samples are to be analysed, in order to prevent a pair of identical samples (although not known to be such) being analysed in immediate succession. Thus the analyst is not alerted to the fact that the samples are actually duplicates. A weakness in this model is that the astute analyst may deduce, after having obtained a number of results, which of the samples are 'blind duplicates' despite the fact that the samples have been randomly coded. Having identified these, the analyst may then be tempted to carry out further determinations (despite specific instructions to the contrary) in those cases where it is assumed that poor duplicate results have been obtained, thereby again introducing an element of analyst bias which the inclusion of 'blind duplicates' was designed to prevent.

To overcome this possibility the 'split-level' model was introduced.[7] In this model two different samples are provided for each analyte concentration level, the concentration of analyte in the two samples differing by some two to three times the expected repeatability standard deviation.

However, the split-level model is not without its critics. In many cases the collaborative study co-ordinator may experience difficulty in preparing samples with the required difference in analyte concentration. Obviously it is quite simple to prepare such samples if the method for study is concerned with the determination of the alcohol content of wine, since the adjustment of the level of alcohol in the wine can be readily made by the addition of a known amount of alcohol (or water). Ensuring that the adjusted sample of wine was homogeneous would not present any difficulty. But with sample matrices of the nature of oils or fats the preparation of split-level samples may be far from simple. It will be appreciated that the preparation of samples of vegetable oils with small, but significantly different levels of tocopherols or sterols as well as making sure that the prepared samples are completely homogeneous, may be quite difficult, if not impossible.

The split-level model has also been criticised on the grounds that it is not possible to calculate the repeatability standard deviation (S_r) from the submitted results without introducing an element of intra-laboratory variance, with the consequence that the estimate of S_r is not as good as that which can be obtained from duplicate results carried out on the same sample. To overcome the latter criticism the Netherlands Normalisation Institute (NNI) introduced what has come to be known as the 'double split-level' model.[8] In this model two identical samples (blind coded) are provided for each of the sublevels.

Some very successful collaborative studies have been organised using this model—one example is the joint ISO/IUPAC study on a method for the determination of iron, copper and nickel in oils and fats by carbon furnace atomic absorption spectrometry.[9] It will be interesting to see whether this model increases in popularity, especially in view of the fact that some critics consider that requiring laboratories to undertake the analysis of twelve samples is too demanding.

The fifth model illustrated in Table 1.3 is the one advocated by the German Federal Office of Health and involves the multiple analysis of each sample provided. Results from up to five replicate determinations on each sample are normally required, the analyst being then required to check if any of the five results is a statistical outlier, and where such occurs to carry out a further three determinations. The main criticism of this model derives from the claim that many analysts will work towards submitting results which are as close together as possible, thereby introducing a replication bias which would inevitably lead to an incorrect estimate of the precision of the method being made.

However, the protocols adopted for the development (by collaborative study) of certified reference materials by the Community Bureau of Reference (BCR)[10] require the multiple analysis of samples (generally not less than five replicate analyses are required to be carried out by each laboratory involved in the certification exercise). But in the BCR studies the objective is not to assess the precision of analytical methods but to determine the concentration level of an analyte to the greatest degree of accuracy possible. During the certification exercise an endeavour is made to obtain, whenever possible, results using more than one method in order to prevent the certified values being 'method dependent'.

The IUPAC-1987 Harmonized Protocol lists in 'approximate order of desirability' the split-level model, a model based on a combination of blind replicates and split-level, and a blind replicates model; a model requiring known replicates is listed last with the caution that such is not to be used unless it is impractical to use one of the foregoing designs. In view of the fact that the split-level model is recommended as the model of choice, it is perhaps somewhat incongruous that the future revision of ISO 5725 is to designate the split-level model as merely an 'alternative' to that of the uniform-level model.

1.3. ASSESSMENT OF THE ACCURACY OF A METHOD

International collaborative studies of methods have generally been confined to the determination of precision only. The reason for this would appear to be that it is not always feasible to assess the accuracy (trueness) of a method. This is especially the case with empirical methods, where the end result is method-dependent and is not an entity which can be defined in absolute terms. An example that could be cited here is the method for the determination of the dilatation of fats, the dilatation being the isothermal expansion, due to change of state from solid to liquid, of a fat which has been solidified under precisely prescribed conditions.

For trace element analysis, it is essential that the text of the method gives not only an indication as to the precision of the procedure, but also information regarding the accuracy (trueness), the limit of detection, and the sensitivity in respect of the analyte to be determined. Furthermore it is essential to state whether the procedure is subject to interference from materials likely to be present in the samples to which the method of analysis may be applied. These requirements could only be met if they were taken into account when the protocol of the collaborative study designed to study the method was drawn up. It would be necessary to provide 'spiked' samples (i.e. samples with known amounts of added analytes), as well as samples containing substances which could be a possible source of interference, in order to establish the accuracy of the method in the presence of interfering materials. In some samples the levels of analyte for determination would need to be close to the expected limit of detection.

In methods published in the United Kingdom by the Standing Committee of Analysts of the Department of the Environment/Water Research Council (DoE/WRC) these important factors are taken into account—for example, see the DoE/WRC method for Arsenic in Potable Waters by AAS (atomic absorption spectrometry).[11] The section on the Performance Characteristics of the Method for this procedure cites the *limit of detection* for arsenic. As to the method's *sensitivity,* the user of the method is informed that under the conditions of the procedure, arsenic, at a level of 2·0 μg/litre, gives an absorbance of approximately 0·10. Interferences, such as copper, silver and selenium are listed in another section of the method, together with an indication of the levels above which these elements, when present, may interfere with the determination of arsenic.

One of the major difficulties that is frequently encountered when attempting to establish the accuracy of a method by collaborative study is the provision of samples in which the concentration of the analytes, required to be determined, can be guaranteed to be stable during the course of the collaborative study; this is particulary true with labile compounds such as tocopherols. The organisation of collaborative studies on methods for microbiological assays presents problems of a kind not normally encountered with chemical methods and for this reason the drafting of guidelines for such studies is now being undertaken by the AOAC. The inherent difficulties in providing samples which will be microbiologically stable will be immediately obvious, and this means in effect that it may be practically impossible for all participants in the study to be provided with samples of identical composition.

It is of interest to read in the report published in *Pure and Applied Chemistry* of the collaborative study of the IUPAC method for butyric acid[12] that one sample with a known amount of butyric acid was provided for analysis in the study, in order to determine both the *trueness* and *precision* of the method. The results reported for this sample were quite remarkable, inasmuch as the mean values obtained by some 13 laboratories, using two different methods, were 1·79% and 1·82% (respectively) for the sample which had been prepared with a level of 1·82% butyric acid (equivalent to approximately 50% butterfat). To check both the precision and trueness at a level of butyric acid equivalent to 5% butterfat a sample of the latter, suitably blended with tallow was also provided. The results obtained indicated that the trueness (as well as the precision) was acceptable at this lower level. It is perhaps unfortunate that some reference to the trueness of the procedure has not been included in the text of the method (which was published simultaneously with the report on the collaborative study) so that the analyst is made aware of the trueness of the method without having to refer to the report on the collaborative study of the method.

Under the heading 'Analysis—accuracy and precision'? a report in *Chemistry in Britain,*[13] which generated a considerable amount of subsequent correspondence, discussed the outcome of a survey conducted by the UK Ministry of Agriculture, Fisheries and Food (MAFF) in which forty UK laboratories took part in determining the lead and cadmium levels in a number of foodstuffs ranging from cabbage to liver. This report 'shows that all is not well' with the quality of analysis being performed by many UK analytical laboratories and the survey had demonstrated that whilst the *precision* of analysis exhibited by individual laboratories was in the main acceptable, the wide disparity in results between laboratories focussed attention on the fact that the *trueness/accuracy* of the results obtained was far from satisfactory.

For example, the report stated that in the case of the results for lead in the cabbage sample 'the spread of results [was] remarkable, covering one and a half orders of magnitude' and, furthermore, that 'unfortunately these results [were] typical of those received during this survey'. When such disparities in results between laboratories arise (assuming that they are analysing identical laboratory samples—in the MAFF survey particular attention had been paid to ensuring that participants received identical material), it has to be assumed that the problem originates from either the inaccuracy of the methods employed or the competence of the analysts using those methods (or both) and, where instrumental techniques are employed, from the use of inferior or poorly calibrated equipment. If standardised validated methods had been used by all laboratories, considerably better agreement between the results could have been expected, but in the MAFF study participating laboratories were free to use their own 'method of choice'. When reviewing the results obtained in the MAFF survey referred to above, it is important to bear this in mind and also that the objective of the survey (as pointed out by the authors of the *Chemistry in Britain* report in a letter published in a later issue of the journal[14]) 'was to assess the accuracy of the data being produced for lead and cadmium levels in foodstuffs in the UK, by analysts known . . . to publish data in this field or to offer a commercial analytical service'. The authors, in answer to criticisms regarding the way in which the study was organised, emphasised that 'the study was not a collaborative trial as defined by official bodies such as the Association of Official Analytical Chemists (AOAC), nor was it an assessment of standardised methods'. What the report clearly showed was the fact that there was a need for standardised methods of acceptable precision and trueness for the determination of lead and cadmium in foodstuffs, and that the procedures laid down in those methods must be meticulously followed by the analyst if satisfactory results are to be obtained.

Referring again to the IUPAC method for butyric acid (which is based on the gas–liquid chromatography of free butryic acid) it is of interest to note that in the report published in *Pure and Applied Chemistry*[12] on the collaborative study, which resulted in the standardisation of the IUPAC method for the determination of butyric acid, it is stated that one of the butterfat samples used in the study was a reference material being developed by the Community Bureau of Reference (BCR). The results obtained in the BCR certification studies indicated a very close correlation between the results obtained for butyric acid in the butterfat sample using three methods which determine this acid as the free acid (mean values of 3·46, 3·47 and 3·52) and the results for butyric acid when chromatographed as the methyl ester (by GLC) and the phenethyl or p-bromophenacyl esters (by HPLC), i.e. mean values of 3·47, 3·50 and 3·48 respectively.

Although the trueness of the IUPAC method has been questioned by some analysts, the BCR studies would appear to confirm that there is little justification for claiming that any one of the methods referred to above yields the best results in terms of trueness. But it should be noted that, as indicated above, the IUPAC free butyric acid method was, during the course of the collaborative study, demonstrated to be capable of very high accuracy at a level of butyric acid of 1·8%. This level corresponds to the level of butyric acid which, on average, would be expected to be found in a 50:50 blend of butterfat with a non-butyric fat. Since butyric acid is frequently used as a criterion for estimating the butterfat content of a product, the accuracy (i.e. both trueness and precision) of the method chosen for the determination of butyric acid is of particular importance. This is emphasised by the fact that, since 100% butterfat contains only some 3·5% butyric acid, an under- or over-estimation of about 0·3% butyric acid could lead to an error approaching 10% for the butterfat content of a product.

The use of certified reference materials can be invaluable when determining the accuracy or trueness of a method, in addition to its precision. Unfortunately the number of biological materials that are available as certified reference standards is somewhat limited, although the BCR has begun to offer reference materials of oils and fats—details of these will be found in Table 1.4. A section on the use of reference materials in determining the trueness of a method forms

TABLE 1.4
BCR oils and fats reference materials

Already available:
Soya/maize oil with certified values for fatty acid profile and indicative values (see Note below) for iodine value and tocopherols
Lard/tallow with certified values for fatty acid profile and indicative values for minor fatty acids
Anhydrous milk fat with certified values for butyric acid content and fatty acid profile; indicative values for triglycerides profile, and vanillin content
In course of development:
Certified values for the sterol profile and sterol content of the soya/maize and anhydrous milk fat, and cholesterol content of the lard/tallow reference materials
Other reference materials proposed for future development:
Cocoa butter, coconut oil, groundnut oil, hydrogenated fish oil, olive oil and rapeseed oil—these materials will be available either individually or in the form of blends

Note: Indicative values are those for which insufficient data are available to provide certified values.

TABLE 1.5
Revision of ISO 5725-1986: accuracy (trueness and precision) of measurement methods and results

Part 1: General Principles and Definitions
Part 2: A Basic Method[a] for the Determination of the Repeatability and Reproducibility of a Standard Measurement Method
Part 3: Measures of Precision intermediate between Repeatability and Reproducibility
    Section 1: Calibration and recalibration
    Section 2: Changes within a laboratory
Part 4: A Basic Method for Determination of the Trueness of a Test Method[b]
Part 5: Test Methods Alternative to the Basic Methods for Determining Trueness and Precision
Part 6: Practical Applications of Trueness and Precision Measures
    Section 1: Introduction
    Section 2: Calculating repeatability and reproducibility limits and other limits
    Section 3: Acceptability of results
    Section 4: Stability of results within laboratories
    Section 5: Assessment of laboratories
    Section 6: Comparison of test methods

[a] The basic method described is that applicable to results obtained from a uniform-level collaborative study.
[b] Includes the use of certified reference materials.

part of the 'long-term' revision of ISO 5725—at the time of writing, part of this revision is expected to be available in 1990; a summary of the contents of the revision is given in Table 1.5—see also Section 1.8, below. At the 3rd International Symposium on Biological Reference Materials (sponsored by IUPAC and the University of Bayreuth and held at Bayreuth, FRG in May 1988), the author presented a paper[15] outlining the availability and application of the BCR certified edible oils and fats reference materials.

1.4. STATISTICAL ANALYSIS OF COLLABORATIVE STUDY RESULTS

The time and resources spent on organising and completing a collaborative study of a method would be wasted if the results obtained from the study were not correctly evaluated by an appropriate statistical method; the same would be true if, when drafting the text of a standardised method, adequate use was not made of the precision parameters determined by a statistical analysis of the results. The 1987-IUPAC Harmonized Protocol outlines an internationally agreed approach to the statistical analysis, stating that 'one way analysis of variance must be applied on a material-by-material basis to estimate the repeatability and reproducibility parameters'. Also outlined in the Protocol is a system of outlier treatment using the Cochran and Grubbs tests and it would appear appropriate here to say something about the nature of these outlier tests, especially as the Grubbs test is not too well known.

1.4.1 Outlier Tests

In any set of collaborative results there are normally one or more results which do not appear to belong to the overall population of results, i.e. they appear to fall outside a normal distribution and statistically are referred to as 'outliers'. Numerous tests have been devised over the years for ascertaining whether results which appear to be somewhat removed from the overall mean of results are truly outliers in statistical terms and therefore should be omitted when the precision estimates of the performance of the method in terms of repeatability and reproducibility are calculated.

Cochran's test, which is quite simple to apply, is designed to identify pairs of test results (obtained from analyses carried out in duplicate under repeatability conditions) which display significantly outlying large differences. The squared difference between a pair of results from duplicate analyses is divided by the sum of the squared differences between the pairs of results obtained by all laboratories, and the resulting quotient compared with the critical values given in tables.

Dixon's outlier test is designed to identify any significantly outlying result (either too low or too high) of an individual laboratory, as compared with the results of all laboratories, and is also relatively simple to apply. Both of these tests, and their application, are fully described in ISO 5725; however Dixon's test has been recommended for replacement by Grubbs' test.[16] The latter has not achieved the popularity enjoyed by Dixon's test, probably because it requires the calculation of standard deviations, but since calculators with standard deviation facilities are now more readily available the major criticism of Grubbs' test is no longer valid. Moreover Grubbs' outlier test takes into account all results whereas Dixon's test simply involves examination of only the results which lie at the extreme ends of the range of results; furthermore Dixon's test is susceptible to the effect of masking, i.e. two results at the extreme end of the range of results may prevent either result being identified as an outlier.

ISO 5725 outlines the procedures involved in statistically determining the precision characteristics of a method and those who are familiar with statistical techniques will have no difficulty in using the ISO standard. Analysts with little or no statistical knowledge may find some sections of the ISO standard difficult to follow, and should this be the case the simplified approach outlined in *Guidelines for the Development of Standard Methods by Collaborative Study*[17] will be found invaluable. This publication provides easy-to-follow calculation schemes applicable to results obtained from any of the five collaborative study models described above. Computer programmes, specially designed for the statistical analysis of collaborative results, have been available for some years now. Programs written in Fortran have been reproduced as an Appendix to BS 5497 in both the 1979 and 1987 editions.[18,19] Other computer programs have recently become available[20] in the form of standard computer disks which contain the programs ready for immediate use. They have the obvious advantage that the somewhat laborious task of transferring the program to a disk using a word processor is not required, and consequently the possibility of errors that inevitably arise when the program is transcribed by word processor are avoided.

But even without computer facilities the use of 'scientific' calculators having standard deviation facilities can considerably reduce the volume of calculations required to be made, thereby keeping to a minimum the possibility of introducing statistical calculation errors. The *Guidelines* publication referred to above comprehensively covers a simplified approach to the rapid calculation of statistical parameters using such calculators.

1.4.2 Application of Statistical Parameters

Statistical parameters which are of the greatest value in assessing the precision of a method are the estimates of the *repeatability/ reproducibility standard deviations* (S_r, S_R). These give an indication of the spread of results experienced when the method was collaboratively studied—see also below under Number of Determinations Clause (1.6.1). In recent years the number of validated standard methods which cite repeatability/reproducibility values (r, R) (or limits) (see Table 1.14 for definitions of repeatability/reproducibility) in the precision clauses of the methods has steadily increased. These values (which are related to a 95% probability criterion) are derived by multiplying the values for S_r, S_R by a factor $2\sqrt{2}$, i.e. 2·83 (now generally rounded to 2·8) (for the derivation of $2\sqrt{2}$ see ISO 5725-1986 Section 5.5). This criterion means that the values for the differences between results obtained from the analysis of a sample would not be expected, in more than 5% (1 in 20) of cases, to be greater than the values for r (for within-laboratory determinations carried out under repeatability conditions), or R (for between-laboratory determinations carried out under reproducibility conditions). Unfortunately the best use of these values in the precision clauses of many published standard methods has not always been made—see Section 1.6.

There has been some discussion, on an international level, regarding the extent to which statistical information can be usefully included in the texts of standard methods. Harmonisation in this area is clearly desirable and if the IUPAC-1987 recommendations are followed we can expect to see in many future published standards a table of statistical parameters similar to the example given in Table 1.6. However, reproducing this table in a validated standard method will only be of value to the analyst if the precision clauses in the text of the method clearly indicate how the statistical parameters are to be interpreted. The role of precision clauses in standardized methods and

TABLE 1.6
Format for a statistical report in the text of a standard method

Results of interlaboratory tests
[*x*] interlaboratory tests, carried out at the international level in [year(s)] by [organisation] in which [*y* and *z*] laboratories participated, each performing [*n*] analyses of each test sample, gave the statistical results (evaluated in accordance with ISO 5725-1986) summarised in the following table:

TABLE

Results expressed in [   ]

Sample	[Description of samples][a]
Number of laboratories	..
Number of results	..
Number of laboratories retained after eliminating outliers	..
Number of outliers (laboratories)	..
Number of accepted results	..
True, or accepted value	..
Mean	..
Repeatability standard deviation (S_r)	..
Repeatability relative standard deviation	..
Repeatability limit (r) [$2\cdot83 \times S_r$]	..
Reproducibility standard deviation (S_R)	..
Reproducibility relative standard deviation	..
Reproducibility limit (R) [$2\cdot83$] $\times S_R$]	..

[a] Data for each sample tabulated with the mean values in increasing order of magnitude.

the most practical way of drafting them will therefore be considered in some detail below (Section 1.6).

1.5. ACCEPTANCE CRITERIA FOR ADOPTION OF VALIDATED METHODS

The adoption of methods by standards organisations would appear to be in many cases very much on an *ad hoc* basis—before recommending adoption some organisations have not always first established whether the method has been adequately studied and its precision shown to meet a required standard. In the absence of guidelines it has often been difficult for standards organisations to decide whether a method warrants adoption or not. A paper by Horwitz[21] reported on the average reproducibility standard deviation coefficient of variation values calculated for a considerable number of AOAC methods that had been collaboratively studied during the past decade or so, and established a mathematical relationship between these values and the concentration of the analyte. This would appear to provide a basis for deciding whether a particular method has the degree of 'ruggedness' expected of a standard method and the application of this criterion is outlined in Table 1.7. Important questions which need to be addressed by a committee responsible for the adoption of collaboratively studied methods will be found in Table 1.8. When considering a method for adoption, the committee may learn that, in the development of the method by collaborative study, the type of samples analysed did not fully cover the range of analyte concentrations and matrices for which the method would likely be used (cf. question 3 in Table 1.8). In such cases the committee will have to decide if a further collaborative study (which takes these factors into account) should be carried out, and the results of the study evaluated, before the method is finally adopted.

Following the adoption of the validated method by the standardising body, the final (and probably one of the most important tasks), is to

TABLE 1.7

Application of coefficient of variation[a] criterion

As a general guide, the value for the reproducibility relative standard deviation (RSD_R) (as determined by the statistical analysis of the collaborative study results) should not significantly exceed the value (indicated in the table below) which corresponds to the order of analyte concentration, as determined by the method subjected to collaborative study.

Analyte concentration	Average values found for RSD_R[b]
1 ppb (10^{-9})	45%
0·01 ppm (10^{-8})	32%
0·1 ppm (10^{-7})	23%
1 ppm (10^{-6})	16%
0·001% (10^{-5})	11%
0·01% (10^{-4})	8%
0·1% (10^{-3})	5·6%
1% (10^{-2})	4%
10% (10^{-1})	2·8%
100%	2%

[a] Relative standard deviation.
[b] Calculated from: $RSD\ (\%) = 2^{(1-0\cdot5\log C)}$, where C is the concentration expressed as powers of 10.

TABLE 1.8
Requirements to be met by a method of analysis prior to adoption as a standard method

(1) Does the method represent an improvement over existing methods in terms of cost, time required for the determinations, etc.? Is the procedure likely to be replaced by an improved method in the near future?

(2) Does the precision of the method represent a significant improvement over any existing standard? Is it intended to be used as a rapid (routine) method or a reference method?

(3) Did the recommended minimum number of laboratories (8) participate in the collaborative study? Did the type and number of samples analysed cover the complete range of analyte concentrations and matrices for which the method would possibly be used?

(4) Does the statistical analysis of the results indicate that the number of outliers exceeds 20% of the total number of participating laboratories?

(5) To what degree of accuracy is it necessary to determine the analyte concentration and does the precision and trueness of the method, as demonstrated in the collaborative study, indicate that the required degree of accuracy could be achieved?

(6) In the case of a method proposed as an international standard, was an adequate number of different countries represented in the study?

ensure that the text of the method is drafted in a format which the user of the method will find the most practical.

1.6. DRAFTING OF THE TEXTS OF STANDARDISED METHODS

An outline of the ISO format for the text of a standard method is reproduced in Table 1.9. This format is based on the recommendations outlined in the ISO Standard 78/2.[22] It is inappropriate to consider here in any detail the drafting of standard methods but three aspects merit particular attention: the *number of determinations clause, expression of results clause* and the *precision clauses*. It should be mentioned that there is little harmonisation at present in the way that these clauses are drafted but the need for harmonisation in their presentation was raised at the May 1987 joint IUPAC/ISO/AOAC Harmonisation Workshop held in Geneva. An outcome of that meeting was the circulation in 1988 of a questionnaire to organisations involved in the publishing of standard methods. On the basis of the

TABLE 1.9
Format of a standard method

1. Scope
   This states briefly what the method determines, e.g. 'This Standard describes a method for the determination of free volatile hydrocarbons'

2. Definition
   Here a precise definition of what is to be understood by the analyte or parameter referred to in the Scope as being determined by the method, e.g. 'The free volatile hydrocarbons content is the quantity of volatile hydrocarbons determined by the method specified, expressed as hexane in milligrams per kilogram (ppm)'

3. Field of application
   This section indicates the type of material to which the method is applicable, e.g. 'The method is applicable to animal and vegetable oils and fats containing volatile solvent residues resulting from the extraction of the oils and fats by means of hydrocarbon based solvents'

4. Principle
   The principle outlines the basic steps involved in the procedure, e.g. 'Desorption of volatile hydrocarbons by heating the sample at 80°C in a closed vessel after the addition of an internal standard, followed by determination of the hydrocarbons in the headspace using packed or capillary column gas–liquid chromatography'

5. Apparatus
   Specific apparatus required for the determination is listed but not that which is to be found in a reasonably well-equipped laboratory

6. Reagents
   Analytical grade reagents are specified where this is considered desirable; it may also be necessary to specify the quality of water, e.g. distilled/deionised that should be used

7. Sampling
   A reference to a sampling procedure is cited here together with recommendations as to the storage of the laboratory sample

8. Procedure
   This section is divided into numbered paragraphs or sub-clauses for the sake of clarity and to allow reference to be made to certain steps in the method at earlier or later stages of the procedure. It should not include a 'number of determinations' clause

9. Expression of results
   This section indicates how the final results are calculated and the units in which the results are to be expressed

10. Notes
    In certain cases it may be preferable to incorporate these in the main text of the method

TABLE 1.9—*contd.*

Annexe
  This will include specific information on analytical quality control, i.e. precision clauses citing repeatability and reproducibility limits, a statistical analysis report on the precision of the method as determined by collaborative study, any necessary figures of apparatus, tables required for calculation of results, etc.

References
  A reference to the report on the collaborative study which led to the standardisation of the method must be included

responses to this questionnaire a number of recommendations were drafted and presented for discussion at an International Workshop on the Harmonization of the Adoption and Presentation of Methods Standardised by Collaborative Study, held in Washington (USA) during April 1989. The objectives before this Workshop were to draft internationally agreed protocols which, if followed by standards organisations, would ensure that: (1) all methods to be published as standards would have met the minimum requirements of specified performance characteristics, and (2) the texts of standard methods, when published, would incorporate practical information regarding the precision and other performance characteristics of the methods. Subsequent to the discussions at the Workshop, protocols were drafted which outline methods adoption criteria and the incorporation of quality control information in standard methods. The finalised protocols have now been published in *Pure and Applied Chemistry*.[23]

1.6.1 Number of Determinations Clause

In a clause which specifies the number of determinations to be carried out (meaning the number of test results to be obtained on each test sample), some standard methods instruct the analyst to perform two analyses on the same test samples. This instruction often takes the following form: 'Carry out two determinations in rapid succession [or simultaneously], using a fresh test portion for each determination'. An instruction to obtain two test results for each sample would apply in particular when just one type of sample material is being analysed in isolation and an indication of the precision of the analyses is required for quality assurance purposes. When several materials of only slightly differing composition are being analysed it may be viewed uneconomic to perform duplicate analyses on every test sample, the assumption being that a check on the precision of the analysis will be given by the test results themselves. However, such an assumption may not be technically justified, and the carrying out of a single analysis only on each test sample in a routine analysis context is a risk that must be carefully measured against quality control considerations. But it would be impracticable for standard methods to take into account all the various situations likely to be encountered in laboratories regarding the most appropriate number of test results that should be obtained from the analysis of each test sample. For this reason the IUPAC-1989 Protocols recommend that the text of a standard method should not include a number of determinations clause. Nevertheless information on analytical quality control should be considered an essential part of a compendium of standard methods and such information should outline the principles on which quality assurance procedures are based.

When an analyst wishes to check his performance of a standardised method, in terms of precision, it is usually recommended that at least five replicate test results are obtained from the same test sample. The standard deviation of the five or more results thus obtained can then be calculated and compared with the value for the repeatability standard deviation (s_r) reported in the table of statistical data which should be included in a section (appended to the text of the standard method) which provides information on analytical quality control specific to that method. The analyst should aim to achieve a value for the repeatability standard deviation of his results which is better (i.e. less) than the cited value for S_r (which will have been derived by a statistical analysis of all accepted results (i.e. outliers excluded) obtained during a collaborative study of the method).

1.6.2 Expressions of Results Clauses

In this clause, a standard method may instruct the analyst to report as the final result the arithmetic mean of two single test results (obtained under the conditions of repeatability) 'provided the requirements of repeatability are met', and if the requirements of repeatability are not met 'to discard the results and obtain two more test results from a further duplicate analysis of the same test sample'. Such an instruction alone is inadequate since it does not inform the analyst of the action to be taken if the results obtained from two further analyses of the test sample do not provide results which meet the requirements of repeatability.

Under these circumstances, and in absence of any appropriate instruction in the method should this situation arise, at least two courses of action are possible:

(a) to report all four results (drawing attention to the fact that the requirements of repeatability were not met by both pairs of results), and to leave the recipient of the test report to decide what is to be accepted as the final result; or

(b) to continue repeating the analyses with the objective of obtaining a pair of results whose agreement does meet the requirements of repeatability.

Neither of these solutions to the problem of poor duplicate results can be considered satisfactory. In the former case the recipient of the report may not be qualified to make a sound judgement as to what should be considered as the final result, and in the latter case there is no certainty that a pair of results which meet the requirements of repeatability will be obtained; furthermore, in the event of a pair of results eventually being obtained which met the requirements of repeatability, this in itself would not necessarily mean that an accurate analysis had been achieved.

Accordingly, when it is found that the results of an analysis do not meet the requirements of the stated repeatability, it is obvious that the analyst should examine his technique and ascertain whether he is strictly adhering to the method (for example, in terms of the reagents being used and the prescribed procedure), to check that the test sample has been properly homogenised, and also to confirm that the method is applicable to the test material. If, as a result of this examination, no factors are discovered that could have contributed to the poor repeatability of the results, a decision may have to be taken as to whether the determinations should be repeated by another competent analyst.

It will be observed that in standard methods of analysis which contain instructions for dealing with poor duplicates, these instructions take quite different forms. Such differences may even be found in methods published by the same standards organisation. Table 1.10 compares the instructions found in the ISO standard methods for the determination of the content of oil, admixture and moisture in oilseeds. The first and third examples, although requiring the final

TABLE 1.10

Comparison of instructions in three ISO methods for dealing with unsatisfactory duplicate analysis results

ISO 659-1979 'Oilseeds—determination of hexane extract called "oil content"'
Take as the result the arithmetic mean of the two determinations, provided that the requirement concerning repeatability is satisfied. Otherwise, repeat the determination on two other test portions. If this time the difference still exceeds 0·4 g per 100 g of sample, take as the result the arithmetic mean of the four determinations carried out

ISO 665-1977 'Oilseeds—determination of moisture and volatile matter content'
Take as the result the arithmetic mean of the two determinations, provided that the requirement concerning repeatability is satisfied. Otherwise, repeat the determination on two other test portions. If this time the difference again exceeds 0·2 g per 100 g of sample, take as the result the arithmetic mean of the four determinations carried out, provided that the maximum difference between the individual results does not exceed 0·5 g per 100 g of sample

ISO 658 'Oilseeds—determination of impurities content'
Take as the result the arithmetic mean of the two determinations, if the conditions of repeatability are satisfied. If the difference is greater than the limit indicated in Table 2, obtain two other test portions, analyse one as before and keep the other for a fourth determination if necessary. In this case, take as the result the arithmetic mean of the result obtained from the third analysis and the nearest result obtained from the previous analyses, provided that the difference does not exceed the allowed limit. Failing this, analyse also the fourth test portion and take as the result the mean of the four determinations

result to be derived in different ways, do at least give clear instructions as to how the final result is to be arrived at. In the second example it will be noted that no provision is made for the eventuality when the maximum difference between the individual results does exceed the figure cited. The possibility of having to report as the final result the mean of four test results (as indicated in the first and third examples), whilst expedient, may not be entirely satisfactory—much would depend on how great a difference was experienced with the results obtained from the first two (duplicate) analyses.

A more satisfactory way of dealing with poor duplicate results is outlined in Section 3 of Part 6 of the long-term revision of ISO 5725. This section is concerned with the determination of the acceptability of results and a practical application of one approach outlined in Section

3 has been adopted in the expression of results clause of at least one standard method recently published by IUPAC in *Pure and Applied Chemistry*.[24] An example is reproduced here:

> If test results in duplicate have been obtained, report as the final result the mean of the two test results, provided the requirements for repeatability (9·1)° are met. If the requirements for repeatability are not met, obtain two more test results by a further duplicate analysis of the test sample.
>
> If the range $(x_{max} - x_{min})$ of the four test results obtained is $\leq 1·3 \times r$, report as the final result the mean of the four test results, otherwise report as the final result the median of the four test results, i.e. the mean of the two intermediate test results.
>
> °(this number refers to the precision clause in the text of the method).

This would appear to be a sound approach to solving the problem faced by the analyst when the difference between duplicate results are greater than the repeatability limit (r), but the 1989-IUPAC Protocols recommend that such procedures should not form an integral part of the calculation and expression of results clause in the text of the method but should be included, along with guidance as to the number of determinations to be carried out, in a section of the compendium of methods which outlines the general principles of quality assurance.

1.6.3 Precision Clauses

Table 1.11 reproduces examples of precision clauses in four standard methods for the determination of moisture by the Karl Fischer procedure. Examination of the format of these clauses shows that there is a complete lack of harmonisation in their presentation.

As will be noted from the examples of precision clauses in Table 1.11 the analyst is not always clearly informed as to the precision of the method and how it is to be applied when interpreting his results. A format adopted some years ago by the IDF (International Dairy Federation) for one of its methods (see Table 1.12) is based on a statistical expression of probability (95%) which, although in itself could not be faulted for correctness, nevertheless is of little practical value to the analyst since he will not normally carry out twenty or more analyses of the one test sample!

Recommended formats for both repeatability and reproducibility clauses have been published with the IUPAC-1989 Protocols and examples are reproduced in Table 1.13. The advantage of these

TABLE 1.11
Comparison of precision clauses in standard methods: example: Karl Fischer
determination of moisture

AOAC 28·003/5 (1984)
Method performance
0·03–3·0% ($S_x = 0·0085 - 0·609$, $S_o = 0·004 - 0·062$)

BS 684: 2.1:1976
Reproducibility
The maximum deviation from the mean of results obtained in different
laboratories should not exceed 0·01% water

AOCS Ca 2e-84
Precision
With average moisture levels of 0·52 and 0·57%:
1. Within laboratory coefficient variation is ±2·7%
2. Between laboratory coefficient variation is ±10·7%

IDF 23:1984
Accuracy of the determination
The maximum deviation between duplicate determinations should not
exceed 0·04% water

TABLE 1.12
Examples of precision clause in an IDF method

(IDF Standard 4A: 1982—Cheese and processed cheese: Determination of the
total solids content (reference method[a]))

9.2 Precision
9.2.1 Repeatability
The difference between two single results found on identical test material by
one analyst using the same apparatus within a short time interval shall
exceed 0·10 g of solids per 100 g of product on average not more than once
in 20 cases in the normal and correct operation of the method

9.2.2 Reproducibility
The difference between two single and independent results found by two
operators working in different laboratories on identical test material shall
exceed 0·20 g of solids per 100 g of product on average not more than once
in 20 cases in the normal and correct operation of the method

[a] This method has now been drafted as ISO 5534-1985 and also published as
BS 770: Part 10: 1986.

TABLE 1.13
Proposed format for precision clauses

Repeatability
When the mean value [*m*] of two independent single test results, obtained with the same method on identical test material in the same laboratory by the same operator using the same equipment within short intervals of time, lies within the range of the mean values cited in the Table, the difference between the two test results obtained should not be greater than the repeatability limit (*r*) deduced by linear interpolation from the Table (see also Note below)

Reproducibility limit (value)
When the values of two single test results, obtained with the same method on identical test material in different laboratories with different operators using different equipment, lie within the range of the mean values cited in the Table, the absolute difference between the two test results should not be greater than the reproducibility limit (*R*) deduced by linear interpolation from the Table

Note: When the results of the interlaboratory test make it possible the value of (*r*) or (*R*) can be indicated as a relative value (e.g. as a percentage of the determined mean value), or as an absolute value, or calculated according to a simple formula—see examples in Table 1.16.

formats is that the analyst is clearly informed as to the maximum difference that should be expected to occur between his own duplicated results and also the maximum difference that should be expected to occur between his final result and the final result of another laboratory using the same method for the analysis of the same laboratory sample.

At this point it would be useful to say something about the use of the reproducibility value (*R*) in the precision clause. Unfortunately there has been some confusion as to what is to be understood by the expression 'final result'. The method may clearly instruct the analyst to report as the 'final result' the mean of two test results obtained by duplicate analysis of the test sample (assuming that the conditions of repeatability have been met). However, if no such instruction appears in the method the 'final result' may be that from just one analysis of the test sample, the mean result of two analyses, or the mean of results obtained from replicate analysis of the test sample.

It is evident that it has not always been fully appreciated that, in order to conform to the ISO definition of reproducibility (see Table 1.14), the reproducibility value *R* (as determined by a statistical analysis of collaborative study results), applies only in the case of the difference between single test results obtained by two laboratories from the analysis of identical laboratory samples. If 'final results' which have been derived from two or more analyses are to be compared, then the reproducibility value (R) has to be adjusted appropriately. An outline of how this calculation is made appears in ISO 5725 but a summary of the principles involved in the calculation is given in Table 1.15.

The note in Table 1.13 refers to the possibility of citing actual values in the repeatability and reproducibility clauses, either in absolute or relative terms, and that in certain cases a formula may be provided which would allow the analyst to calculate the appropriate values for r and R. This is to be preferred if (as the note states) 'the results of the interlaboratory test make it possible', since this would not necessitate reference to a statistical results table. Further examples of clauses in which the precision is cited in absolute terms, and also by reference to formulae, are given in Table 1.16.

TABLE 1.14
ISO 5726-1986 Definitions of repeatability and reproducibility

---

Repeatability: Precision under repeatability conditions
Repeatability limit[a]
  The value less than or equal to which the absolute difference between two single observed values or test results obtained under repeatability conditions[b] may be expected to be with a probability of 95%

Reproducibility: Precision under reproducibility conditions
Reproducibility limit[a]
  The value less than or equal to the absolute difference between two single observed values or test results obtained under reproducibility conditions[c] may be expected to be with a probability of 95%

---

[a] Previously termed repeatability/reproducibility 'value'.
[b] Repeatability conditions: conditions where independent test results are obtained with the same method on identical test material in the same laboratory by the same operator using the same equipment within short intervals of time.
[c] Reproducibility conditions: conditions where test results are obtained with the same method on identical test material in different laboratories with different operators using different equipment.

TABLE 1.15
Calculation of reproducibility values

It is extremely unlikely that two laboratories would compare results that have been derived from single test results (i.e. one analysis only having been carried out on identical laboratory samples). However it is both possible, and necessary, to derive from the calculated values (R) for the reproducibility the equivalent values for the 95% confidence limit critical difference $[CrD_{95}]$ which are applicable when the final result reported by a laboratory represents:

(a) the mean of two test results obtained by the duplicate analysis of the test sample under repeatability conditions, or

(b) either of the laboratories (whose results are being compared) has carried out two or more analyses of an identical laboratory sample, and reported as their final result the mean of the tests results from those replicate analyses.

For the general case (i.e. as is specified in many standard methods of analysis), where the final result has been derived from the mean of two test results, the critical difference (95% probability) between the means should be derived from:

$$CrD_{95}(y_1 - y_2) = \sqrt{[R^2 - r^2/2]}$$

where y_1, y_2 are the final results derived from the means of duplicate test results obtained by the first and second laboratories, respectively, and $CrD_{95}(y_1 - y_2)$ is the value below which the absolute difference between the final results, obtained in different laboratories from the analysis of identical laboratory samples, may be expected to lie with a probability of 95%.

When the final results are derived from the means of more than one test result in either laboratory, the critical difference (95% probability) between the means may be derived from:

$$CRD_{95}(y_1 - y_2) = \sqrt{\left[R^2 - r^2\left(1 - \frac{1}{2n_1} - \frac{1}{2n_2}\right)\right]}$$

where n_1, n_2 are the number of test results obtained by the first and second laboratories, respectively

1.7. NAMAS† ACCREDITATION OF METHODS

In order to conform to the NAMAS (UK) requirements any non-standard methods employed by a laboratory must have been validated 'in-house'. Such validation could take the form of establishing the accuracy of the method by comparing results it returns with those

† National Measurement Accreditation Service—formed as a result of merging the National Testing Laboratory Accreditation Scheme (NATLAS) with the British Calibration Service (BCS).

TABLE 1.16
Examples of Precision Clauses Citing Absolute Values

Repeatability value
   The absolute difference between two independent single test results,
   obtained with the same method on identical test material in the same
   laboratory by the same operator using the same equipment within short
   intervals of time, should not be greater than 0·5 mg/kg

Reproducibility value
   The absolute difference between two single test results, obtained with the
   same method on identical test material in different laboratories with
   different operators using different equipment, should not be greater than
   0·8 mg/kg

Formulae[a] for the calculation of values for (r) and (R)

Copper in oil: $r = 0·0102 + 0·1397m$

$R = 0·0085 + 0·3584m$

where m is the corresponding mean value

[a] Reproduced from IUPAC method 2.631—Determination of copper, iron and
nickel by direct graphite furnace atomic absorption spectrometry.

obtained using a standard method or with results obtained by
analysing a reference material. It would be expected that a monitoring
programme would be followed in order to document the performance
of the method. This would involve recording, on a regular basis,
results obtained when analysing a standard material whose matrix is
similar to that of the samples being analysed. For example, if a
laboratory were to use semi-automatic equipment such as the
SOXTEC for the determination of the oil content of oilseeds it would
be incumbent on that laboratory to include in each series of analyses a
standard oilseed. This could be a sample of oilseed whose oil content
had been established accurately by replicate analyses using a standard
method (such as ISO 659-1979) for the determination of oil content. A
very informative and concise account of what is involved in the
NAMAS accreditation of laboratories has been published in
International Analyst.[25]

1.8. REVISION OF ISO 5725

This revision has already been referred to above and it would appear
appropriate, before concluding this chapter, to summarise in some

detail what the revision covers and in particular its value when determining the trueness and precision of a method.

(a) The situation often arises when it is desired to compare the precision of one method with that of another and Section 6 of Part 6 of the Standard (Comparison of Test Methods) outlines how this comparison can be made.
(b) The use of reference materials (referred to in Section 1.3 above) in determining the accuracy/trueness of the method is fully described in Part 4 of the Standard (A Basic Method for Determination of the Trueness of a Test Method).
(c) Assessment of laboratories, with particular reference to the degree of accuracy to which they are obtaining results, is covered in Section 5 of Part 6.
(d) In Part 3, methods for monitoring any changes in the precision performance of laboratories over a period of time, due to changes in staff and/or equipment are considered in some detail.

Undoubtedly the new version of ISO 5725 will prove to be invaluable for assessment of the accuracy of analytical methods and of the performance of laboratories using those methods.

1.9. CONCLUSION

In this chapter reference has been made to the fact that many standard methods are lacking in precision data. That all future methods published as standards will have some indication of their precision is reassuring but the question remains as to what can be done regarding existing standards. It would indeed be a mammoth undertaking to arrange collaborative studies for all of these standards in order to provide data on which precision characteristics could be based. Nevertheless there is available a considerable amount of data in existence for many of the well-established standard methods, data which understandably have been viewed as a possible source of estimating precision.

The AOCS (American Oil Chemists' Society) Smalley programme, which has been conducted since about 1915 with the purpose of 'stimulating the upgrading of analytical work through public acknowledgement of individual excellence', has generated a great deal of analytical data. The programme involves the distribution of a series of check samples of uniform quality and each participant in the programme analyses the sample using AOCS methods and notifies AOCS of the results. A final compilation of results of all participants indicates to a participant whether his/her analysis was accurate, thus providing laboratories with a means for checking the proficiency of their analytical techniques.

Whether these data can be used to provide precision data for the methods themselves is open to question however, since the protocols for the studies were designed to be a test of the proficiency of laboratories rather than for the purposes of determining the precision of the methods.

FOSFA International (Federation of Oils, Seeds and Fats Associations Ltd) likewise runs a programme which was introduced primarily to obtain information on the ability of laboratories to carry out analyses as required under the terms of FOSFA contracts. However, the programme provides valuable information not only on the performance of laboratories but also on the methods used. Again, careful scrutiny of the data, which have been obtained during the course of this programme over many years, will be necessary before it can be determined whether such data can be used to provide precision data for the methods used in the studies.

Reference has been made to the role of quality assurance in analytical laboratories. The view has been expressed that many laboratories are not paying sufficient attention to ensuring that the quality of their results meets the highest standard possible. There also continues to be some confusion among recipients of analytical reports as to the effect of experimental error on the accuracy of analysis. Whilst reproducibility values may be used to give an indication of what are meaningful differences between results reported by different laboratories for the analysis of identical material, this indication is obviously only valid when the laboratories have used the same standard method. When different methods are used, the source of any differences in results may be attributable to the method rather than the expertise of the laboratories. If the trueness (as opposed to the precision) of methods having similar precision characteristics has not been determined experimentally, then it will be impossible to determine which laboratory is providing the most accurate assessment of the analyte concentration.

The 'customer' who is meeting the cost of the analytical work naturally expects that the results he receives will be the most accurate achievable, but unless he is fully cognisant with the meaning of precision and trueness (as defined in the ISO standard) he may be confused when apparent differences in results become issues in litigation. It is therefore vital that laboratories, when submitting analytical reports, clearly inform the recipient of the reports as to how the results should be viewed in light of the accuracy (trueness and precision) of the method used to obtain them. It would be more satisfactory if all laboratories, instead of simply reporting values for the concentration of an analyte without any qualification as to their possible accuracy, also provided in the report an indication of how much these results may differ from the true values for the analyte concentration. But, again, such information can only be given if the laboratory uses methods for which precision (and trueness) data are available.

In the analytical report it may also be advisable to comment on the nature of the procedure. This is especially important when what is actually determined by the procedure is not specifically defined. For example, when analysing an oilseed for 'oil content', what is determined using the hexane extract procedure is not exactly equivalent to that determined by an NMR (Nuclear Magnetic Resonance) procedure, since the former determines all material present in the oilseed which is soluble in hexane (for this reason the standard method ISO 659:1979 includes in its title the phrase 'called "oil content"'), whereas the NMR method determines the hydrogen bonding content (hence the need for careful pre-drying) which is directly related to the oil content and the result obtained using this method may be influenced to some degree by the protein and other components in the seed as is evident by the residual NMR signal given by defatted seed.

For too long, many analysts have been somewhat complacent about the quality (and validity) of their analytical results, and the introduction of laboratory accreditation schemes, for this reason alone, can be viewed as being long overdue. Such schemes will undoubtedly be welcomed by the laboratories that take a pride in the quality of their analytical expertise.

Despite the shortcomings highlighted in this chapter regarding the lack of data in the texts of standard methods as to their performance characteristics, it must be recognised that considerable progress has been made in the past few years in the way in which standard methods have been validated. The same is true for the way in which validated and standardised methods are now being drafted, as well as the availability of information regarding their precision (and trueness). The introduction of the NAMAS scheme in the UK can be expected to make a significant contribution to improving the performance of UK analytical laboratories.

Progress in all these areas continues and there are sufficient grounds now for optimism, so that before the advent of the 21st century we can expect analysts to be as well-equipped with validated, standardised methods (together with full information regarding the accuracy of the procedures), as they will be with sophisticated laboratory equipment. On the assumption that the protocols drafted at the 1989 Workshop (see Section 1.6 above) will be adopted by organisations publishing standard methods, we can be optimistic that the majority of standard methods to be published in the future will have met the specified requirements as to their performance characteristics, and will be accompanied by comprehensive information outlining all the essential aspects of analytical quality control.

To return to the analogy of the hospital and the analytical laboratory (referred to at the beginning of this chapter)—should we find ourselves unfortunately lying on the operating table, drowsy from anaesthetic, hopefully we can lapse into unconsciousness with full assurance in the outcome of the operation. Our confidence can be based on the fact that, not only has the surgeon the highest of qualifications, and that we are in a modern theatre with the most up-to-date equipment, but also on the knowledge that a validated surgical technique will be employed. Likewise, in our analytical laboratories, as our staff apply their expertise, using quality equipment and reagents, we can rest assured that their results of analysis will be completely reliable.

But confidence in our laboratory results will only be justified if we ensure that our analysts are both using validated methods and adhering meticulously to the procedures outlined therein. We do well to continually monitor our own attitude (and likewise that of our staff) to the requirements of analytical quality control.

REFERENCES

1. AOAC, *Official methods of analysis*, 15th edn, Association of Official Analytical Chemists, Inc., Arlington VA, 1989.

2. ISO/TC 34/SC 11, committee document N 836, International Organisation for Standardisation, Geneva, July 1986.
3. BSI, *British Standard methods of analysis of fats and fatty oils*, BS 684, British Standards Institution, London, 1976.
4. *IUPAC Standard methods for the analysis of oils, fats and derivatives*, 6th edn, Pergamon Press, Oxford, 1979.
5. *IUPAC Standard methods for the analysis of oils, fats and derivatives*, 7th edn, Blackwell, Oxford, 1987.
6. IUPAC-1987 Harmonized protocol for the design, conduct, and interpretation of collaborative studies resulting from the IUPAC Workshop on the Harmonization of Collaborative Analytical Studies, Geneva, Switzerland, 4–5 May 1987, prepared by William Horwitz, Food and Drug Administration, Washington, DC, *Pure Appl. Chem.*, **60**(6) (1988) 855–64.
7. Youden, W. J. and Steiner, E. H., *Statistical Manual of the AOAC*, 3rd printing, Association of Official Analytical Chemists, Arlington, VA, 1982.
8. *Vegetable and animal oils and fats—Determination of repeatability and reproducibility of methods of analysis by interlaboratory tests*, Netherlands Normalisation Institute standard NEN 6303, April 1988.
9. IUPAC method 2.631: Determination of copper, iron and nickel by direct graphite furnace atomic absorption spectrometry, *Pure Appl. Chem.*, **60**(6) (1988) 893–900.
10. Community Bureau of Reference (BCR), Commission of the European Communities, rue de la Loi 200, B-1049 Bruxelles, Belgium.
11. *Arsenic in potable waters by atomic absorption spectrophotometry—methods for the examination of waters and associated materials*, Department of the Environment/Water Research Council, Her Majesty's Stationery Office, London, 1982.
12. IUPAC method 2.310: *Determination of butyric acid in fats containing butterfat, Pure Appl. Chem.*, **58**(10) (1986) 1419–28.
13. *Chemistry in Britain* **21**(10) (1985) 1019–21.
14. *Chemistry in Britain*, **22**(7) (1986) 627.
15. Pocklington, W. D., *Fresenius Z. Anal. Chem.*, **322** (1988) 674–8.
16. Grubbs, P. E., *Technometrics*, **11**(1) (1969) 1–21.
17. Pocklington, W. D., *Guidelines for the development of standard methods by collaborative study*, 5th edn, – Laboratory of the Government Chemist, Teddington, Middx, 1990.
18. BSI, British Standard BS 5497: Part 1: *Precision of test methods Part 1. Guide for the determination of repeatability and reproducibility for a standard test method by inter-laboratory tests*; British Standards Institution, London, 1969.
19. BSI, BS 5497: Part 1, 1987.
20. Computer Program for the statistical analysis of results from a uniform-level Collaborative Study, Laboratory of the Government Chemist, Teddington, Middx, 1987.
21. Horwitz, W., *Anal. Chem.*, **54**(1) (1982), 67A–76A.
22. International Standard ISO 78/2, *Layouts for standards–Part 2: Standard*

for chemical analysis, 1st edn 1982-04-01, International Organisation for Standardisation, Geneva, 1982.
23. Pocklington, W. D., *Pure Appl. Chem.*, **62**(1) (1990) 149–162.
24. IUPAC method 2.432: *Determination of tocopherols and tocotrienols in vegetable oils and fats by high performance liquid chromatography, Pure Appl. Chem.*, **60**(6) (1988) 877–92.
25. Mesley, R. J., *International Analyst*, **1**(6) (1987) 15–18.

2

Analysis and Properties of Oilseeds

J. L. R. PRITCHARD

*Lingmoor, 7 Hisnams Field, Bishops Cleeve, Cheltenham,
Glos. GL52 4LQ, UK*

2.1. INTRODUCTION

Most plant seeds contain some oil but the commercial importance of a seed as an oilseed is based on the quantity and composition of the triglyceride oil present. In value terms the oil is the most highly priced constituent. The value of the residual meal, after removal of the oil, largely depends on the level of toxins and anti-nutritional factors. When these are present to the extent that the meal cannot be used as a feedstuff then the value is degraded, usually to that of fertiliser. Value as human or animal feed is largely dependent on the protein and fibre levels and, in a lesser degree, the compositions of amino acids and trace elements assume importance.

Millions of tonnes of commercial oilseeds are grown and traded annually on contracts and/or rules and specifications issued by Trade Associations such as FOSFA International (Federation of Oils, Seeds & Fats Associations Ltd), NSPA (National Soyabean Processors' Association) and NOFOTA (Netherlands Oils, Fats & Oilseeds Trade Association). Residues are handled by separate associations and their analysis and properties are dealt with in Chapter 3.

Oilseed contracts contain clauses covering sampling and analysis and usually define the methods to be used. There is also a warranty assuring good merchantable quality, in which authenticity and damaged seed are important. The main analytical indices specified are oil, moisture and volatile matter, impurities and free fatty acid (FFA) content and the major reference methods for assessment are discussed in this chapter. These methods are issued in the main by FOSFA International, International Organisation for Standardisation (ISO), National Standards Organisations such as the British Standards Institution (BSI), the American Oil Chemists' Society (AOCS), the International Union of Pure and Applied Chemistry (IUPAC), and the International Association of Seed Crushers (IASC).

In addition to reference methods some discussion is also included on instrumental and rapid methods for oil and moisture used in plant breeding and quality assurance. Major topics, i.e. glucosinolates and mycotoxins are the subjects of separate chapters by specialist authors. The importance of using a specified method will become evident when studying the text, as in some cases different methods lead to different results or a wider spread of results. Indeed it is improvement in precision—repeatability and reproducibility—which is the focus of attention at international level within ISO, where the methods for oil content and FFA are under study.

In considering properties of oilseeds, toxic and anti-nutritional factors are listed together with trace element composition. Common commercial oilseeds are described together with details of source and composition. The more important changes in composition which have resulted from seed breeding by the plant geneticist, together with future targets, are also mentioned and referenced. It was apparent at the outset that space would not allow a comprehensive number of oilseeds to be treated in this way. This has been partly counter-balanced by some reference to new oilseeds, so that the reader will, hopefully, appreciate the widespread effort in research and development which is taking place.

2.2. ANALYSIS OF OILSEEDS

2.2.1 Oil

It is perhaps not surprising that oil content plays a major part in assessing product value when oilseeds are traded. It is the amount, the fatty acid composition and quality of the oil which account for its value in world markets. Quality factors which reduce this value, e.g. free fatty acid (FFA), moisture and impurities, admixture, etc., are the subject of allowances, together with factors which influence ease of processing, such as bleached colour, phosphatide level and so on. The meal residue after removal of oil is less important and, with certain exceptions (such as soyabeans and cottonseed), does not make a major contribution to value. Naturally the protein content and amino-acid composition and the presence or otherwise of deleterious substances such as glucosinolates, gossypol, hydrocyanic acid, etc., will influence meal value and be reflected in the oil/meal price differential and the oilseed price. Nevertheless, most international contracts for the sale of oilseeds feature oil content as a basis, with a scale of allowances either side of the contractual figure to buyer or seller.

2.2.1.1 Definition

The hexane extract (or light petroleum extract) is called the 'oil content' and is defined as the whole of the substances extracted under the operating conditions specified and expressed as a percentage by mass of the product as received. On request it may be expressed relative to the dry matter.[1]

The above definition of the International Standards Organisation (ISO) is also adopted in the FOSFA International Official Method,[2] with the additional proviso that the oil content may be expressed as a percentage on the cleaned seed.

The EEC adopts a similar definition but restricts the use of the solvent to hexane only.[3,4] The method of the IASC adopts a similar definition to ISO and FOSFA International.[5] Analyses of oilseeds shipped under contracts based on the Trading Rules of NIOP[6] are generally based on the methods of the AOCS[7] where the definitions are similar to the ISO method, with the restriction of solvent to petroleum ether.

There is an exception here with safflowerseed, where oil content is defined as the substances extracted by petrol ether under the conditions of the NIOP test. Another exception occurs in the trading of corn germ, where the Standard Analytical Methods of the Corn Refiners' Association[8] are used both in the USA and other countries where US subsidiary companies are based. In this instance the definition includes extraction with carbon tetrachloride for 16 h.

The basic premise of the definition is that, provided that the method is sufficiently detailed and followed by the analyst, the figure obtained can be related by seedcrushers to the potential industrial yield of oil from a particular parcel of seed. In other words, the method may arguably be regarded as a factory simulation method, although the extraction must be exhaustive in order to meet the requirements of precision.

A further consequence of the definition is that any changes introduced into the 'operating conditions specified' are likely to meet with stiff opposition from interested trade organisations, particularly if the end result is changed, unless change can be justified on grounds of improved precision. An example of such a change was the introduction and specification of a microgrinder for interstage grinding when ISO 659, 1979 was published. The original intention was to eliminate the tedium of grinding with a mortar and pestle and improve the method, particularly for rapeseed, which has a slow rate of extraction. On the other hand, the extractable matter was increased by at least 0·4% using the microgrinder. Although the ISO standard still retained the option of a mortar and pestle grind with certain provisos, the adoption of the method by FOSFA International did not include this option. This met with considerable criticism from crushers who felt disadvantaged by the change, particularly as no evidence was forthcoming from collaborative ring tests to justify any claim for improved precision. A further example occurred with the inclusion of waxes in the estimation, which will be discussed later under sunflowerseed.

It will be apparent that reported oil content very much depends on the definition and method of analysis, and one cannot foresee a solvent extraction reference method being replaced by an instrumental reference method in the present century. Much research has been carried out on instrumental techniques, as described later, but little productive research has been devoted to the reference method until recently under the sponsorship of FOSFA International. Moreover, the need for revision and research on AOCS methods also has recently been highlighted.[9]

2.2.1.2 Precision

There has been some concern recently in the trade over the reproducibility of results on rapeseed, expressed in a paper by Cooper.[10] This concern gave rise to subsequent discussion in papers by Kershaw[11] and Pritchard.[12] The present ISO method gives a repeatability within a laboratory of 0·4%, whereas the FOSFA method widens this to 0·6%. AOCS method Aa-4-38 for cottonseed lists a repeatability of 0·42% and single determinations in two different laboratories shall not differ by more than 0·73%. The only other data, listed in AOCS method Ai-3-75, concern sunflowerseed (or dehulled kernels) with a repeatability of 1·93% and an interlaboratory difference of 2·19% on single determinations.

Although FOSFA International regularly carries out world-wide ring tests amongst their member analysts, so far publication of results has been on a restricted basis—a policy which could change in the future. Published data on reproducibility between laboratories is, therefore, limited to results from the AOCS Smalley check series.[13] These results are listed in Table 2.1 and refer to the AOCS method, except for safflowerseed and rapeseed where the NIOP method was also used. Reproducibility between laboratories and the associated statistics are discussed in Chapter 1; at this stage, however, it is apparent that sunflowerseed, rapeseed and safflowerseed presented most of the problems. Data for cottonseed and sunflowerseed support the precision figures listed in the AOCS methods.

Consideration of these factors, together with results of internal ring tests led FOSFA International to sponsor research work on the method, culminating in the issue of an amended protocol in 1986. The revised method and the background to it are described below.

2.2.1.3 FOSFA International reference method

2.1.1.3(a) Basic method. The method involves the following stages, of which the first in particular is tailored to meet the requirements of a particular oilseed or contract.

(i) Preparation of the test sample, involving separation of impurities (when contractually required), pre-drying, grinding, dc-linting (cottonseed). Two separate determinations are performed on the same test sample.

(ii) Extraction with refluxed solvent for 4 h in a thimble located in an approved extraction apparatus.

(iii) Solvent removal from the meal, regrinding and further extraction for 2 h.

(iv) Repeat of stage (iii); removal of solvent and impurities from the extract; weighing.

2.2.1.3(b) Revised method. In further development of the reference method it is important to note that the hexane (petroleum ether) extract is a complex mixture of glycerides, free fatty acids, glycerol, colouring matter, natural anti-oxidants, sterols, trace elements, phospholipids and their complexes and waxes. The degree and rate of extraction of the individual components can vary if the particle size of

TABLE 2.1
Oil content (%) AOCS methods—precision data from Smalley checks

	No. of samples	No. of analyses	No. of outliers	Mean	Ranges		
					SR	RSD	R
Cottonseed (a)	10	194	11	14·83–18·58	0·152–0·288	0·82–1·71	0·43–0·82
Soyabeans (b)	10	537	28	18·37–19·93	0·241–0·542	1·39–2·95	0·68–1·53
Sunflowerseed (c)							
clean seed	8	166	7	28·42–43·43	0·478–0·803	0·80–2·30	0·96–2·27
as is	8	166	5	27·52–41·92	0·380–0·810	0·98–2·94	1·08–2·29
Peanuts (d)	7	125	10	45·64–48·15	0·285–0·567	0·60–1·24	0·81–1·60
Safflowerseed (e)	8	87	5	41·31–43·59	0·241–0·817	0·56–1·93	0·68–2·31
Rapeseed							

AOCS Method: (a) Aa 4-38; (b) Aa 2-38; (c) Ba 6-84; (d) Ab 3-49; (e) Ai 2-75 (or NIOP).
SR = reproducibility standard deviation.
RSD = reproducibility relative standard deviation.
R = reproducibility value $(2·83 \times SR)$.
(See Chapter 1 and *J. Assoc. Off. Anal. Chem.*, **71** (1988) 161 for definition of statistical terms.)

the ground seed and the physical conditions during the test are varied. In this context it is perhaps useful to draw a parallel with experience in crushing plants, where the seed pre-treatment, initial grinding/rolling of the seed, composition of the solvent, the extraction temperature and percolation rate of solvent through the seed bed all contribute to the yield of oil. Heat conditioning of the seed prior to extraction can double the amount of phospholipids extracted, for example. The effect of heat is also demonstrated by a slight rise in oil content between the extractor and toaster discharge. On soyabeans this increase is normally assessed at 0·2–0·4% oil (dry basis),[14] whereas personal experience has shown that this does not occur with palm kernels, probably due to the absence of phosphatides. It is also possible to grind the seed too finely as further collapse of seed structure can occur during extraction, producing an impervious layer of very fine particles which can restrict the percolation of solvent through the seed bed. A number of these factors, notably temperature, particle size and percolation through the sample bed are not well detailed in the reference texts. Admittedly their significance in the laboratory requires research, some of which has been carried out and the results introduced into the method, but further work is still required.

2.2.1.3(c) Particle size. For small seeds such as rapeseed, linseed, etc., the procedure was to weigh a test portion of whole seed and grind before extraction. This has the benefit of ensuring the correct hull/meal proportions and removed any significance concerning loss of moisture on grinding. It did, however, have the disadvantage that large sized admixture such as wheat, unevenly distributed in the analysis sample, could have a marked effect on the end result. For example, a single wheat seed in a test portion of 10 g can reduce the oil content from 40% to 39·85%. The AOCS method overcomes this problem by analysis of cleaned seed but most seeds are traded and processed with some admixture present. On balance, therefore, the merits of grinding a large test sample (100 g) favoured its adoption. An ultracentrifugal mill (UCM) was preferred for grinding as the screen defines the maximum particle size. It was found that such a mill released oil from rapeseed and a domestic coffee grinder gave preferable results. The UCM mill was adopted for sunflowerseed where it had been previously shown[15,16] that the tough fibrous hulls make it difficult to obtain a uniform finely ground sample with other methods of grinding.

A new clause is included in the method detailing mills found suitable for specific seeds. As an example, a much expanded separate clause is included on milling of palm kernels as a consequence of FOSFA contracted research.[17] Shell and dirt (impurities) are present with the kernels and the hard shell is difficult to mill. It is almost impossible to obtain a homogeneous sample if kernels and impurities are milled together. The impurities in this instance must be separated and milled in a microgrinder. This is a particularly good example of a seed which, on the plant scale, can be readily straight extracted with a fast extraction rate to a very low oil content yet has given problems in the laboratory in the precision of both oil content and free fatty acid determination. Similar attention to detail has been focussed on the interstage grinding in a microgrinder, in which cup and ball size and duration of grind are specified for the Dangoumau Prolabo analytical grinder.

2.2.1.3(d) Temperature. The extraction temperature in the method is the resultant of the temperatures of the solvent vapour and condensed solvent, rate of reflux and heat loss to the room. Towards the end of the extraction the miscella in the flask is more concentrated and boils at a higher temperature. A siphoning Soxhlet extractor not only induces a lower extraction temperature but can cause flooding of the condenser due to rapid volatilisation when cool solvent siphons over to the concentrated hot miscella in the receiver. On this basis use of the Soxhlet has been excluded from the method.

Temperature still remains an uncontrolled variable, compensated to some extent by an extended extraction time. A wide area of research needs to be covered in re-defining both the extraction temperature and extraction time.

2.2.1.3(e) Miscellaneous. A number of minor changes have been introduced into the text with the intention of clarification and definition of the procedures. The method has now been submitted to BSI and ISO for consideration in the revision of ISO 659. It is noteworthy that special mention of sunflowerseed waxes has now been omitted. In France the AFNOR method for oil disregards the wax content of the hull, which is said to be as high as 0·4%.[18] The current ISO method includes waxes as oil and representations were made by France, through the EEC Seed Crushers' and Oil Processors' Federation (FEDIOL) to amend the ISO method. On further investigation lower levels of wax were found in the hulls than claimed and this proposal has now been dropped.

The reasoning for the adoption of a fixed three-stage extraction cycle of $4 + 2 + 2$ h is based on practical considerations rather than ultimate technical accuracy. A detailed investigation of the ISO 659 method and the current FOSFA method was carried out in 1983. The main differences between the methods, i.e. an extended extraction cycle and use of hexane, as compared with petroleum ether, in the ISO method, were evaluated on groundnuts, rapeseed, soyabeans and sunflowerseed.[19] Significant differences between the results were obtained mainly on groundnuts and soyabeans, attributed to the regrinding rather than solvent difference. The small increases in the amount of oil extracted in the extended cycle by the ISO method were judged to be outweighed by the savings in operator time and in reducing the risk of errors. The boiling point range of the light petroleum spirit allowable as an alternative to technical n-hexane has now been widened to 40–70°C as compared with 50–70°C in the ISO method.

The petroleum fraction referred to as 'hexane' can vary in the range 45–90% n-hexane depending on the regional source of the crude oil. Other major constitutents are 2- and 3-methyl pentane, methyl cyclopentane and cyclohexane.[20] Methyl pentane increases the solvency power, i.e. the degree and rate of extraction, as also does the presence of naphthenes in commercial extraction solvents. As mentioned previously, extraction temperature is an uncontrolled variable and closer control, together with improved solvent specification, in my personal opinion, could lead to improved precision in the method.

2.2.1.4 Instrumental methods

The attraction of instrumental methods in analysis is that they require less operator time with a possibility of automation, less elapsed time, are less subject to operator error and possibly give improved precision. They must, however, give a result the quality of which is fit for the purpose intended! For contractual measurement of oil content, correlation with the solvent extraction reference method must be established unless the instrumental method has been approved in the contract. For example, the determination of oil content of sunflowerseed by wide line nuclear magnetic resonance (NMR) in the USA by the USDA Federal Grain Inspection Service (FGIS) is an approved method for domestic and export lots.[21] Low resolution NMR is also approved for the analysis of colza and rapeseed by the European Economic Commission (EEC)[22] for intervention purposes. A new NMR method has recently been developed and approved by FOSFA for the determination of oil in rapeseed. This differs from existing standard texts mainly in terms of multipoint calibration and regression analysis rather than single point calibration. The determination of oil, moisture and volatile matter and protein content in soyabeans by infra-red reflectance is an official method of FOSFA International. Use of this method for FOSFA contracts is restricted to instruments which have been calibrated by FOSFA and, in this instance, the calibration shall have a multiple correlation coefficient of not less than 0·8 between the solvent extraction reference method and the instrumental method. Moreover, the standard error of results from the calibration line shall not be more than 0·3 for oil content.

It will be appreciated that not all analysis is carried out for the settlement of contracts. Instrumental methods figure largely in quality assurance programmes and in plant breeding. In the latter case the non-destructive analysis of a single seed can be carried out[23] and germination potential retained by low temperature drying before analysis.

2.2.1.4(a) Continuous wave wide line nuclear magnetic resonance (NMR): (i) standard and contractual methods. There is no doubt that NMR is the most widely used instrumental method and standard methods have been published by ISO,[24] BSI,[25] and IUPAC.[26] It is important to note that the definition of oil in these standards, namely: 'Oil is the whole of the organic substances contained in the oleaginous seeds and fruits which are liquid at the measuring temperature, basically (20°C)' limits the field of application and, under normal conditions of use, does not apply to shea, palm, illipe, cocoa, etc. The definition differs from that of the solvent extraction reference method of each standardising body and there is no published information to compare the methods and relate the results. As discussed later, publications relate NMR to non-standard, non-contractual solvent extraction procedures. It is understood that experience in France has demonstrated, on rapeseed in particular, that the IUPAC methods give closely similar results.

The instrument most widely used is the Oxford 4000 NMR Analyser (Fig. 2.1) and the underlying theory of the technique has been previously described.[27,28] Basically a sample of seed is placed inside a

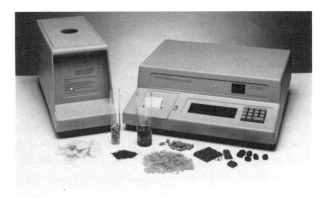

Fig. 2.1. Oxford 4000 Analyser—Wide line NMR.

coil located between the poles of a magnet. The coil is connected to an oscillator and detector circuit. When the radio frequency of the oscillator in kilohertz is just over four and a quarter times the strength of the magnetic field in gauss the hydrogen nuclei in the sample absorb energy from the oscillator and this resonance can be detected, amplified and measured in the Oxford Analyser.

The magnetic field is scanned over a range of field centred about the resonance frequency. A supplementary electromagnet of variable field is used and the series of absorption resonances produced is integrated for a fixed time interval to give a readout signal proportional to the number of protons in the sample. The close proximity of other atoms in a solid matrix causes a change in the local magnetic field called the 'chemical shift'. As a consequence protons in solid fat or protein resonate at a different frequency from protons in liquid oil or water. The 'liquid signal' is, moreover, a sharp peak as opposed to the broad flat 'solid signal'. The two signals are separated by a 'gate' usually set at two thirds of the scan amplitude.

In the EEC, BSI/ISO and IUPAC procedures the NMR signal is compared with the signal from an oil reference after drying the seed at $103 \pm 2°C$. A correction is made for the defatted meal response which is assumed to be independent of the sample for a given seed. The failure to take account of any sample to sample variation in the response of the meal signal is one of the weaknesses of the single point calibration methods. In the case of colza (rape) the response is equivalent to 0·30% oil, about the same as the repeatability of the solvent extraction reference method. The precision of the meal correction is lower than that of measurement on the seed and the magnitude is dependent on the extent of defatting and the gatewidth setting. More scatter is introduced into the results although accuracy is improved.

Special attention must also be given to the removal of free iron (upwards of 30 ppm affects the results) and to temperature control, where a change of 1°C is equivalent to 0·30% oil. It is not clear in the methods how impurities (other than particulate iron) affect the result. Impurities are removed before analysis in the EEC and FGIS methods and thereafter ignored. On the other hand the IUPAC, ISO and BSI methods do not raise the issue.

Whereas the standard methods use oil and defatted meal reference standards for calibration and do not claim to relate to the solvent extraction method, the EEC method is similarly calibrated and does imply this relation by discarding NMR results differing by more than 0·4 g of oil per 100 g of sample from the EEC extraction method. This must be followed by re-calibration. The FGIS method uses sunflowerseed as a standard for calibration purposes together with relaxed water (containing cupric nitrate) which is used for checking the instrument settings for optimum precision. The oil content of the sunflowerseed is presumably determined by solvent extraction, using the AOCS method Ai-3-75 which has an allowable difference between duplicates of 1·93%. It is to be noted that the final result is based on clean seed and corrected to 10% H_2O. An updated calibration has recently (22 February 1989) been introduced and is expected to reduce the reported oil content by 0·36% (Federal Register February 22nd 1989, p. 7575).

Robertson and Morrison[29] carried out development work on this method, studying the effect of depth of sunflowerseed in the 130 ml sample tube. They concluded that the largest possible sample within the length of homogeneous radio frequency field gave the greatest accuracy. They also found a decrease of 0·2% oil for each 1°C increase from the calibration temperature between 15 and 32°C, somewhat lower than quoted in the ISO standard. The effect of change in linoleic acid content of the seed was also measured and it was found that at 50% oil content a 10% difference between sample and standard was equivalent to 0·5% oil by NMR.

This work was followed by a comparative study of methods for oil content of sunflowerseed using NMR, near infra-red reflectance (NIR) and the AOCS solvent extraction method Ai-3-75.[15] Eight samples of seed, each sub-divided into five sub-samples, were analysed by the different methods and NMR was found to give the best precision. The NIR results were quite variable, with a higher coefficient of variation. It is noted, however, that mean results of sample 2 in the series differed by 0·9% oil measured by extraction and NMR. At this particular time (1981) the NMR method was already in use for domestic trading of sunflowerseed.

2.2.1.4(a) (ii) Methods used in plant breeding and quality assurance. Conway and associates were the first to report the application of the method to plant breeding, firstly on corn[30] and later on a number of species.[31] Moisture content of corn was reduced to less than 5% over 5 days at 65°C so as not to reduce germination. On soyabeans moisture was reduced to less than 4% over 5 days at 52–55°C. Two probes were used on a Varian PA-7 unit, one for 20–25 g seed and one for 1–3 g. Samples of 25 g were matched by NMR scan and pairs selected, one of each pair for a standard and one for oil analysis by a specially developed technique. The work was later extended on soyabeans[32] and corn,[33] with the conclusion that single kernel analysis overestimates oil content and each seed crop must be compared with standard samples selected from the crop. A more sophisticated technique using high resolution Carbon 13 NMR spectra has been used in breeding for study of fatty acid composition of the oil.[34] Madsen[35] described a comprehensive study of NMR and traditional solvent extraction methods in use at the Danish State Seed Testing Station on rapeseed. First he compared the solvent extraction oil content on seed (telle quelle) against oil content determined on dried seed calculated back to a telle quelle basis, and claimed good agreement up to a moisture content of 26%. He also found good reproducibility of results by NMR which correlated well with the extraction results. Unfortunately the method used for solvent extraction was not one which merits universal acceptance, and restricts the wider application of the conclusions.

2.2.1.4(b) Pulsed wide line nuclear magnetic resonance. There is considerable current interest in applying pulsed NMR in the form of the Bruker Minispec instrument to measurement of oil content. The model used is a modified version of the p20i instrument used for determination of solid fat content of fats, as larger sized samples must be used due to lack of homogeneity. This instrument is also based on wide line low resolution NMR, as is the Oxford Analyser, but in this instance the high frequency energy is applied in a pulsed sequence rather than a continuous wave. Early work[36] was carried out by Tiwari *et al.* using a Yugoslavian instrument and was followed by research using the Bruker Minispec p20i.[37] In principle protons in all phases are excited by a high frequency energy pulse and emit an NMR signal on returning to their ground state when the pulse is completed. The signal intensity is a maximum when all the protons have been rotated by 90° with reference to the static magnetic field and the instrument is adjusted to give this '90° pulse'. The initial amplitude of the signal is proportional to the total content of hydrogen nuclei but the rate of decay (relaxation time) differs in accordance with physical phase or chemical state. For example solid phase signals decay far more rapidly (microseconds) than those from liquids (milliseconds to seconds). This principle is used in the measurement of solid fat content in fats, where the signal amplitude is normally measured at 15 µms (includes solid and liquid) and 70 µs (liquid only).

In the application to oilseeds the amplitude measurement at 70 µs is proportional to the total liquid, i.e. oil and moisture, provided that the moisture content is less than 12% (rapeseed). This principle can be used for oil content measurement after drying the seeds at 105°C. For non-destructive seed analysis, drying at 60°C and signal measurement at 110 µs is used. A linear calibration curve with extracted oil from the same crop enables oil content to be calculated after weighing the seed. A further successful application has been to the estimation of fat content in chocolate powder where a measurement temperature of 60°C gives more reliable results.

The more recent exciting advance in this field has been the application to simultaneous measurement of oil and moisture content of oilseeds and residues.[38] In this instance a further radio frequency pulse is applied (a 180° pulse) after a set interval and a 'spin-echo' signal produced. The amplitude of this signal, 7–10 ms after the initial pulse, is proportional to the oil content only.

In practice rapeseed with less than 12% moisture is subjected to a 'spin-echo' pulse sequence and signal amplitudes measured at 70 µs and 7–10 ms after the initial pulse. The 7–10 ms signal corresponds to the oil content of the seed whilst the 70 µs signal corresponds to oil plus moisture, the difference between the two corresponding to the moisture content. Suitable data processing features can be included to give immediate digital read-out or print-out of the oil and moisture figures. The instrument is modified to accommodate a 40 mm diameter sample tube with a fill height such that a 20–25 g sample can be used. The probehead, the device seated between the poles of the magnet and which delivers the energising pulses to the sample, is also modified. The signal per gram of sample is measured and related to calibration values from seeds analysed by reference tests for contractual purposes. So far the method has been applied to rapeseed, sunflowerseed, soyabeans and residues from these seeds which contain liquid oils. It is in the early stages of assessment by ISO and much work needs yet to be done, but the advantages of such a method for trade and industrial use would be quite extensive. The technique has also been applied to measurement of oil content in cottonseed[39] and the effect of hydrogen content of the oil on the oil content has also been evaluted.[40]

2.2.1.4(c) Near infra-red reflectance. Whereas NMR is concerned with energy of radio wave frequency the near infra-red has wavelengths from 750 to 2500 nm and it is the absorption of energy corresponding to these wavelengths which causes molecular vibrations. Some of these involve movement of small groups of atoms while others involve chemical bonds so that the resulting absorption bands can identify the molecule. In this wavelength area the frequencies are generally too high for fundamental vibrational states but are overtones (2 × fundamental frequency) or combination frequencies. In addition, rotational energy states contribute 'fine structure', adding to the complexity of the spectra and, in particular, the constituent oil, moisture and protein spectra overlap. The spectra are, however, simpler in the wavelength region of the weaker overtone bands and, by careful selection and measurement of reflectance at a number of wavelengths, followed by mathematical analysis, it is possible to determine the separate constituents. It is, of course, necessary to calibrate with samples of predetermined properties to compute constants for the mathematical equations involved. Reflectance has the advantage of simpler spectra and cell packing than required for transmission spectra although measurements are still sensitive to the sample grinding and packing technique.

A comparative study of three different instruments, namely Infra-matic 8100, Infra-Alyser 300 and Instalab 800, available commercially, has recently been reported by Ribaillier.[41] This paper includes a description of the instruments and their performance in the measurement of oil, moisture and protein on ground samples of rapeseed, sunflowerseed, soyabeans and rapeseed meal. The method was found acceptable for quality control provided that careful calibration is carried out, which is frequently checked. Sunflowerseed gave the most problems, with a standard deviation of 1·2–1·3 for oil content, probably due to difficulty in sample preparation. A yearly review of the calibration is advised, otherwise there was little to choose between the instruments as regards performance.

The variability of results on sunflowerseed has been referred to previously[15] and is also shown in further work[42] comparing NIR and wideline NMR for the analysis of oil and moisture. Again the variability from sampling was a problem, with errors four to five times those in either method. The NMR method was found to be more precise but requires a dried sample whereas NIR is fast and generates data on constituents other than oil, such as moisture, protein and fibre. In this work the samples for NIR analysis were ground with an equal weight of Hyflo Super Cel for 2·5 min in a high speed grinder.

Much attention has been devoted to analysis of rapeseed by NIR due to the intensive crop breeding programmes in different countries. Tkachuk[43] in Canada has applied the technique to both oil and protein analysis of whole rapeseed kernels. Although reflectance spectra were better resolved in ground seed, acceptable comparative results with reference methods were also found for whole seeds although this was not the case for chlorophyll and glucosinolate analyses which were also investigated. Interestingly seed colour, which was specifically researched using yellow and brown admixtures of seed, did not increase the standard deviation of results although the inclusion of grey stripe samples of sunflowerseed had affected results in a previous paper.[42] In Sweden, Bengtsson,[44] in a study in which an extraction reference method was compared with NMR and NIR, used whole seeds dried at 105°C/15 h for NMR analyses, and whole seeds for NIR. In the latter experiments duplicate readings were taken on two separate packings of the cell. A bias of 0·5% on the high side for oil content was found by NMR, coupled with a standard deviation (SD) of 0·88 and a coefficient of variation (CV) of 1·81. Comparable NIR statistics were SD 0·74 and CV 1·60. In this instance, therefore, NIR is favoured for plant breeding, particularly as the protein analyses were also found to be acceptable. From personal experience of using NIR on ground rapeseed it has been satisfactory for quality assurance work, provided that sufficient safeguards are built into the operation. These include grinder and bias checks, weekly extraction checks, repeats of analyses outside a given range and a thorough calibration check with new crop rapeseed. The calibration for home produced rapeseed was not satisfactory on imported seed. Problems in grinding the seed uniformly are thought to be due to the 'case hardening' (for want of a better term) which can occur during industrial seed drying.

Presently this technique has been approved by FOSFA International for contractual analysis on soyabeans only. The method is applied to the analysis of oil, moisture and volatile matter and protein content and is restricted to calibrations carried out by FOSFA. The calibration must achieve a multiple correlation coefficient of not less than 0·8 against the reference method and the standard error of results about the calibration line shall be not more than 0·3 for oil content. The level of impurities is restricted to 2·0% max. and every tenth sample analysed must be checked by the reference method. The experimental work leading to the adoption of this method devoted special attention to choice of grinder, temperature and cell packing of the sample and careful calibration.[45] So far research on rapeseed has not produced results acceptable for contractual work on this oilseed.

2.2.1.4(d) Miscellaneous rapid methods. There are a number of rapid methods of long standing, in which the seed is comminuted and the oil extracted with solvent in a grinder extractor. This is followed by filtration and measurement of dielectric properties (constant and/or loss) or density of the resulting miscella. The most well researched method is the USDA rapid dielectric method for soyabeans and cottonseed.[46] Silica gel is added during grinding of the seed to remove water, which has a high dielectric constant of 81 compared with the solvent and soyabean oil (10 and 3 respectively). The time required for analysis is 10 min compared with 9 h for the reference AOCS method. In a comprehensive statistical survey the standard error of the variation in oil content compared with the AOCS method was 0·19–0·25% for soyabeans and 0·237–0·260% for cottonseed. The nature of the solvent—orthodichlorobenzene—would restrict its use in many countries.

Measurement of density of tetrachloroethylene miscella by means of a magnetic float cell is used in the Foss–Let method. This instrument has been used satisfactorily on oven dried peanuts[47] and, with perchloroethylene, on olives in Spain.

2.2.2 Moisture and Volatile Matter

2.2.2.1 Reference methods

The method described in ISO 665-1977,[48] also published by BSI as a dual standard,[49] is used by FOSFA for contractual work and the definition contained therein, i.e. 'the loss in mass measured under the operating conditions specified' is sufficiently wide-reaching to cover all other methods. The most important of these methods are probably the methods of AOCS and IASC.

In the ISO method the principle is to carry out the determination on a test portion of the material as received (pure seed and impurities), or, if required, of the pure seed alone. Drying is carried out at $103 \pm 2°C$ in an oven at atmospheric pressure until practically constant mass is reached. The description of the apparatus focusses attention on a number of important areas which can influence the results of a test which, on the surface, appears to be quite straightforward. The mechanical mill, for example, used in preparing most test samples, should allow the seed to be ground without heating and without appreciable change in moisture, volatile matter and oil content. The depth of the test portion in the flat-bottomed vessel used for drying is controlled to some extent by spreading at an amount of $0·2$ g/cm². The electric oven should have good natural ventilation capable of being regulated so that the temperature of the air and of the shelves in the neighbourhood of the test portion lies between $101°C$ and $105°C$ in normal operation. The desiccator should contain an efficient desiccant and be provided with a metal plate to allow rapid cooling.

In the preparation of the test sample all seeds are ground (copra is grated) with certain notable exceptions. These comprise small seeds such as linseed, colza, sesame, hemp, etc., as well as safflowerseed, sunflowerseed and cottonseed with adherent linters. Possible reasons for these exceptions are susceptibility to oxidation with a high content of unsaturated and polyunsaturated fatty acids present (which would increase the mass) and difficulties in grinding in particular without appreciable change in moisture. The heating period is a minimum of 3 h followed by 1 h intervals until a weight difference less than $0·005$ g (on a test portion of 5 g) between successive weighings is achieved. In the case of cottonseed the initial period is 12–16 h.

The IASC method is similar in many respects to the ISO method but is less clearly detailed. In this instance all seeds are ground (or grated in the case of copra) without exception. The method also notes that, in some countries on certain contracts, it is customary to dry the sample at 130°C for exactly 1 h. This loss in mass is then the final result. AOCS methods differ in detail for different seeds but are all carried out using cleaned whole seed and dried in a forced draught oven at 130° ± 3°C for a fixed period, e.g. 3 h for sunflowerseed and 4 h for castor beans.

2.2.2.2 Sources of error

At this stage it is worth considering the sources of water and the nature of the non-aqueous volatile matter. Pixton[50] refers to water in oilseeds as:

(a) Bound water or water of constitution in chemical union.
(b) Adsorbed water physically linked with the adsorbent.
(c) Absorbed water loosely held.

It is the sum of (b) and (c), often called 'free water', which is normally determined as moisture content but some bound water may be released in the determination. According to Wittka[51] water may be released by an intermolecular reaction of proteins during heat treatment and this loss in mass is greater for seeds high in protein, such as groundnuts.

Volatile matter other than water has not been quantified and identified. Free oil could be present in some seeds, particularly with a high impurities content. This could result in loss of low molecular weight fatty acids with seeds containing lauric acid, such as palm kernels, especially if the free fatty acid content is higher than 6·0%. Protein denaturation can occur with loss of hydrogen sulphide from sulphur-containing amino acids. This is more likely with seeds such as groundnuts and soyabeans if the moisture content is high and the test is carried out at 130°C. Decomposition of carbohydrates can yield furfural derived from pentosans. Furfural is normally produced from the hulls of cottonseed. A number of reactions take place on heating different seeds involving different constituents, usually changing toxic into non-toxic compounds. There is, however, no evidence to suggest loss in mass in the process.

In the case of soyabeans and linseed it has been demonstrated that the volatile matter condenses to an oily liquid which gave a positive test for phosphate and was presumed to be phospholipids.[52] Precipitates of barium carbonate were also obtained with barium hydroxide. Moreover the presence of volatile contaminants is always a possibility after transportation. The summation of all these changes which occur on heating contribute to the difficulty in achieving constant mass. Hunt and Neustadt[53] in discussing factors affecting the precision of moisture measurement, prefer forced draught ovens since they recover their fixed temperature more rapidly. Nevertheless the input of samples to the oven should be carefully regulated. The relative humidity (RH) of the room air up to 100% RH had no effect on moisture levels of samples dried at 130°C but reduced values of samples dried at 103°C by 0·2%. It is advisable to keep ovens which operate at the lower temperatures in a room at 40–50% RH. Dried seed is extremely hygroscopic and special attention should be given to the desiccant with activated alumina beads or $\frac{1}{16}$ inch pellets of molecular sieves (type 4A) preferred.

Moisture loss in sample preparation can be quite significant,[54] particularly in grinding softer seeds, and the type of grinder must be carefully selected for the particular seed. Lowest losses established on grain such as wheat and barley ranged from 0 to 0·6% at moisture levels of 6–10%. The Hobart 2040 coffee grinder and the Buhler laboratory grinder achieved these results out of the four grinders evaluated. Studies on sunflowerseed have shown that approximately 7·5% of the moisture is lost from samples at 8·9% moisture content (i.e. 0·67%).[42] Although this loss will vary with the particular seed and the room conditions it is a factor which has implications for other analyses where a ground analysis sample is used, such as moisture and oil content by NIR (see below).

2.2.2.3 *True water by Karl Fischer titration*

The Karl Fischer reagent titration method is specific for water and involves a chemical reaction with iodine, sulphur dioxide and pyridine in the presence of methanol. The generally accepted course of the reaction is as follows:[55]

$$I_2 + SO_2 + 3C_5H_5N + H_2O \rightarrow 2C_5H_5NHI + C_5H_5NSO_3$$

$$C_5H_5NSO_3 + CH_3OH \rightarrow C_5H_5NHSO_4CH_3$$

The general procedure for the determination of water is to mix the sample with anhydrous methanol in a suitable flask and to titrate with Karl Fischer (KF) reagent to a cherry red colour. Alternatively an excess of reagent can be added, followed by back titration with a standard solution of water in methanol. An electrical method may be used to determine the end-point.

A notable adaptation of this procedure to the determination of free water in linseed, soyabeans and other grain used a short extraction time (5 min) and a low temperature (64·5°C) in order to keep the release of bound water to a minimum.[52] The sample of whole seed was ground under methanol in a grinder-extractor and, after settling, a measured aliquot titrated by the back titration method. The end-point was measured by an electrical method depending on the fact that a measurable current will pass between two platinum electrodes immersed in the solution, provided iodine is present in excess. The iodine removes hydrogen which collects on the cathode and when all the iodine is removed by titration the cathode polarises and the resistance of the cell increases sharply. This method of detection is known as the 'deadstop' principle and is applied in both manual and automatic KF titrations.

The results furnished good agreement with the air oven method (103°C) for grain such as corn, pea and kidney beans. Low results with differences of 0·62 (linseed) and 0·56 (soyabeans) against the AOCS method (3 h at 130°C) were found, probably due to the factors previously outlined.

More recently Robertson and Windham[56] have applied a development of the method, using automatic KF titration, to sunflowerseed. The automatic method used electrometric end-point detection and titrators with a motor driven burette. A ball mill was used to grind whole seed in methanol for 30 min during which time the temperature rose to near the boiling point of methanol. After cooling and settling, an aliquot of the extract was titrated and corrected for a blank titration. A correction was also applied for the residue, i.e. oil and soluble carbohydrates which are partly extracted with the methanol. The results revealed a significant decrease in KF moisture levels, with increase in sample size, and 3–4 g of sample gave closest agreement with the AOCS oven method for moisture and volatile matter in the range 5·5–10·5%. It was also reported that workers from the USDA have found KF and AOCS oven values for soyabeans agree within ± 0·3% at moisture levels of 11·5%, whereas at 12·5% and 17·0% KF values were lower by 0·5% and 1·0% respectively. True water by KF analysis and moisture and volatile matter by oven methods were compared for whole and ground sunflowerseed and the results are listed in Table 2.2. In this instance there is no significant difference

TABLE 2.2

Comparison of Karl Fischer (KF) moisture analyses with oven methods on whole and ground sunflowerseed

Sample No.	Whole seed			Ground seed	
	130°C[a]	100°C[b]	KF	100°C[b]	KF
1	5·40	5·43	5·48	5·13	5·58
2	7·99	8·01	7·90	7·23	7·36
3	10·41	10·31	10·27	9·94	9·73
4	12·67	12·84	12·63	11·33	11·48

[a] AOCS official method Ai 2-75, 130°C for 3 h.
[b] Vacuum oven 100°C, 20 h, 30 mm Hg.

between methods but a marked difference between whole and ground seed, with the exception of sample No. 1. A loss of moisture in grinding is indicated and this could affect the precision of results in other analyses in which the seed is ground before analysis, as referred to earlier.

2.2.2.4 Instrumental methods

2.2.2.4(a) Moisture meters. Meters are used extensively for quality control purposes from farm to factory, including crop drying installations where these are necessary. It is important at all stages that the meter used is reliable and properly calibrated against the requisite contractual test. Most meters in use measure an electrical property related to moisture content such as resistance, capacitance, conductivity or dielectric constant. These parameters are not only affected by moisture content but also by the chemical and physical properties of the sample. Particle size, shape and bulk density affect the cell packing, and temperature control is also important. Electrical properties are found to vary from season to season and are also area dependent, therefore the calibration chart represents an average which should be frequently checked. It is believed that the main source of electrical property variations is the distribution of water in the seed. 'Free' water contains dissolved mineral salts, will conduct electricity and is therefore measured by all types of meter, whereas 'bound' water, physically linked within the molecular structure of the seed, does not conduct electricity and is responsive only to meters based on dielectric properties. Meters such as the Marconi, Weston and Universal, conductance type meters, measure 'free' water only and a constant value for 'bound' water is added, on the assumption that it is a fixed amount. The Motomco and Steinlite meters, dielectric type meters, measure both 'free' and 'bound' water and the Motomco is found to be least subject to electrical variations and shows least deviation from oven methods.[53] In the USA the Motomco 919 meter (Seedburo Equipment Co., Chicago) as shown in Fig. 2.2 is used in the approved dielectric meter method for grain.[57] In Canada the NEL model 919 meter (Nuclear Enterprises Ltd, Winnipeg) is approved.

Main features of the Motomco meter are the use of a large weighed 250 g sample in a cell with a centre post which corrects for bulk density variation. It contains a scaled standard for calibration which, in turn, is calibrated against a master standard held by the USDA Grain Divison. In the use of the instrument temperatures of the seed, other than 77°F (the standardisation temperature) are corrected for in the calibration

Fig. 2.2. Motomco 919 moisture meter.

charts. Calibration of the meter is carried out by the USDA Federal Grain Inspection Service.

In the UK the Sinar Datatec moisture analysers (Fig. 2.3) have achieved popularity for use on rapeseed. The design concept is based on simultaneous sensing of capacitance, a wide weight range (20–240 g) and temperature of the sample. These signals are fed into a micro-processor and the results of the computation compared with a pre-programmed calibration curve. An oscillating weight balance uses frequency measurement to determine mass. The frequency is detected through an electromagnetic system and processed by the computer. The P25 model holds 25 calibration curves, the P6 can hold 6, which can be transferred via a cable from the P25. This enables a number of P6 analysers to be identically standardised. The RS232 data output on the P25 can feed a printer giving readout facilities for moisture content, weight (g), temperature (°C) and specific weight.

A cautionary note on the determination of moisture content of sunflowerseed with various moisture meters (including the Motomco) has been expressed by Zimmerman.[58] Rapid drying of the hull surface of the seed can produce a moisture differential between hull and meat which was shown to indicate a moisture content 0·2–2·0% lower than the oven dried values. It is therefore advisable to equilibrate the sample by placing it in a sealed container for 12 h.

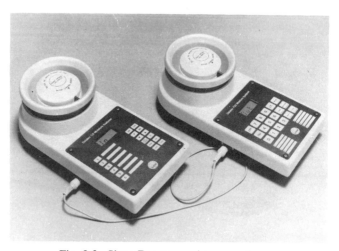

Fig. 2.3. Sinar Datatec moisture analyser.

2.2.2.4(b) Calibration of moisture meters. A measure of the importance attached to the calibration of moisture meters is the recent publication of an International Standard[59] devoted to the calibration of meters for cereals. This is now being used as a basis for an International Standard for oilseeds, which is at the draft stage, and will also be published by the British Standards Institution as a dual numbered standard. The intention is that it will be applicable to colza (rape) seed, sunflowerseed and soyabeans.

As with cereals the preparation of samples involves cleaning the seeds and, in addition, selecting seeds of low acidity. A moisture range of samples is selected, preferably from naturally occurring samples. Failing this, a large sample is dried gradually at 30°C and sub-samples placed in air tight bottles. The calculated amount of water is then added to achieve a particular moisture level and the sub-samples carefully equilibrated, conditioned and the moisture and volatile matter determined by the ISO reference method using oven drying. Maximum permitted errors (difference between oven and meter results) are tabulated, with a suggested tolerance of 0·7 for Class I meters and 0·8 for Class II meters for moisture contents below 10% (m/m). Tolerance for sunflowerseed will be slightly higher, as will all tolerances above 10% (m/m) moisture.

A very useful practical guide to moisture meter calibration has been detailed by Gough[60] in which it is advised to carry out the tests close to the normal temperature of the commodity. Examples are given of temperature corrections, e.g. +0·08% moisture per 1°C rise for soyabeans. The increase is lower (+0·05%) for oilseeds of intermediate oil content such as groundnuts. For temperatures below the calibration value a fall in moisture content of a similar level is predicted.

2.2.2.4(c) Near infra-red reflectance. In the discussion of NIR as applied to the measurement of oil content many of the references and general observations also apply to the measurement of moisture. The method is approved by FOSFA for use in contractual work on soyabeans, provided that the correlation coefficient with the oven drying reference method is not less than 0·8 and the standard not more than 0·2. It is to be noted that in this application the seed is ground for the reference method and for the determination by NIR. Where this is not the case, for example in the application of NIR to moisture analysis of sunflowerseed by Robertson and Barton,[42] the results are not so good. These workers carried out the calibration using data obtained by oven drying whole sunflowerseed by the AOCS moisture method. In the NIR analyses the sample was ground with an equal weight of Hyflo Super Cel and the assumption made that moisture loss was uniform from day to day. The average NIR moisture value found was 0·59% lower than the average oven value and the standard deviation of ±0·57% unacceptably high. Comparison with the precision of the oven method from Smalley checks[61] shows that a standard deviation of 0·255% can be achieved.

Tkachuk[43] has shown that the NIR spectra of whole and ground rapeseed are similar, although some of the peaks in the ground sample are better resolved. Moisture in whole rapeseed forms a very prominent peak at 1930 nm and it is suggested that the reflectance value of this peak, corrected for background reflectance at 2100 nm, can be used for accurate analysis although no results are presented for study. On ground rapeseed a correlation coefficient of 0·95 with oven moisture at 103°C (15 h) and a standard deviation of 0·20–0·30 has been found when comparing different NIR instruments.[41] The standard deviations for sunflowerseed and soyabeans were somewhat higher at 0·3–0·4 and 0·4–0·5 respectively. In these experiments the main reflectance peak was at 1940 nm.

2.2.3 Impurities

Admixtures, foreign matter, shell and dirt, and dockage are terms which are used, in addition to impurities, all of which may be synonymous. On the other hand, in some countries when defining grades of particular seeds they have a different meaning. Impurities, admixture and shell and dirt are terms used in FOSFA contracts and in Europe generally, where they have the same meaning as given by ISO in the definition of impurities. Foreign matter and dockage are terms more generally used in the USA and Canada. The USDA defines foreign material with a separate definition for each seed, whereas the AOCS outlines a method of determination of foreign matter but does not define it, except in the case of cottonseed. Dockage is a term usually reserved for material which can be removed from the seed by designated screening and aspiration equipment. In Canada dockage is removed from the seed when assessing grade, then added back to commercially clean rapeseed, for example, at a level of 2·5%. It will be apparent that special considerations apply to different oilseeds and special care must be taken that detail of the analysis complies with contract requirements. This is particularly important when other analyses are required on a cleaned seed basis.

2.2.3.1 Methods

2.2.3.1(a) Oilseeds: Determination of Impurities Content ISO 658 1980. [62] This method is also published as a British Standard (BS 4289-Part 2-1981) and includes the following definitions:

Impurities—all foreign matter, organic and inorganic, other than seeds of the species under consideration.

Fines—the particles passing through the sieves for which the aperture sizes are given for different species. In the case of groundnuts meal from the seeds, contained in the fines, is not regarded as an impurity.

Non-oleaginous impurities—non-oleaginous foreign bodies, (bits of wood, pieces of metal, stones, seeds of non-oleaginous plants, etc.), fragments of stalks, leaves and all other non-oleaginous parts belonging to the oleaginous seed analysed (e.g. bits of shell, loose or adhering to palm kernels), retained by sieves of the aperture sizes tabled. In the case of seeds sold in their shells, such as sunflower-seeds or pumpkin seeds, the loose shells are regarded as impurities only if their proportion is larger than that corresponding to the kernels in the same sample.

Oleaginous impurities—oilseeds other than those of the species under consideration.

The method involves separation by sieving and sorting into the three classes of impurities, using sample sizes ranging from 1000 g for copra to 200 g for small seeds such as rapeseed. For other seeds of intermediate size 500–600 g is specified. The separation of non-oleaginous impurities in small seeds, which are about the same size as the seeds, as well as small foreign oleaginous seeds, is a meticulous, laborious job. It is usually carried out on an aliquot portion of the partially sorted seeds. The results depend on attention to detail and the creation of fines during sieving must be avoided. It is worth consulting the relevant ISO method for test sieving[63] ensuring that the sieves are not overloaded and also checking that the sum of 'retained' and 'throughs' differ by not more than 2% of the charge mass.

In the special cases of groundnut fines the proportion originating from the seed is determined by analysis of oil content of fines and cleaned seed. Contractual samples of seeds such as palm kernels tend to be oily, with fines adhering to whole kernels. The special problems with these samples (in which the impurities are named 'shell and dirt') has recently been discussed.[64]

An indication of the precision of the method can be obtained from a recent ring test carried out by FOSFA International. Thirteen laboratories in nine different countries analysed two samples of French sunflowerseed derived from the same bulk sample, with the results given in Table 2.3.

2.2.3.1(b) Determination of impurities—IASC Method. This method retains the main features of the ISO method in less precise terms but omits the special case of groundnuts. The definition of impurities is the same and reference to the ISO method is advisable when the detail is obscure.

2.2.3.1(c) EEC provisions for rapeseed[65] and sunflowerseed. The method for rapeseed is based on the ISO method with two important differences. First, a second sieve with circular holes of 3·0 mm is used to separate impurities larger than rapeseed. Second, special mention is made of charlock, or wild mustard seed (sinapis arvensis), which is an undesirable foreign seed contaminant in rapeseed, allowable at a restricted level. It is similar to rapeseed in appearance and size (although some concentration of charlock occurs by screening through a 1·80 mm sieve). It can be distinguished microscopically with a lens of magnifying power 4 but requires considerable expertise and botanical skills. Other methods available are based on fluorescence of charlock when treated with KOH or, more recently, by temperature programmed gas–liquid chromatographic identification of the glucosinolates. Charlock contains sinalbin (4-hydroxybenzyl glucosinolate) which is absent from rapeseed. Chemical differentation by surface wax analysis[66] is the most promising technique.

TABLE 2.3
FOSFA ring test—% impurities in sunflowerseed

	No. of laboratories	Mean	Standard deviation	Max.	Min.
Sample A	13	2·73	0·413	3·57	2·20
Sample B	13	2·74	0·436	3·36	2·15

Some confusion has arisen within the EEC when applying the method for impurities which is detailed in EEC Regulation 1470/68 (23 September 1968) to sunflowerseed. Medium sized seeds have been classified, in error, with copra, so that for official purposes a 2·0 mm screen should be used for removal of fines. A resolution was drafted by the recent meeting of ISO/TC34/SC2 held in London (1987) with the intention of correcting the error so that a 1·0 mm screen will be used in future.

2.2.3.1(d) Determination of foreign matter and dockage—US methods. It is necessary to refer to the USDA FGIS grain inspection handbook for the US Standards definition of foreign matter which varies for different seeds and is not included in AOCS methods except in the case of cottonseed. For example:

Sunflowerseed:[67]
All matter other than whole sunflowerseeds containing kernels which may be removed from a test portion of the original sample by use of an approved device and by hand picking a portion of the sample in accordance with procedures prescribed in the grain inspection handbook.
Hull (husk): the ovary wall of the seed.
Kernel: the interior contents surrounded by the hull.
Whole sunflowerseed: includes intact seeds and broken seeds with kernels. Seeds that do not contain kernels are not considered as whole seeds.

Soyabeans:[68a]
All matter including soyabeans and pieces or soyabeans which will readily pass through an $\frac{8}{64}$ in sieve and all matter other than soyabeans remaining on the sieve after sieving.

Flaxseed:[66b]
All matter other than flaxseed which can be removed readily from a portion of the original sample using an approved device. Also underdeveloped, shrivelled and small pieces of flaxseed, removed in separating the material other than flaxseed, and which cannot be recovered by properly re-screening or re-cleaning.
Note: In this instance the definition refers to dockage and the method of analysis equates with the AOCS method for foreign matter.

The FGIS method for sunflowerseed involves mechanical separation from a 600–650 g sample, using a Carter Dockage Tester, followed by hand picking a 75 g aliquot of clean seed. The corresponding AOCS method (Ai 1-80) is closely similar but also incorporates the alternative of hand cleaning. The FGIS method for soyabeans requires removal of coarse foreign material by hand picking from a 1000–1050 g sample, followed by hand or mechanical sieving of a 125 g aliquot of residue and hand picking. Interestingly soyabean hulls retained by the $\frac{8}{64}$ in screen are not considered as foreign matter even above normal proportions. Broken soyabeans are classed as foreign matter.

The AOCS method (Ac 1-45) does not define foreign matter but it is clear from the procedure that the FGIS definition is used. In this case the whole sample (1250–1700 g) is screened and hand picked.

The FGIS method for flaxseed includes dockage separated mechanically with the Carter Dockage Tester on a 1000–1050 g sample, also reclaiming flaxseed from the separated fractions with hand sieves. This is followed by hand picking a 15 g aliquot of mechanically cleaned seed. The AOCS method (Af 1-54) is closely similar to the FGIS method but includes additional detail for use with samples which contain excessive amounts of pigeon grass seed and/or other weed seeds of similar size and shape.

Foreign matter in cottonseed is dealt with in the Trading Rules of the National Cottonseed Products Association.[69] In this instance a preliminary sorting of the whole 50 lb bulk sample is carried out on a shaker/cleaner. Dirt, sand, stones, hulls, leaves, sticks, grabbots flues and lint cotton are removed and the amount recorded. A reduced sample of 1000 g is then analysed by the AOCS method (Aa2-38). This method involves removal of large foreign matter with a 6 mesh sieve and hand picking the 'throughs'. Special attention has lately been devoted towards the amount of bract and leaves in cottonseed for environmental health reasons in crushing plants in the USA.[70]

A visual system of photographic slides approved by the USDA is now available for grade classification of soyabeans, flaxseed and other grain. A number of slides are included which can assist identification of foreign seeds and moulds, e.g. crotalaria, castor beans, cockleburr, yellow star thistle, velvet leaf, sclerotinia, etc.[71]

2.2.3.1(e) Determination of foreign matter and dockage—Canadian methods. The Canadian methods are given by the Canadian Grain Commission[72] who, fortunately, have interpreted the definition of foreign material given in the Canada Grain Act—1970. The definition is in two parts, namely:

Dockage is material that, by use of approved cleaning equipment, can and must (with a few specific exceptions) be removed from grain in order that the grain can be assigned to the grade for which it qualifies.

Foreign material is material, other than grain, of the same class, which remains in the sample after cleaning.

As with the US methods it is necessary to refer to the individual oilseed grade interpretation to appreciate the separate classifications. For example, dockage in rapeseed[73] is defined as all readily removable foreign material plus 1% maximum inseparable foreign material, plus up to 2·5% soft earth pellets hand picked from the cleaned sample, plus all material removed for grade improvements. The methods, in general, are similar to those of the USA involving mechanical separation with a Carter Dockage Tester, sometimes preceded by manual sieving and finally hand picking an aliquot of the cleaned seed.

2.2.4 Acidity or Free Fatty Acid Content

The acidity or free fatty acid (FFA) content of a fat is normally a measure of the extent to which hydrolysis has liberated the fatty acids from their ester linkage with the parent triglyceride molecule. In oilseeds the acidity is expressed on the oil extracted from the seed and, as discussed later, methods differ in the manner in which the extraction is carried out. Acidity may also be determined on the seed as received (pure seeds and impurities) or on pure seeds and separated impurities. Calculation of FFA entails an assumption of the molecular weight of the fatty acids. FFA content of lauric oils is expressed as lauric acid (mol. wt 200), of palm oil as palmitic acid (mol. wt 256), and of most other oils as oleic acid (mol. wt 282). The term 'FFA' is invariably used in the oilseeds trade and in contracts: acidity may, however, also be expressed as 'acid value'. This latter term is more commonplace in the refining industry where blends of oils are concerned and it not clear which molecular weight to take for the calculation. *Acid value* (AV) is defined as the number of milligrams of potassium hydroxide required to neutralise the free fatty acids in 1 g of oil.

At the outset it is worth a brief mention of the causes of acidity development during seed deterioration as these factors have a bearing on the handling, storage, sub-division of samples and the analysis. Damage to the seed by bruising, disease, breakage or other stress factors or by germination makes demands on the storage energy of the seed as represented by the neutral triglycerides. In the course of supplying energy due to damage or exposure to certain micro-organisms, fat degradation reactions occur which are catalysed by endogenous enzyme systems or by enzymes from the micro-organisms. Lipase and lipoxygenase are the two principal enzymes involved and it is the lipases which catalyse hydrolysis to FFA. Lipoxygenase, on the other hand, catalyses the oxidation of polyunsaturated fatty acids present in the seed. It is the acid lipase, with maximum activity at pH 4·0–5·0, which is most active in damaged or diseased seeds and retains this activity at storage temperatures up to 75°C.[74] Lipases can also be produced by micro-organisms and these fungal or microbial lipases are mostly found under the seed coat and can be more active than endogenous seed lipases. Growth of micro-organisms in or on seeds is accelerated by conditions of high moisture and temperature, but storage at sub-zero temperatures reduces growth. Deterioration due to these factors can also occur in the field if harvest is delayed by wet weather. The microbial lipases hydrolyse all three positions on a triglyceride quite rapidly. In contrast, seed lipases hydrolyse the one and three positions rapidly and the two position much more slowly. It will be apparent, therefore, that the composition of the free fatty acids is not necessarily the same as the esterified fatty acids and in practice differences equivalent to three units of Iodine Value have been found.

From the above it follows that the precautions below are necessary in preparation for this analysis:

(i) Damaged seed (broken, discoloured, etc.) must be evenly distributed before sub-sampling, and this may be a problem where there is a large difference in size, e.g. palm kernels. In this instance FFA in damaged kernels may differ by as much as a factor of 20 from undamaged seed.

(ii) There is a rapid rise in FFA of impurities, especially in the case of copra and groundnuts. The rise is likely to accelerate when impurities are separated from the pure seed and they should be analysed straightaway or stored in a deep freeze.

(iii) Temperature rise during sample grinding should be minimal. Even with pure seed, oil should be extracted immediately afterwards as grinding brings enzymes into contact with the oil.

With some seed, such as copra, it is advisable to store the samples at a temperature of $-15°C$ and mill the frozen sample.[75]

2.2.4.1 Methods

2.2.4.1(a) Oilseeds—determination of acidity of oils ISO 729-1988. The fundamental point of this method is that it is based on the analysis of the oil extracted in the determination of oil content by ISO 659-1979. This is important from two standpoints, namely FFA is a degrading factor in quality and economic terms and should therefore be related to the product produced by industry. Also in fractional extraction of the oil the later fractions have higher acidity. This is particularly so when polyphenols are present, although the prime example—cottonseed—which contains a polyphenolic binaphthaldehyde (gossypol) is excluded from the standard if the sample contains adherent linters. Black cottonseed is, however, covered by the method. Copra and palm kernels are also excluded due to the poor results from inter-laboratory tests on these seeds, recently carried out by sub-committee SC2 of ISO Technical Committee ISO/TC34-Agricultural Food Products. In view of the fact that ring tests on the determination of acidity in coconut and palm kernel oils (rather than oil from seeds) were very satisfactory it is clear that the poor results on the seeds arose from sample handling and oil extraction. The matter is still under consideration by committee ISO/TC 34/SC2.

The method entails taking the whole of the extract obtained during the determination of oil content and dissolving it in a neutralised mixture of diethyl ether and ethanol. The cold stirred solution is then titrated, with alcoholic potassium hydroxide solution (0·1 or 0·5 mol/litre) to a pink colour change, with added phenolphthalein indicator. For highly coloured oils an alternative indicator—alkali blue 6B—or a potentiometric method can be used.

The AV and FFA are expressed as

$$AV = 56 \cdot 1 \, VP/M$$

$$FFA = CVP/M\%$$

where V is the volume of the standard alkali used (ml),
       P is the molarity of the potassium hydroxide (mol/litre),
       M is the mass of the test portion (g),
       C is one tenth of the molar mass (g/mol) of the appropriate
       fatty acid, i.e. 20·0 for lauric, 25·6 for palmitic, 28·2 for oleic,
       the factor 56·1 is the molar mass (g/mol) of potassium
       hydroxide.

For most oils where FFA is expressed as oleic acid the FFA is equal to
the AV divided by 1·99, which is usually rounded to 2·0.

The selection of diethyl ether/ethanol as the primary solvent has
been the source of some controversy due to the hazards in handling
diethyl ether. The problem in the choice of solvent stems from the
difference in solubility of the fatty acids and neutral triglycerides.
There is also the possibility of saponification, particularly of mono-
and diglycerides at elevated temperatures. Moreover, when a single
solvent is used, such as ethanol, the titration has to be done at near
boiling point, with vigorous shaking, or phase separation occurs. In
these circumstances the end-point can be influenced by absorption of
atmospheric carbon dioxide. Glycerides and fatty acids are both
soluble in higher alcohols such as propanol and butanol but again the
possibility of glyceride saponification exists.

The use of diethyl ether/ethanol in analysis of FFA in fats has also
been approved by the International Union of Pure and Applied
Chemistry, Dutch and German Standards Authorities and the IASC.
In the UK industrial practice utilises ethanol or industrial methylated
spirits, carrying out the titration at boiling point with aqueous sodium
hydroxide, which is not subject to the stability problems experienced
with alcoholic potassium hydroxide.

The ISO standard does, in fact, allow the following secondary
choices, all of which are used at room temperature:-

solvent: diethyl ether/ethanol, toluene/ethanol, diethyl ether/2-
           propanol, toluene/2-propanol;
alkali: alcoholic or aqueous potassium hydroxide, aqueous
           sodium hydroxide.

The method also takes account of the separate analysis of impurities, ensuring that 10 g is taken for oil extraction to obtain
sufficient oil for a reasonable titration. Details of precision are also
included from inter-laboratory tests on colza, sunflowerseed and
soyabeans.

2.2.4.1(b)   FOSFA International Official Method. As with method (a) the trade method is based on the oil extracted in the determination of oil content by the FOSFA official method. The extracted oil is then analysed using the British Standard method for oils.[76] The special case of cottonseed with adherent linters has still to be resolved. In the current ISO method for determination of oil content cottonseed is dried for 2 h at 130°C, cooled, then treated by fuming with hydrochloric acid. This treatment makes the seed and the fibres friable and assists the grinding operation which follows. It is not technically possible to do this when subsequently measuring FFA in the oil and the FOSFA revised method omits treatment with hydrochloric acid. It is found, however, that heating the seed darkens the colour of the oil and necessitates the use of Alkali Blue 6B as indicator. There are also other drawbacks which are under discussion.

The freezing of copra before milling is another feature of the FOSFA method for oil content which has implications for FFA content previously discussed.

2.2.4.1(c)   American Oil Chemists' Society Official Methods. These methods are interesting inasmuch as the problems are approached in a different manner. For example, the method for cottonseed and sunflowerseed (Aa 6-38) is carried out on dehulled seed followed by partial extraction of oil from the ground separated meats with petroleum ether at room temperature. Heating is involved only if it is necessary to dry the seed and then only for 30 min at 100–105°C. Ethyl or isopropyl alcohol and aqueous sodium hydroxide are used in the titration, with phenolphthalein as indicator. On the other hand, the methods for oil content of these seeds are designed to be more exhaustive in extraction with the same hot solvent, i.e. petroleum ether.

A similar instance applies in the case of castorseed. In this case FFA is determined on oil partially extracted with methyl alcohol at room temperature, whereas oil content is determined by a more exhaustive extraction with hot petroleum ether.

2.2.4.1(d)   Canadian Grain Commission Method.[77] A different approach is made by extraction of the oil from ground seed in Swedish extraction tubes with petroleum ether at room temperature. Steel balls are added to the tube and the whole is shaken longitudinally for 2 h. This is then followed by further extraction of the separated meal in a

Goldfisch extraction apparatus. After solvent removal the titration is carried out in ethanol: chloroform: 2-propanol solvent (ratio 1:2:2), with aqueous sodium hydroxide, using Alkali Blue 6B indicator.

2.2.5 Damaged Seed

Damaged seed is of increasing importance in international trade[18] and is included in grading standards for oilseeds in the USA and Canada. FOSFA International Manual also includes methods for damaged seed in cottonseed, soyabeans and castor seed in its list of Official Methods for use in the assessment of Good Merchantable Quality in contracts. It is an index of paramount importance in insurance claims and a focus of attention in arbitrations and it is thought worthwhile including some general guidance to the analyst who is asked for factual evidence.

The assessment is usually carried out on the sample of clean seed remaining after determination of impurities, i.e. approximately 200 g for small seeds (rape, linseed, sesame, etc.), 500–600 g for other seeds and 1000 g for copra. It is then necessary to hand pick the clean seed, weigh and classify the damaged seeds. Criteria of damage differ with variety of oilseeds and source country. Some of these are listed in Table 2.4.

In the USA the types of damage are identified and evaluated by a

TABLE 2.4
Criteria for assessment of damage in oilseeds

Country	Oilseed	Types of damage
USA	Soyabeans	Ground, badly weathered, diseased, frost, heat, insect-bored, mould, sprout, stink bug stung, immature or materially damaged soyabeans, and pieces of soyabeans
	Sunflower	Heat, sprout, frost, bad weather, mould, diseased or materially damaged seed and pieces of seed
Canada	Rapeseed	Shrunken, frost-shrivelled, mould-discoloured, rimed or weathered, distinctly green, sprouted, heated or otherwise damaged seeds
EEC	Rapeseed	Immature, sprouted, mechanical, empty or split seeds

colour slide number, as previously mentioned.[71] Heat damaged soyabeans, for example, are materially discoloured seeds identified by reference to a slide and also supported by odour which is similar to smoke and called 'sour'. They are separately classified from beans which are heating from excessive respiration and have a high temperature.

In Canada damage in rapeseed is separated into four categories, i.e. visual, distinctly green, heated and other. A canola damage guide prepared by the grading department of the Alberta wheat pool is available from the Canola Council of Canada. It includes a colour chart for assessing distinctly green seeds of high chlorophyll content by making a colour comparison with a number of crushed strips of seeds. There is also a heated seed colour guide with sub-divisions into light tan, brown, dark brown and black. Distinctly heated seeds are compared with the guide. Light tan or brown seeds are considered heated only if supported by a heated odour. All these criteria are based on visual assessment and may have to be supported by data on the extracted oil and residual meal. Such data would include colour degradation, high chlorophyll levels, high refining losses, oxidation, poor taste and lack of stability in the refined, deodorised oil. Denaturation of protein in the meal would have to be demonstrated by chemical tests, possibly supported by feeding trials.

2.3. PROPERTIES OF OILSEEDS

Oilseeds, apart from their natural employment in germination and propagation of the species, are an extremely important commodity of commerce. The demand is based on their value as a human or animal foodstuff, either in the raw state or as the separated components—oil and meal residue. The oil contributes energy and essential fatty acids to the diet. Vegetable proteins in the residual oilcakes and meal are converted to animal protein when used in animal feedstuffs and contribute essential nutrients and variety to the human diet. The economy of many countries depends, to some extent, on cultivation, processing or export of vegetable oilseeds. As an example, in the UK in recent years industries using products supplied by the seed crushing and oil processing industry accounted for some 22% of total employment in manufacturing.

The physical characteristics of the seeds vary widely, possibly more so nowadays than in the past, with an increasing contribution from the plant geneticist. Take, for example, the reduced hull content of sunflowerseed, which has dropped to less than 20% by weight of the seed from levels of 42%. The oil content has increased accordingly from 25–32% to 40–49%. In this instance the hull is much thinner and lighter in weight. Another example in which the colour of the seed is changed from black or reddish brown is in the development of yellow hulled varieties of Canola. This development is to render this particular Canadian rapeseed variety more competitive with soyabeans by reducing the fibre content of the meal residue from 11·0% closer to that of soyabean meal, 6·0%. There are, however, generalities which one can apply to oilseeds, apart from drupes such as olive and oil palm fruit; they consist of a meat surrounded by a relatively hard protective coating known as the hull, husk or shell which is fibrous in nature. The level of hull can range widely from 91% in babassu nuts to 7% in soyabeans. Chemically the seeds are highly complex and the elucidation of this complexity is a continuing field of research. The major components of the seed are lipids, of which the more important are the triglycerides, waxes and the complex phospholipids and glycolipids. Non-lipid components are the proteins, carbohydrates, crude fibre, moisture and inorganic matter. More minor components are pigments, vitamins, antioxidants, enzymes and anti-nutritional factors, all of which can exercise a major effect on the processing and use of the seed and its separated components. In general parlance the hull consists mostly of carbohydrates and crude fibre while the meat is mainly protein and oil. On the other hand, it is the shape and size of starch grains in the meat which are of systematic importance in determining structure and seed identity microscopically.[78]

2.3.1 Toxic Constituents and Anti-nutritional Factors

Some species of raw oilseeds contain toxic constituents and complex chemical compounds which have a profound effect on the processing of the seed and use of the separate constituents. In general it is the residual meal and its potential use as an animal feedstuff which is the problem. In cases where an undesirable constituent is extracted with the oil, e.g. gossypol in cottonseed, this is readily removed in the refining process.

A number of the more common deleterious components are listed below and will be referred to later under the seed in question.

Alkaloids: ricinine in castor seed; lupin seed, safflowerseed
Cyanogens: linseed
Chlorogenic acid: sunflowerseed
Gossypol: cottonseed
Flatus producing factors: oligosaccharides in soyabeans
Glucosinolates: rapeseed, mustard seed
Metal binding constituents: phytates, phosphorus (in most seeds)
Oxalates: sesame seed
Protease inhibitors: trypsin inhibitor in soyabeans
Saponins: soyabeans, teaseed
Tannins: groundnut skins
Theobromine: cocoa beans

A comprehensive treatise on this subject can be found elsewhere.[79]

2.3.2 Trace Element Composition

Oilseeds contain trace elements of varying levels depending on many factors such as species, variety, stage of maturity, environment, etc. They are present both in the meat and the hull and are important when the seed is used as a human foodstuff, either raw, e.g. groundnut and sesame, or as an oilseed flour, e.g. soyabean, or as a protein isolate. They are also valuable when assessing the potential use of the meal residue, either as animal feed or fertiliser.

The main trace elements are phosphorus, calcium, magnesium and iron, supported by lower levels of zinc, manganese, copper, molybdenum and chromium. In a recent update of levels[80] in sesame, mustard, groundnut, safflower and coconut, values found were in the listed order, with sesame highest.

2.3.3 Common Oilseeds

A number of oilseeds are featured in Fig. 2.4 to illustrate the wide variation in size and shape between the different species. The following list describes the main features of the more common seeds and includes the genetic developments which are proceeding with the different species. The analytical properties and fatty acid composition of the oils obtained from these seeds are given in Chapter 7. The nomenclature used is derived from the International Standard.[81]

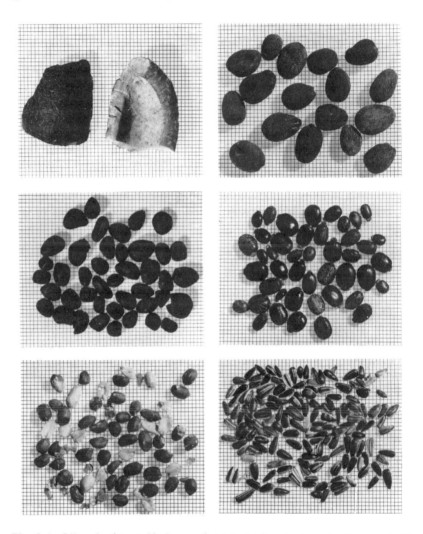

Fig. 2.4. Oilseeds. (top left) Copra; (middle left) palm kernels; (bottom left) groundnuts; (top right) sheanuts; (middle right) castorseed; (bottom right) sunflowerseed.

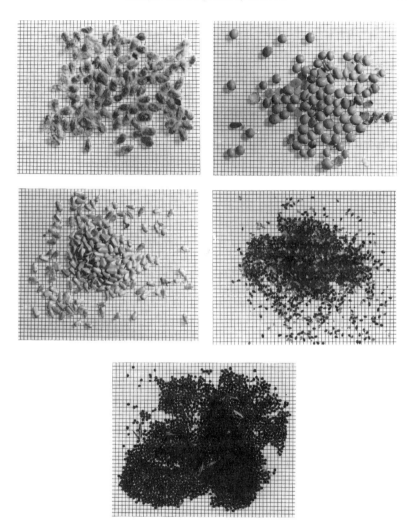

Fig. 2.4—*contd.* (top left) cottonseed (partially delinted); (middle left) safflowerseed; (top right) soyabeans; (middle right) sesame seed (mixed); (bottom centre) rapeseed.

Babassu (nuts)   *Attalea speciosa* Martius (synonym: *Orbignya speciosa* Martius, Barbosa Rodrigues)
Source: Brazil, Mexico, British Honduras
Description: Species of palm with fruits (drupes) in bunches of 200–400. Starchy edible mesocarp. Hard endocarp 4–8 kernels/fruit, each oblong in shape, 20 × 10 cm. Kernel weight approximately 3 g.
Composition: The thick-shelled kernels represent 9% by weight of the fruit and contain 63–70% oil. Residue contains 23% protein.
Babassu oil: An edible oil similar to palm kernel oil.
Castor beans   *Ricinus communis* Linnaeus
Source: India, Brazil, China
Description: Seed is oval, laterally flattened and of variable length, 8–20 mm. Colour is variable, with distinctive mottling of several colours from white, brown, purple or red. Seed weight 0·3 g (range 0·1–1·0 g)
Composition: Hull 25–35%, kernel 65–75%, oil 46–52%, moisture 3·0–7·0%, protein 17–25%. Contains a toxic protein—ricin (1·5% of oil-free residue) an alkaloid—ricinine and a potent allergen.
Castor oil: A liquid oil with mainly industrial uses based on the high level of ricinoleic acid.
Cocoa beans   *Theobroma cacao* Linnaeus
Source: West Africa, South America, Sri Lanka, Indonesia, etc.
Description: The fruit comprises a husk surrounding about 20 seeds embedded in a mucilagenous pulp. The seed is oval shaped, 15–30 mm long and consists of chocolate coloured cotyledons (nibs) covered with a thin red/brown shell (11–12% by weight). Seed weight is 1·0–1·3 g.
Composition: Typical analysis (%)—moisture 5·0; oil 54·0; protein 12; ash 1·46; theobromine 1·09; caffeine 0·44. Beans from Mexico and Indonesia have lower oil levels (53%) than from West Africa (58%) on a dry basis. Disposal of meal and shell residues is restricted due to the presence of theobromine.
Cocoa butter: A hard, brittle, edible fat, highly valued due to its distinctive taste and triglyceride composition.

Copra (kernels)   *Cocos nucifera* Linnaeus[82]
Source: Philippines, Sri Lanka, Papua New Guinea
Description: The coconut fruit is a fibrous drupe which weighs
                        2–3 lb. Removal of the fibrous husk yields a
                        kernel with a hard shell which is filled with a
                        watery liquid. In mature nuts the liquid forms a
                        firm white endosperm or meat about 1 cm thick.
                        The brown 'paring' on the shell side of the
                        endosperm is removed only when producing de-
                        ssicated coconut. The meat is removed from the
                        shell and dried from 50% to 5·7% moisture and
                        represents the cup copra of commerce. (When
                        dried in the shell it is called ball copra.) It consists
                        of large, irregular pieces which are discoloured in
                        handling and drying.
Composition: Oil 65–68%, moisture 4–7%, protein 21·2% (in
                        cake residue at 7% oil, 10% moisture).
Coconut oil: A semi-solid dual-purpose fat. Oil from 'parings'
                        differs in composition. Industrial use based on the
                        presence of short chain fatty acids. Edible uses
                        based on the sharp melting characteristics.
Corn/maize germ   *Zea mays* Linnaeus
Source: USA, Belgium, France
Description: The most widely grown variety is called Dent,
                        yellow or Indian corn. The kernel is tooth-like in
                        shape with an indentation at the top of the fruit.
                        The kernel consists of four major components:
                        starch (59·4%), fibre, protein (gluten) and germ
                        and contains 4·0–4·8% oil. The pure germ (8% by
                        weight of the kernel) contains 50–54% oil. The
                        separated germ is encased in a sheath which has
                        the appearance and shape of an insect wing. It is
                        light to dark brown in colour depending on the
                        drying process used.
Composition: The germ is a separated by-product from the wet
                        milling of corn for starch or dry milling for
                        brewers or maize grits. The purest commercial
                        germ is from wet milling and is typically 42–50%
                        oil, 5% max. moisture, 8% max. residual starch.
                        The meal residue—corn germ meal—contains
                        20·0—24·2% protein. The dry milled germ frac-

	tion can be 20–25% oil or 15–17% oil depending on the process.
Corn/maize oil:	A relatively stable liquid edible oil high in poly-unsaturated acids. Oil from South African white corn is less unsaturated. Commercial corn hybrids at 6–7% oil have been developed in the USA,[83] breeding for lipid composition is also being carried out.[84]
Cottonseed	*Gossypium* spp.
Source:	USA, Australia, China, West Africa, Brazil
Description:	Cotton is grown for its lint (seed-coat hairs). The seed of commerce usually retains some lint and is called woolly/white (8–12% lint) or black, with residual short hairs, linters or fuzz. It is pear-shaped, up to 12 mm long, 5 mm wide, coloured brown, red or black when bald. Hull content including fibres 35–40%. Weight of 100 seeds 7 g.
Composition:	Average values for US cottonseed[85] moisture 9·9%, oil 19·5%, protein 19·4%, crude fibre 22·6%, ash 4·7%. Oil content ranges from 15 to 24%. Meats contain 1–2% by weight of pigment glands which, uniquely, contain a yellow toxic pigment—gossypol—at levels of 0·4–1·7%. It is toxic to non-ruminants.
Cottonseed oil:	Crude oil contains 0·1–1·3% cyclopropenoid fatty acids, causing a positive Halphen test, which are destroyed on processing. Used as a liquid edible oil. Commercial varieties of glandless, gossypol-free cottonseed have been developed and grown in significant trial quantities in the USA.[86,87] The incentives for widespread adoption have yet to be clarified.
Crambe seed	*Crambe abyssinica* Hochstetter ex R. E. Fries[88]
Source:	Canada, USA
Description:	The seed is laterally flattened, 2–3 mm long and olive green to yellow in colour. The seed weight is 7·0–7·5 g per 1000 seeds. Hull (pod) content 14–20% by weight (Canada) and 25–40% (USA).

Composition:

	Moisture	Oil	Protein ($N \times 6 \cdot 25$)	Ash
Seed + hull%	7·1	33·3	17·1	5·3
Dehulled seed%	4·6	45·6	24·2	4·2

Contains glucosinolates. The principal glucosinolate in the meal residue is epi-progoitrin which is deleterious to non-ruminants. Recent approval for toasted meal by USDA at a level of 4·2% in cattle feed.[89]

Crambe oil: An industrial liquid oil with uses based on the high erucic acid content, which at levels of 55–60% is more consistently higher than in high erucic rapeseed oil.

Chinese tallow tree fruit *Sapium sebiferum* (Linnaeus) Roxburgh

Source: China, USA (trial plantings)

Description: The fruit has a brown outer husk with a white mesocarp which yields the tallow fat. The seed or kernel is oval-shaped, about 10 mm long with a smooth creamy-white surface. It is enclosed within a hard brown shell (endocarp). Weight of 100 fruits 20 g. The kernel represents 30% of the fruit.

Composition: Fruit contains 27–33% Chinese vegetable tallow. Seed contains 50–55% stillingia oil at 3·5% moisture.

Chinese vegetable tallow: A hard brittle fat with edible and industrial uses.

Stillingia oil: A highly unsaturated liquid drying oil, with an unusual fatty acid composition. Used industrially.

Grapeseed *Vitis vinifera* Linnaeus

Source: Italy, Argentina

Description: A by-product of the wine industry the brown seed is pear shaped and about 6·5 mm long.

Composition: Seeds contain 11·5% oil, 8·5% protein, 47·1% fibre and 9% moisture. Oil content ranges from 6% (black grape varieties) to 20% (white grape varieties). Residue is of low nutritional value in feeds.

Grapeseed oil:	A liquid oil high in polyunsaturated fatty acids. Used for soap, as a drying or edible oil.
Groundnuts (peanuts)	*Arachis hypogaea* Linnaeus
Source:	USA, China, Argentina, West Africa, Brazil
Description:	The pod (syn: shell, hull, husk) which matures underground, is cream-coloured, size 1·0–0·5 cm to 6·0–2·5 cm and contains 2 or 3 seeds. Each seed (kernel, nut) contains two cotyledons, between which lies the germ, and the whole is covered by a light tan or red skin (testa or cuticle), 2–4% of the seed by weight. The nuts, which weigh 0·2–2·0 g, range from spherical to cylindrical, with a flattened, rounded or pointed end. The larger nuts, e.g. Virginia variety in the USA, are selected for edible purposes. Traded product may be in shells or decorticated (shell removed). Shell weight is 25–35% of the whole pod.
Composition:[90,91]	The kernels contain 46–52% oil, 4·0–6·0% moisture, 25–30% protein and 2·5–3·0% ash. The skin contains tannins and is removed for some food uses because of its bitter taste. This also applies to the germ, where the taste precursor is not known. The meal residue is used for animal feed.
Groundnut oil:	A soft liquid oil with widespread uses for edible purposes. Higher oil content varieties are being developed in the United States.[92]
Hazelnuts	*Corylus avellane* Linnaeus
Source:	Europe, USA, North Africa
Description:	The nut is roughly spherical, about 20 mm long and consists of a kernel enclosed within a thick reddish brown shell.
Composition:	Kernels contains 50–68% oil, 18% protein. Residue is used as flour for human food or for animal feed.
Hazelnut oil:	A liquid edible oil with food and pharmaceutical uses.

Illipe (a)	Borneo illipe nut *Shorea macrophylla* (De Vries) Ashton Shorea illipe nut *Shorea stenoptera* Burck
Source:	Borneo, Sumatra, Sarawak, Malaysia
Description:	The fruit (60 g weight) has a characteristic winged calyx enclosing the hard, brittle shell of the nut. Inside is the black or reddish brown kernel (seed). Weight of 100 seeds 150 g.
Composition:	Kernels contain 45–70% oil. Meal has a low protein level, 10–15%, and is used in animal feed.
Illipe butter	(Borneo tallow): A hard edible fat with a sharp melting range between 30 and 35°C. Used mainly as a cocoa butter substitute.
Illipe (b)	Indian illipe seed *Bassia longifolia* Linnaeus
Source:	India
Description:	The seed is spindle-shaped, 1·25–4 cm long, enclosed in a husk. Weight of 100 seeds, including 25% husk, 157 g.
Composition:	Seed contains 52–60% oil. Meal residue contains a toxic saponin (mowrin) and is used as a fertiliser.
Indian illipe butter (Mowrah or Bassia butter):	A softer fat than Borneo tallow, with different properties. Used mainly in India for edible or soap-making purposes. The fat is sometimes called 'true illipe' by botanists but this term can mislead as it is a minor item of trade in comparison to Borneo tallow illipe.
Linseed (Flaxseed)	*Linum usitatissimum* Linnaeus
Source:	Argentina, Uruguay, Canada
Description:	Linseed is cultivated both for the fibre (flax) and, a shorter stemmed separate plant, for the seed. The oval lentil-shaped seed is 4–6 mm long, 2–5 mm wide, pointed at one end. It is smooth, sometimes shiny, and greenish brown in colour. Average seed weight is 4·5–7·5 mg/seed. Density is 1·10–1·15 g/ml. The seed coat uniquely contains a water dispersible carbohydrate (2–7% by weight of dry seed). The fibrous hull is 40–50% by weight of the seed.

Composition: Contains 35–44% oil, 24% protein at 10% moisture. Immature seed contains a small amount of a toxic cyanogenic glucoside—*linamarin*—and an enzyme—*linase*—which, under certain conditions, can act to release hydrocyanic acid.

Linseed oil: A highly unsaturated liquid oil, used industrially as a drying oil due to the high level of linolenic acid. Genetic developments have produced strains at 1% linolenic acid and increased linoleic acid levels which are more suited for edible use.[93]

Mustard seed

(a) *Brassica juncea* (Linnaeus) Czernajew et Cosson.[94] Brown or Indian mustard

(b) *Brassica nigra* (Linnaeus) W, D, J. Koch Black mustard

(c) *Sinapis alba* (Linnaeus) White or yellow mustard

(d) *Sinapis arvensis* (Linnaeus) Charlock or wild mustard

Source: Europe, USA, India, China (white and black); India, Far East, Canada (brown); Europe, USA (charlock)

Description: The seeds are broadly oblong, spherical or slightly flattened laterally, with a pitted or stippled surface. Brown mustard (a) is spherical, 1·9 mm diameter, and yellow seeded varieties exist in India. Black mustard (b) is spherical, reddish brown and 1·3 mm diameter. The white species (c) is flattened laterally, 2·5 mm long and is commonly yellow. Charlock is similar to rapeseed and is difficult to separate from it. Weight of 100 seeds 0·37 g.

Composition: Oil content 24–40%. Protein in meal residue 25–35%. Contains glucosinolates and the presence of allyl and parahydroxybenzyl isothiocyanate is responsible for the pungent smell, with minimum limits of 0·70% (a) and 1% (b) of allyl and 2·3% (c) of parahydroxybenzyl isothiocyanate given in the ISO specification.[95]

Mustard seed oil: The liquid oil is similar in composition to high erucic rapeseed oil. It is used for edible purposes, particularly in India.

Olive *Olea europaea* Linnaeus

Source: Mediterranean countries

Description: The fruit of the olive is an egg-shaped, elongated drupe, 2–3 cm long, often ending in a point. The colour varies from green or white to blue, purple, red and black. It comprises an outer skin (1·5–3·5% by weight), mesocarp or pulp (65–83% by weight) and an endocarp or stone (13–30% by weight) which contains one seed (3% by weight of the drupe).

Composition: The whole fruit comprises 40–50% water, 20–25% oil, and 25–40% solid residue. The oil distribution ratio is 96–98% (pericarp) and 2–4% (seed). Olive pomace, the residue containing the stones after milling, is used in animal feeds but has a low feeding value and may also be used as fuel or fertiliser. Alternatively the oil may be extracted with solvent for industrial use.

Olive oil: A liquid edible oil with an unusually high content of oleic acid. The composition of the kernel oil is similar to that from the pulp. The various grades of unrefined olive oil obtained from the pulp are used for edible or pharmaceutical purposes. Some grades of refined extracted oil are for edible use, lower grades are for industrial outlets.

Palm *Elaeis guineensis* N. J. Jacquin

Source: West Malaysia, Sabah, Indonesia, Nigeria

Description: The palm fruit is a drupe and is grown in large bunches with fruit representing 60–65% of the bunch weight. Average bunch weight is 10–20 kg in Malaysia, with 1500 fruits/bunch. The fruit is oval shaped, 2·5–5 cm long, 2·5 cm in diameter and varies in weight from 3 to 30 g. Colour is not uniform and varies from deep orange to reddish brown and black. The pulp encloses the seed or kernel which, with its surrounding hard shell, is called the nut. The kernels have an oily greyish

Composition:[96]

Palm oil:

Palm kernel oil:

Poppy seed
Source:

Description:

white flesh with a hard dark brown coat and occasionally there are two per nut. Weight of 100 kernels is 85–120 g and represents about 10% (weight) of the fruit. In Malaysia in recent years palm kernel production is about 27% of the palm oil output (by weight). Palm oil content of pulp averages 56%. Mean composition of Malaysian palm kernels is: oil 48·6% (46·4–50·8) moisture 6·5% (5·9–8·5) fibre 6–7%. Protein content of extracted palm kernel meal is 17–19%. Important minor components are the carotenoids present in the pulp and in palm oil at levels of 500–1000 mg/kg.

A semi-solid edible oil containing a high level of palmitic acid. Mainly edible uses and can be fractionated to provide a cocoa butter substitute.

A semi-solid edible oil containing a high level of lauric acid. Melts at a lower temperature than palm oil and more closely resembles coconut oil in composition. Used as an edible or industrial oil, the solubility and lathering properties of lauric acid are utilised in soaps and detergents. Developments in Malaysia have concentrated on increasing the yield of palm oil by the introduction of the weevil *Elaeidobius* (Pantzaris, T. P., Palm Oil Dev., Palm Oil Res. Inst. Malaysia No. 3, 1985).

A separate species of palm—*Elaeis melanococca* auctores von Gaertner—is grown in South America.[97] The palm oil from this species is more unsaturated and the palm kernel meal has a lower protein content than *E. guineensis*. Some work on hybrids of the two species is being carried out.[98]

Papover somniferum Linnaeus

Grown in India and Asia for opium and in Asia and Europe for seed oil

The seeds are contained in a capsule and opium is obtained from the juice in unripe capsules. They are white, yellow, brown, bluish or black coloured, kidney shaped and 1–1·5 mm long.

Composition: Seeds contain 40–55% oil, 9% moisture. Cake residue contains 36% protein (11% moisture, 8·6% oil). It is suitable for cattle feed providing opium alkaloids are absent.

Poppyseed oil: A highly unsaturated liquid edible oil. Seeds and oil are used for edible purposes.

Rape (Colza) seed *Brassica napus* Linnaeus
                  Brassica rapa (*campestris*) Linnaeus

Source: Canada, France, Denmark, Poland, UK, India, China

Description: The reddish brown seeds are elliptical, spherical or slightly flattened laterally, about 2 mm diameter. The hull (15–20% of seed weight) is reticulated and its structure is the main identifying feature between individual *brassica* and *sinapis* species. Weight of 100 seeds 0·37 g.

Composition: Average values are 39–45% oil, 6–9% moisture, 24% protein.[99] Polish winter rapeseed (typical values) are 40–50% ether extract, 19–25% protein.[100] Canadian seed (canola and rapeseed) over a 5 year period, 1977–1982, averaged 43·3–47·2% oil (dry basis) and 39·2–45·6% protein (oil free dry basis).[101] The seeds contain glucosinolates at levels ranging from 76 (campestris) to 123 (napus) μmol/g dried oil extracted meal. Lower glucosinolate varieties are being developed as explained later.
The hulls contain 36% fibre, 26% carbohydrate, and are high in polyphenols.

Rapeseed oil: A liquid edible oil with uses based on the erucic acid content. Oil with high levels of erucic acid (greater than 40%) is mostly used industrially whereas, in the Western world, oil for edible use is controlled by legislation at a level of 5% maximum.

Developments: (a) Canada
A rapeseed variety named Canola is now officially approved in the amendments to the Canada Seeds Regulations; grades were introduced in 1987. The term 'Canadian rapeseed' will now apply to seed

for industrial use. Canola is defined as 'the seed of *B. napus* or *B. campestris* the oil component of which seed contains less than 5% erucic acid and the solid component less than 30 μmol of any one or any mixture of 3-butenyl glucosinolate, 4-pentenyl glucosinolate, 2-hydroxy-3-butenyl glucosinolate and 2-hydroxy-4-pentenyl glucosinolate per gram of air dry oil free solid'. (GLC method of the Canadian Grain Commission.)

It should be noted that indolyl glucosinolates are excluded from the definition.[102] All export shipments of rapeseed from Canada since December 1983 have been of canola quality. Yellow coated low fibre varieties are being developed (hull analysis 8% fibre, 38% carbohydrate, low levels of polyphenols) to improve the meal residue quality.

(b) EEC

In the EEC all rapeseed grown for edible purposes is the low erucic acid variety. In 1990 the so-called double low varieties (low erucic acid, low glucosinolates) only will qualify for intervention purposes. Indolyl glucosinolates will be included in the definition. The intervention aid level in 1988 was set at 35 μmol glucosinolates per gram of seed. High erucic acid rapeseed (40% minimum C22:1) continues to be grown for industrial purposes.

Rice
Source:
Description:

Oryza sativa Linnaeus
Asia, USA, Africa
A main cereal crop, the rice grain is enclosed in a hard brown siliceous husk. It is of oblong shape about 8 mm long. Milling involves dehulling to give 'brown rice'. Bran and germ are then removed and the polished milled rice is used for human food.

Composition:

Hull 20%, kernel 52%, stalk 15%, bran 10%. Rice bran includes the germ, some hulls, pericarp and some endosperm. It contains oil 9–23%, protein 7–12%, fibre 6–27%, ash 8–22%. The lipase enzymes present produce a rapid rise in

FFA and oxidation of the oil on milling. Consequently the bran is either used in animal feed or the extracted oil used for soap manufacture. More recently pre-milling inactivation of the enzymes has extended the potential use as an edible oil.

Rice bran oil: A liquid oil containing vitamin E and with large volume potential as an edible oil.

Safflower seeds *Carthamus tinctorius* Linnaeus[103,104]

Source: Mexico, India, USA, USSR, Spain, Portugal, Australia, China

Description: The fruit is an achene (dry, one-seeded with a thin hull). It resembles sunflowerseed but is smaller in size and coloured cream, white or grey. It is irregularly pear-shaped, smooth and shiny and up to 10 mm long. The hull is thick and fibrous and can vary in proportion to the meat (48% in India, 40% USA). Weight of 100 seeds 4·7 g.

Composition: Oil 25–40%, protein 20–22% (in meal residue), 60% (in fully decorticated meal residue). High fibre content of hull (60%) restricts use of meal to ruminants, unless it is removed.

Safflower oil: A highly unsaturated liquid edible oil with a high content of linoleic acid. Two oil types are grown—high linoleic safflower oil (C18:1/15%, C18:2/77%), Iodine Value 138–145; high oleic safflower oil with the levels reversed (C18:1/77%). Spanish and Australian oil has less than 74% of linoleic acid, whereas Chinese has more than 80%. Originally an industrial oil, both types are now used for edible purposes.

Sesame seeds *Sesamum indicum* Linnaeus

Source: China, Mexico, Sudan, India

Description: The pear-shaped seeds are 2·5–3·0 mm long, 1·5 mm wide and coloured black, brown or white. Seed use, such as in the Asian confection 'Tahina', is based on colour. Here the appearance of the ground white seed mixed with sugar is preferred. The hull is 15–20% weight of the seed and contains calcium oxalate crystals. Weight of 100 seeds 0·30 g.

Composition: Oil 53% (40–60), higher levels from the Sudan,
                          lower levels from India.[105] Protein 20–25%, fibre
                          12%, ash 6·6% (dry basis). Contains very high
                          levels of trace elements and high levels of potas-
                          sium magnesium phytate. Meal residue used in
                          animal feeds.
Sesame oil: A stable liquid edible oil, rich in oleic and linoleic
                          acids. A previous relation between seed colour,
                          oil content and quality has not been
                          substantiated.[106]
Sheanuts   *Vitellaria paradoxa* C. F. Gaertner
                          (syn. *Butyrospermum paradoxum, Butyrosper-
                          mum parkii*)
Source: West Africa, Central Africa
Description: The shea tree grows wild and the nut, after
                          removal of the edible yellowish-green pulp from
                          the fruit, consists of a brownish-red crinkled shell
                          enclosing a chocolate coloured seed composed of
                          two cotyledons. It is oval shaped, 3·5–4·5 cm
                          long, similar to a plum in size. The shell content,
                          typically 35–38%, can vary up to 49%. Colour can
                          vary widely, as well as kernel weight, which can
                          differ by a factor of ten within a sample. Mean
                          weight of 100 kernels 400 g.[107]
Composition: Oil 50% (40–60), moisture 5–8%, protein 7–9%,
                          fibre 3–6%. Special features are the low protein
                          and fibre content and the presence of tannin and
                          latex in the flesh of the kernels. The latex is the
                          main constituent in the very high unsaponifiable
                          matter in the oil. The meal residue is used in
                          animal feeds.
Sheanut oil (shea   A semi-solid edible fat, difficult to process due to
  butter):   the high level of latex present. Used for edible
                          purposes. A stearine produced by solvent frac-
                          tionation is very useful in cocoa butter substitute
                          manufacture.
Soyabeans   *Glycine max.* Linnaeus (Merrill)
Source: USA, Argentina, Brazil, China
Description: The seed is ovoid, nearly spherical, up to 12 mm
                          long, similar in size to a small pea. It is generally

yellow but a number of coloured varieties—green, brown, black, etc., exist. The meat consists of two cotyledons enclosed by a light hull (8% by weight). Weight of 100 seeds 8·5 g.

Composition: Oil 17–21%, moisture 13·0%, protein 35% min., fibre 4·8%. Oil and protein levels vary inversely. In the USA average variety differences of 3·9% oil and 5% protein are found. Contains an enzyme—*lipoxygenase*—producing beany flavors by oxidation of the oil, and a trypsin inhibitor in the meal residue. Meal residue used in animal feeds or isolates for human food.

Soyabean oil: An unsaturated liquid edible oil. Mainly edible uses with some industrial applications.

Developments: Breeding aims to reduce linolenic acid content to 1%. So far 3% is the lowest achievement. Strains with 50% reduction in *lipoxygenase* are available.

Sunflower seeds *Helianthus annuus* Linnaeus

Source: Argentina, France, USA, Hungary

Description: The fruits (achenes or seeds of commerce) are of two varieties—large seeded types for confectionery (25–35% oil) and small seeded types for oilseeds (40–50% oil). The large seeded types have longitudinal striping ranging from black to white. The seeds can be up to 18 mm long, pointed at the base, rounded at the top and roughly four-sided. The oilseeds are black or dark brown and consist of an outer husk, 23–36% by weight (43% on large seeded type), enclosing the meat which has two cotyledons. Weight of 100 seeds 5·7–6·75 g.

Composition: Oil 40–50%, moisture 6–9%, protein 16·5–19·6%, fibre 13·2–15·7%, ash 2·8–3.9%.

The hulls are high in fibre and contain waxes which are extracted with the oil.[108] The meat contains 1·5–2·0% chlorogenic acid which causes darkening in colour of the meal residue when oxidised.

Sunflower oil: An unsaturated liquid edible oil containing high levels of linoleic acid. Used for edible purposes.

Developments: The high linoleic acid content of the oil (31·8–
                  75%), which is nutritionally desirable, can lead to
                  stability problems. Selective breeding has resulted
                  in commercial production of high oleic sunflo-
                  werseed with oleic acid levels in excess of 80%
                  and less than 10% linoleic acid. Apart from the
                  fatty acid composition and related analyses such
                  as iodine value, saponification value and refractive
                  index the composition of the oil is similar to
                  normal sunflowerseed oil. There is little change in
                  composition of the meal residue.[109]

Tea seeds   Camellia sinensis Linnaeus (O. Kuntze)
                  Camellia sasanqua Thunberg
                  Although all species have oily seeds it is camellia
                  sasanqua which is grown especially for oil
Source: China, Japan, Assam, North Vietnam
Description: An orange-brown, plano-convex seed, 10–22 mm
                  in diameter, with a hard, pitted shell.
Composition: Oil content 51–60%. The meal residue contains
                  saponins and is generally unsuitable for animal
                  feeds. It is used in insecticides in China.
Teaseed oil: A liquid edible oil, similar in fatty acid composi-
                  tion to olive oil. Used for edible purposes.

Tung nuts   Aleurites fordii Helmsley
                  Vernicia montana Loureiro (syn. Aleurites mont-
                  ana Wilson)
Source: There are two species of tung tree. A. fordii, the
                  main species, is grown in China, USSR, USA and
                  Argentina. Exports are mainly from China, Arg-
                  entina and Paraguay. A. montana is grown in
                  South East Asia.
Description: The fruit is the size of a large walnut, 14–35 mm
                  long, surrounded by a fibrous outer hull about
                  6 mm thick. It contains four or five seeds. The
                  seed consists of a kernel in a thin hard shell, with
                  a mottled fawn surface and is roughly triangular.
                  Weight of 100 seeds 3–4 g.
Composition: The kernels contain about 66% oil (dry basis).
                  The meal residue contains a toxic protein. It is
                  unsuitable for animal feeds and is used as a
                  fertiliser.

Tung oil (Chinese wood oil):	The oil is extracted from the seed and oil from both species has similar characteristics. It is a liquid, highly unsaturated, industrial drying oil. Contains high levels of the conjugated acid—α-eleostearic acid (*cis*-9-*trans*-11-*trans*-13-octadecatrienoic).
Tobacco seeds	*Nicotiana tabacum* Linnaeus
Source:	USA, India, South Africa
Description:	The seed is brown, oval shaped and minute, about 0·5 mm long, with a reticulated surface. Weight of 100 seeds 5–10 mg.
Composition:	Seeds contain 33–43% oil, 18–21% protein, 3·3–4·5% moisture, 15% fibre. Residue is used as cattle feed or fertiliser.
Tobaccoseed oil:	A highly unsaturated liquid oil. Used in soap manufacture and other industrial outlets. Also refined for edible use.
Walnuts	*Juglans regia* Linnaeus
Source:	France, Italy, Romania, China, California
Description:	A roughly spherical, light brown nut about 35 mm long, consisting of a kernel within a smooth shell with two distinct ridges.
Composition:	Kernels contain 60–64% oil and the cake residue, used in animal feed, contains (dry basis) 40·4% protein 14·1% oil.
Walnut oil:	A polyunsaturated liquid edible oil, also used in high grade artists' colours.

2.3.4 New Oilseeds

It is appreciated that the preceding list of oilseeds is selective and by no means comprehensive. One has also to be selective in describing new oilseeds: first because they are not new in the accepted meaning of the word, but rather from the viewpoint that developing knowledge in nutrition and other fields, together with the expertise of the plant geneticist, have highlighted their potential. Secondly, the research effort devoted to this topic is so extensive. For example, the USDA alone screened over 6500 species of wild plants in a search for new industrial oils. An AOCS monograph[110] serves as an introduction to the subject, and a more recent article[111] deals with new oilseed crops for oleo-chemical raw materials. A selection of some of the more developed crops is given below.

2.3.4.1   Oleo-chemical raw materials

Meadowfoam seed *Limnanthes alba*
Contains up to 33% of a wax with a high concentration of fatty acids (above 90%) of chain length greater than C18:0.

Jojoba seed[112] Contains 41·5–56·6% of a wax which, in the same way as meadowfoam, can be used to replace waxes derived from sperm whale oil.

Vernonia Contains 40–42% of a triglyceride oil which
galamensis seed includes 72–78% of vernolic acid (*cis*-12, 13-epoxy-*cis*-9-octadecenoic acid). The oil is suitable for baked films or coatings.

Stokes aster seed *Stokesia laevis* Hill (Greene)
Contains 30–40% oil which can also be used as a source of vernolic acid.

Cuphea seed The oil has been identified as a source of C8, C10 and C12 saturated fatty acids.[113] Seeds from different genus contain 33–36% oil (dry basis) and over 70% of one of these acids.

2.3.4.2   Edible oil source materials

2.3.4.2(a)   General purpose

Lupin seed *Lupinus albus, Lupinus mutabilis*[114]
Seeds contain 10–15% oil (*albus*), 13–23% oil (*mutabilis*) with a fatty acid composition similar to soyabean oil. Protein 34–41%, although use is restricted by the presence of alkaloids.

Winged bean *Psophocarpus tetragonolobus*[115]
Grown in South East Asia the seeds contain 15·0–20·4% oil. Fatty acid composition is similar to groundnut oil. Protein 29·3–39%. Potential use is seen as similar to soyabeans, apart from the oil which, with its lower level of unsaturated fatty acids, will appeal mainly as a cooking oil.

2.3.4.2(b)   Nutritional purpose

Evening primrose *Oenothera* spp[116,117]
seed
Seeds contain 24% of oil rich in linoleic acid (65–80%) and also 7–14% of gamma linolenic

acid, with α-linolenic acid absent. Species of the *Boraginaceae* family have also been shown to contain appreciable amounts of gamma linolenic acid.

Blackcurrant seed[118]   The seed or pip contains 27–30% oil. The fatty acids include 12–14% alpha and 15–19% gamma linolenic acids together with 45–50% linoleic acid.

There is a demand for these oils for clinical and pharmaceutical use.

2.3.4.2(c)   Cocoa butter substitutes. In addition to fats already discussed which are used in cocoa butter substitutes (palm oil, illipe and shea butters), fats from aceituno, kokum and sal seeds, as well as mango kernel fat, are now being used for this purpose.

Aceituno seed   *Simarouba glauca* Quassia[119]
Grown in Central and South America the fruit is similar to an olive, with a sweet edible pulp. The seed consists of a kernel (30%) surrounded by a shell (70%). Oil content of the kernel is 55–65% (dry basis).

Kokum seed   *Garcinia indica*[120]
Grown in India and the East Indies the small black ovoid seeds are enclosed in a fleshy fruit about 35–50 mm diameter. The seeds contain 23–26% oil.

Sal seed   *Shorea robusta* Dipterocarpaceae[121]
Grown in North India the winged seed comprises 60% kernel, 12–20% shell and 20–30% wings. Oil content of the kernel is 14–18%.

Mango kernels   *Mangifera indica* Anacardiaceae
The mango stone (seed) forms 6·5% of the dry weight (15% wet basis) of mango fruit grown in India. The dry stone consists of equal weights of shell and kernel. The dry kernel contains 9–13% of mango kernel fat.

ACKNOWLEDGEMENTS

Thanks are due to the undermentioned for supplying information and photographic prints with permission for publication.

British Food Manufacturing Industries Research Association, Leatherhead, UK.
Bruker Analytische Messtechnik GMBH, Rheinstetten, FRG.
Oxford Analytical Instruments Ltd, Abingdon, UK.
Seedburo Equipment Co., Chicago, USA.
Sinar Technology, Weybridge, UK.

REFERENCES

1. International Standards Organisation, *Oilseeds—Determination of hexane extract (or light petroleum extract) called 'oil content'*, ISO, Geneva, Standard No. 659, 1979.
2. Federation of Oils Seeds & Fats Associations Ltd., *Manual*, FOSFA International, London, 1982, p. 283.
3. European Economic Community, Regulation No. 1470/68, *Off. J. European Communities* (1) (28 September 1968) 239, 2.
4. European Economic Community, Regulation No. 1881/70, *Off. J. European Communities* (1) (15 July 1970) 154/13, 415.
5. International Association of Seed Crushers, *IASC Handbook*, 4th edn, IASC, London, 1988, p. 28.
6. National Institute of Oilseed Products, *Trading Rules 1980–81* (corrected and revised to April 1988), NIOP, San Francisco, 1980.
7. American Oil Chemists' Society, *Official & Tentative Methods of the AOCS*, 3rd edn, (corrected and revised to 1986), ed. R. O. Walker, AOCS, Champaign, 1978.
8. Corn Refiners' Association, *Standard Analytical Methods*, CRA, Washington, DC, 1978.
9. Daun, J. K. and Snyder, H., *J. Am. Oil. Chem. Soc.*, **66** (1989) 1074.
10. Cooper, A., *Oilseeds*, **3**(2) (1985) 16.
11. Kershaw, S. J., *Oilseeds*, **3**(4) (1986) 10.
12. Pritchard, J. L. R., *Oilseeds*, **4**(2) (1986) 8.
13. Firestone, D., Ridlehuber, J. and Berner, D., *J. Am. Oil Chem. Soc.*, **65** (1988) 1427.
14. Karnofsky, G., *J. Am. Oil Chem. Soc.*, **26** (1949) 564.
15. Robertson, J. A. and Windham, W. R., *J. Am. Oil Chem. Soc.*, **58** (1981) 993.
16. Kochhar, S. P. and Rossell, J. B., *J. Am. Oil Chem. Soc.*, **64** (1987) 865.
17. Kershaw, S. J. and Hardwick, J. F., *Oleagineux*, **40** (1985) 557.
18. Pritchard, J. R., *J. Am. Oil Chem. Soc.*, **60** (1983) 322.
19. Cooke, M. V. and Lewins, S. C., *Determination of 'oil content' of oilseeds. A study of two solvent extraction methods*, BFMIRA report No. 430, Leatherhead (available to non-members), 1983.
20. Johnson, L. A. and Lusas, E. W., *J. Am. Oil Chem. Soc.*, **60** (1983) 229.
21. Federal Grain Inspection Service, *FGIS Instruction 918-37*, USDA, Washington, DC, 1984.

22. European Economic Community, Regulation No. 3519/84 Annex VII, *Off. J. European Communities* (15 December 1984) L358/12.
23. Collins, F. I., Alexander, D. E., Rodgers, R. C. and Silvelas, L., *J. Am. Oil Chem. Soc.*, **44** (1967) 708.
24. International Standards Organisation, *Oilseeds—Determination of oil content—low resolution nuclear magnetic resonance spectrometric method*, ISO, Geneva, Standard No. 5511, 1984.
25. British Standards Institution, *Oilseeds Part 7, Determination of oil by nuclear magnetic resonance*, BSI, London, Standard No. 4289, 1985.
26. International Union of Pure & Applied Chemistry, *Standard methods for the analysis of oils, fats & derivatives*, 7th edn, Blackwell, Oxford, Method No. 1. 123, 1987.
27. Oxford Analytical Instruments Ltd, *Measurement of oil in seeds by NMR—applications report*, Oxford Analytical Instruments Ltd, Abingdon, 1984.
28. Waddington, D. In: *Fats & Oils, Chemistry & Technology*, ed. R. J. Hamilton and A. Bhati, Applied Science Publishers, London, 1980.
29. Robertson, J. A. and Morrison, W. H., *J. Am. Oil Chem. Soc.*, **56** (1979) 961.
30. Conway, T. F., *Proceedings of symposium on high oil corn*, University of Illinois, Illinois, 1960, p. 29.
31. Conway, T. F. and Earle, F. R., *J. Am. Oil Chem. Soc.*, **40** (1963) 265.
32. Collins, F. I., Alexander, D. E., Rodgers, R. C. and Silvelas, L., *J. Am. Oil Chem. Soc.*, **44** (1967) 708.
33. Alexander, D. E., Silvelas, L., Collins, F. I. and Rodgers, R. C., *J. Am. Oil Chem. Soc.*, **44** (1967) 555.
34. Schaefer, J. and Stejskal, E. O., *J. Am. Oil Chem. Soc.*, **52** (1975) 366.
35. Madsen, E., *J. Am. Oil Chem. Soc.*, **53** (1976) 467.
36. Tiwari, P. N., Gamhir, P. N. and Rajan, T. S., *J. Am. Oil Chem. Soc.*, **51** (1974) 104.
37. Tiwari, P. N. and Burk, W., *J. Am. Oil Chem. Soc.*, **57** (1980) 119.
38. Kroll, E., Private communication, 1986, Unimills, VDO Mannheim, FRG.
39. Srinivasan, V. T., Singh, B. B., Chidambareswaran, P. K. and Sindaram, V., *J. Am. Oil Chem. Soc.*, **62** (1985) 1021.
40. Srinivasan, V. T., Singh, B. B., Chidambareswaran, P. K. and Sindaram, V., *J. Am. Oil Chem. Soc.*, **63** (1986) 1059.
41. Ribaillier, D., *Rev. franc. corps gras*, **31** (1984) 181.
42. Robertson, J. A. and Barton, F. E., *J. Am. Oil Chem. Soc.*, **61** (1984) 543.
43. Tkachuk, R., *J. Am. Oil Chem. Soc.*, **58** (1981) 819.
44. Bengtsson, L., *Fette Seifen Anstrich*, **87** (1985) 262.
45. Cooke, M. V., *Infra-red methods of analysis for oilseeds*, FOSFA technical seminar report, FOSFA International, London, 1979, p. 13.
46. United States Department of Agriculture, (a) *Rapid method for determining oil content of soyabeans*, Technical Bulletin 1296. (b) *Rapid method for determining oil content of cottonseed*, Technical Bulletin 1298, USDA, Washington, 1963.

47. Heinis, J. L. and Saunders, M. M., *Oleagineux,* **29** (1974) 91.
48. International Standards Organisation, *Oilseeds—determination of moisture and volatile matter content.* Standard No. 665, ISO, Geneva, 1977.
49. British Standards Institution, *Oilseeds, Part 3—Determination of moisture and volatile content,* BSI, London, Standard No. 4289, part 3, 1978.
50. Pixton, S. W., *Tropical Stored Products Information* (43) (1982) 16.
51. Wittka, F., *Olearia,* **8** (March–April 1955) 55.
52. Hart, J. R. and Neustadt, M. H., *Cereal Chem.,* **34** (1957) 26.
53. Hunt, W. H. and Neustadt, M. H., *J. Assoc. Off. Anal. Chem.,* **49** (1966) 757.
54. Williams, P. C. and Sigurdson, J. T., *Cereal Chem.,* **55** (1978) 214.
55. Mitchell, J. and Smith, D. M., *Aquametry,* Interscience, New York, 1948.
56. Robertson, J. A. and Windham, W. R., *J. Am. Oil Chem. Soc.,* **60** (1983) 1773.
57. American Association of Cereal Chemists, *Approved methods of American Association of Cereal Chemists, Vol. 2,* Method 44-11, 1969, AACC St Paul, Method approved 10 August 1976.
58. Zimmerman, D. C., *J. Am. Oil Chem. Soc.,* **53** (1976) 548.
59. International Standards Organisation, *Check of the calibration of moisture meters, Part 1, Moisture meters for cereals,* ISO, Geneva, Standard No. 7700/1, 1984.
60. Gough, M. C., *Tropical Stored Products Information,* **46** (1983) 17.
61. Ridlehuber, J., *J. Am. Oil Chem. Soc.,* **59** (1982) 888A.
62. International Standards Organisation, *Oilseeds: determination of impurities content,* ISO, Geneva, Standard No. 658, 1980.
63. International Standards Organisation, *Test sieving,* ISO, Geneva, Standard No. 2591, 1973.
64. Kershaw, S. J. and Hardwick, J. F., *Oleagineux,* **40** (1985) 397.
65. European Economic Community, Regulation 1470/68 Annex VIII, *Off. J. European Communities* (1 Aug. 1986) L210/56.
66. Andrew, M., Hamilton, R. J. and Rossell, J. B., *Fett Wissenschaft Technologie,* **89** (1987) 7.
67. Federal Grain Inspection Service, *Grain Inspection Handbook, Book II,* Transmittal No. 23, 1984, USDA, Washington, 1984, pp. 11–12.
68. (a) Transmittal No. 32, 1984, USDA, Washington, 1985, pp. 6–16. (b) Issuance Change No. 46, USDA, Washington, 1986, pp. 9–19.
69. National Cottonseed Products Association, *Trading Rules 1981–82,* NCPA, Memphis, 1981, p. 113.
70. Morey, P. R. and Bethea, R. M., *J. Am. Oil Chem. Soc.,* **59,** (1982) 473A.
71. S. J. Systems Co., 647 W. Virginia St, Milwaukee, Wisconsin.
72. Canadian Grain Commission, *Official Grading Guide,* CGC, Winnipeg, 1980.
73. Canadian Grain Commission, *Grain Grading Handbook for Western Canada,* CGC, Winnipeg, 1986.
74. Angelo, A. J. S. and Ory, R. I., *Phytopathology,* **73** (1983) 315.

75. Federation of Oils, Seeds & Fats Associations, Ltd., (FOSFA International), *Manual,* FOSFA International, London, Contract No. 1, 1982, p. 105.
76. British Standards Institution, *Fats & fatty oils. Determination of acidity acid value & mineral acidity,* BSI, London, BS 684 Section 2.10 Method 1, 1976.
77. Canadian Grain Commission, *Grain Research Laboratory Methods,* CGC, Winnipeg, 1981.
78. Vaughan, J. G. In: *The structure and utilisation of oilseeds,* Chapman & Hall, London, 1970, p. XIII.
79. Liener, I., *Toxic constituents of plant foodstuffs,* 2nd edn, 1980, Academic Press, New York.
80. Deosthale, Y. G., *J. Am. Oil Chem. Soc.,* **58** (1981) 988.
81. International Standards Organisation, *Oilseeds—nomenclature,* ISO, Geneva, Standard No. 5507, 1982.
82. Cornelius, J. A., *Trop. Sci.,* **15** (1973) 15.
83. Fitch Haumann, B., *J. Am. Oil Chem. Soc.,* **62** (1985) 1524.
84. Weber, E. J. and Alexander, D. E., *J. Am. Oil Chem. Soc.,* **52** (1975) 370.
85. Tharp, W. H. In: *Cottonseed & Cottonseed Products,* ed. A. E. Bailey, Interscience Publishers, New York, 1948, p. 128.
86. Lusas, E. W. and Jividen, G. M., *J. Am. Oil Chem. Soc.,* **64** (1987) 839.
87. United States Department of Agriculture, *Proceedings of conference on glandless cotton,* ARS, Beltsville, Md, 1977.
88. Lessman, K. J. and Powell Anderson, W. In: *New sources of fats & oils,* ed. E. H. Dryde, L. H. Princen and K. D. Mukherjee AOCS, Champaign, 1981, p. 223.
89. United States Federal Register, 5 June 1981, p. 30081.
90. Rosen, G. D. In: *Processed plant protein foodstuffs,* ed. A. M. Altschul, Academic Press, New York, 1958, p. 425.
91. Taira, H., *J. Am. Oil Chem. Soc.,* **62** (1985) 699.
92. Hammons, R. O., *Proc. Am. Peanut Res. Ed. Assoc.,* **13** (1981) 12.
93. Green, A. G., *J. Am. Oil Chem. Soc.,* **63** (1986) 404.
94. Abidi, A. B., *J. Am. Oil Chem. Soc.,* **58** (1981) 947.
95. International Standards Organisation, *Mustard seed—specification,* ISO, Geneva, Standard No. 1237, 1981.
96. Tang, T. S. and Teoh, P. K., *J. Am. Oil Chem. Soc.,* **62** (1985) 254.
97. Hartley, C. W. S., *The Oil Palm,* Longmans, London, 1967, p. 65.
98. McFarlane, M., Alaka, B. and MacFarlane, N., *Oil Palm News* (20) 1945, 1.
99. Thomas, A., *Oil Palm News,* **59** (1982) 1.
100. Krzymanski, J., *Fette Seifen Anstrich.,* **86** (1984) 468.
101. Daun, J. K., *Cereal Foods World,* **29** (1984) 291.
102. Daun, J. K., *J. Am. Oil Chem. Soc.,* **63** (1986) 639.
103. Weiss, E. A., *Castor, Sesame & Safflower,* Leonard Hill, London, 1971.
104. Knowles, P. F., *J. Am. Oil Chem. Soc.,* **52** (1975) 374.
105. Yermanos, D. M., Hemstreet, S., Saleeb, W. and Huszar, C. K., *J. Am. Oil Chem. Soc.,* **49** (1972) 20.

106. Tinay, E. L., Khattab, A. H. and Khidir, M. O., *J. Am. Oil Chem. Soc.*, **53** (1976) 648.
107. Kershaw, S. J. and Hardwick, E., *J. Am. Oil Chem. Soc.*, **58** (1981) 706.
108. Morrison, W. H., *J. Am. Oil Chem. Soc.*, **60** (1983) 1013.
109. Purdy, R. H., *J. Am. Oil Chem. Soc.*, **63** (1986) 1062.
110. American Oil Chemists' Society, *New sources of fats & oils*, AOCS Monograph 9, ed. E. H. Pryde, L. H. Princen and K. D. Mukherjee, AOCS, Champaign, 1981.
111. Hirsinger, F., *Oleagineux*, **41** (1986) 345.
112. Yermanos, D. M. and Duncan, C. C., *J. Am. Oil Chem. Soc.*, **53** (1976) 80.
113. Miller, R. W., Earle, F. R., Wolff, I. A. and Jones, Q., *J. Am. Oil Chem. Soc.*, **41** (1964) 279.
114. Hudson, B. J. F., *Plant Fds hum. Nutrition*, **29** (1979) 245.
115. Sri Kantha, S. and Erdman, J. W., *J. Am. Oil Chem. Soc.*, **61** (1984) 515.
116. Hudson, B. J. F., *J. Am. Oil Chem. Soc.*, **61** (1984) 540.
117. Wolf, R. B., Kleiman, R. and England, R. E., *J. Am. Oil Chem. Soc.*, **60** (1983) 1858.
118. Bracco, U., Private communication. Nestlé Research Centre, Switzerland, 1987.
119. Bolley, D. S. and Holmes, R. L. In: *Processed plant protein foodstuffs*, ed. A. M. Altschul, Academic Press, New York, 1958, p. 852.
120. Williams, K. A., *Oils, fats & fatty foods*, Churchill, London, 1966.
121. Gunstone, F. D., Harwood, J. L. and Padley, F. B. In: *The Lipid handbook*, Chapman & Hall, London, 1986.

3

Oilseed Residues—Analysis and Properties

J. L. R. Pritchard

*Lingmoor, 7 Hisnams Field, Bishops Cleeve, Cheltenham,
Glos. GL52 4LQ, UK*

3.1. INTRODUCTION

Considering the seed as a beginning then the residue is the end of the beginning, and it is the end use of the residue with which analyses and properties are directly concerned. The bulk of residues remaining after processing for oil is used for animal feeding stuffs and as such they are subject to governmental controls in most developed countries. These are necessary as various animals are concerned, ranging from farm animals used for human food to domestic pets and racehorses! In the UK these take the form of the Feeding Stuffs Regulations.[1] Under these regulations oilseed residues are called 'straight feeding stuffs' and, to conform, certain analyses must be given on the label. Declaration of protein and fibre levels is compulsory, sometimes oil content, and, in addition, the seller has the option of declaring ash and moisture content. The regulations are also concerned with permitted additives, such as binders, anti-oxidants, etc., and the content of deleterious substances. Sampling and analysis methods are given.[2] The main UK methods are discussed in this chapter and, where appropriate, are compared with methods of the International Standards Organisation (ISO) and the American Oil Chemists' Society (AOCS).[3]

Nutritional value and protein denaturation are also important in the use of residues in feedstuffs and, to this end, additional analyses such as urease activity and cresol red index are included. Safety in transportation and storage are also considerations, and methods for determining residual solvent are discussed. Physical properties are important as particle size of meals and durability of pellets affect the bulk density and costs of handling. Moreover dust is not infrequently a cause of explosions in storage silos, particularly with solvent extracted residues, so that in this instance safety and economy are both achievable by pelletising suitable materials.

Many oilseeds contain deleterious substances and/or anti-nutritional factors which may be present in the residue. A lot of attention and research effort have been focussed on two of these, namely mycotoxins and glucosinolates, and the reader is referred to Chapters 4 and 5 where these topics are covered in detail. The level and the effect of some of the remaining problem areas can be reduced by proper processing, e.g. gossypol in cottonseed and trypsin inhibitor in soyabeans. Attention is drawn to these residues and others when discussing properties. The main concern, however, must lie with nutritional properties as their value must outweigh other considerations. Animal nutrition is a highly specialised subject and no claim is made to membership of its expert apostles. Some knowledge is, however, desirable in interpreting the composition of residues. This interpretation is attempted when discussing energy, protein, fibre, amino-acid and mineral element composition using the values for ruminants, which are listed in tables.

Some residues, such as defatted soya flour, are used for human food and microbiological and protein solubility analyses are required. These are not detailed in this chapter. At the lowest value level is the use of the residue as a fertiliser, either directly or when part excreted by the animal. Manurial values can be calculated from analyses of nitrogen, phosphoric acid and potash levels.[4]

3.2. OILSEED RESIDUES—DEFINITION

The definition used in International Standards is 'oilseed residues include meals, extraction, expeller cakes or slab cakes resulting from the production of crude vegetable oil from oil seeds by pressure or solvent extraction. It does not include compounded products'. Slab cakes are further defined as cakes of oilseed residues produced by hydraulic presses, a typical mass being about 10 kg.[5]

It is necessary to include an accurate full description of the processing undergone in the product title. The best examples are found in the UK Regulations. For example four products are listed for residues from groundnuts: decorticated groundnut expeller, extracted decorticated groundnut expeller, partly decorticated groundnut expeller and extracted partly decorticated groundnut. The meaning is also given: 'by product of oil manufacture obtained by pressing (or extraction) from decorticated (or partly decorticated) groundnuts (species *arachis hypogaea* and other species of *arachis*). Decortication is the removal of shell or husk and has the effect of increasing the protein level and decreasing the level of fibre. Expelled or pressed residues have part of the seed oil removed and are normally in a compressed form referred to as cake. Extracted residues have the bulk of the oil removed by solvent (normally to less than 2% residual oil). They are usually referred to as 'meals' in the crushing industry although quite often nowadays they may be pelletised for ease of transport and storage. In the feed industry pellets and compacted material are referred to as cake.

It is usual to heat (cook) the comminuted oilseed prior to expelling or pressing. Mostly closed steam is used and some analytical changes in protein solubility and toxins occur. High pressure expelling reduces oil levels to a 5–10% range in traded residues and is usually applied to seeds of oil content greater than 20%. Solvent extraction of the seed (or expelled cake for seeds of high oil content) is used in most developed countries. Heat is then subsequently applied to remove solvent using open steam in a desolventiser or toaster. It is at this stage that major changes in the chemical composition of the residue take place. Some protein denaturation occurs—protein solubility is greatly reduced and detectable amounts of hydrogen sulphide have been found by the author in toaster vapours from groundnuts and soyabeans. On the other hand, anti-nutritional factors such as trypsin inhibitor are much reduced in activity. This is most important in the case of soyabeans and is recognised in the UK Regulations by the name 'extracted toasted soya' and the meaning given 'by product of oil manufacture obtained by extraction and appropriate heat treatment from soyabeans'. In the USA cottonseed residues are designated as 36 or 41% protein meal and in this instance the seed is sometimes decorticated and hulls added back at the end of the process to control the protein level.

3.3. ANALYSES

3.3.1 Sample Preparation

The care necessary in preparation of test samples (samples for analysis) from laboratory samples is well documented.[2,5] The method involves grinding when required and sub-division by quartering or dividing apparatus such as described in Chapter 11. Special precautions are necessary for certain analyses such as moisture and volatile matter and residual solvent. These are detailed under the respective methods. Samples of high moisture content are likely to be encountered and these require drying before grinding and sub-division.

3.3.2 Oil and Ether Extract

There are several internationally recognised procedures for oil analysis using variations in extractor types, extraction times and differing solvents. Similar variations are encountered as with oilseeds. The major difference with residues lies between the use of hydrocarbon solvents such as hexane or petroleum ether and the use of diethyl ether. The former solvents are used by ISO,[6] AOCS for cottonseed meal and in the UK Feeding Stuffs Regulations. Diethyl ether is used in a separate ISO method[7] and by the Association of Official Analytical Chemists[8] in the USA. In the latter two cases the extract may be called ether extract although the term has also been applied to petroleum ether extract. As diethyl ether is the more polar solvent more phospholipids and polar compounds are extracted and the results will differ. As the boiling point is also much lower (34·5°C as compared with 62–68°C) the extraction temperature will also be lower, affecting the results by a slower extraction rate. Ether extract results are favoured by farmers and compounders as a library of information has been accumulated on results which, together with digestibility coefficients, are used for energy calculations in formulating feeds to meet the nutrient requirements of the animal. On the other hand, when assessing the efficiency of the industrial extraction process the seed crusher requires methods for oilseeds and residues which are related and employ a solvent similar to the solvent used in the plant—mostly some form of commercial hexane.

3.3.2.1 *Hexane extract/oil*

The ISO method is closely similar to the method for oilseeds detailed in Chapter 2. The initial oil content, however, is much lower and reduced extraction times are required. In Europe it is more than likely that the test sample will have to be pre-dried as the marketed moisture levels are in the range 12–14%. It is important to note that the intermediate microgrind, followed by a second extraction, is retained in the method. This is necessary as size reduction in the plant process is not always at peak efficiency and it is not unknown for whole seeds, particularly small seeds such as rape and sesame, to be found in the residue.

The methods of the AOCS and UK Regulations do not include this precautionary measure. This may be understandable insofar as the Regulations are concerned as the required accuracy is not very demanding. For example, if an oil content of 2·0% is declared on the product label the allowable limits (to include process and analytical variations) are 1·4–3·2%. The precision of the AOCS method from Smalley check data is given in Table 3.1

3.3.2.2 Diethyl ether extract

The ISO method consists of a preliminary grind of the sample through a 1 mm screen before weighing a 5–10 g test portion. Pre-drying is carried out by mixing the test portion with anhydrous sodium sulphate or drying at 60°C and atmospheric pressure in the case of residues rich in volatile acids (copra, palm kernel, etc.). After extracting for 4 h, diethyl ether solvent is removed from the meal before adding sand and re-grinding with a pestle and mortar. Use of a microgrinder without added sand is preferred. Then follows a further 2-h extraction using the same flask, solvent removal from the extract (at 75°C in a vacuum oven) and weighing. The flask is then re-heated and re-weighed, with an allowable difference in weight of 0·01 g. It should be noted that this weight difference is equivalent to 0·2% extract on a 5 g test portion and this is the allowable repeatability for two determinations given in the method. A weight difference of 0·01 g on a 10 g test portion would be more desirable.

3.3.2.3 Instrumental methods

Many of the methods described for oilseeds in Chapter 2 have also been applied to residues with varied amounts of success. In the control of solvent extraction plants the author used the Foss–Let method for a number of years calibrated against the reference method. Latterly infra-red reflectance gave good results with soyabean, palm kernel and rapeseed meals. It has not been so successful with maize germ meal due to large variations in colour. In Germany an NMR method has been used successfully for the simultaneous determination of moisture and oil content. A method is currently being studied by ISO/TC34/SC2, which is the sub-committee dealing with oilseeds and oilseed residues.

TABLE 3.1
AOCS methods—precision data from Smalley checks (%)

Meal residue	Moisture			Oil			Nitrogen			Crude fibre		
	Mean	SD	CV	Mean	SD	CV	Mean	SD	CV	Mean	SD	CV
Soybean	8·71	0·193	2·22	1·42	0·086	6·06	7·96	0·064	0·804	4·53	0·237	5·23
Peanut	8·24	0·202	2·45	1·04	0·062	5·96	7·89	0·058	0·735	11·39	0·398	3·49
Cottonseed	7·97	0·234	2·94	5·54	0·104	1·88	6·73	0·060	0·891	12·56	0·606	4·82
Coconut	8·77	0·355	4·05	4·54	0·085	1·87	3·63	0·040	1·10	11·74	1·555	13·2
Safflower	6·31	0·182	2·88	1·31	0·067	5·11	4·97	0·070	1·41	28·55	1·033	3·62
Rapeseed	8·19	0·206	2·52	3·64	0·110	3·02	5·99	0·056	0·935	11·37	0·676	5·95

SD = standard deviation.
CV = coefficient of variation (%).

3.3.3 Moisture and Volatile Matter

Much of the detail under this heading in Chapter 2 concerning oilseeds is relevant to oilseed residues. There are, however, several important differences, especially when dealing with solvent extracted residues. When desolventising the residue after oilseed extraction, moisture is lost by azeotropic distillation and it is common practice in the seed crushing industry to add back moisture to a moisture level in equilibrium with the prevailing atmosphere. The amount of free water present is therefore much higher than in oilseeds, and it may well be unevenly distributed. There is also the distinct possibility of the volatile matter containing residual solvent, particularly with rapeseed residues which desorb solvent slowly, as discussed later.

Some account is taken of these factors in the ISO protocol for preparing test samples which involves mixing and, if necessary, grinding of laboratory samples to pass a 1 mm sieve. A correction factor C is used if either of these operations is likely to involve loss of moisture with moist samples.[5] The factor is given by the equation

$$C = \frac{100 - U_o}{100 - U_1}$$

U_o is the moisture and volatile matter content, expressed as a percentage by mass determined on the partly prepared sample, i.e. mixed or, in the case of very coarse samples ground through a 2·80 mm screen. U_1 is the moisture and volatile matter content expressed as a percentage by mass of the prepared test sample. The factor C is used as a multiplication factor to relate the analysis result on the test sample to the original condition as regards moisture and volatile matter content.

The ISO method of analysis[9] on the test sample is closely similar to that used for oilseeds and drying to constant weight is carried out at $103 \pm 2°C$ in an oven at atmospheric pressure. The repeatability allowed is 0·2% maximum.

The regulatory method in the UK is based on the principle of drying to constant weight in an oven at 100°C and is referred to as moisture content. This method also takes account of the problems encountered with appreciably moist samples by correcting the analysis of the test sample for moisture loss in sample preparation. The differences between this method and the ISO method include the amount of exposed surface area of the test portion in the drying oven, the drying temperature and the definition of constant weight. It is therefore not advisable to use the ISO method for the purpose of establishing conformity with the regulations.

Sampling of residues in the AOCS methods (method Ba1-38) incorporates size reduction, with the exception of meals, to a coarsely ground meal. The screen size is not defined and the residues are used for the determination of moisture and volatile matter (method Ba2-38) without further grinding. The method entails drying the sample for a fixed period of 2 h at 130°C. Under these circumstances the design and loading of the oven will have a large influence in the absence of a constant weight check.

Many of the instrumental methods used on oilseeds are also applicable to meal residues and near infra-red reflectance has proved useful for quality control purposes.[10] Future progress in this area is likely to arise from the application of pulsed nuclear magnetic resonance.

3.3.4 Protein

As a consequence of its importance in trade and nutrition the estimation of protein content has been subjected to scientific scrutiny on a worldwide basis. Carbohydrates and oil contain only three chemical elements—carbon, hydrogen and oxygen. In addition to these, proteins contain about 16% of nitrogen, together with small percentages of sulphur and phosphorus. The percentage of crude (or total) protein in a residue is conventionally found by determining the total nitrogen content using the Kjeldahl method and multiplying by the factor 6·25 ($\frac{100}{16}$). The principle of this method is to digest the organic matter with sulphuric acid in the presence of a catalyst, render the reaction product alkaline, then distil and titrate the liberated ammonia.

Not all the nitrogen in feeding stuffs, however, is in the form of true protein and may be present as simpler nitrogenous compounds such as asparagine, amino acids, ammonium compounds, etc. These are usually grouped together as 'amides' and, in feeding stuffs such as roots, tubers, silage and kale, etc., where their proportion is high, have the effect of reducing the feeding value. The level in oilseed residues is usually quite small. True protein is estimated by boiling a known weight of the residue with a copper reagent, such as copper sulphate, under specified conditions. The true protein is precipitated, separated by filtration and the amount estimated by the Kjeldahl method. The amides are left in solution and are calculated arithmetically by difference:

$$\% \text{ crude protein} = \% \text{ true protein} + \% \text{ amides}$$

Until recently there was a separate ISO standard (ISO 3099-1974) for the specific application of the Kjeldahl method to oilseed residues. This has now been superseded by a general standard applicable to all animal feeding stuffs.[11] Other methods include AOCS method Ba4-38, applicable to cottonseed, soyabean, groundnut and linseed residues, and the Corn Refiners' Association method,[12] applicable to corn germ residues. The mandatory method in the UK is given in the Feeding Stuffs (Sampling and Analysis) Regulations 1982. It is similar to the ISO method except for the use of a mercuric oxide catalyst for the oxidation and sulphuric acid as the collecting medium in distillation.

The most critical stage in the procedures is the digestion, and precautionary notes are given in the methods. For high protein levels mercury is the preferred catalyst and is used as the oxide in the UK and AOCS methods, whereas copper is also allowed in the ISO method and is used as the selenite on corn germ residues. Higher fat levels in the sample may require higher levels of sulphuric acid in the digestion, and in all cases potassium sulphate is added to raise the boiling point. As a check ISO advise a parallel test using acetanilide or tryptophan. Acetanilide, together with a small quantity of sucrose, is recommended for easily digested samples, otherwise dry tryptophan and sucrose are used. Precision data for the AOCS method is given in Table 3.1. Ring tests are presently in progress to revise and extend precision data given in the ISO method. Although the factor 6·25 is most frequently used in the calculation of crude protein content alternatives have been listed[13] which are more specific for particular oilseeds and residues. For raw oilseeds these include soyabeans: 5·71, groundnuts: 5·46, coconuts, cottonseed, linseed, sesameseed, sunflowerseed and castorbeans: 5·30. More recently the Nitrogen Committee of the AOAC has reviewed the conversion factors and concluded that the factor 6·25 should be used for oilseed residues. For groundnuts and brazil nuts: 5·46 and tree nuts and coconuts: 5·30.[14] In practice the factor used should always be quoted in a report.

Foremost amongst the rapid methods of analysis which have been developed have been instrumental methods based on near infra-red reflectance (NIR). In the evaluation of this technique Ribaillier and Maviel[10] have compared the performance of three NIR instruments on rapeseed, sunflowerseed, soyabeans and extracted rapeseed meal.

Results on the residue were poorer than results on the seed although judged satisfactory for quality control purposes. An excellent example of the precautions necessary in using this technique can be found in the instructions for protein analysis of wheat under the US Grain Standards Act.[15] Pulsed nuclear magnetic resonance is also being researched as a rapid method, with development as yet at an early stage.[16] The semi-automated Kjel–Foss method which is a rapid version of the Kjeldahl method, using a much reduced digestion time of 9 min, is in widespread use. The digestion mixture is sulphuric acid/hydrogen peroxide/potassium sulphate with mercuric oxide as a catalyst and the final end-point in the ammonia titration is measured photometrically. The method has been adopted by FOSFA International and is an official method for use on oilseeds and their products.[17] It has also been approved by the AOAC and more recently by the AOCS Uniform Methods Committee for use on soyabean meal,[18] using a copper catalyst and a digestion time of 90 min.

3.3.5 Fibre

The determination of crude fibre has received much attention in the USA with the impetus arising from the control of dehulling of soyabeans and the production of high protein soyabean meal and soya flour with low fibre content. The method has also been progressed by the Crude Fibre Liaison Committee of the AOCS and AOAC since their earlier findings in 1959[19] that all of the methods in existence were inadequate for checking specification limits set up by the National Soyabean Processors' Association for soyabean meal. As a consequence of this attention the AOCS method (Ba6-84 revised 1985) is probably the most detailed method and also includes precision data from tests carried out by the AOAC. The method is a variant of the more general ISO method[20] which does not include precision data. Both methods are based on the Weende method and include grinding and defatting of the test portion, boiling with sulphuric acid solution (0·255 mol/litre), separation and washing of the insoluble residue. The residue is then boiled with sodium hydroxide (0·313 mol/litre) followed by separation, washing, drying and weighing of the insoluble residue and determination of the loss in mass on ashing. In the AOCS method a number of important points are taken into account. The test portion is weighed out from a finely ground sample after determination of the moisture loss on grinding. Fat is then extracted if the level is greater than 1%. In the post-digestion separation process the filtration is carried out on a modified California state Buckner funnel covered with a 200 mesh stainless steel screen. Separate procedures are detailed for asbestos, ceramic fibre or glass wool, all of which can be used to coat the filter screen. In the case of asbestos and ceramic fibre the aid to filtration is added prior to acid digestion, whereas glass wool is formed into a mat on the funnel. The filtration media and retained residue are quantitatively transferred for the second digestion with alkali. In the asbestos procedure the second separation is the same and the asbestos mat plus residue is dried and weighed before determining the loss in weight on ashing at 600°C. A Gooch crucible coated with glass wool or a glass fibre filter disc is used for the other procedures. The asbestos technique is to be deleted shortly by the AOCS for safety reasons.

The ISO method raises further points of interest. De-fatting for example is recommended, but not essential, up to a level of 10% fat and must be quantitatively assessed. The presence of carbonates above a level of 1% requires pre-treatment with hydrochloric acid before acid digestion with sulphuric acid, which would otherwise be diluted in effect. It should be noted that calcium carbonate is used as a carrier for trace element mixtures which may be added to the residues by compounders of feeding stuffs. Separation techniques used include centrifuging, asbestos, sea sand, filter cloth and filter paper. The regulatory method in the UK is similar in principle to the above methods but necessarily limits the choice of procedure. In particular the separation media allowed is restricted to filter paper and special treatment is required for levels of calcium carbonate above 3%.

The importance of the method lies in the fact that the crude fibre mainly consists of substances such as cellulose, lignin, etc., which are not readily digestible and, as such, devalue the use of the residues as feedstuffs. Most countries have their own national regulatory methods. The methods are lengthy and quick methods for quality control using NIR[10] and photomicroscopy[21] have been reported.

A new approach—the Van Soest analysis—is gaining ground in the USA.[22] The method separates the dry matter into two fractions by boiling with neutral detergent solution,[22,23] then filtering. The more soluble (and digestible) fraction in the filtrate, called neutral detergent solubles (NDS), consists of lipids, sugars, starches and protein. The residue, called neutral detergent fibre (NDF), consists of plant cell wall constituents, i.e. hemicellulose, cellulose, lignin, silica and some protein. It is claimed that NDF represents the total fibre fraction more closely than the classical analysis of crude fibre.

3.3.6 Ash

Ash content is the residue obtained after incineration at a defined temperature under certain operating conditions. It is indicative of the level of inorganic material present and can arise from naturally occurring trace elements, residual impurities and residual processing aids, such as filter aid, bleaching earth or pelletising aids. In some countries soapstock is fed back to the toaster when desolventising the meal and the presence of soap can contribute to the ash content. Mineral supplementation of straight residues is not normal practice.

In the AOCS method (Ba5-49) a 2 g test portion taken from a ground analysis sample is weighed into a porcelain combustion capsule and ashed at 600°C for a fixed period of 2 h. The ISO method[24] has recently (1987) been confirmed without amendment and involves incineration of a 2 g test portion at 550°C until constant mass is achieved. It is necessary to heat carefully over a gas flame or on a hot plate until the material carbonises before placing in an electric muffle furnace. The visual appearance of the ash determines the length of the initial ashing. It is usually 2–3 h. After cooling and weighing it is then heated for further 1 h periods until a weight difference less than 0·002 g is achieved.

The UK mandatory method is closely similar to the ISO method but does not include the proviso of ashing to constant mass. The appearance of the ash is the criterion used for assessing complete incineration. It should be white, light grey or reddish and free from carbonaceous particles. When necessary a few drops of ammonium nitrate may be added to assist the ashing process. The addition is followed by oven drying before resuming the calcination. Care should be taken when lead or zinc are present as these metals are slightly volatile and may be gradually lost. Platinum dishes are advised for the operation but should not be used if lead is present as this may alloy with the platinum.

3.3.7 Ash Insoluble in Hydrochloric Acid

It is important nutritionally and economically to differentiate between fractions of the total ash which are of value to the animal. Most elements, with the exception of silver, lead and mercury, form soluble chlorides and the fraction of the total ash which remains undissolved after treatment with hydrochloric acid is of little value.

Both the ISO method[25] and AOCS method (Ba5a-68) employ the same procedure as used for ash content apart from using larger test portions. The total ash is then heated in the presence of concentrated hydrochloric acid, filtered and washed free from chloride with boiling water. The residue is then ashed in the same manner as previously described. The UK regulatory method requires treatment with hydrochloric acid prior to and after the procedure used for total ash content provided that the acid insoluble ash content is above 1·0%. At lower values the sequence is similar to the ISO method. As an example of the expected order of magnitude of results more than 0·5% acid insoluble ash in linseed meal is considered excessive in the USA.

3.3.8 Residual Solvent

In recent years there have been considerable developments in methods for residual solvent analysis focussed particularly on hexane. This attention has been stimulated by rising price trends and stricter safety and environmental regulations. In Europe the growth in production and processing of rapeseed has highlighted particular problems due to the high retention and slow release of solvent from the residue. More than 80% of the solvent losses in a rapeseed extraction plant can be attributed to the meal residue due to inefficient desolventisation.[26] A post treatment has been proposed[27] to compensate for the slow desorption and seed dehulling suggested as a means of further improvement.[28]

Earlier methods used explosimeters for the measurement of hexane, recent methods use gas chromatography. One of these early methods differentiated between free and adsorbed hexane.[29] After measurement and removal of free solvent the remainder was released by adding aqueous detergent to the meal and heating under vacuum on a boiling water bath.

A method for total residual hexane, used in France,[30] has recently been published by ISO[31] and is also available as British Standard 4325 part 10. In principle hexane is desorbed by heating with water (about 50% of the sample weight) at 110°C in a flask sealed with a septum. Subsequently hexane is determined in the head space vapour by gas chromatography using capillary or packed columns. Special precautions are detailed for handling the laboratory sample for this determination. It must be packed in a sealed container such as a crimped metal box (not plastic) and stored at −20°C or below. For analysis it is brought to room temperature and a test portion taken without further preparation. It is worth noting that hexane vapour is heavier than air and test portions from different depths in the container should be analysed. After heating with water for 90 min a sample of the head space gas is taken through the septum with a gas syringe and injected into the chromatograph. The sum of the peak areas of hexane and various hydrocarbons present in the technical extraction solvent (2-methyl pentane, 3-methyl pentane, methylcylopentane, cyclohexane, etc.) is determined. Volatile oxidation products such as hexanal, which may give peaks, are excluded from the measurement. Calibration is carried out with technical *n*-hexane or light petroleum of composition similar to that used in the industrial extraction of oilseeds. A statistical analysis of the results of two interlaboratory tests is presented in the standard. The coefficient of variation of reproducibility was 18% for soyabean and 20% for rapeseed residues.

A method similar in principle used in the USA cottonseed industry has been described by Cherry.[32] A method for free hexane used in Germany, which is similar in principle apart from the desorption step, has not yet received international approval. IUPAC have published a method and it is probable that ISO will adopt the same protocol. Conclusions regarding safety matters from the results of this test should not be derived in isolation from results for total residual hexane.

3.3.9 Urease Activity

The urease enzyme present in soyabeans and soyabean residues will liberate ammonia from urea and it is this property which is used in the determination. The ISO method[33] defines urease activity as: 'the amount of ammoniacal nitrogen liberated per minute by the product, under the specified conditions of operation expressed as milligrams of nitrogen per gram of the product as received or related to the dry material'. The definition in the UK regulatory method of analysis is closely similar to the ISO method whereas AOCS method Ba9-58 requires the results to be expressed as a pH difference between a test solution and a blank.

In each method the sample is finely ground and a small test portion of 0·20 g weighed out. A phosphate buffered urea solution (pH 7·0) is then added, vigorously mixed by shaking, and kept at 30°C for a timed interval of 30 min. The AOCS method requires a blank test to be carried out in the same manner with the omission of the urea followed by pH measurement of test and blank samples. The ISO method follows the timed reaction with acidification using 0·1 N hydrochloric acid, cooling to 20°C and titration of excess acid with 0·1 N sodium hydroxide to pH 4·7. In this instance a blank test is carried out repeating the whole procedure with the proviso that the acid is added before the heating period. A potentiometric titration apparatus or a pH meter sensitive to 0·02 pH unit is necessary for the method where results to the nearest 0·01 mg are required. It is also advisable to defat samples which have more than 10% oil by cold extraction. The test is important in assessing the degree of cooking, particularly under-cooking, of soyabean meal as the decrease in urease activity parallels the decrease of trypsin inhibitor. Soyabeans have an activity in excess of 0·50 mg N/min/g. Opinion differs on acceptable limits for properly cooked meal. For example the European Feed Manufacturers' Federation defines limits of 0·05–0·20 mg N/min/g whereas the FAO Protein Advisory Committee recommends 0·02–0·30 mg N/min/g. Raw soyabeans give an increase in pH of 1·8–2·1 using the AOCS method and properly cooked soyabean meal gives an increase in the range 0·05–0·30.[34,35] It is common trade practice to target for a range of 0·05–0·12 to cover all eventualities, such as use of the residue mixed with added urea and molasses, as ammonia can be released in these circumstances, leading to an unpalatable feed. Urease activity is one of a number of factors to be considered in assessing the quality of soyabean meal.[36] The ultimate test lies in feeding tests with the animals concerned.

3.3.10 Cresol Red Index

The ISO method[37] is based on the property of cresol red dye to combine with the cationic groups in proteins, particularly soyabean protein. Denaturation of the proteins increases the number of cationic groups able to react with cresol red. The quantity of cresol red combined enables the degree of cooking of the product to be assessed. It is especially useful for detecting over-cooking, taken in conjunction with other criteria.

The index is defined as: 'the amount of cresol red bound by the product under the specified conditions of operation expressed in milligrams and related to 500 mg of crude protein ($N \times 6·25$)'. In the method special attention is directed towards test sample preparation, which must be ground to pass a sieve of 200 μm aperture without significant heating. A ball mill is proposed for grinding although one should bear in mind that it is moist heat which will have the biggest effect on denaturation. If the fat content exceeds 10%, defatting is carried out by grinding in the presence of hexane and desolventising at less than 60°C.

A test portion of size equivalent to 100·0 mg of crude protein is taken and agitated for 30 min with a standard cresol red solution in dilute hydrochloric acid. After centrifuging an aliquot is rendered alkaline with sodium hydroxide and the absorbance measured at 570 nm. This is then compared with the absorbance of the standard cresol red solution diluted under the same conditions. The cresol red index expressed in milligrams of cresol red combined per 500 mg of crude protein (N × 6·25) is equal to $[(A_2 - A_1)/A_2] \times 10$ where A_1 is the absorbance of the test solution and A_2 is the absorbance of the diluted standard cresol red solution. The method is based on the method of Glomucki and Bernstein[38] who correlated their results on soyabean meal with chick feeding tests. Properly heated meal was classified by the index range 3·8–4·3 mg dye absorbed per gram of meal. (At a mean value of 4·22 the coefficient of variation was 1·2%.)

3.3.11 Castor Oil Seed Husks (*Ricinus communis*)

Castor oil seed is an undesirable substance in animal feeds and residues and its detection by identification of the husks has merited an International Standard.[39] The standard is applicable to oilseed residues, straight and compound animal feeding stuffs, with a limit of detection of 5 mg/kg. The method involves boiling a test portion successively with nitric acid solution and sodium hydroxide solution, followed by washing and separation of the residue by decantation. It is then examined under a stereomicroscope or binocular lens of magnification ×10–15 and the husk fragments isolated. After drying the fragments are identified using a microscope, by comparing them with similarly treated castor oil seed husks. The husks, which are weighed, have a particular characteristic pitted surface, colour and shape and are well illustrated in the Standard. The method is preferably carried out by an experienced specialist.

The husks are non-toxic and a specific test for the presence of *ricin*, which is the toxic protein present in castor seed meal, has been devised.[40] It is based on the fact that serum from an immune animal will neutralise the toxicity of *ricin*. The presence of castor seed is indicated if an extract of the residue or feedstuff, mixed with normal serum, causes greater mortality when injected into rodents than a mixture of extract with immune serum. The lowest level of determination is 1 mg/kg *ricin*.

3.3.12 Free and Total Gossypol

Gossypol is the predominant yellow toxic pigment, with the formula $G_{30}H_{30}O_8$, occurring in cottonseed. It is a very reactive polyphenolic binaphthaldehyde and exhibits strongly acidic properties. Gossypol can therefore be readily removed from cottonseed oil by alkali washing. The aldehyde groups react with amines to form Schiffs' bases in a condensation reaction. It is present in the pigment glands of cottonseed and, during processing of the seed, the glands may be ruptured by water, aqueous solvents or the shearing forces of rolls and presses. When present in glands or released, gossypol is designated as free and is toxic. Certain processing conditions can cause the free gossypol to react with protein amino groups, thus binding the gossypol in a non-toxic form. The difference between the total gossypol and free gossypol content is called bound gossypol. Knowledge of the levels of free gossypol (which is toxic to mono-gastric species such as pigs and poultry), total and bound gossypol (which are indicative of nutritional quality), are important in animal nutrition. In the Feeding Stuffs Regulations 1982 of the UK, free gossypol is classified as an undesirable substance with restrictive limits of 1200 mg/kg at 12% moisture for cottonseed cake and meal and 20 mg/kg for other residues.

AOCS method Ba7-58 for free gossypol is based on aqueous acetone extraction, which also extracts gossypol-like pigments such as diamino-gossypol and gossypurpurin. An aliquot is diluted with alcohol and treated with aniline, with which both gossypol and like pigments give a coloured reaction product. The gossypol content is then estimated from the absorbance at 440 nm. Pure gossypol can now be purchased for calibration purposes; alternatively it can be prepared from separated pigment glands and purified by re-crystallisation as gossypol acetic acid. Acceptable absorptivity values for standards are included in the method. Bound gossypol is released by hydrolysis and, when in the free state can be estimated with the free gossypol as total gossypol.

In the earlier AOCS method Ba8-55 the hydrolysis was carried out with 0·1 M oxalic acid in methyl ethyl ketone–water azeotrope solution at a temperature of 75°C for a minimum period of 6 h. The method is time consuming, is not applicable to chemically treated meals and, moreover, in mixed feeds there is incomplete removal of free gossypol. Also feed constituents may be extracted which interfere with the subsequent spectrometric determination. The latest (1983) AOCS method Ba-78 uses 3-amino-propan-1-ol as a complexing agent. A test portion sized to contain 0·5–5·0 mg of gossypol is weighed out from the prepared sample and treated at 100°C for 30 min with a solution of 3-amino-propan-1-ol in N,N-dimethylformamide. After cooling to room temperature a mixture of propan-2-ol and hexane is added, an aliquot treated with aniline and the absorbance measured at 440 nm. Precision data from a collaborative study are included in the method. There is also an important note which draws attention to an increase in absorption during a period of up to 1 h after the aniline addition before stable readings are attained.

The UK mandatory method, applicable to straight residues and mixed feeds, is similar to the latest AOCS method for total gossypol but also includes specific absorbances for free and total gossypol which can be used to calculate the result in place of a calibration curve. The method for free gossypol uses 3-amino-propan-1-ol alone as a complexing agent and a mixture of propan-2-ol and hexane for extraction. The lower limit of both determinations is 20 mg/kg.

Pons[41] has reviewed methods of analysis and Berardi and Goldblatt[42] have described the occurrence, analysis and physiological effects of gossypol. In the USA some cottonseed meal is detoxified by binding the gossypol with aniline. Qualitative and quantitative tests for dianilinogossypol are then necessary before assessment of the residual gossypol.[43]

3.4. PROPERTIES

3.4.1 Physical Properties

The physical form of oilseed residues is largely the result of the type of process used in extraction of the oil. When low pressure methods are used, such as hydraulic presses, the residue will be in the form of a slab cake. Residues from high pressure expelling processes are usually smooth-sided sheets, curved in shape, of thickness upwards of 3 mm dependent on the setting of the restrictive choke fitted to the barrel discharge. The pieces of cake may be hand-sized or smaller. Cottonseed cake, for example, is marketed in the USA as nut, sheep, pea and pebble size after breaking and grading on screens.[44] Residues from solvent extraction processes are usually in the form of meal but for transport and storage purposes, to reduce costs and dust problems, they may be in pelletised form. Palm kernel, maize germ and soyabean meal are often handled in this manner. For example soyabean meal in the USA is marketed as cubes, pellets, pea size, cake, chips or flakes but each product is identified by the method of manufacture, i.e. 'hydraulic', 'expeller' or 'solvent extracted'.[45]

The results of a survey of the particle size of solvent extracted soyabean meal from various source countries, carried out by the author, are listed in Table 3.2. The Canadian meal in these experiments was more finely ground than other meals. Meals of this particle size can be used as straight feeding stuffs or mixed with other ingredients as compound feeding stuffs, without further grinding.

When the residue has been pelleted for transportation and storage purposes it is, of course, highly desirable that the physical form is retained to the destination. Hardness and durability of the pellets are important criteria which have been used in this connection. Hardness indicates the ability to withstand compression without fracture and can be measured in different ways. As an example, the weight necessary to disintegrate the pellet with a specially designed knife is one suitable method. Durability indicates the ability of the pellets to withstand attrition and impact without breaking down to meal. A method of measurement developed in the USA consists of removal of fines, subjecting the pellets to a mechanical tumbling action for a timed interval then re-screening. A durability index is then calculated representing the percentage of pellets which have resisted breakdown. Minimum levels of 96% are desirable in practice for bulk transport, somewhat lower levels may be acceptable for bagged material.

One of the methods used in the UK measures pellet breakdown after pneumatic handling in the Holmen Pellet Tester.[46] This method, together with a hardness test method and other physical properties of residues such as abrasiveness, are described in the useful handbook.[46]

TABLE 3.2
Particle size analysis of extracted soyabean meal from various source countries
(% retained)

BSS sieve	UK	Denmark	Canada	Germany
5	nil	nil	nil	nil
10	54	54	2	31
36	44	41	72	51
36 (thro')	2	5	26	18

Colour of the residue is a complaint occasionally raised in commerce. Light or excessively dark colours in soyabean meal, for example, may be indicative of the degree of toasting and confirmation by chemical analysis should be sought. Colours of other residues, such as cottonseed, are influenced by the degree of decortication as the hulls are darker in colour. From experience it is simpler to distinguish burn damaged material by colour and smell than material which has simply been overcooked or overtoasted. It is large changes in colour of a particular source material which require further investigation.

3.4.2   Deleterious Substances and Antinutritional Factors

There are a number of substances present in certain residues which are either toxic to animals or to certain species such as poultry or, more generally, non-ruminants. There are, in addition, substances which impair the feeding value of the residue in some way by restricting the availability to the animal of the desirable constituents present in the residue. These substances are known as antinutritional factors. This whole area of knowledge is quite complex as not only are the substances of complex chemical composition and usually present in very small amounts but also the restrictive effects can be reduced in some cases by processing or elimated in others by seed breeding programmes.

3.4.2.1   Toxic factors

In the UK experience to date has resulted in legislation with limits for the presence of undesirable substances in oilseed residues used in feeding stuffs. The limits for use of the residue as a straight feeding stuff are given in Table 3.3. Aflatoxins and glucosinolates are areas in which a great deal of research has recently been applied, devoted principally to analysis and seed breeding. Both these areas were judged sufficiently important to merit separate chapters in the present book. Further progress in glucosinolate legislation will, no doubt, follow from the recent phased introduction of double zero rapeseed in the EEC and the acceptance of the importance of indolyl glucosinolates which concentrate in these varieties. In the interim the current legislation takes account of egg taint by stipulating a lower tolerance for laying hens. A limit for the castor oil plant is necessary as the seed meal contains the toxic protein ricin, a harmless alkaloid, ricinine and an extremely potent allergen referred to as castor bean allergen. Detoxification of the meal from castor beans is carried out in

TABLE 3.3

Limits in mg/kg[a] for undesirable substances in oilseed residues used in feeding stuffs (UK Feeding Stuffs Regulations)

Substance	Maximum level
Aflatoxin B_1	0·05
Castor oil plant *Ricinus communis* L.	10[b]
Crotalaria L. spp	100
Free gossypol	20, 1200[c]
Hydrocyanic acid	50, 350[d]
Theobromine	300
Vinylthio-oxazolidone	1000,[e] 500[f]
Trace elements	Arsenic 2·0, fluorine 150, lead 10, mercury 0·1
Seeds/residues prohibited	Apricot, bitter almond, black mustard, camelina, croton, Chinese yellow mustard, Ethiopian mustard, mowrah, bassia, madhuca, physic nut, sareptian mustard, unhusked beech mast

[a] Maximum content in mg/kg at 12% moisture, Ref. 1 regulation 15.
[b] Expressed in terms of castor oil plant husks.
[c] Level for cotton cake or meal.
[d] Level for linseed cake or meal.
[e] Level for poultry feed.
[f] Level in feed for laying hens.

the USA but limits are necessary due to the risk of adventitious contamination of other oilseeds. Immature linseed contains a small amount of the cyanogenetic glucoside *linamarin* and an associated enzyme *linase*. At certain temperatures (40–50°C) and conditions of acidity and moisture, hydrocyanic acid can be released enzymatically from *linamarin*. Under normal conditions of crushing linseed the enzyme is destroyed by the high temperature used. It is, however, possible for unchanged enzyme and glucoside to be present in the meal. Toxic hydrocyanic acid can be released if such a meal is fed to animals under moist conditions.

Cottonseed meal can be fed to ruminants in unrestricted proportions but free gossypol in the meal is toxic to mono-gastric species such as pigs and poultry. Again the levels in the meal can be controlled by adjustment of the seed crushing conditions but variability will be experienced with different source materials. The tolerance level of 1200 mg/kg is quite generous even for decorticated cottonseed meal. Crotalaria is a small, heart-shaped toxic weed seed which raised particular problems with soyabeans in the USA due to difficulties in separation. The principal source of theobromine will be in cocoa bean residues.

Trace element levels are restricted as a starting point in the chain of human food legislation in addition to protection of the animal. The seeds and residues which are prohibited exhibit toxicity associated with different constituents. For example in the various mustards and camelina it is with isothiocyanates, in mowrah and bassia with poisonous saponins, and in unhusked beech mast with an alkaloid, fagine.

3.4.2.2 Antinutritional factors

Protease inhibitors, which restrict the proteolytic (protein break-down) activity of some enzymes, are present in many residues, in particular soyabean, groundnut and winged bean.[47] These inhibitors can be inactivated in many cases by adjustment of heating conditions when processing the seed. For example the digestive enzyme trypsin in soyabeans is inhibited by protease in the raw material. The heating of the meal in the presence of open steam when removing solvent post-extraction can reduce the anti-trypsin activity to very low levels and improve the nutritional value of the meal provided that the protein is not denatured by overheating. Analytical tests such as urease activity, cresol red index and protein solubility are used for guidelines in quality control.

Metal binding constituents such as phytic acid—the hexaphosphate derivative of myoinositol—are present in most plant seeds. Most of the phosphorus present is in this form and sesame seed contains one of the highest levels found in nature. Sesame seed meal contains 1·44% phytate on a dry basis,[48] corn germ also contains high levels. The problem is that complexes formed with metallic ions such as iron, zinc and calcium are insoluble, not readily absorbed from the intestinal tract and are therefore unavailable to the animal. On the other hand the metals in complex form are available for the germination of seedlings, therefore the problem is best approached by mineral supplementation of the meal rather than seed breeding.

A number of other factors which need to be taken into account by the animal nutritionist include:

flatus producing factors: oligosaccharides in soyabean meal,
chlorogenic acid in sunflowerseed meal,
alkaloids in meal from castor seed, lupin seed and safflowerseed,
saponins in meal from soyabeans and teaseed,
oxalates in sesame seed meal,
tannins from groundnut skins.

3.4.3 Composition and Nutritional Value

The following discussion is restricted to properties of residues useful in formulating animal feedstuffs where most residues, apart from separated hulls, are used as protein concentrates. This is the major outlet for products which contribute energy, mineral elements and B vitamins to the ration in addition to essential amino acids from proteolysis. Deleterious and anti-nutritional factors have already been described and more positive comment on nutritional properties is necessary to restore the balance. More detailed information on animal nutrition can be found in the references.[49–51]

3.4.3.1 Feeding standards and nutrient requirements

Many countries have manuals of feeding standards setting out the dietary needs of the animal for a particular function such as maintenance, weight gain, milk production, pregnancy, etc. The standards are set in terms of daily requirements, requirements per kilogram of milk or as a percentage of the diet. For example with dairy cows separate standards are given for maintenance, lactation and pregnancy whereas for beef cattle standards are given for maintenance and production or weight gain. In the case of poultry, which are fed to appetite, the standards are given as a percentage of the diet. The maintenance standards represent the feed allowance needed to keep the animal healthy without change of body weight. They are precisely defined in energy terms and include an allowance for voluntary muscular activity.

The criteria used in the standards for nutritive purposes are energy, protein, protein quality, mineral and trace elements and vitamin content. It is important that the same criteria expressed in the same units are used for the evaluation of ration components such as oilseed

TABLE 3.4
Composition/nutritional data: UK, ruminants (dry basis)[a]

Oilseed residue	Chemical analysis						Digest. coeff.				Nutritive value	
	CP[b] (%)	EE (%)	Ash (%)	CF (%)	NFE (%)	GE (MJ/kg)	CP (%)	EE (%)	CF (%)	NFE (%)	ME (MJ/kg)	DCP (%)
Beech mast cake shelled	40·6	9·4	8·8	7·6	33·7	20·4	88	90	24	76	12·6	35·7
Beech mast cake unshelled	21·7	10·1	5·6	29·9	32·8	20·6	75	91	16	51	8·9	16·2
Castor bean meal (detoxicated)	32·4	1·6	7·1	41·2	17·7	19·0	81	93	09	43	6·2	26·3
Coconut cake	23·6	8·1	6·6	12·7	49·1	19·7	78	97	63	83	13·0	18·4
Coconut cake meal	22·0	7·6	7·2	15·3	47·9	19·5	79	97	63	83	12·7	17·4
Cotton cake Bombay	23·1	5·4	6·6	24·8	40·1	19·3	77	94	20	54	8·5	17·8
Cotton cake Brazilian	30·4	6·1	5·0	28·0	30·4	20·1	77	93	21	54	8·9	23·4
Cotton cake Egyptian	26·3	5·7	6·6	24·2	37·2	19·5	77	92	21	54	8·7	20·3
Cotton cake (decorticated)	45·7	8·9	7·4	8·7	29·3	20·8	86	94	28	67	12·3	39·3
Cotton cake (semi-decorticated)	42·6	6·9	6·6	14·3	29·7	20·4	86	93	27	66	11·4	36·6
Groundnut cake (decorticated)	50·4	6·7	6·3	7·2	29·3	20·7	89	90	08	85	12·9	44·9
Groundnut cake (undecorticated)	33·7	10·1	6·3	25·6	24·3	20·9	92	90	11	84	11·4	31·0
Groundnut meal (decorticated extd)	55·2	0·8	6·3	8·8	28·9	19·6	89	86	08	85	11·7	49·1
Groundnut meal (undecorticated extd)	34·3	2·1	4·7	27·3	31·6	19·5	92	79	11	69	9·2	31·6

Hempseed cake	34.4	9.7	8.7	26.8	20.4	20.5	74	90	08	58	9.0	25.5
Hempseed meal	39.4	1.9	10.6	29.1	19.0	18.6	75	77	08	53	6.9	29.6
Kapok cake	31.3	8.1	7.3	29.9	23.3	20.3	74	91	20	50	8.7	23.2
Linseed cake English made	33.2	10.7	5.9	10.2	40.0	20.9	86	92	49	80	13.4	28.6
Linseed cake foreign	35.4	7.7	6.2	10.4	40.2	20.3	86	93	50	80	12.9	30.5
Linseed meal (extd)	40.4	3.6	7.3	10.2	38.4	19.4	80	90	50	80	11.9	34.8
Niger cake	36.4	6.6	10.4	20.3	26.2	19.4	80	81	27	84	10.5	29.2
Olive cake	7.1	20.1	6.1	33.8	32.9	22.1	97	95	33	70	12.7	6.9
Palm kernel cake	21.6	6.8	4.4	15.0	52.2	19.8	91	88	38	85	12.8	19.6
Palm kernel meal (extd)	22.7	1.0	4.4	16.7	55.2	18.5	90	89	50	88	12.2	20.4
Poppy seed cake	40.8	10.8	15.2	9.2	24.0	19.6	79	93	49	64	11.3	32.2
Rape cake	38.8	10.6	13.6	9.1	28.0	19.8	83	79	08	80	11.4	32.2
Rape meal (extd)	41.3	3.4	8.2	10.4	36.6	19.2	83	77	11	80	10.9	34.3
Sesame cake English	49.1	13.1	9.8	4.9	23.1	21.5	90	90	31	56	13.0	44.2
Sesame cake French	41.2	12.1	9.7	18.7	18.3	21.1	90	90	31	56	11.7	37.1
Sesame meal (extd)	49.3	2.6	11.4	8.2	28.4	18.8	90	92	31	56	10.4	44.4
Soyabean cake	50.4	6.6	6.2	6.0	30.8	20.7	90	91	72	77	13.3	45.4
Soyabean meal (extd)	50.3	1.7	6.2	5.8	36.0	19.5	90	93	71	77	12.3	45.3
Sunflower cake (decorticated)	41.3	15.2	7.4	13.4	22.6	22.1	90	88	30	71	13.3	37.2
Sunflower cake (undecorticated)	20.6	8.0	8.0	32.3	31.1	19.6	90	88	18	71	9.5	18.5
Sunflower meal (extd)	42.3	1.1	7.2	18.1	31.2	19.0	90	90	30	71	10.4	38.1
Walnut cake	40.4	14.1	5.9	7.7	31.9	22.0	90	95	25	85	14.7	36.4

[a] All contain 90% dry matter: further data in Ref. 54.
[b] For abbreviations, see Table 3.8 and text.

residues and presently this applies only on a national basis. Internationally differences exist, particularly in energy assessment, for example, values for net energy are considerably lower in the USA than those calculated from metabolisable energy in the UK. The Agricultural Research Council has published standards for ruminants, pigs and poultry in the UK[52] and the National Research Council in the US.[53] These standards have been revised in the last 8 years, keeping pace with advances in research on energy requirements. The properties of oilseed residues required to compute the formulation of a ration on an additive basis to meet the required standards are next described. Compositional data for ruminants used in the UK, mostly on imported residues,[54] are given in Table 3.4. Data used in the USA[55] are given in Tables 3.5, 3.6 and 3.7.

3.4.3.2 Energy

The classical or proximate analysis of feed constituents attributed to Henneberg and Stohmann is over 100 years old and is still in use for legal requirements in the UK. The analyses and chemical components are given in Table 3.8. These analyses, together with digestibility data, formed the basis of Kellner's energy system of starch equivalents. In this system the net energy value of a feed for fattening is expressed relative to the net energy value of starch. Energy values can be calculated using a net energy value of starch as $2 \cdot 36 \, kcal/g$. This approach, widely used until recently, and still used in East Germany, has now been superseded and more complex systems introduced with no uniform approach.

For further understanding of developments it is necessary to define certain energy terms which have been set out by the National Research Council.[56]

Gross energy (GE). Total heat of combustion as measured by a bomb calorimeter.

Digestible energy (DE). Gross energy minus the energy content of faeces. It includes metabolisable energy, energy of urine and gaseous products of digestion (mostly methane).

Metabolisable energy (ME). DE minus energy of urine and gaseous digestion products. It is used by the animal for work, growth, fattening, foetal development, milk production and/or heat production.

Net energy (NE). The difference between ME and heat increment. Net energy for maintenance (NE_{main}) is the energy used to keep the

Composition/nutritional data: US, ruminants (as fed)

Oilseed residue	Dry matter (%)	CP (%)	EE (%)	Ash (%)	CF (%)	TDN (%)	DE (Mcal/kg)	ME (Mcal/kg)	NE_{main} (Mcal/kg)	NE_{main} (Mcal/kg)	NE_{lac} (Mcal/kg)
Copra meal (solvent extd.)	91·0	21·3	3·5	6·0	14·0	68·0	3·01	2·63	1·57	1·01	1·56
Cottonseed hulls	91·0	3·7	1·5	2·6	43·3	41·0	1·80	1·41	0·88	0·05	0·89
Cottonseed meal (exp., 36% pro.[a])	92·0	38·6	4·2	6·7	14·3	67·0	2·97	2·58	1·54	0·97	1·54
Cottonseed meal (exp., 41% pro.[a])	93·0	41·0	4·6	6·1	11·9	72·0	3·19	2·80	1·68	1·10	1·66
Cottonseed meal (solvent extd, 41% pro.[a])	91·0	41·2	1·4	6·5	12·1	70·0	3·06	2·68	1·60	1·04	1·59
Linseed exp.	91·0	34·3	5·4	5·7	8·8	74·0	3·28	2·90	1·76	1·18	1·71
Groundnut hulls	91·0	7·1	1·8	3·8	57·3	20·0	0·89	0·48	0·70	—	0·38
Groundnut meal (solvent extd)	92·0	48·1	1·3	5·8	9·9	71·0	3·12	2·74	1·64	1·07	1·63
Rapeseed meal (solvent extd, decorticated)	91·0	37·0	1·7	6·8	12·0	63·0	2·77	2·39	1·41	0·86	1·43
Safflowerseed meal (solvent extd, decorticated)	92·0	43·0	1·3	7·5	13·5	67·0	2·95	2·57	1·53	0·97	1·53
Sesameseed meal exp.	93·0	45·5	6·9	11·2	5·7	71·0	3·15	2·76	1·66	1·08	1·64
Soyabean hulls	91·0	11·0	1·9	4·6	36·4	70·0	3·09	2·71	1·63	1·06	1·61
Soyabean meal (solvent extd, 44% pro.[a])	89·0	44·6	1·4	6·5	6·2	75·0	3·31	2·94	1·79	1·20	1·73
Soyabean meal (solvent extd) (dehulled, 49% pro.[a], decorticated)	90·0	49·7	0·9	5·8	3·4	78·0	3·46	3·09	1·89	1·28	1·81
Sunflowerseed meal (solvent extd, decorticated)	93·0	46·3	2·9	7·6	11·4	60·0	2·67	2·27	1·34	0·76	1·37

[a] pro. = protein.

TABLE 3.6

Amino acid composition of oilseed residues (% dry basis)[55]

Oilseed residue	Arginine	Glycine	Histidine	Isoleucine	Leucine	Lysine	Methionine	Cystine	Phenylalanine	Tyrosine	Serine	Threonine	Tryptophan	Valine
Copra meal, extracted	2·65	1·14	0·41	0·91	1·59	0·66	0·35	0·27	0·95	0·63	—	0·73	0·22	1·14
Maize germ meal	1·43	1·20	0·76	0·76	1·97	0·98	0·64	0·44	0·98	0·76	1·09	1·19	0·21	1·31
Cottonseed meal (exp. 41% pro.)	4·51	2·06	1·15	1·56	2·50	1·73	0·62	0·78	2·35	1·01	1·84	1·44	0·57	2·05
Cottonseed meal (extd, 41% pro.)	4·62	2·17	1·21	1·67	2·56	1·86	0·64	0·85	2·46	1·13	1·92	1·52	0·61	2·06
Linseed meal (exp.)	3·10	1·80	0·71	1·86	2·11	1·30	0·64	0·67	1·53	1·06	2·09	1·25	0·56	1·77
Groundnut meal (extd)	4·95	2·56	1·03	1·91	2·94	1·93	0·46	0·79	2·22	1·65	3·37	1·26	0·52	2·04
Palm kernel meal (extd)	2·36	0·81	0·32	0·64	1·19	0·54	0·33	0·28	0·79	0·47	—	0·61	0·20	0·82
Rapeseed meal (extd)	2·26	1·97	1·09	1·48	2·74	2·18	0·78	0·33	1·55	0·87	1·72	1·72	0·47	1·96
Safflowerseed meal (decorticated, extd)	3·98	2·54	1·16	1·70	2·68	1·38	0·74	0·76	1·91	1·17	—	1·42	0·65	2·54
Soyabean meal (extd, 44% pro.)	3·38	2·03	1·19	2·27	3·65	2·99	0·58	0·83	2·36	1·48	2·36	1·85	0·71	2·25
Soyabean meal (extd, dehulled, 49% pro.)	4·07	2·68	1·35	2·73	4·14	3·52	0·79	0·83	2·71	1·86	—	2·15	0·77	2·82
Sunflowerseed meal (decorticated, extd)	4·75	3·03	1·32	2·42	4·12	2·06	1·25	0·79	2·54	1·49	2·37	2·07	0·65	2·80

—Indicates no result available.

TABLE 3.7
Mineral element composition of oilseed residues (dry wt basis)[55]

Oilseed residue	Ca (%)	Cl (%)	Mg (%)	P (%)	K (%)	Na (%)	S (%)	Co (mg/kg)	Cu (mg/kg)	I (mg/kg)	Fe (mg/kg)	Mn (mg/kg)	Se (mg/kg)	Zn (mg/kg)
Copra meal (extd)	0·19	0·03	0·36	0·66	1·63	0·04	0·37	0·14	10·0	—	750·0	72·0	—	—
Cottonseed hulls	0·15	0·02	0·14	0·09	0·87	0·02	0·09	0·02	13·0	—	131·0	119·0	—	22·0
Cottonseed meal (exp., 36% pro.)	0·20	—	0·58	1·04	1·46	0·05	0·28	0·16	20·0	—	197·0	25·0	—	—
Cottonseed meal (exp., 41% pro.)	0·21	0·05	0·58	1·16	1·45	0·05	0·43	0·17	20·0	—	197·0	24·0	—	69·0
Cottonseed meal (extd, 41% pro.)	0·18	0·05	0·59	1·21	1·52	0·05	0·28	0·17	22·0	—	228·0	23·0	—	68·0
Groundnut meal (extd)	0·29	0·03	0·17	0·68	1·23	0·08	0·33	0·12	17·0	0·07	154·0	29·0	—	22·0
Linseed meal (exp.)	0·45	0·04	0·64	0·96	1·34	0·12	0·41	0·46	29·0	0·07	194·0	42·0	0·89	36·0
Palm kernel meal (extd)	0·21	—	—	0·50	—	—	—	—	—	—	—	164·0	—	53·0
Rapeseed meal (extd)	0·67	0·11	0·60	1·04	1·36	0·10	1·25	—	—	—	—	—	1·07	—
Safflowerseed meal (decorticated, extd)	0·38	0·18	1·11	1·40	1·19	0·05	0·22	2·15	9·0	—	528·0	43·0	—	36·0
Soyabean hulls	0·49	—	—	0·21	1·27	0·01	0·09	0·12	1810·0	—	324·0	11·0	—	24·0
Soyabean meal (extd, 44% pro.)	0·34	0·04	0·30	0·70	2·20	0·04	0·47	0·10	25·0	0·15	133·0	32·0	0·34	48·0
Soyabean meal (extd, dehulled, 49% pro.)	0·29	0·05	0·32	0·70	2·30	0·03	0·48	0·07	22·0	0·12	148·0	41·0	0·11	61·0
Sesameseed meal (exp.)	2·17	0·07	0·50	1·46	1·35	0·04	0·35	—	—	—	100·0	52·0	—	108·0
Sunflowerseed meal (decorticated, extd)	0·44	0·11	0·77	0·98	1·14	0·24	—	—	4·0	—	33·0	20·0	—	—

TABLE 3.8
Classical analysis of feed components

Analysis	Components	
1. Moisture	Water, volatile acids, residual hydrocarbons	
2. Ash	Essential elements:	
	Major:	calcium, potassium, magnesium, sodium, sulphur, phosphorus, chlorine
	Trace:	iron, manganese, copper, cobalt, iodine, zinc, molybdenum, selenium, chromium
	Non-essential elements:	silicon, nickel, titanium, aluminium, vanadium, boron, lead, tin
3. Crude protein (CP)	Protein, amino acids, amines, nitrates, nitrogenous glycosides, glycolipids, nucleic acids, B vitamins	
4. Ether extract (EE) (diethyl ether)	Fats, oils, waxes, fatty acids, pigments, sterols, phosphatides, vitamins A, D, E, K	
5. Crude fibre (CF)	Cellulose, hemi-cellulose, lignin	
6. Nitrogen-free extractives (NFE) (calculated)	Cellulose, hemi-cellulose, lignin, sugars, fructans, starch, pectins, organic acids, resins, tannins, pigments, water-soluble vitamins	

% NFE = 100 − (ash + moisture + CP + EE + CF) %.

animal in equilibrium without loss or gain in weight. Net energy for production or gain (NE_p or NE_{gain}) is the net energy required in addition to NE_{main} for work, tissue gain or production of milk, eggs, wool, fur, etc. For dairy cows a separate component—net energy for lactation—NE_{lac} is used.

Heat increment (HI). The increase in heat production following consumption of feed when the animal is in a thermo-neutral environment. It is the difference between ME and NE and is incidental to nutrient digestion and metabolism.

Energy systems for ruminants in the UK use ME as the basis for feed components and the enery requirements of the animal are expressed as NE. A complex system of equations, attributed to Blaxter, is used to relate the two values in terms of maintenance, growth and lactation. In the USA both animal requirements and feeds were evaluated in terms of *Total Digestible Nutrients* (TDN) defined by the equation

$$TDN = \%DCP + \%DCF + \%DNFE + 2 \cdot 25\%DEE$$

DCP = digestible crude protein,
DGF = digestible crude fibre,
DNFE = digestible nitrogen-free extract,
DEE = digestible ether extract.

Digestibility is the proportion of the feed not excreted in faeces and is assumed to be absorbed by the animal. It is expressed on dry matter as a digestibility coefficient or a percentage. A decision was taken in 1958 to phase out this system and replace it with a net energy system in which each feed is given separate NE values, i.e. NE_{main}, NE_{gain} and NE_{lac} calculated from ME values. For pigs and poultry there is less emphasis on energy systems as the range of feed components used is more limited due to the comparative inability to digest fibre and the supply of rations is not restricted. ME and DE values are used for the formulation of rations and assessing the relative quality of residues. It can be seen from energy values in Table 3.5 that GE ranges from 18·8 to 22·1 MJ/kg (extracted sesame meal → olive cake). ME values are much lower with a much wider range: 6·2–13·4 MJ/kg (hempseed cake → linseed cake).

3.4.3.3 Protein

In addition to the value of protein in contributing to the energy levels, the protein composition in terms of essential amino acids must be taken into account. These amino acids are eseential for growth and cannot be synthesised or produced from other amino acids by the animal. They include arginine, histidine, isoleucine, leucine, lysine, methionine, phenylalanine, threonine, tryptophan and valine. Furthermore, unless the protein is digested by the animal and broken down into simpler substances it will not be absorbed. Digestible crude protein (DCP) levels are, therefore, the necessary attributes. For pig and poultry rations available lysine and methionine are valuable, as cereals, which form the basis of the ration, are deficient in these amino acids. Moreover the digestive system of these non-ruminants is such that the amino acids released from the proteins can be absorbed, unchanged, by the animal.

The assessment of protein value for ruminants is not quite so straightforward as the action of micro-organisms in the rumen facilitates the synthesis of essential and non-essential amino acids. The mixture of amino acids, finally absorbed, does not relate to the amino-acid content of the ration. For practical purposes therefore, the value of the feed protein for ruminants is based on the crude protein content. These values are recorded in Tables 3.4 and 3.5 for a number of oilseed residues and the amino-acid composition of USA residues is given in Table 3.6. Inspection of Tables 3.4 and 3.5 shows that the residues, apart from hulls, are deservedly called protein concentrates, and removal of hulls by decortication increases the protein content and reduces fibre levels. In general terms the residues have a deficit of at least one essential amino acid and are low in cystine and methionine with a variable low lysine content.

Soyabean meal in the USA is graded as 44% and 49% protein. The hulls removed in manufacture may be used in part to adjust the protein levels. This is common practice in the USA where it is also used on cottonseed and groundnut meals. It is regarded as the top class oilseed residue as it is high in protein, contains all the essential amino acids and is low in fibre content. Groundnut and cottonseed are both low in cystine and methionine: in formulations, however, the first limiting amino acid is lysine.

Both sesame seed and sunflowerseed residues are comparable to soyabean meal in protein levels but the protein quality is not so good when decorticated. The protein levels of sunflowerseed and sesame seed meals are increased but they are still low in lysine content. Rapeseed meal has an acceptable balance of amino acids but the protein content and its digestibility are lower than soyabean meal.

Coconut and palm kernel meals have relatively low levels of protein but the restriction on use to ruminants is based on the high fibre levels. In feed formulations all oilseed residues are supplemented with other protein sources to achieve the required amino-acid balance.

3.4.3.4   Ether extract (EE)

As previously described (Table 3.8) a number of compounds are included under this heading. The nomenclature is an unfortunate choice and may refer to hexane extract (or oil content), petroleum ether extract, diethyl ether extract or even carbon tetrachloride extract. It is therefore necessary to establish the method of analysis used when comparing data in the tables. The nutritional interest stems mainly from the free fatty acid and triglyceride content providing a source of energy and essential fatty acids. Decorticated, expelled or hydraulic pressed residues have the highest levels of residual EE. Solvent extracted residues in normal efficient processing have levels less than 2·0% but with some difficult materials higher levels may be encountered. The essential fatty acids (EFA), all polyunsaturated, are regarded as linoleic, linolenic and arachidonic acids. The fatty acid composition of fats present in the various residues can be found in Chapter 7 and need no further comment. The residues are an important source of EFA for pigs and poultry but with ruminants there is a reduction in EFA in the rumen due to hydrogenation. The level and nature of the EE can also influence the yield and fat content of milk in dairy cows and groundnut, coconut and palm kernel residues are used in this connection.

3.4.3.5 Fibre

The amount and chemical composition of the crude fibre affects digestibility of the residues. Most of the fibre is present in the hull of the seed and much lower levels can be achieved by decortication. There is however a certain amount of fibre present in the meat. For example virtually complete removal of hulls from soyabeans still leaves a level of 3·0–3·5% in the extracted meal. Crude fibre contains only part of the cellulose, hemicellulose and lignin from the cell walls. Lignin is resistant to enzyme digestion and the level increases with age of the plant tissues. According to Van Soest[22] the neutral detergent fibre corresponds more closely to the total fibre fraction in a feed although it is not a uniform chemical entity. It is suggested that lignin be determined separately and allowed for when predicting digestibility of the neutral detergent fibre. As a start some data on cell walls, hemicellulose, cellulose, lignin and acid detergent fibre contents have been published.[55] It is also possible that data from this procedure can be used to predict energy values.

The importance of fibre digestibility, even for ruminants, can be seen in Table 3.4 with a range of digestibility coefficients from 8% for rape cake to 72% for soyabean cake.

3.4.3.6 Minerals

Nutritionally mineral elements are important for a number of purposes of which the most well known will be calcium and phosphorus for skeletal development such as teeth and bone. A number of other elements have also been identified as essential including major amounts of potassium, sodium, chlorine, sulphur, magnesium and trace amounts of iron, zinc, copper, manganese, iodine, cobalt, molybdenum, selenium and chromium. There are losses of minerals from the body in urine, faeces, skin, etc., which require replacement.

Special consideration has also to be given to dairy cows, quickly growing pigs, young cattle and laying hens, where losses may occur in milk and eggs, and higher levels are required in the short term. Minerals present in the rations must also be in a chemical form available to the animal. As an example phosphorus present in the form of phytates is of limited availability to non-ruminants but is available to ruminants. Other essential elements also form insoluble complexes with phytic acid. The mineral element composition of oilseed residues is listed in Table 3.7 from which it can be seen that, although the phosphorus levels are satisfactory, the calcium to phosphorus ratio is less than 1 in most residues. For nutritional purposes a ratio of between 1:1 and 2:1 (Ca:P) is regarded as desirable.

Although phytates are present in most residues their influence on quality is variable, as in some instances protein solubility is affected as well as metal complexation. Problems with rapeseed meal,[57] sesame seed meal[58] and corn germ meal[47] have been discussed. As previously mentioned sesame seed meal contains the highest level of phytate, where it is complexed with magnesium. In this instance binding of zinc has been found when it is used in chick rations although the effect on protein solubility is minimal.

3.4.3.7 Vitamins

Oilseed residues are generally accepted as good sources of the B vitamins but are low in carotene and vitamin E. Ruminants are able to synthesise B vitamins by the action of micro-organisms in the rumen. The B vitamin content of residues is therefore mainly of interest in rations for pigs, poultry and young calves.

REFERENCES

1. Her Majesty's Stationery Office, *The Feeding Stuffs Regulations 1982*, Statutory Instrument No. 1143, HMSO, London.
2. Her Majesty's Stationery Office, *The Feeding Stuffs (Sampling and Analysis) Regulations 1982*, Statutory Instrument No. 1144, HMSO, London.
3. American Oil Chemists' Society, *Official and Tentative Methods of the AOCS. Section B Sampling and Analysis of Oilseed By-products*, 3rd edn, ed. R. O. Walker, AOCS, Champaign, 1978 (corrected and revised to 1986).

4. Woodman, H. E., *Rations for livestock*, Bulletin No. 48, HMSO, London, 1952, p. 9.
5. International Standards Organisation, *Oilseed residues—Preparation of test samples*, ISO, Geneva, Standard No. 5502, 1983.
6. International Standards Organisation, *Oilseed residues—Determination of hexane extract (or light petroleum extract) called 'oil content'*, ISO, Geneva, Standard No. 734, 1979.
7. International Standards Organisation, *Oilseed residues—Determination of diethyl ether extract*, ISO, Geneva, Standard No.736, 1977.
8. Association of Official Analytical Chemists, *Official Methods of Analysis*, 14th edn, AOAC, Arlington, Method No. 7.062, 1984.
9. International Standards Organisation, *Oilseed residues—Determination of moisture and volatile matter content*, ISO, Geneva, Standard No. 771, 1978.
10. Ribaillier, D. and Maviel, M. F., *Rev. franc. corps gras*, **31** (1984) 181.
11. International Standards Organisation, *Animal feeding stuffs—Determination of nitrogen and calculation of crude protein content*, ISO, Geneva, Standard No. 5983, 1979.
12. Hunt, W. G. In: *Seminar Proceedings—Products of the corn refining industry in food*, Corn Refiners' Association, Washington, DC, 1978, p. 70.
13. Watt, B. K. and Merrill, A. L. In: *Composition of foods*, Agriculture Handbook No. 8, USDA, Washington, DC, 1963, p. 159.
14. AOAC Nitrogen Committee, *J. Am. Oil Chem. Soc.*, **58** (1981), 415A.
15. Federal Grain Inspection Service, *FGIS Notice 82-11*, USDA, Washington, DC 1982.
16. Coles, B. A., *J. Am. Oil Chem. Soc.*, **57** (1980) 202.
17. Federation of Oils Seeds & Fats Associations, Ltd (FOSFA International), *Manual*, FOSFA International, London, 1982, p. 277.
18. AOCS Uniform Methods Committee, *J. Am. Oil Chem. Soc.*, **64** (1987) 1516.
19. Holt, K. F., *J. Am. Oil Chem. Soc.*, **36** (1959) 549.
20. International Standards Organisation, *Agricultural food products—Determination of crude fibre content—general method*, ISO, Geneva, Standard No. 5498, 1981.
21. Anderson, R. E. and Holt, K. F., *J. Am. Oil Chem. Soc.*, **44** (1967) 583.
22. Van Soest, P. J., *J. Assoc. Off. Anal. Chem.*, **49**, (1966) 546.
23. Goering, H. K. and Van Soest, P. J. In: *Forage fibre analysis*, USDA ARS Agricultural Handbook No. 379, Government Printing Office, Washington, DC 1970.
24. International Standards Organisation, *Oilseed residues—Determination of total ash*, ISO, Geneva, Standard No. 749, 1977.
25. International Standards Organisation, *Oilseed residues—Determination of ash insoluble in hydrochloric acid*, ISO, Geneva, Standard No. 735, 1977.
26. Dahlen, J. A. H. and Lindh, L. A., *J. Am. Oil Chem. Soc.*, **60** (1983) 2009.
27. Knuth, Van M., *Fette Seifen Anstrichmittel*, **86** (1984) 497.
28. Schneider, Von F. H. and Rutte, U., *Fette Seifen Anstrichmittel*, **84** (1984) 331.

29. Pritchard, J. R. L., Farmer, S. N. and Wong, D. R., *Chemy Ind.*, (50) (1964) 2062.
30. Wolff, J. P., *J. Am. Oil Chem. Soc.*, **60** (1983) 220.
31. International Standards Organisation, *Oilseed residues—Determination of total residual hexane*, ISO, Geneva, Standard No. 8892, 1987.
32. Cherry, J. P., *J. Am. Oil Chem. Soc.*, **60** (1983) 360.
33. International Standards Organisation, *Soyabean products—Determination of urease activity*, ISO, Geneva, Standard No. 5506, 1978.
34. Diser, G. M., Technical paper No. 233, Archer Daniels Midland Co., Minneapolis, USA.
35. Diser, G. M., *Milling* (April 10, 1964) 373.
36. Wright, K. N., *J. Am. Oil Chem. Soc.*, **58** (1981) 294.
37. International Standards Organisation, *Soyabean products—Determination of cresol red index*, ISO, Geneva, Standard No. 5514, 1979.
38. Glomucki, E. and Bernstein, S., *J. Assoc. Off. Anal. Chem.*, **43** (1960) 440.
39. International Standards Organisation, *Animal feeding stuffs—Determination of castor oil seed husks—microscopical method*, ISO, Geneva, Standard No. 5061, 1983.
40. Clark, E. C., *J. Pharm. Pharmac.* (1953) 458.
41. Pons, W. A. Jr, *J. Assoc. Off. Anal. Chem.*, **60** (1977) 252.
42. Berardi, L. C. and Goldblatt, L. A. In: *Toxic constituents of plant foodstuffs*, ed. I. E. Liener, 2nd edn, Academic Press, New York, 1980.
43. National Cottonseed Products Association, *Trading Rules 81–82*, NCPA, Memphis, TN, 1981, p. 137.
44. National Cottonseed Products Association, *Trading Rules 81–82*, NCPA, Memphis, TN, 1981, p. 90.
45. American Soyabean Association, *Soya blue book 81*, ASA, St Louis, Mo., 1981, p. 175.
46. MacMahon, M. J. and Payne, J. D. In: *Holmen pelleting handbook*, Holmen Chemicals Ltd, Basingstoke, UK, 1981.
47. Liener, I. E. and Kakade, M. L. In: *Toxic constituents of plant foodstuffs*, ed. I. E. Liener, 2nd edn, Academic Press, New York, 1980, p. 7.
48. O'Dell, B. L. and De Boland, A., *J. Agric. Fd Chem.*, **24** (1976) 804.
49. McDonald, P., Edwards, R. A. and Greenhalgh, J. F. D., *Animal nutrition*, 3rd edn, Longmans, London, 1981.
50. Cullison, A. E. and Lowrey, R. S., *Feeds and feeding*, 4th edn, Prentice-Hall, N.J., 1987.
51. Bondi, A. A., *Animal nutrition*, John Wiley, Chichester, 1987.
52. Agricultural Research Council, *The nutrient requirements of farm livestock, No. 1 Poultry, No. 2 Ruminants, No. 3 Pigs*, HMSO London, 1963 (No. 1), 1980 (No. 2), 1960 (No. 3).
53. National Academy of Sciences, National Research Council: *Nutrient requirements of poultry*, Pub. No. ISBN 0 309 01867 7, 1971; *Nutrient requirements of swine*, Pub. No. ISBN 0 309 01599 5, 1968; *Nutrient requirements of dairy cattle*, Pub. No. ISBN 0 309 01916 8, 1971; *Nutrient requirements of beef cattle*, Pub. No. ISBN 0 309 01754 8, 1970; *Nutrient requirements of sheep*, Pub. No. ISBN 0 309 01693 2, NRC, Washington, DC 1968.

54. Agricultural Development & Advisory Service, *Nutrient allowances and composition of feeding stuffs for ruminants,* Advisory Paper No. 11, 2nd edn, HMSO, London, 1976.
55. National Academy of Sciences, National Research Council, *US/Canadian tables of feed composition,* 3rd edn, NRC, Washington, DC, 1982.
56. Committee on Animal Nutrition of the Agricultural Board, *Nutrient requirements of domestic animals—glossary of energy terms,* Pub. No. 1040, National Academy of Sciences, NRC, Washington, DC, 1962.
57. Siy, R. D. and Talbot, F. D. F., *J. Am. Oil Chem. Soc.,* **59** (1982) 191.
58. Lyon, C. K., *J. Am. Oil Chem. Soc.,* **49** (1972) 245.

4

Methods of Analysis for Mycotoxins—an Overview

P. M. Scott

Health and Welfare Canada, Health Protection Branch, Bureau of Chemical Safety, Ottawa, Canada K1A 0L2

4.1. INTRODUCTION

Mycotoxins are toxic substances produced by fungi and can be classified according to their fungal origin, chemical structure and biological activity. Occurrence of these toxins in human foods is mainly as a result of direct contamination of the agricultural commodity and their survival of food processing to some extent. Oilseeds such as groundnuts (peanuts), cottonseed, maize (corn), and copra are particularly favourable substrates for aflatoxin formation. Environmental factors affecting mycotoxin production in general include moisture, temperature, interference by other fungi, the gaseous environment, preservatives and, for growing plants, varietal resistance and plant stress. Transmission of mycotoxins or their metabolies to edible animal products such as milk may result from use of contaminated feeds. Foods may also be sometimes contaminated by direct mould growth, particularly at the consumer level. Specific regulations pertaining to mycotoxins, particularly aflatoxins, exist in many countries for foods and foodstuffs and in several countries for animal feeds; in some cases the maximum tolerated level is based on the detection limit of the analytical method used. The detection and reliable measurement of mycotoxins is essential for control at all stages of food production, survey work, and research on stability, toxicology, metabolism, production and epidemiology of mycotoxins. Analytical methods for mycotoxins may be divided generally into quantitative (or semiquantitative) assays and rapid screening tests; into methods for a

141

specific mycotoxin (or group of closely related mycotoxins) and multimycotoxin methods; into physicochemical determinations and bioassays; and into presumptive and confirmatory methods. Determination techniques include thin layer chromatography (TLC), high performance liquid chromatography (HPLC), gas chromatography (GC), mass spectrometry (usually in combination with GC), and immunoassay. Methods have been specifically developed for determination of aflatoxins (and in some cases other mycotoxins) in various oilseeds, oils and fatty foods.

Mycotoxins may be defined as secondary metabolites produced by filamentous fungi and having harmful biological effects in animals and, in some cases, man. Some would extend this definition to include toxins produced by higher fungi and fungal metabolites that have a toxic effect on any living organism or on cells in culture. However, for evaluation of possible human health hazard from the presence of mycotoxins in food, data from animal studies are essential. These must be coupled with a knowledge of concentrations of mycotoxins in agricultural commodities used as foodstuffs (and what happens to the mycotoxins during food processing) or of the concentrations in foods themselves as consumed. Epidemiological studies must also be considered.

It is apparent that quantitative and qualitative analyses of foods and foodstuffs for mycotoxins constitute a critical reference point for both research and regulatory work. The problem is not simple. Although considerable progress has been made since the first methods for detection and determination of aflatoxins appeared in the early 1960s, we are generally limited to analyses using authentic mycotoxin standards and available methods. To put this into perspective, there were about 3000 structurally characterized fungal metabolites known in 1983[1,2] and many of these have never been tested for mammalian toxicity. In 1985, Watson[3] estimated that 432 fungal metabolites known in the literature were toxic in the widest sense and of these over a hundred possessed mammalian toxicity. Cole and Cox[4] in 1981 compiled the properties of over 270 mycotoxins from filamentous fungi (although some of these were related metabolites and had no known toxicity). In 1983, Ciegler and Vesonder[5] listed about 170 fungal toxins, including some toxins from higher fungi, while van Egmond[6] in 1984 gave a figure of over 200 different known mycotoxins. These figures are all on the low side—more recent work has raised the number of the trichothecenes alone to over 100.[7,8] In any event, and considering the wide variety of chemical structures, the

TABLE 4.1
Mycotoxins from *Alternaria* spp.[15–17]

Tenuazonic acid
Alternariol
Alternariol monomethyl ether
Altenuene
Altertoxins I, II, III

number of known mycotoxins is just too large for screening by presently available analytical methods. The analyst must be selective, particularly for quantitative determinations, and take into account the particular situation. He should ideally be guided by the type and incidence of toxigenic fungal species reported to contaminate the foodstuff,[9] by knowledge of the known mycotoxins these fungi are able to produce[4,9,10] and by toxicological information,[8,11,12] with emphasis on chronic toxic effects for most regulatory analyses of human foodstuffs. Take as an example the potential for mycotoxin contamination of sunflower seeds. Two studies in Argentina and India showed that fungi of the *Alternaria, Cladosporium, Aspergillus, Penicillium, Fusarium, Rhizopus* and *Curvularia* genera were isolated from sunflower seeds in highest incidence.[13,14] Of the Aspergilli, *A. wentii* and *A. flavus* occurred most frequently in whole seeds while other species of fungi known to be toxigenic included *A. ochraceus, F. moniliforme* and *P. citrinum*; the Penicillia were not identified as to species in the Argentinian study. Just to consider the *Alternaria* and the first two *Aspergillus* species, whose known mycotoxins are listed in Tables 4.1–4.3, and *P. citrinum* it appears that multimycotoxin analysis that includes tenuazonic acid, emodin, wentilactone A, aflatoxins, sterigmatocystin, α-cyclopiazonic acid and citrinin could initially be carried out. However, extension of this reasoning to include other fungal species would lead to a long list of known mycotoxins to be analysed in sunflower seeds, for many of which

TABLE 4.2
Mycotoxins from *Aspergillus wentii*[2,4,18]

β-Nitropropionic acid
Kojic acid
Emodin
Physcion
Wentilactone A

TABLE 4.3
Mycotoxins from *Aspergillus flavus*[2,19,20]

Aflatoxins B_1, B_2, G_1 and G_2
Aflatoxins M_1 and M_2
Aflatoxin B_3 (parasiticol)
Aflatoxicols I and II
Aspertoxin
Versicolorin A
Sterigmatocystin
Physcion (parietin)
Kojic acid
Aflatrem
Paspalinine
β-Nitropropionic acid
Aspergillic acid
Flavoglaucin
α-Cyclopiazonic acid
Flavutoxin (metal chelate of α-cyclopiazonic acid)
Palmotoxins B_0 and G_0

methods are unavailable or poorly developed. In practice, sunflower seeds have only been analysed for aflatoxins.[13,14,21–24] To submit a sample of any foodstuff for 'mycotoxin analysis' is indeed wishful thinking!

This review will cover mainly the mycotoxins that have been selected for method development because of a potential for harm to human health from low level ingestion and because at least one of the producing fungi is of common occurrence on the foodstuff. In a study of the global significance of mycotoxins (which covered animal as well as human health aspects), a poll of mycotoxin research workers in 30 countries indicated that the aflatoxins were considered the most important mycotoxins, followed by the trichothecenes[25] and that the commodities with the most mycotoxin problems were maize (corn) and groundnuts (peanuts). The General Refereeship on Mycotoxins of the Association of Official Analytical Chemists (AOAC) at present covers aflatoxin M, aflatoxin methods, *Alternaria* toxins, citrinin, cyclopiazonic acids, ergot alkaloids, ochratoxins, penicillic acid, secalonic acids, sterigmatocystin, mycotoxins in tree nuts, trichothecenes, xanthomegnin and related naphthoquinones, and zearalenone; two recently added topics are emodin and related anthraquinones, and *Penicillium islandicum* toxins.[26] Compounds selected for inclusion in a volume on selected methods of analysis for mycotoxins published by the International Agency for Research on Cancer (IARC) were aflatoxins, sterigmatocystin, ochratoxin A, patulin, penicillic acid, citrinin, T-2 toxin, rubratoxin B, and luteoskyrin.[27]

4.2. CLASSIFICATION OF MYCOTOXINS

4.2.1 By Fungal Origin

In general, mycotoxins can be produced both in the field before harvest and during storage, although the responsible fungi may be dfferent in each case. Important genera of fungi that may produce mycotoxins on living plants include *Fusarium, Alternaria, Aspergillus, Claviceps* and *Trichoderma*. In stored products, *Aspergillus* and *Penicillium* are major spoilage fungi and many mycotoxigenic species are known.[28–32] In addition, metabolites and residues of mycotoxins in food producing animals have to be considered. Mycotoxins may thus be classified as to their fungal origin, compare the list of *Aspergillus flavus* toxins identified in Table 4.3; indeed the name 'aflatoxin' itself is derived from this species.

It should be noted that usually not all strains of a given toxigenic species produce the expected mycotoxin(s) or mycotoxin profile.[33] There may also be considerable overlap in originating fungi for a given mycotoxin, e.g. sterigmatocystin is known to be formed by *Aspergillus versicolor* and several other *Aspergillus* and *Eurotium* species, *Bipolaris sorokiniana, Chaetomium* spp., *Monocillium nordinii, Talaromyces luteus* and *Emericella* spp.[31,34–36]

4.2.2 By Structure

Mycotoxins may also be classified according to their chemical structure and biosynthetic origin.[1,2,37] The chemical structures of some major mycotoxins known to contaminate oilseeds are shown in Fig. 4.1. Small changes in chemical structure, however, may result in a large difference in potency. Thus aflatoxin B_1 is much more carcinogenic than aflatoxin M_1, a hydroxylated derivative formed by mammalian metabolism.[38,39] Turner and Aldridge[1,2] grouped the secondary metabolites of fungi, included in which of course are the mycotoxins, according to their biosynthetic origin (Table 4.4).

Two classifications of mycotoxins (and other fungal metabolites) that are relevant to their detection by mass spectrometry are by

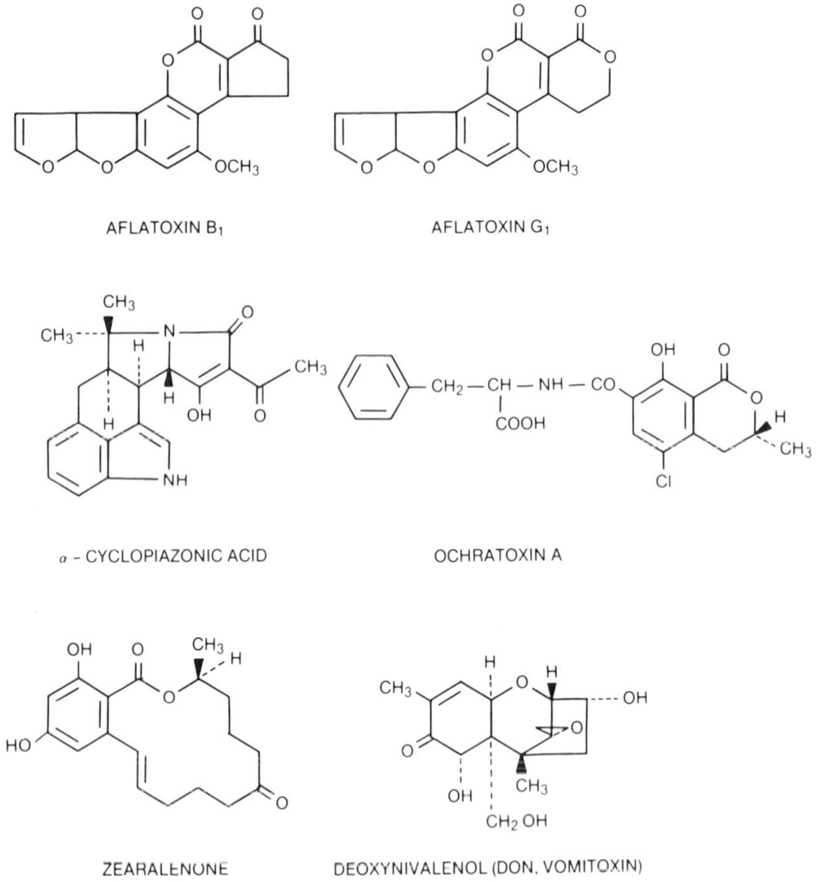

Fig. 4.1. Chemical structure of some mycotoxins known to occur in oilseeds.

molecular formulae[1,2,4,5] and by molecular weight.[4] Indeed classification by mass spectrum itself in a computer library is a working reality for at least one group of mycotoxins, namely the trichothecenes.[7]

4.2.3 By Biological Effect

Finally, mycotoxins may be classified as to their biological effects.[8,11,12] They usually at least possess acute toxicity to laboratory animals. Acute toxicity from mycotoxins has been associated with

TABLE 4.4
Classification of secondary fungal metabolites by biosynthetic origin[1,2]

Biosynthetic derivation	Mycotoxin example(s)
Without the intervention of acetate (e.g. from intermediates of the shikimic acid pathway)	Xanthocillin X
From fatty acids	Epicladosporic acid
Polyketides[a]	Aflatoxins, sterigmatocystin, ochratoxins, patulin, citrinin, alternariol, zearalenone
Terpenes (and steroids)	Trichothecenes, penitrems, PR toxin
From intermediates of the tricarboxylic acid cycle	Tenuazonic acid, rubratoxins
From amino acids	Ergot alkaloids, cyclopiazonic acids
Miscellaneous	Cytochalasins, moniliformin

[a] Tetraketides, pentaketides, hexaketides, heptaketides, octaketides, non-aketides and decaketides.

death or illness in humans, for example, an outbreak of hepatitis in India in 1974 was attributed to contamination of maize with *Aspergillus flavus* and aflatoxins.[40] However, the sub-acute properties of mycotoxins are usually of greater importance with regard to human health, in view of the low levels of toxin expected to be found as contaminants in foods. The carcinogenic potential of aflatoxin B_1 and several other mycotoxins has been demonstrated in laboratory animals and there is epidemiological evidence that aflatoxins are a factor in some primary human liver cancers.[41,42] Structurally related to the aflatoxins, sterigmatocystin is also a potent hepatocarcinogen in rats and mice but no evidence from food surveillance indicates a potential human health hazard, except possibly from its occurrence in cheese.[43] In addition, aflatoxins G_1 and M_1, cyclochlorotine, (−)-luteoskyrin, (+)-rugulosin, griseofulvin, ochratoxin A, zearalenone and T-2 toxin are animal carcinogens by oral administration.[44] Patulin and penicillic acid produced local tumours in rodents by subcutaneous injections, which has prompted analytical method development for these myco-toxins. Many other mycotoxins are mutagenic in the *Salmonella* assay (Ames test) but no tests for carcinogenicity have been carried out (e.g. altertoxins, emodin, kojic acid and fusarin C).[44]

   In addition to the foregoing toxin types, mycotoxins may also be regarded as nephrotoxins, e.g. ochratoxin A;[45] teratogens, e.g.

aflatoxin B_1, cytochalasins A–E, ochratoxin A, rubratoxin B and T-2 toxin;[46] oestrogens, e.g. zearalenone;[47,48] hepatotoxins, e.g. aflatoxins and sporidesmins;[49] neurotoxins/tremorgens, e.g. α-cyclopiazonic acid, citreoviridin and the penitrems;[50] cardiotoxins, e.g. xanthoascin and deoxynivalenol;[51,52] depressors of the immune response, e.g. T-2 toxin and other trichothecenes, which are now considered responsible for human Alimentary Toxic Aleukia (ATA);[8,53] and other toxin types or combinations, e.g. the ergot alkaloids, which have central nervous, neurohumoral and peripheral-muscular effects.[54]

Ueno[8] classified the important mycotoxins according to their major affinity to cellular organelles and summarized their toxicities with special reference to metabolic transformation. Thus citreoviridin, luteoskyrin, xanthomegnin and moniliformin, for example, are inhibitors of energy production; the trichothecenes and ochratoxin A are inhibitors of protein synthesis; griseofulvin, cytochalasins and cyclochlorotine are cytoskeleton modifiers. Oestrogenic, tremorgenic and carcinogenic mycotoxins complete this classification.

4.3. TYPES OF FOOD AT RISK

Any commodities susceptible to fungal growth, including oilseeds, fats and fatty foods, are likely candidates for mycotoxin contamination. These have often been determined by experiments involving the culture of toxigenic fungi on various foodstuffs in the laboratory or by inoculation in the field. However, it should be borne in mind that experimental demonstration of mycotoxin formation is not evidence of natural occurrence.

It is worthy of note that some foodstuffs, in particular certain spices such as mustard and cinnamon, inhibit mycelial growth of toxigenic fungi[55–58] and that aflatoxins have not been detected in commercial mustard and cinnamon.[59,60] Of some importance to the oilseed industry are observations that soyabeans are not as good a substrate in general for aflatoxin production as some other agricultural commodities such as peanuts, corn, and other grains;[61–63] it appears that seed coat integrity controls the colonization of soyabeans by *Aspergillus flavus*.[64] Nevertheless aflatoxins have been found in soyabeans affected by heavy rainfall in the field.[65] Natural occurrence of a mycotoxin in a foodstuff has in some cases been demonstrated directly even before carrying out inoculation experiments. The aflatoxins were in fact first isolated directly from Brazilian groundnut meal that was a component of feed lethal to young turkeys on English farms in 1960;[66] and deoxynivalenol (vomitoxin) was first obtained in Japan and the USA from barley and corn infected in the field with *Fusarium graminearum*.[67,68] Many foodstuffs have been shown to contain mycotoxins by direct analysis once contamination in other commodities has been demonstrated. For example, Brazil nuts are frequently naturally contaminated with aflatoxins,[69] but inoculation studies with Brazil nuts do not appear to have been done. Ciegler and Vesonder[5] listed some representative findings of mycotoxins in foods and feeds, of which the most important known natural occurrences in oilseeds and nuts are aflatoxins in groundnuts (peanuts) and processed products, tree nuts (especially Brazil nuts and pistachio nuts), maize (corn) and cottonseed; and ochratoxin A, zearalenone and trichothecenes (including deoxynivalenol) in maize. α-Cyclopiazonic acid has also been found in maize and groundnuts.[70,71] Other oilseeds in which aflatoxins have been detected include copra,[72,73] soyabeans, as mentioned previously,[65] olives,[74] palm kernels[75] and shea-nuts[76] (both of which are sources of cocoa butter equivalents),[77] and sunflower seeds.[21,78] A report on the natural occurrence of trichothecenes in safflower seeds in India is of particular interest.[79] With reference to fatty foods, the occurrence of aflatoxins in cheese and other dairy products as a result of direct mould contamination is very rare; however, important findings of sterigmatocystin in cheese stored in warehouses in the Netherlands[80] and of various *Penicillium* mycotoxins—ochratoxin A, citrinin, patulin, penicillic acid, penitrem A, mycophenolic acid, roquefortine and α-cyclopiazonic acid—in mouldy or mould-ripened cheese have been documented.[81–83]

Major sources of human exposure to known mycotoxins and the types of mycotoxins themselves may vary from country to country and also within individual countries. Thus peanuts and corn are clearly the major sources of dietary aflatoxins in the USA but the contribution of aflatoxins from corn products, while virtually the total contribution in the southeast USA, is not significant in the north and west USA.[84] Also human exposure to ochratoxin A is significant in parts of Yugoslavia[85] but not, so far as is known, in the USA and Canada.[45]

It is important to distinguish between agricultural products used for animal feed[86] and those destined for human consumption when evaluating the mycotoxin risk from a particular oilseed. Often the presence of mycotoxins or potential for their occurrence will determine the end use of a commodity if the mycotoxin is regulated in that commodity. As an important example, the segregation-3 programme of the US peanut industry requires that all lots of farmers' stock peanuts found to contain kernels with visible *Aspergillus flavus* growth be kept separate from edible stocks (with no *A. flavus* kernels) during the official grading operation at the first marketing point.[87-89] These segregation-3 peanuts are crushed for oil and the meal is not used for food or feed purposes, but has been used for fertilizer.[90] Segregation-2 peanuts, containing no visible *A. flavus* but more than 2% damaged kernels or more than 1% concealed damage due to moulds, rancidity or decay, are crushed for oil and the meal is used for animal feed if aflatoxin analysis does not indicate otherwise. The remaining peanuts (segregation-1) are shelled and used for food purposes if aflatoxin levels are acceptable. It should be noted that lots of segregation-1 peanuts may still contain aflatoxins, but contamination is much less than that of segregation-3 peanuts.[89]

Contamination of animal feeds by mycotoxins may result in carry-over of the mycotoxin into edible animal products. Of most concern is the occurrence of aflatoxin M_1, a metabolite of aflatoxin B_1, in milk and other dairy products, including cheese.[91] The percentage of aflatoxin B_1 in feed that is excreted in cow's milk varies from 0·2 to 3·2 and depends to some extent on the nature of the feed.[92] Feed/tissue ratios for aflatoxins B_1 and M_1 have been established experimentally for cattle, poultry and swine.[93,94] Cow kidney has the lowest ratio of 79. In general, however, there is little or no public health concern from carryover of aflatoxins and certain other mycotoxins, such as the trichothecenes, into the muscle tissue of animals. An important exception is transfer of ochratoxin A into pig liver and kidney, for which regulatory control is necessary in Denmark.

Most of the data on mycotoxin contamination apply to agricultural products such as small grains and oilseeds whereas human health concerns relate to dietary intake. For epidemiological studies, foods may be analysed as eaten on the plate. However, this approach is not practical for regulatory purposes. Thus knowledge of the effects of food processing, storage and home cooking on mycotoxin retention is important. The various mycotoxins that have been studied have wide differences in stability.[95] Considerable research on the processing of oilseeds and other foodstuffs containing aflatoxins has been carried out, including studies on the sorting and roasting of peanuts, the cleaning and milling of corn, the manufacture of vegetable oils, cooking of maize products, breadmaking, and the processing of milk and cheese containing aflatoxin M_1. Aflatoxins are moderately stable during roasting processes and carry through into finished foods such as peanut butter but one would not expect to find any significant concentrations of aflatoxin in a refined vegetable oil. Only a small percentage of aflatoxin B_1 present in oilseeds passes into extracted or pressed oil and refining and bleaching operations essentially eliminate it.[96] Natural occurrence of aflatoxins in crude vegetable oils such as cottonseed oil (up to 65 ng aflatoxin B_1/g), groundnut oil (up to 500 ng aflatoxin B_1/g), coconut oil (up to 1000 ng aflatoxin B_1/g), olive oil (up to 32 ng total aflatoxins/g) and even palm oil (up to 250 ng aflatoxin B_1/g) has nevertheless been reported.[72–74,97–102] Transfer of 2–4% of the *Alternaria* toxins alternariol and alternariol methyl ether (but not tenuazonic acid or altenuene) from contaminated olives into olive oil has been demonstrated experimentally.[103] In studies on the effect of oil refining processes alone,[104,105] no aflatoxins, deoxynivalenol, nivalenol or zearalenone added to crude vegetable oil at 0·1–10 µg/g levels could be detected in the refined oil. Nevertheless, sub-ng/g concentrations of aflatoxins were found in commercial edible peanut oils by Kamimura *et al.*[105] Also aflatoxin B_1 has been detected at levels up to 24 ng/g in refined groundnut, cottonseed and palm oils in Nigeria.[97,102] Aflatoxin B_1 appears to be moderately stable in oil during cooking[100,106,107] but exposure to sunlight (coconut oil) can significantly reduce aflatoxin levels.[107] The effects of food processing on ochratoxin A, zearalenone and deoxynivalenol (vomitoxin) have received considerable study.[95] For example, cooking of artificially moulded maize grits in the first stage of making 'cornflakes' resulted in a 45% loss of ochratoxin A;[108] fermentation of maize naturally contaminated with zearalenone resulted in little destruction of the mycotoxin and the recovered solids contained about twice the zearalenone concentration of the original maize;[109] and zearalenone, deoxynivalenol (and nivalenol) were stable in the baking process.[110]

4.4. REGULATIONS

Protection of the consumer from foods contaminated with hazardous substances is a prime concern of health authorities. Specific regulations pertaining to mycotoxins in foods and foodstuffs are in force in many countries and some countries have legislation for animal feeds and feedstuffs.[111–114] In some cases tolerances or action levels are based on the detection limit of the analytical method.

The following countries regulate aflatoxins in all foods: Australia, Austria, Belgium, Brazil, Czechoslovakia, Finland, France, German Democratic Republic, Hungary, India, Ireland, Japan, Malaysia, Mauritius, Mexico, the Netherlands, New Zealand, Nigeria, Norway, Poland, Portugal, Romania, Singapore, South Africa, Sweden, Switzerland, Thailand, the USA and the USSR. Tolerances range from zero to 35 ng/g, usually based on total aflatoxins (B_1, B_2, G_1 and G_2). Many countries, namely Argentina, Australia, Canada, Republic of China, Colombia, Cuba, Denmark, the Dominican Republic, Federal Republic of Germany, Hong Kong, Hungary, Israel, Italy, Jordan, Kenya, Luxembourg, Malawi, Mauritius, the Netherlands, New Zealand, Norway, Peru, the Philippines, Portugal, Surinam, Switzerland, the UK, Yugoslavia and Zimbabwe specify particular oilseeds, to which tolerances or action levels of zero to 50 ng/g (B_1) apply.[111] For example, in Canada, the tolerance is 15 ng/g for total aflatoxins in nuts and nut products on a shelled basis. It should be noted that the US Delaney Clause does not apply to aflatoxins. Kenya has a tolerance (20 ng/g, total aflatoxins) for edible vegetable oils themselves.

Countries of the European Economic Community, as well as Austria, Brazil, Canada, Republic of China, Dominican Republic, India, Israel, Ivory Coast, Japan, Jordan, Nigeria, Norway, Oman, Peru, Poland, Romania, Senegal, Sweden, Switzerland, the USA, Uruguay and Yugoslavia have regulations for animal feeds and feedstuffs with tolerances, where specified, of zero to 1000 ng/g of aflatoxin B_1 or total aflatoxins. While control of feedstuffs is aimed at yielding low levels of aflatoxins in animal products, several countries— Argentina, Austria, Belgium, Brazil, Czechoslovakia, France, Federal Republic of Germany, Nigeria, the Netherlands, Romania, Sweden, Switzerland, the USA and the USSR—have action levels or tolerances (0–1 ng/g) for aflatoxin M_1 specifically, in either fluid milk only or milk and milk products; for example, in Switzerland a maximum of 50 ng aflatoxin M_1/kg is permitted in milk and certain milk products, enforcement of which presents an analytical challenge.

Countries that export susceptible oilseed commodities have to establish limits that conform to controls in the importing countries. Thus Brazil has a tolerance of 50 ng/g for total aflatoxins in exported groundnut meal, which conforms to the level of 50 ng/g for aflatoxin B_1 permitted by the countries of the European Economic Community in 'straight feedstuffs' but not the permitted level in feedstuffs for dairy cattle (10 ng/g); and India allows 120 ng/g for aflatoxin B_1 in peanut meal exported for feed while Japan has a (high) tolerance of 1000 ng aflatoxin B_1/g in peanut meal imported for feed.

Of the other mycotoxins that are currently regulated—ochratoxin A, patulin, sterigmatocystin, phomopsin, zearalenone and trichothecenes (in particular deoxynivalenol[115])—few tolerances exist specifically for their presence in oilseeds, fats and fatty foods. Regulations in Brazil cover ochratoxin A and zearalenone in maize; in the USSR there are tolerances for zearalenone in grains, fats and oils and for T-2 toxin in grains, while the following countries cover all foods; Belgium (patulin, ochratoxin A, sterigmatocystin, and zearalenone),[113,114] Czechoslovakia (patulin and ochratoxin A), Finland (patulin) and Romania (patulin, ochratoxin A, zearalenone and, for feeds only, deoxynivalenol, stachybotryotoxin and chetomyotoxin).[111]

4.5. ENVIRONMENTAL FACTORS AFFECTING PRODUCTION OF MYCOTOXINS

Fungal growth is of course a prerequisite for mycotoxin production, although it should be noted that optimum conditions for fungal growth may differ from those for mycotoxin production and that the toxins may persist after growth has occurred and the fungus has died. The isolation of a toxigenic fungus from a foodstuff does not necessarily mean that mycotoxins are present, particularly if the fungus was not growing. The main factors affecting toxin production by fungi on a given substrate in a given time period are strain variation in the fungus, interference by other microorganisms, moisture, temperature, pH, the gaseous environment and preservatives. Additionally, in growing crops, plant stress and varietal resistance are important factors affecting fungal invasion and hence subsequent formation of mycotoxins.[116] The whole subject of formation and control of mycotoxins in foodstuffs has been reviewed in detail by Hesseltine[62] and, more recently, by Bullerman *et al.*[61] The factors referred to above will be discussed briefly here.

Different fungal strains of the same species that colonize a particular oilseed or other agricultural product may differ in which mycotoxins they produce, to what extent they produce a particular mycotoxin under defined conditions, and also in their response to changes in the conditions affecting mycotoxin production by each strain. Studies that define optimum culture conditions for mycotoxin formation using one strain only are therefore merely indicative. In the real world, fungi usually occur as mixed populations of species and cultural competition can reduce mycotoxin yields. Thus Hill *et al.*[117] showed that when the ratio of *Aspergillus flavus* to *A. niger* in drought stressed edible grade groundnut kernels exceeded 19:1 aflatoxin was present, but when it was below 9:1 no aflatoxin could be detected. *Penicillium oxalicum*, a producer of secalonic acid D, inhibited aflatoxin production by *A. flavus* on maize and the yield of secalonic acid D was also lower in the mixed culture.[118] Storage experiments using naturally-infected non-sterilized commodities have good predictive value, e.g. Abramson *et al.*[119] demonstrated ochratoxin A production and development of *P. verrucosum* var. *cyclopium* in maize adjusted to 21% moisture content and stored for up to 52 weeks.

Moisture and temperature have a critical effect on mould growth and mycotoxin production, as numerous studies have shown. Moisture is measured as moisture content (MC), relative humidity (RH), or, as index of water available for fungal growth, water activity (a_w). The a_w not only influences fungal growth but also affects mycotoxin production. It can be measured by the salt crystal liquefaction test,[120] which may be regarded as an indirect screening test to indicate the possibility of the presence of mycotoxins. The minimum a_w for growth of a toxigenic fungus and the generally higher minimum a_w for toxin production may be dependent on the temperature. Optimum temperatures and temperature ranges for mycotoxin production are important parameters. The aflatoxins, for example, are produced in the highest amounts at about 25°C by *Aspergillus flavus* and most strains of this species do not grow below 13°C. However, members of the *Penicillium* and *Fusarium* genera can grow below 5°C, even at −2°C to −10°C for strains of *F. poae* that were associated with ATA in the USSR, with greater eventual toxin accumulation than at room temperature.[53]

Toxigenic fungi require oxygen for growth. Results of studies by several workers[61] show that atmospheres with high carbon dioxide and low oxygen concentrations appear to have promise in preventing mould growth and toxin production in stored peanuts and other commodities.

The use of chemicals to control mycotoxin formation in stored grains and foods and in grain growing in the field has been studied. Organic acids such as sorbic acid, benzoic acid and propionic acid or their salts have antifungal properties and the antibiotic natamycin (pimaricin) is used in cheese to inhibit mould growth and mycotoxin production. Certain pesticides, such as fonofos, carbaryl and maneb, have also been shown to inhibit mycotoxin production significantly in corn cultures and in corn that was field-inoculated with *Fusarium roseum*.[121]

Insect damage, mechanical damage, moisture, and drought stress, can all affect the invasion of oilseeds by *Aspergillus flavus* and other fungi. Insects may also serve as fungal vectors. Other damage may be from birds, hail or agricultural machinery. Under drought conditions there are more immature groudnut kernels, which appear to be colonized more readily by *A. flavus*. Drought stressed maize is also more susceptible to fungal attack. Varieties of groundnuts and maize can be developed that are more resistant to *A. flavus*[122,123] and inbreds and hybrids of maize that have some resistance to zearalenone formation by *Fusarium graminearum* have been differentiated.[124]

4.6. METHODS FOR DETECTION AND DETERMINATION OF MYCOTOXINS

Analysis for mycotoxins is an essential aspect of research and regulation in this area. We need accurate methods to carry out surveys for mycotoxins in foods and feeds and hence determine human and animal exposure; to exercise control at commodity, manufacturing and retail levels; to assess stability of mycotoxins during storage, processing and decontamination; to conduct toxicological studies with naturally contaminated materials; and to carry out other research such as metabolism studies, microbiological production and epidemiology.

Considerable effort has been devoted to the development of methods for the analysis of mycotoxins since 1961. Even by 1979, the US Food and Drug Administration's computerized data base contained over 1300 methods, 38% of which concerned aflatoxins.[125] The interest in method development is ongoing and a more recent analysis of the data base for the years 1980 and 1984 showed respectively 105 and 71 papers on physicochemical methods of analysis plus 16 and 7 papers on biological methods.[25] These are out of totals of 669 and 653 papers altogether on mycotoxins for those 2 years, indicating physicochemical analytical methodology is one of the leading areas of mycotoxin research. According to a 1985 poll of scientists engaged in research on mycotoxins in 30 countries, the most needed mycotoxin research is on rapid and improved analytical methods but one of the areas of least needed research is complex analytical methodology.[25]

4.6.1 Sampling

It must be realized that the reliability of any assay of an agricultural commodity for mycotoxins is largely dependent on sampling. The difficulty in obtaining a representative sample of a foodstuff for mycotoxin analysis depends on whether the product is in the form of whole kernels (and on their size), flour, paste or liquid and on the type of mycotoxin. Sampling of oilseeds before aflatoxin analysis has been reviewed in detail in Chapter 11. In the case of aflatoxins in ground-nuts, a single nut may contain up to 0·1% of aflatoxins and only a few kernels per thousand may be contaminated. Thus if a lot of ground-nuts contained 20 ng/g (ppb) of aflatoxins distributed heterogeneously, it has been determined that sampling contributes at least two thirds of the total variability of measuring the aflatoxin content, subsampling of the ground sample about 20%, and the actual analysis only the remaining 13%.[126] Similar studies have been carried out with aflatoxin in shelled maize and it was found that in this case the sampling variation was about the same as the analytical variation.[127] Sampling procedures employed by the US Food and Drug Administration for analysis of mycotoxins in a variety of foodstuffs have been tabulated.[128,129] The minimum total sample size is the product of the minimum number of sample units and the minimum unit size; this varies from 10–12 lb for peanut butter and small grains to 50 lb for bulk in-shell pistachio nuts and 60 lb for bulk cottonseed and large lots of Brazil nuts. The design of all sampling plans must be based on the critical mycotoxin concentration (tolerance), definitions of acceptable and rejectable lots, and the consumer's and producer's risks. Clearly the cost of the sample may be high. Protocols for surveys, sampling, post-collection handling, and analysis of grain samples relative to mycotoxin problems have been recommended by Davis et al.[130]

Comminuting (grinding), blending and then subsampling are further important operations that must be carried out before obtaining the actual analytical sample, which is usually 50 g.[129,131] The Dickens–Satterwhite subsampling mill automatically gives a subsample during comminution but further subdivision of the subsample is desirable. One way of doing this is by slurrying the 1100 g subsample (peanuts) with water, if the presence of water in the analytical sample is compatible with the extraction solvent of the analytical method.[132,133]

4.6.2 Types of Methods

Analytical methods for mycotoxins may be broadly divided in four ways: into quantitative (or semi-quantitative) assays and rapid screening tests; into methods for individual mycotoxins and multimycotoxin methods; into physicochemical determinations and bioassays; and into presumptive and confirmatory techniques. Methodology for mycotoxins in general has been reviewed by Stoloff,[134] Park and Pohland,[135] Pohland *et al.*,[125] Hunt[136] and, more recently, by Bullerman,[137] Cole,[138] Gorst-Allman and Steyn,[139] Coker,[140] van Egmond,[6] van Egmond and Paulsch,[141] Crosby,[142] Romer[143] and Shotwell.[90] Methods for particular mycotoxins—aflatoxins;[144-147] *Alternaria* toxins;[16] and *Fusarium* toxins,[148] including deoxynivalenol,[149] other trichothecenes[150,151] and zearalenone[152]—have been thoroughly reviewed separately, as have mycotoxin methods employing specific detection and quantitation techniques—thin layer chromatography (TLC),[153-156] minicolumn chromatography,[157] gas chromatography (GC),[158] high performance liquid chromatography (HPLC),[159,160] immunoassay[161-164] and biological tests.[165-169] In addition, reviews of methods for mycotoxins have been included in some cases as part of more general reviews, as for example methods for ochratoxins.[45] The choice of methods is indeed great. For a continuing update of mycotoxin methodology the annual reports of the AOAC General Referee for Mycotoxins should be consulted.[26,170-174]

Methods for aflatoxins B_1, B_2, G_1, G_2 and M_1, ochratoxins, patulin, sterigmatocystin and zearalenone that have been validated in inter-laboratory studies are described in Chapter 26 of the Official Methods of Analysis of the AOAC.[132] The most recent edition of this compilation of 'Official Methods' was published in 1990 and it is updated on an ongoing basis. Five new methods, including two for deoxynivalenol (vomitoxin), have been approved as AOAC official methods;[26,172] official methods applicable to oilseeds and fatty foods now cover aflatoxins B_1, B_2, G_1 and G_2 in corn, peanuts and peanut products, coconut and copra, cottonseed products, pistachio nuts, cocoa beans and soybeans, aflatoxin M_1 in dairy products, and zearalenone and α-zearalenol in corn. Other organizations, such as the American Oil Chemists' Society (AOCS), American Association of Cereal Chemists (AACC) and the International Union of Pure and Applied Chemistry (IUPAC), have method validation programmes. These four organizations, whose approved methods have been discussed by Park and Pohland,[175] are represented on the Joint Mycotoxin

Committee.[176] The purpose of this Committee is to coordinate the efforts of these societies in the development and validation of analytical methods and to provide a forum for the exchange of ideas and dissemination of information on all aspects of mycotoxin research, including the identification of new problems.

At present there are no official AOAC methods for multimycotoxin analysis. Multimycotoxin methods are numerous and becoming increasingly sophisticated in their judicious use of TLC systems, so that up to 22 mycotoxins can be detected in foods, grains and feeds.[153,156,177–181] However, such methods are rarely quantitative and information on the type of fungal contamination of the commodity would assist in reducing the number of mycotoxins to be looked for. Even so, analytical challenges remain for particular groups of mycotoxins and there is much current interest in developing a quantitative multitrichothecene analysis (see, for example, Eller and Sobolev[182]). While many of the multimycotoxin methods available include oilseeds such as groundnuts and maize as substrates, it should be noted that such methods have also been applied to the analysis of vegetable oils[183] and fatty foods such as cheese and milk.[180,184,185]

Many biological methods (bioassays) have been developed to detect mycotoxins.[165–168] They have generally not been used in food or feed surveillance—one notable exception is the rabbit skin test that was used with extracts from toxic grain responsible for outbreaks of ATA in the USSR.[53] Other examples are the demonstration by the chicken embryo test of unidentified mycotoxins in market samples of mouldy cheese in Yugoslavia[186] and detection of trichothecenes in maize by using protein synthesis inhibition in cultured fibroblasts.[187] The major value of bioassays has been in the initial identification and purification of mycotoxins from fungal cultures or mouldy agricultural products and this approach has been historically quite successful;[166] a notable example was the isolation of aflatoxins from toxic Brazilian groundnut meal during 1960–62 using a duckling bioassay.[66] The types of bioassay that have been used may be classified as microbial assays (e.g. *Bacillus thuringiensis, B. megaterium* for a bacterial bioluminescence assay[169] and, for mutagenesis testing, *Salmonella typhimurium*), invertebrate assays (e.g. brine shrimp larvae and insects), cytological assays (HeLa cells, tracheal organ cultures, etc.), vertebrate assays (e.g. chicken embryo, zebra fish, ducklings and mice) and plant bioassays (wheat coleoptiles, intact plants and pollen). One system that has been extensively tested by mycotoxin research workers is the brine shrimp

(*Artemia salina*), whose larvae have been used to detect trichothecenes in selected animal feedstuffs[188] and mycotoxins on thin layer chromatograms by the technique of bioautography.[189] However, in general, it appears that false positives are a problem in feed analysis.[190] The chicken embryo assay was employed in the past by the US Food and Drug Administration for confirmation of toxicity of aflatoxin B_1 but has been replaced by physicochemical tests.

4.6.3 Sequence of the Physicochemical Analysis[6,139–141,143]

Following the sampling and sampling preparation procedures, an analytical method is usually composed of extraction, clean-up, qualitative assessment (for screening tests), quantitative determination and confirmatory steps. The first two steps will also usually be used in a bioassay, although the clean-up step may be quite simple.

Mycotoxins are extracted from foods by blending or shaking the sample with organic solvents such as methanol, acetonitrile, ethyl acetate, acetone, chloroform or dichloromethane, often together with water and sometimes in the presence of acids (e.g. for ochratoxin A) or at alkaline pH (for ergot alkaloids). A correction should be made, as in the CB (Contaminants Branch) method (AOAC Method I), for fat and other materials extracted from peanuts and other oilseeds that add volume to the extract aliquot used for clean-up; also where clean-up in a method is inadequate there is also a small error caused by these extractives in the volume of final extract solution used for TLC.[191] Recently supercritical fluids such as carbon dioxide have been tested for extraction of trichothecenes from wheat.[192] Extraction procedures should be tested on naturally contaminated samples as well as spiked samples to establish extraction times required for maximum recovery of the mycotoxin(s).[193] Refinements made to the BF (Best Foods) method for aflatoxins in peanuts (water slurry modification, AOAC Method II) have resulted in optimum methanol concentration and peanut/solvent ratios that extract 12% more aflatoxin.[194]

Sample clean-up to remove lipids and other extraneous substances is an essential step in the analysis of foods for trace levels of mycotoxins. Techniques used include defatting with a solvent such as hexane, either in a Soxhlet extractor or by liquid–liquid partitioning; solid phase extraction columns; chemical adsorption by addition of such reagents as copper carbonate or ferric gel to the extract solution; dialysis, used in one of the multimycotoxin assays; and various forms of chromatography. One type of solid phase extraction tube, or Chem-tube™, is filled with an inert hydrophilic matrix so that partition of the mycotoxin from an aqueous phase to a percolating organic solvent can occur, effectively replacing a separatory funnel. Chromatographic clean-up may involve open glass columns packed in the laboratory; small disposable prepacked columns that are now commercially available; and preparative HPLC or preparative TLC, which are usually used only if necessary and also for isolation of a mycotoxin for confirmation by a technique such as mass spectrometry.[195] Silica gel is the most frequently used adsorbent for column chromatography; alumina, florisil, charcoal, ion exchange resins and gel permeation materials are also employed in some mycotoxin methods. There is a trend towards the use of smaller clean-up columns than specified in, for example, the CB method for aflatoxins in peanuts and peanut products (AOAC Method I); reduction in the amount of silica gel, aliquot of extract and volume of elution solvents by a factor of 5 gives equivalent results and the procedure is applicable to corn, peanuts, soybeans, coconut and pistachios.[196] The small disposable clean-up columns, e.g. Sep-pak® and Baker®, reduce cost and time and are now being widely used; a wide range of adsorbents is offered by the manufacturers. One example of a method that incorporates these columns is that of Trucksess et al.[197] for determination of aflatoxins in corn and peanut butter; another is the method of Hurst et al.[198] for analysis of aflatoxins in raw peanuts.

The detection-quantitation step in mycotoxin analysis has classically been TLC[153] using visual or densitometric determination. TLC is a simple, inexpensive technique that is usually carried out in one dimension on a silica gel layer with a single developing solvent but where the extra separation is required, two-dimensional TLC is a powerful technique.[6] TLC is still the most commonly used procedure in routine mycotoxin analysis and is especially useful for aflatoxin analyses down to the low ppb range, for screening purposes, and as previously mentioned, for multimycotoxin analysis. In addition to the official AOAC methods for aflatoxins (and zearalenone) in various oilseeds and nuts, there are TLC methods for other mycotoxins, including ochratoxin A, sterigmatocystin, rubratoxin B, citrinin and α-cyclopiazonic acid in maize and other commodities.[155,156,199–201] There are also TLC methods for mycotoxins in fatty foods, e.g. for aflatoxins in vegetable oils[104,202,203] and dairy products,[132,204,205] ochratoxin A in olive oil,[202] sterigmatocystin in cheese,[43,206] and α-cyclopiazonic acid, mycophenolic acid and roquefortine in mould-ripened cheeses.[207–210]

In the last 10–15 years there has been increased research and application of HPLC, with its advantages of increased sensitivity, precision and potential for automation. As many as 134 mycotoxins and other secondary fungal metabolites have been chromatographed in a single gradient elution reverse phase system[10] although application of HPLC to food analysis has only been for a very limited number of mycotoxins. For some mycotoxin analyses, e.g. ergot alkaloids in flour,[211] HPLC is clearly preferable over TLC for separation, sensitivity (down to a few ppb for individual alkaloids) and the ease of exclusion of light during the analysis. Microbore HPLC has found little application to mycotoxin separation. HPLC of aflatoxins may be carried out under normal or reverse phase conditions with detection by ultraviolet (UV) absorption or fluorescence. For reverse phase HPLC, derivatization of aflatoxins B_1 and G_1 is necessary for fluorescence measurements;[159,212] a flow cell packed with silica gel may be used to intensify fluorescence of the aflatoxins under normal phase conditions.[159] Shepherd[160] has recently discussed the chromatographic conditions required to optimize resolution of the aflatoxins. Detection limits of overall HPLC methods for determination of aflatoxins B_1, B_2, G_1 and G_2 in oilseeds,[212] aflatoxin B_1 in edible oils,[104,203] and aflatoxins M_1 and M_2 (in addition to B_1, B_2, G_1 and G_2) in dairy products[132,213–215] are generally below 1 ng/g for each aflatoxin. Other mycotoxins for which HPLC methods have been developed for their analysis in oilseeds include trichothecenes, sterigmatocystin, patulin, penicillic acid, α-cyclopiazonic acid, ochratoxin A, citrinin, citreoviridin, moniliformin and zearalenone.[150,159,160,174,216–219] Detection limits in the low ng/g range are usually possible. HPLC methods for mycotoxins other than aflatoxins in fatty foods include methods for *Alternaria* toxins in olive oil;[103] roquefortine in blue cheese;[220] patulin, penicillic acid, zearalenone, sterigmatocystin and ochratoxin A in cocoa beans;[221] and ochratoxin A and B in nuts and other foods.[222]

Most mycotoxins are not volatile, so the use of GC for their detection has been limited. Derivatization is usually necessary. For one important group of mycotoxins, the trichothecenes, determination by HPLC[150] is difficult as they are not fluorescent and many trichothecenes do not absorb UV light at a useful wavelength. GC is therefore widely used for analysis of trichothecenes as their trimethylsilyl

(TMS), heptafluorobutyryl or trifluoracetyl derivatives with electron capture or mass spectrometric detection.[143,151,223] Methods for GC determination of some other mycotoxins—zearalenone, aflatoxins B_1 and B_2, patulin, penicillic acid, slaframine, swainsonine, and certain *Alternaria* toxins—have been developed.[143,158] GC methods developed for oilseeds in particular include ones for patulin and penicillic acid in soyabeans, penicillic acid in maize, zearalenone in maize and trichothecenes in maize.[158]

The foregoing analytical chromatographic techniques are the main ones used for the quantitative or semiquantitative determination of mycotoxins in foods and for enforcement purposes. TLC can also be used for screening commodities, with the advantage that up to 20 samples can be run on just one TLC plate. A particularly useful TLC screening method for aflatoxins in maize is that of Dantzman and Stoloff,[224] which incorporates a clean-up step on the TLC plate itself using anhydrous diethyl ether. Another screening technique is minicolumn chromatography, in which the mycotoxin (usually aflatoxin) is adsorbed on the column packing material and observed under UV light. Minicolumn methods[157,225,226] are simple, economical, and can be used by inexperienced personnel. Official AOAC minicolumn methods for aflatoxins are applicable to maize, almonds, pistachios, groundnuts and groundnut products, cottonseed meal and mixed feeds.[132] Other minicolumn methods include the detection of aflatoxins in cheese and other dairy products,[227] olives and olive oil,[228] and several other oilseeds and nuts;[229] they have also been developed for ochratoxin A[230,231] and for zearalenone.[232] The ultimate rapid test is detection of the presence of fluorescing particles in maize (bright greenish-yellow [BGY] fluorescence),[233] cotton seed (greenish-yellow fluorescence),[234] pistachio nuts (BGY fluorescence)[235,236] and almonds (violet-purple fluorescence).[237] This is indicative of aflatoxin[238] although it is based on detection of another *Aspergillus flavus* metabolite, presumably kojic acid in the cotton seed at least.[239] A procedure has been developed for efficient extraction of BGY fluorescent material from ground maize and its determination by fluorimetry.[240]

Immunoassay is an important technique for detection of mycotoxins. It depends on the interaction of a given mycotoxin with a specific antibody obtained from a laboratory animal, such as the rabbit, which has previously been immunized with a mycotoxin–protein conjugate. The two main types of immunoassay are radioimmunoassay (RIA) and enzyme-linked immunosorbent assay (ELISA); the latter has gained ground over RIA as radioactivity is not involved. Among the various types of ELISA method, heterogeneous competitive ELISA is used in mycotoxin analysis. This consists of two systems: direct competitive ELISA, which involves coating the wells of a microtitre plate with specific antibodies and incubation of the sample with a mycotoxin–enzyme conjugate; and indirect competitive ELISA, where a mycotoxin–protein (or polypeptide) conjugate is coated on the wells of the microtitre plate in which the sample is incubated with the specific antibody.[164] Antibodies against several mycotoxins, e.g. aflatoxin B_1, aflatoxin M_1, ochratoxin A, zearalenone, deoxynivalenol (as triacetate), diacetoxyscirpenol, T-2 toxin and group A trichothecenes in general have been prepared[161–164,241-248] and procedures developed for immunoassay of these toxins in foodstuffs. The methods are specific, sensitive and easy to use. Clean-up is usually minimal. ELISA methods are generally useful for screening but quantitation is also possible. Some examples of ELISA methods are the determination of aflatoxin B_1 in naturally contaminated maize, cottonseed and peanut butter[249,250] and in vegetable oils,[104] determination of aflatoxins B_1 and G_1 together in peanut butter,[251] quantitation of aflatoxin M_1 in milk,[252] screening of maize for zearalenone[253] and quantitation of T-2 toxin in corn and wheat.[254]

Chromatographic techniques for detection and determination of mycotoxins in foods should generally be used after sufficient clean-up to remove interferences that could give rise to false positives. Identification remains presumptive, however, until a confirmatory test has been carried out, although confidence is usually adequate (unless the sample is involved in regulatory or economic dispute) if the commodity has a history of contamination by a particular mycotoxin and no interfering compounds are expected. Derivatives of aflatoxins B_1, G_1 and M_1 may be formed directly on the TLC plate with trifluoroacetic acid.[132] This type of test has been extended to other mycotoxins using pyridine and/or acetic anhydride as reagent to change ochratoxin A, citrinin, penicillic acid and zearalenone to derivatives with different TLC mobility.[255] Mass spectrometric (MS) confirmation of identity offers promise as a highly specific confirmation test for mycotoxins.[145,195,256,257] The high degree of specificity of MS confirmatory techniques compared to the less specific chemical confirmatory tests has to be balanced against their higher cost. Other confirmation procedures[145] include use of two different detectors for

HPLC, e.g. UV absorbance and fluorescence detectors; determining the mycotoxin by UV or fluorescence at different wavelengths (HPLC); and bioassay. A novel confirmation test for light sensitive substances, which has been applied to fusarin C, is to irradiate the sample and observe loss of intensity of the HPLC peak.[258]

An essential component of the analytical method is the mycotoxin reference standard. Many of these are now available commercially[259] and this situation is continually improving. Radiolabelled (^{3}H) aflatoxins B_1 and B_2 and aflatoxin B_1 and ochratoxin A conjugates with bovine serum albumin (BSA) can also be purchased from certain biochemical companies. Purity criteria for selected mycotoxin standard reference materials have been published[132] and compilations of physicochemical data on many more mycotoxins[4,260] are most useful. The UV spectrum and extinction coefficient at maximum absorbance are the physical constants most commonly used to assess purity (in conjunction with analytical chromatography) and also to determine concentrations of standard solutions of, for example, aflatoxins,[132] zearalenone,[261] α- and β-zearalenols[262] and trichothecenes.[262] Mass spectrometry has also been used to indicate impurities in mycotoxin standards, e.g. demethylated derivatives of T-2 toxin, HT-2 toxin and T-2 triol.[263] Nuclear magnetic resonance spectroscopy can give a particularly critical indication of purity if sufficient material is available. The analysis is only as good as the reference standard used, not only the primary crystalline material but also the final analytical standard solution.[264] Criticism of analytical results obtained with a 20-month old aflatoxin standard solution was recently publicized.[265]

4.6.4 Assessment and Use of Quantitative Methods for Mycotoxins

Reference has been made above to sampling errors in mycotoxin determination and to interlaboratory collaborative study of the actual analytical methods. The analytical errors occurring within one laboratory for aflatoxin determination in raw peanuts by fluorodensitometric TLC have been shown to be largely due to the TLC quantification step (coefficient of variation 18·6%).[266] Methods for aflatoxin analysis (at the 10 ng/g order of contamination) that achieved coefficients of variation of 30–40% in interlaboratory collaborative studies (with recoveries of $\geq$70%) were considered eligible for 'referee status' by Schuller *et al.*[267] However, even larger coefficients of variation have been obtained in international check sample programmes. These programmes are laboratory proficiency studies, however, rather than collaborative method studies[264,268] and serve to aid the analyst in checking the quality of his analysis. For example, for determination of aflatoxin B_1 in peanut butter by the CB and BF methods (AOAC Official Methods I and II) at the 4–6 ng/g level, coefficients of variation were 72 and 101%, respectively, in one check sample series.[269] Clearly some individual laboratory performances needed improvement and training was called for. However, the situation has been getting better. Not only has the repeatability (within laboratory variation) improved in the AOCS Smalley Aflatoxin Check Sample Programme to a level expected from collaborative studies of the methods used, but also some improvement in the overall reproducibility and the between-laboratory variation has been noted.[264] Sufficient data from check sample programmes have now made some method comparison possible and it has been noted that the BF method gave lower results for aflatoxins B_1 and B_2 in peanut butter and de-oiled peanut meal, but not raw peanut meal, than the CB method.[269,270]

In general, no significant differences in overall variability or accuracy have been apparent between laboratories using HPLC and those using TLC for aflatoxin analysis.[268,270] Furthermore, in a collaborative study on aflatoxins in cottonseed products,[271] within-laboratory error (repeatability) was essentially the same by both TLC and HPLC, although use of the latter did reduce the between-laboratory error component. Chu[161,164] reported that ELISA data from his laboratory for aflatoxin B_1 in maize, groundnuts and defatted groundnut meal and for aflatoxin M_1 in milk agreed well with results obtained by other methods in a check sample programme sponsored by the IARC. An intralaboratory comparison and critical evaluation of six published methods for aflatoxin M_1 in liquid milk has been carried out by Shepherd *et al.*[272] Assessment of method use and performance for mycotoxins other than aflatoxins is more limited. In addition to the collaborative studies carried out by AOAC, IARC has conducted a check sample programme on ochratoxin A in feed.[268] Currently available quantitative methods for determination of trichothecenes were evaluated for accuracy, precision and limits of detection as reported in the literature;[151] a short list of the more accurate and sensitive GLC and immunoassay methods were discussed further with respect to convenience and the number of trichothecenes analysed. Since this review,[151] two methods for deoxynivalenol have been collaboratively studied but no interlaboratory studies for other trichothecenes are planned as yet.

It is important to use the right method for the commodity being analysed since they are sometimes matrix dependent. For example, the BF method for aflatoxins in peanuts gave much lower recoveries than the CB method when applied to corn.[273] Also the AOAC method for sterigmatocystin in barley and wheat[132] is not satisfactory for cheese.[206] Even after the choice of method is made—and it is apparent that analytical methods for mycotoxins of varying degrees of reliability, accuracy and sensitivity exist in profusion—the analyst must ensure that he is using the method correctly. Quality control is necessary when applying methods. The check sample programmes referred to previously and the recent development of certified mycotoxin-containing reference materials[274] are a part of this. The extraordinary interest in recent years in the analysis of biological samples associated with yellow rain[275] and samples from the Persian Gulf war[276] has highlighted certain other aspects of 'quality assurance'. These are the need for correct identification of toxins (by MS), frequent use of blank samples, spiked controls, an awareness of possible toxin degradation related to sample handling, and the need for confirmation of important results by two or more laboratories. Even though most analytical results will not have to stand up to world opinion with the same intensity as the analyses of yellow rain and related samples, they do have to reliably show what is being measured with a definable limit of detection.[277] Awareness of the need for quality assurance can be expected to increase in the future.

The analyst should be aware of hazards that may be involved in handling mycotoxins, particularly carcinogens such as the aflatoxins. Appropriate precautions, including special handling of crystalline mycotoxins, use of gloves and other protective clothing, clean-up of spills, and disposal of contaminated wastes have been outlined by Trenholm and Young.[278] Procedures for destruction of aflatoxins in laboratory wastes using either sodium hypochlorite or potassium permanganate have been given in detail by Castegnaro *et al.*[279] The trichothecenes diacetoxyscirpenol and T-2 toxin are also destroyed by sodium hypochlorite,[280,281] although the possible toxicity of any chlorinated reaction products[282] should be studied. Hazards associated with mycotoxins are not to be confined to the laboratory. Airborne dust generated during the handling, shipping and processing of mycotoxin contaminated agricultural commodities such as maize and groundnuts may contain aflatoxins or zearalenone.[283–285] Such dust could be a potential inhalation hazard to workers and in fact there is epidemiological evidence for increased rates of cancer for workers exposed to aflatoxin in a peanut processing plant in the Netherlands.[286–288]

4.7. MORE RECENT DEVELOPMENTS IN MYCOTOXIN METHODOLOGY

In the area of chromatographic techniques, there are several that have been known for some time but only recently have been incorporated into methods of analysis for mycotoxins. High performance TLC (HPTLC) coupled with densitometry has been compared to HPLC for determination of aflatoxins in peanut butter.[289] With respect to precision, accuracy, sensitivity, recovery, and linearity of response, the two techniques appeared to be similar, and quantitative HPTLC is better than HPLC with respect to reduction of total analysis time for multiple samples. Coker[140] estimated that the average time required for quantification of each of 60 aflatoxin-containing peanut extracts was ≤ 3 min. HPTLC also has other advantages over HPLC such as reduction in solvent volumes used. A multimycotoxin HPTLC method suitable for analysis of cereals, legumes, and oilseeds for aflatoxins, ochratoxin A, zearalenone and sterigmatocystin has been published.[179] The advantages of another TLC variation, reverse phase TLC, are referred to by Romer,[143] but no overall mycotoxin method applications have yet been reported. Improved means of detecting aflatoxins by conventional TLC include use of laser-induced fluorescence (detection limit 10 pg for each aflatoxin)[290] and, as a means of confirmation of aflatoxin on the TLC plate, bromination followed by fluorescence detection.[291]

For HPLC, there have been recent developments in the application of post-column derivatization for mycotoxin analysis, e.g. the iodination of aflatoxin to enhance sensitivity of aflatoxins B_1 and G_1 in reverse phase HPLC,[292,293] which has been used in surveys of nuts and nut products for aflatoxins.[294] A diode array UV detector has been used to detect 182 mycotoxins and other fungal metabolites[295] and has been tested for determination of a number of mycotoxins—patulin, ochratoxin A, zearalenone, deoxynivalenol, diacetoxyscirpenol, HT-2 toxin and T-2 toxin—in wheat;[296] 250–1000 ng/g of toxin could be reliably determined. Microbore HPLC has not yet been incorporated into a mycotoxin method but may find application because of shorter analysis times and the smaller volumes of solvent used; it has been coupled with a mass spectrometer.[297]

Capillary column GC is becoming more widely used in analytical laboratories, especially since the introduction of fused silica columns and chemically bonded stationary phases. Capillary GC is becoming an essential means of separating trichothecenes when analysing for several of them in a given sample and has been used, for example, in Hungary for simultaneous determination of deoxynivalenol, fusarenon-X, diacetoxyscirpenol, T-2 toxin and HT-2 toxin (in addition to zearalenone) in maize and wheat;[298] in Finland for routine analysis of cereals and feeds for deoxynivalenol, nivalenol, fusarenon-X, diacetoxyscirpenol, T-2 toxin, and HT-2 toxin;[299] and in Canada and the Federal Republic of Germany for determination of deoxynivalenol and nivalenol in cereals and cereal products.[300,301]

Mass spectrometric (MS) techniques are being increasingly used in analysis of mycotoxins, particularly trichothecenes,[7,223] for both quantitation and confirmation. The classical method is GC–MS with selected ion monitoring (SIM) and electron impact (EI) ionization.[302] This was the technique used for determination of deoxynivalenol in Canadian grains (as heptafluorobutyryl derivative),[303] in a UK survey of barley and maize (as TMS ether)[304] and in the Finnish survey of trichothecenes (as TMS ethers) in feeds referred to above.[299] GC–MS (SIM) has also been used for selective determination of sterigmatocystin (underivatized) in grains,[305] several other mycotoxins as TMS ethers, i.e. zearalenone in corn flakes,[306] tenuazonic acids in tomato paste,[307] patulin in apple juice,[308] and for confirmation of aflatoxins B_1 and B_2 (underivatized) in peanut extracts.[309] GC of aflatoxin B_1 was itself only fairly recently accomplished.[310] Another MS ionization mode is negative ion chemical ionization (NICI),[311–313] which has been used more so than positive chemical ionization, particularly for trichothecenes; sample introduction by GC has been made with underivatized[256,312] as well as derivatized trichothecenes.[313,314] Sample introduction into the mass spectrometer by HPLC, including microbore HPLC,[149,297,315] has been reported for aflatoxins,[315,316] ochratoxin A,[317] ergot alkaloids,[318] patulin,[296] zearalenone[296,319] and trichothecenes,[149,296,319] and an increase in activity in this area of mycotoxin analysis can be expected. The most versatile and popular interface for HPLC–MS is thermospray, which can handle large amounts of water and volatile buffers. Thermospray HPLC–MS shows great potential for the analysis of mycotoxins[296,319] and concentrations of patulin, ochratoxin A, zearalenone and several trichothecenes in the low ng/g range have been measured for wheat.[296] The NICI mass spectrum of aflatoxin B_1 isolated from peanut butter was recorded for confirmatory purposes not only by direct probe introduction into the mass spectrometer (a collaboratively studied procedure and AOAC official method[195,320]) but also following capillary GC with on column injection.[197,321] Fast atom bombardment MS[322] and capillary column supercritical fluid chromatography/MS[323] are two recent MS techniques to be studied for *Fusarium* mycotoxins in particular. The latter technique has been applied to determine trichothecenes in wheat[192] but no applications to mycotoxins in oilseed matrices have yet been reported. Supercritical fluid chromatography offers advantages over both GC and HPLC. It should be particularly useful for mycotoxins that are heat sensitive or not volatile enough for GC analysis.

The technique of tandem mass spectrometry (MS/MS) has generated considerable interest as a method for rapid trace analysis.[324] One stage of mass separation selects the compound to be analysed (usually as the molecular ion, protonated molecular ion, or molecular anion depending on the ionization mode) and the second MS stage is used for analysis after collision induced dissociation (CID) with a target gas. Derivatization and sample clean-up are not necessary and preliminary applications to determination of aflatoxin B_1, zearalenone and trichothecenes in grains and ergot alkaloids in tall fescue forage have been reported.[324–328]

Another instrumental method recently applied to mycotoxin analysis is Fourier transform infrared (FTIR) spectroscopy. Chen *et al.*[329] interfaced a gas chromatograph with a matrix isolation and FTIR detection system for analysis of TMS derivatives of trichothecenes. This method has excellent potential for confirmation of identity, particularly for isomers. A newer analytical development is TLC–FTIR, but this has not yet been applied to mycotoxins.

The high cost of specificity in the mass spectrometric and FTIR methods referred to above makes the development of simple kits for immunoassay (ELISA) of mycotoxins quite exciting. Kits for aflatoxin B_1 and zearalenone are already available commercially and others, for aflatoxin M_1, zearalenone, deoxynivalenol and T-2 toxin, should be available soon. A simple card test kit used for certain drugs has been redesigned and applied to aflatoxins.[330] Since foods usually would contain higher concentrations of aflatoxin B_1 than aflatoxins B_2, G_1 or

G_2, a kit for this aflatoxin would be ideal for screening purposes in terms of both convenience and specificity.

Monoclonal antibody affinity columns have shown great promise for the specific cleanup of aflatoxins in biological fluids;[41,163] determination is then carried out by standard procedures such as HPLC. Affinity column chromatography has also been applied to T-2 toxin and the guanine adduct of aflatoxin B_1 but no application to analysis of foods has yet been reported.

Advances in biotechnology have also found application in a bioassay for aflatoxin B_1:[331] the SOS-Chromotest is based on a bioengineered strain of *Eschericia coli* that reacts to the damage to its DNA caused by this genotoxic mycotoxin by synthesizing β-galactosidase, which produces a measurable colour with a suitable substrate.

Automation of mycotoxin methods has been discussed by Coker[140] and it is to be expected that the trend will parallel advances in other areas of trace analysis. Most automation to date has taken place with the quantitation step[332] but its introduction into other procedures of analytical methods for mycotoxins may not be too far off. Laboratory robotics capable of performing such tasks as grinding, weighing, centrifuging, manipulating filters and solid phase extraction columns, and other operation of sample preparation is already showing steady growth in other areas of analysis.[333]

REFERENCES

1. Turner, W. B., *Fungal metabolites*, Academic Press, London, 1971.
2. Turner, W. B. and Aldridge, D. C., *Fungal metabolites II*, Academic Press, London, 1983.
3. Watson, D. H., *CRC Crit. Rev. Fd Sci. Nutr.*, **22** (1985) 177–98.
4. Cole, R. J. and Cox, R. H., *Handbook of toxic fungal metabolites*, Academic Press, New York, 1981.
5. Ciegler, A. and Vesonder, R. F. In: *Handbook of foodborne diseases of biological origin*, ed. M. Rechcigl, Jr, CRC Press, Boca Raton, Florida, 1983, pp. 57–166.
6. Egmond, H. P. van. In: *Developments in food analysis techniques—3*, ed. R. D. King, Elsevier Applied Science Publishers, London, 1984, pp. 99–144.
7. Mirocha, C. J., Pathre, S. V., Pawlosky, R. J. & Hewetson, D. W. In *Modern methods in the analysis and structural elucidation of mycotoxins*, ed. R. J. Cole, Academic Press, Orlando, Florida, 1986, pp. 353–92.
8. Ueno, Y., *CRC Crit. Rev. Toxicol.*, **14** (1985) 99–132.

9. Betina, V. In: *Mycotoxins—production, isolation, separation and purification,* ed. V. Betina, Elsevier, Amsterdam, 1984, pp. 3–12.
10. Frisvad, J. C., *J. Chromatogr.,* **392** (1987) 333–47.
11. Betina, V. In: *Mycotoxins—production, isolation, separation and purification,* ed. V. Betina, Elsevier, Amsterdam, 1984, pp. 25–36.
12. Ueno, Y., *Pure Appl. Chem.,* **58** (1986) 339–50.
13. Dalcero, A. M., Chulze, S. and Varsavsky, E. In: *Proc. Int. Symp. Mycotoxins, Cairo, Egypt, Sept 6–8, 1981,* ed. K. Naguib, M. M. Naguib, D. L. Park and A. E. Pohland, Food and Drug Administration, Washington, DC and National Research Centre, Cairo, 1983, pp. 437–41.
14. Vijayalakshmi, M. and Rao, A. S. In: *Trichothecenes and other mycotoxins,* ed. J. Lacey, John Wiley, Chichester, 1985, pp. 33–45.
15. King, A. D., Jr and Schade, J. E., *J. Fd. Protect.,* **47** (1984) 886–901.
16. Schade, J. E. and King, A. D., Jr, *J. Fd Protect.,* **47** (1984) 978–95.
17. Stack, M. E., Mazzola, E. P., Page, S. W., Pohland, A. E., Highet, R. J., Tempesta, M. S. and Corley, D. G., *J. Nat. Prod.,* **49** (1986) 866–71.
18. Dorner, J. W., Cole, R. J., Springer, J. P., Cox, R. H., Cutler, H. and Wicklow, D. T., *Phytochemistry,* **19** (1980) 1157–61.
19. Kirksey, J. W. and Cole, R. J., *Appl. Microbiol.,* **26** (1973) 827–8.
20. Bassir, O. and Emerole, G. O., *FEBS Lett.,* **40** (1974) 247–9.
21. Casper, H. H., Backer, L. F. and Kunerth, W., *J. Ass. Off. Anal. Chem.,* **64** (1981) 228–30.
22. Karmelic, J., Israel, M., Benado, S. and Leon, C., *J. Ass. Off. Anal. Chem.,* **56** (1973) 219–22.
23. Llewellyn, G. C. and Eadie, T., *J. Ass. Off. Anal. Chem.,* **57** (1974) 858–60.
24. Monacelli, R., Aiello, E., Di Muccio, A., Salvatore, G. and Dattolo, G., *Riv. Soc. Ital. Sci. Aliment.,* **5** (1976) 259–61.
25. Hesseltine, C. W. In: *Mycotoxins and phycotoxins,* ed. P. S. Steyn and R. Vleggaar, Elsevier, Amsterdam, 1986, pp. 1–18.
26. Scott, P. M., *J. Ass. Off. Anal. Chem.,* **69** (1986) 240–6.
27. Egan, H., Stoloff, L., Scott, P., Castegnaro, M., O'Neill, I. K. and Bartsch, H. (Eds), *Environmental carcinogens—selected methods of analysis. Vol. 5—Some mycotoxins,* International Agency for Research on Cancer, Lyon, 1982.
28. Mislivec, P. B. In: *Aspergillosis,* ed. Y. Aldoory and G. E. Wagner, Charles C. Thomas, Springfield, Illinois, 1985, pp. 257–68.
29. Moss, M. O. In: *Genetics and physiology of Aspergillus, Vol. 1,* ed. J. E. Smith, Academic Press, London, 1977, pp. 499–524.
30. Scott, P. M. In: *Proc. Int. Symp. Mycotoxins, Cairo, Egypt, Sept. 6–8, 1981,* ed. K. Naguib, M. M. Naguib, D. L. Park and A. E. Pohland, Food and Drug Administration, Washington, DC and National Research Centre, Cairo, 1983, pp. 87–110.
31. Frisvad, J. C. In: *Modern methods in the analysis and structural elucidation of mycotoxins,* ed. R. J. Cole, Academic Press, New York, 1986, pp. 415–57.
32. Scott, P. M. In: *Mycotoxic fungi, mycotoxins, mycotoxicoses. An*

encyclopedic handbook. Vol. 1. Mycotoxic fungi and chemistry of mycotoxins, ed. T. D. Wyllie and L. G. Morehouse, Marcel Dekker, New York, 1977, pp. 283–356.

33. Frisvad, J. C. and Filtenborg, O., *Appl. Environ. Microbiol.,* **46** (1983) 1301–10.
34. Ayer, W. A., Pena-Rodriguez, L. and Vederas, J. C., *Can. J. Microbiol.,* **27** (1981) 846–7.
35. Davis, N. D., *J. Fd Protect.,* **44** (1981) 711–14.
36. Yamazaki, M., Horie, Y., Maebayashi, Y., Suzuki, S., Terao, K. and Nagao, M., *Proc. Jpn Ass. Mycotoxicol.* (1980) 17–19.
37. Steyn, P. S. (Ed.), *The biosynthesis of mycotoxins: a study in secondary metabolism,* Academic Press, New York, 1980.
38. Cullen, J. M., Ruebner, B. H., Hsieh, L. S., Hyde, D. M. and Hsieh, D. P., *Cancer Res.,* **47** (1987) 1913–17.
39. Hsieh, D. P. H. and Ruebner, B. H. In: *Toxigenic fungi—their toxins and health hazard,* ed. H. Kurata and Y. Ueno, Kodansha Ltd, Tokyo and Elsevier, Amsterdam, 1984, pp. 332–8.
40. Krishnamachari, K. A. V. R., Bhat, V. R., Nagarajan, V., Tilak, T. B. G. and Tulpule, P. G., *Ann. Nutr. Alim.,* **31** (1977) 991–6.
41. Groopman, J. D., Busby, W. F., Jr, Donahue, P. R. and Wogan, G. N. In: *Biochemical and molecular epidemiology of cancer,* ed. C. C. Harris, Alan R. Liss, New York, 1986, pp. 233–56.
42. Patterson, D. S. P. In: *Handbook of foodborne diseases of biological origin,* ed. M. Rechcigl, Jr, CRC Press, Boca Raton, Florida, 1983, pp. 323–51.
43. Egmond, H. P. van, Paulsch, W. E., Deijll, E. and Schuller, P. L., *J. Ass. Off. Anal. Chem.,* **63** (1980) 110–14.
44. Stoltz, D. R. In: *Carcinogens and mutagens in the environment,* Vol. III, ed. H. F. Stich, CRC Press, Boca Raton, Florida, 1983, pp. 129–36.
45. Harwig, J., Kuiper Goodman, T. and Scott, P. M. In: *Handbook of foodborne diseases of biological origin,* ed. M. Rechcigl, Jr, CRC Press, Boca Raton, Florida, 1983, pp. 193–238.
46. Hood, R. D. and Szczech, G. M. In: *Handbook of natural toxins, Vol. 1. Plant and fungal toxins,* ed. R. F. Keeler and A. T. Tu, Marcel Dekker, New York, 1982, pp. 201–35.
47. Price, K. R. and Fenwick, G. R., *Fd Addit. Contamin.,* **2** (1985) 73–106.
48. Kuiper-Goodman, T., Scott, P. M. and Watanabe, H., *Regul. Toxicol. Pharmacol.,* **7** (1987) 253–306.
49. White, E. P., Mortimer, P. H. and diMenna, M. E. In: *Mycotoxic fungi, mycotoxins, mycotoxicoses. An encyclopedic handbook, Vol. 1. Mycotoxic fungi and chemistry of mycotoxins,* ed. T. D. Wyllie and L. G. Morehouse, Marcel Dekker, New York, 1977, pp. 427–47.
50. Cole, R. J. In: *Antinutrients and natural toxins in foods,* ed. R. L. Ory, Food & Nutrition Press, Westport, CN, 1981, pp. 17–33.
51. Takahashi, C., Sekita, S., Yoshihara, K. and Natori, S., *Chem. Pharm. Bull.,* **24** (1976) 2317–21.
52. Robbana-Barnat, S., Loridon-Rosa, B., Cohen, H., Lafarge-Frayssinet, C., Neish, G. A. and Frayssinet, C., *Fd Addit. Contamin.,* **4** (1987) 49–55.

53. Joffe, A. Z. In: *Handbook of foodborne diseases of biological origin*, ed. M. Rechcigl, Jr, CRC Press, Boca Raton, Florida, 1983, pp. 353–495.
54. Lorenz, K., *CRC Crit. Rev. Fd Sci. Nutr.*, **11** (1979) 311–54.
55. Azzouz, M. A. and Bullerman, L. B., *J. Fd Protect.*, **45** (1982) 1298–301.
56. Hitokoto, H., Morozumi, S., Wauke, T., Sakai, S. and Ueno, I., *Mycopathologia*, **66** (1978) 161–7.
57. Llewellyn, G. C., Burkett, M. L. and Eadie, T., *J. Ass. Off. Anal. Chem.*, **64** (1981) 955–60.
58. Mabrouk, S. S. and El-Shayeb, N. M. A., Z. *Lebensmittelunters u.-Forsch.*, **171** (1980) 344–7.
59. Suzuki, J. I., Dainius, B. and Kilbuck, J. H., *J. Fd Sci.*, **38** (1973) 949–50.
60. Scott, P. M. and Kennedy, B. P. C., *Can. Inst. Fd Sci. Technol. J.*, **8** (1975) 124–5.
61. Bullerman, L. B., Schroeder, L. L. and Park, K.-Y., *J. Fd Protect.*, **47** (1984) 637–46.
62. Hesseltine, C. W. In: *Mycotoxins and other fungal related food problems*, ed. J. V. Rodricks, American Chemical Society, Washington, DC, 1976, pp. 1–22.
63. Shotwell, O. L., Vandegraft, E. E. and Hesseltine, C. W., *J. Ass. Off. Anal. Chem.*, **61** (1978) 574–77.
64. Stössel, P., *Appl. Environ. Microbiol.*, **52** (1986) 68–72.
65. Bean, G. A., Schillinger, J. A. and Klarman, W. L., *Appl. Microbiol.*, **24** (1972) 437–9.
66. Goldblatt, L. A. In: *Aflatoxin, Scientific background, control, and implications*, ed. L. A. Goldblatt, Academic Press, New York, 1969, pp. 1–11.
67. Yoshizawa, T. In: *Trichothecenes, Chemical, biological and toxicological aspects*, ed. Y. Ueno, Kodansha, Tokyo and Elsevier, Amsterdam, 1983, pp. 195–209.
68. Vesonder, R. F. In: *Trichothecenes, Chemical, biological and toxicological aspects*, ed. Y. Ueno, Kodansha, Tokyo and Elsevier, Amsterdam, 1983, pp. 210–17.
69. Woller, R. and Majerus, P., *Lebensmittel. Gericht. Chem.*, **33** (1979) 115–16.
70. Gallagher, R. T., Richard, J. L., Stahr, H. M. and Cole, R. J., *Mycopathologia*, **66** (1978) 31–6.
71. Lansden, J. A. and Davidson, J. I., *Appl. Environ. Microbiol.*, **45** (1983) 766–9.
72. Kumari, C. K., Nusrath, M. and Reddy, B. N., *Ind. Phytopathol.*, **37** (1984) 284–7.
73. Samarajeewa, U. and Arseculeratne, S. N., *J. Natn Counc. Sri Lanka*, **11** (1983) 225–35.
74. Gracian, J. and Arévalo, G., *Grasas Aceites*, **31** (1980) 167–71.
75. Cornelius, J. A. and Maduagwu, E. N. In: *Spoilage and mycotoxins of cereals and other stored products*, ed. B. Flannigan, CAB International, Slough, UK, 1986, pp. 95–101.
76. Kershaw, S. J., *Appl. Environ. Microbiol.*, **43** (1982) 1210–12.

77. Rossell, J. B., *Fd Flav. Ingr. Proc. Pack.*, **5** (Oct. 1983) 28–32, 57.
78. Gelda, C. S. and Luyt, L. J., *Ann. Nutr. Alim.*, **31** (1977) 477–83.
79. Ghosal, S., Chakrabarti, D. K. and Chaudhary, K. C. B., *Experientia*, **33** (1977) 574–5.
80. Northolt, M. D., Egmond, H. P. van, Soentoro, P. and Deijll, E., *J. Ass. Off. Anal. Chem.*, **63** (1980) 115–19.
81. Josefsson, E., *Vår Föda*, **33** (1981) 237–48.
82. Pfleger, R., *Milchwirtschaft. Ber. Bundesanstalt Wolfpassing Rotholz*, **85** (1985) 297–301.
83. Tantaoui-Elaraki, A. and Khabbazi, N., *Lait*, **64** (1984) 46–71.
84. Stoloff, L., *Nutr. Cancer*, **5** (1983) 165–86.
85. Pepeljnjak, S. and Cvetnić, Z., *Mycopathologia*, **90** (1985) 147–53.
86. Scott, P. M., *J. Fd Protect.*, **41** (1978) 385–98.
87. Dickens, J. W. In: *Mycotoxins in human and animal health*, ed. J. V. Rodricks, C. W. Hesseltine and M. A. Mehlman, Pathotox, Park Forest South, Illinois, 1977, pp. 99–105.
88. Dickens, J. W., *J. Am. Oil Chem. Soc.*, **54** (1977) 225A–8A.
89. Dickens, J. W. and Satterwhite, J. B., *Oléagineux*, **26** (1971) 321–8.
90. Shotwell, O. L. In: *Modern methods in the analysis and structural elucidation of mycotoxins*, ed. R. J. Cole, Academic Press, Orlando, Florida, 1986, pp. 51–94.
91. Egmond, H. P. van, *Fd Chem.*, **11** (1983) 289–307.
92. Frobish, R. A., Bradley, D. B., Wagner, D. D., Long-Bradley, P. E. and Hairston, H., *J. Fd Protect.*, **49** (1986) 781–95.
93. Pestka, J. J. In: *Advances in meat research, Vol. 2*, ed. A. M. Pearson and T. R. Dutson, Avi Publ. Co., Westport, Connecticut, 1986, pp. 277–309.
94. Rodricks, J. V. and Stoloff, L. In: *Mycotoxins in human and animal health*, ed. J. V. Rodricks, C. W. Hesseltine and M. A. Mehlman, Pathotox, Park Forest South, Illinois, 1977, pp. 67–79.
95. Scott, P. M., *J. Fd Protect.*, **47** (1984) 489–99.
96. Parker, W. A. and Melnick, D., *J. Am. Oil Chem. Soc.*, **43** (1966) 635–8.
97. Abalaka, J. A., *Fd Chem. Toxicol.*, **22** (1984) 461–3.
98. Abalaka, J. A. and Elegbede, J. A., *Fd Chem. Toxicol.*, **20** (1982) 43–6.
99. Ling, K.-H., Tung, C.-M., Sheh, I-F., Wang, J.-J. and Tung, T.-C., *J. Formosan Med. Assoc.*, **67** (1968) 309–14.
100. Dwarakanath, C. T., Sreenivasamurthy, V. and Parpia, H. A. B., *J. Fd Technol. (Mysore)*, **6** (1969) 107–9.
101. Arseculeratne, S. N. and De Silva, L. M., *Ceylon J. Med. Sci.*, **20** (1971) 60–75.
102. Okonkwo, P. O. and Nwokolo, C., *Nutr. Rep. Int.*, **17** (1978) 387–95.
103. Visconti, A., Logrieco, A. and Bottalico, A., *Fd Addit. Contamin.*, **3** (1986) 323–30.
104. Isohata, E., Toyoda, M. and Saito, Y., *Eisei Shikensho Hokoku*, **104** (1986) 138–42.
105. Kamimura, H., Nishijima, M., Tabata, S., Yasuda, K., Ushiyama, H. and Nishima, T., *J. Fd Hyg. Soc. Jpn*, **27** (1986) 59–63.

106. Peers, F. G. and Linsell, C. A., *Trop. Sci.*, **17** (1975) 229–32.
107. Samarajeewa, U., Arseculeratne, S. N. and Bandunatha, C. H. S. R., *J. Natn. Sci. Counc. Sri Lanka*, **5** (1977) 1–12.
108. Wood, G. M., Cooper, S. J. and Chapman, W. B. In: *Proc. V. Int. IUPAC Symp Mycotoxins and Phycotoxins, Vienna, Austria, Sept. 1–3*, Austrian Chemical Society, Vienna, 1982, pp. 142–5.
109. Bennett, G. A., Lagoda, A. A., Shotwell, O. L. and Hesseltine, C. W., *J. Am. Oil Chem. Soc.*, **58** (1981) 974–6.
110. Tanaka, T., Hasegawa, A., Yamamoto, S., Matsuki, Y. and Ueno, Y., *J. Fd Hyg. Soc. Jpn*, **27** (1986) 653–5.
111. Egmond, H. P. van, *Joint FAO/WHO/UNEP Second Int. Conf. Mycotoxins*, Sept. 28–Oct. 3, Bangkok, Thailand, 1987.
112. Labuza, T. P., *J. Fd Protect.*, **46** (1983) 260–5.
113. Schuller, P. L., Egmond, H. P. van and Stoloff, L. In: *Proc. Int. Symp. Mycotoxins, Cairo, Egypt, Sept. 6–8, 1981*, ed. K. Naguib, M. M. Naguib, D. L. Park and A. E. Pohland, Food and Drug Administration, Washington, DC and National Research Centre, Cairo, 1983, pp. 111–29.
114. Schuller, P. L., Stoloff, L. and Egmond, H. P. van. In: *Environmental carcinogens—selected methods of analysis. Vol. 5—Some mycotoxins*, ed. H. Egan, L. Stoloff, P. Scott, M. Castegnaro, I. K. O'Neill and H. Bartsch, International Agency for Research on Cancer, Lyon, 1982, pp. 107–16.
115. Scott, P. M. In *Toxigenic fungi—their toxins and health hazard*, ed. H. Kurata and Y. Ueno, Kodansha Ltd, Tokyo and Elsevier, Amsterdam, 1984, pp. 182–9.
116. Moss, M. O. and Frank, J. M. In: *Trichothecenes and other mycotoxins*, ed. J. Lacey, John Wiley, Chichester, 1985, pp. 257–68.
117. Hill, R. A., Blankenship, P. D., Cole, R. J. and Saunders, T. H., *Appl. Environ. Microbiol.*, **45** (1983) 628–33.
118. Ehrlich, K., Ciegler, A., Klich, M. and Lee, L., *Experientia*, **41** (1985) 691–3.
119. Abramson, D., Sinha, R. N. and Mills, J. T., *Sci. Aliments*, **5** (1985) 653–63.
120. Northolt, M. D. and Heuvelman, C. J., *J. Fd Protect.*, **45** (1982) 537–40.
121. Draughon, F. A. and Churchville, D. C., *Phytopathology*, **75** (1985) 553–6.
122. Mixon, A. C., *J. Am. Oil Chem. Soc.*, **58** (1981) 961A–6A.
123. Lillehoj, E. B. and Zuber, M. S., *J. Am. Oil Chem. Soc.*, **58** (1981) 970A–3A.
124. Shannon, G. M., Shotwell, O. L., Lyons, A. J., White, D. G. and Garcia-Aguirre, G., *J. Ass. Off. Anal. Chem.*, **63** (1980) 1275–7.
125. Pohland, A. E., Thorpe, C. W. and Nesheim, S., *Pure Appl. Chem.*, **52** (1979) 213–23.
126. Horwitz, W. and Albert, R., *Ass. Fd Drug Officials Quart. Bull.*, **46** (1982) 14–24.
127. Whitaker, T. B. and Dickens, J. W., *J. Am. Oil Chem. Soc.*, **56** (1979) 789–94.

128. Campbell, A. D., Whitaker, T. B., Pohland, A. E., Dickens, J. W. and Park, D. L., *Pure Appl. Chem.*, **58** (1986) 305–14.
129. Dickens, J. W. and Whitaker, T. B. In: *Environmental carcinogens—selected methods of analysis. Vol. 5—Some mycotoxins*, ed. H. Egan, L. Stoloff, P. Scott, M. Castegnaro, I. K. O'Neill and H. Bartsch, International Agency for Research on Cancer, Lyon, 1982, pp. 17–32.
130. Davis, N. D., Dickens, J. W., Freie, R. L., Hamilton, P. B., Shotwell, O. L., Wyllie, T. D. and Fulkerson, J. F., *J. Ass. Off. Anal. Chem.*, **63** (1980) 95–102.
131. Dickens, J. W. and Whitaker, T. B. In: *Modern methods in the analysis and structural elucidation of mycotoxins*, ed. R. J. Cole, Academic Press, Orlando, Florida, 1986, pp. 29–49.
132. Stoloff, L. and Scott, P. M. (chapter Eds). In: *Official methods of analysis of the Association of Official Analytical Chemists*, 14th edn, ed. S. Williams, Association of Official Analytical Chemists, Arlington, Virginia, 1984, pp. 477–500.
133. Whitaker, T. B., Dickens, J. W. and Monroe, R. J., *J. Am. Oil Chem. Soc.*, **57** (1980) 269–72.
134. Stoloff, L., *Clin. Toxicol.*, **5** (1972) 465–94.
135. Park, D. L. and Pohland, A. E. In: *Trace organic analysis: a new frontier in analytical chemistry*, National Bureau of Standards, Gaithersburg, Maryland, 1979, pp. 321–31.
136. Hunt, D. C. In: *HPLC in food analysis*, ed. R. Macrae, Academic Press, London, 1982, pp. 271–84.
137. Bullerman, L. B. In: *Food and beverage mycology*, 2nd edn, ed. L. R. Beuchat, AVI, Westport, Connecticut, 1987, pp. 571–98.
138. Cole, R. J. (Ed.), *Modern methods in the analysis and structural elucidation of mycotoxins*, Academic Press, New York, 1986.
139. Gorst-Allman, C. P. and Steyn, P. S. In: *Mycotoxins—production, isolation, separation and purification*, ed. V. Betina, Elsevier, Amsterdam, 1984, pp. 59–85.
140. Coker, R. D. In: *Analysis of food contaminants*, ed. J. Gilbert, Elsevier Applied Science Pubishers Ltd, Barking, Essex, 1984, pp. 207–63.
141. Egmond, H. P. van and Paulsch, W. E., *Pure Appl. Chem.*, **58** (1986) 315–26.
142. Crosby, N. T., *Fd Addit. Contamin.*, **1** (1984) 39–44.
143. Romer, T. In: *Food constituents and food residues. Their chromatographic determination*, ed. J. F. Lawrence, Marcel Dekker, New York, 1984, pp. 355–93.
144. Nesheim, S. In: *Trace organic analysis: a new frontier in analytical chemistry*, National Bureau of Standards, Gaithersburg, Maryland, 1979, pp. 355–72.
145. Nesheim, S. and Brumley, W. C., *J. Am. Oil Chem. Soc.*, **58** (1981) 945A–9A.
146. Rottinghaus, G. E. In: *Diagnosis of mycotoxicoses*, ed. J. L. Richard and J. R. Thurston, Martinus Nijhoff Publishers, Dordrecht, 1986, pp. 239–55.

147. Stubblefield, R. D., In: *Diagnosis of mycotoxicoses*, ed. J. L. Richard and J. R. Thurston, Martinus Nijhoff Publishers, Dordrecht, 1986, pp. 257–69.
148. Gilbert, J. In: *The applied mycology of fusarium*, ed. M. O. Moss and J. E. Smith, Cambridge University Press, Cambridge, 1984, pp. 175–93.
149. Pohland, A., Thorpe, C. and Sphon, J. In: *Toxigenic fungi—their toxins and health hazard*, ed. H. Kurata and Y. Ueno, Kodansha, Tokyo and Elsevier, Amsterdam, 1984, pp. 217–30.
150. Pohland, A. E., Thorpe, C. W., Trucksess, M. W. and Eppley, R. M. In: *Diagnosis of mycotoxicoses*, ed. J. L. Richard and J. R. Thurston, Martinus Nijhoff Publishers, Dordrecht, 1986, pp. 271–81.
151. Scott, P. M., *J. Ass. Off. Anal. Chem.*, **65** (1982) 876–83.
152. Bennett, G. A. and Shotwell, O. L., *J. Am. Oil Chem. Soc.*, **56** (1979) 812–19.
153. Betina, V., *J. Chromatogr.*, **334** (1985) 211–76.
154. Nesheim, S. and Trucksess, M., In: *Modern methods in the analysis and structural elucidation of mycotoxins*, ed. R. J. Cole, Academic Press, Orlando, Florida, 1986, pp. 239–64.
155. Scott, P. M. In: *Thin layer chromatography: quantitative environmental and clinical applications*, ed. J. C. Touchstone and D. Rogers, Wiley Interscience, New York, 1980, pp. 251–74.
156. Scott, P. M. In: *Advances in thin layer chromatography, Clinical and environmental applications*, ed. J. C. Touchstone, John Wiley, New York, 1982, pp. 321–42.
157. Holaday, C. E., *J. Am. Oil Chem. Soc.*, **58** (1981) 931A–4A.
158. Beaver, R. W. In: *Modern methods in the analysis and structural elucidation of mycotoxins*, ed. R. J. Cole, Academic Press, Orlando, Florida, 1986, pp. 265–92.
159. Scott, P. M. In: *Trace analysis, Vol. 1*, ed. J. F. Lawrence, Academic Press, New York, 1981, pp. 193–266.
160. Shepherd, M. J. In: *Modern methods in the analysis and structural elucidation of mycotoxins*, ed. R. J. Cole, Academic Press, Orlando, Florida, 1986, pp. 293–333.
161. Chu, F. S., *J. Fd Protect.*, **47** (1984) 562–9.
162. Chu, F. S. In: *Diagnosis of mycotoxicoses*, ed. J. L. Richard and J. R. Thurston, Martinus Nijhoff Publishers, Dordrecht, 1986, pp. 163–76.
163. Chu, F. S. In: *Mycotoxins and phycotoxins*, ed. P. S. Steyn and R. Vleggaar, Elsevier Science Publishers, B.V., Amsterdam, 1986, pp. 277–92.
164. Chu, F. S. In: *Modern methods in the analysis and structural elucidation of mycotoxins*, ed. R. J. Cole, Academic Press, Orlando, Florida, 1986, pp. 207–37.
165. Cole, R. J. In: *Mycotoxins—production, isolation, separation and purification*, ed. V. Betina, Elsevier, Amsterdam, 1984, pp. 45–58.
166. Cole, R. J., Cutler, H. G. and Dorner, J. W. In: *Modern methods in the analysis and structural elucidation of mycotoxins*, ed. R. J. Cole, Academic Press, Orlando, Florida, 1986, pp. 1–28.
167. Watson, D. H. and Lindsay, D. G., *J. Sci. Fd Agric.*, **33** (1982) 59–67.

168. Yates, I. E. In: *Diagnosis of mycotoxicoses*, ed. J. L. Richard and J. R. Thurston, Martinus Nijhoff Publishers, Dordrecht, 1986, pp. 333–78.
169. Yates, I. E. and Porter, J. K. In: *Toxicity screening procedures using bacterial systems*, ed. D. Liu and B. J. Dutka, Marcel Dekker, New York, 1984, pp. 77–88.
170. Stoloff, L., *J. Ass. Off. Anal. Chem.*, **66** (1983) 355–63.
171. Scott, P. M., *J. Ass. Off. Anal. Chem.*, **67** (1984) 366–9.
172. Scott, P. M., *J. Ass. Off. Anal. Chem.*, **68** (1985) 242–8.
173. Scott, P. M., *J. Ass. Off. Anal. Chem.*, **70** (1987) 276–81.
174. Scott, P. M., *J. Ass. Off. Anal. Chem.*, **71** (1988) 70–6.
175. Park, D. L. and Pohland, A. E. In: *Foodborne microorganisms and their toxins, Developing methodology*, ed M. D. Pierson and N. J. Stern, Marcel Dekker, New York, 1986, pp. 425–38.
176. Scott, P. M., Bernetti, R., Bowers, R. H., Dickens, J. W., Henderson, J. C., Pohland, A. E., Shotwell, O. L., Stubblefield, R. D. and Waltking, A. E., *J. Ass. Off. Anal. Chem.*, **70** (1987) 357–8.
177. Gimeno, *J. Ass. Off. Anal. Chem.*, **62** (1979) 579–85.
178. Grabarkiewicz-Szczęsna, J., Goliński, P., Chelkowski, J. and Szebiotko, K., *Nahrung*, **29** (1985) 229–40.
179. Miguel, J. A. and de Andres, V., *Ann. Inst. Nac. Invest. Agrar. Ser.: Agric. (Spain)*, **27** (1984) 89–99.
180. Nowotny, P., Baltes, W., Krönert, W. and Weber, R., *Chem. Mikrobiol. Technol. Lebensmittel.*, **8** (1983) 24–8.
181. Steyn, P. S., *Pure Appl. Chem.*, **53** (1981) 891–902.
182. Eller, K. I. and Sobolev, V. S., *Zh. Anal. Khim.*, **38** (1983) 903–7.
183. Hagan, S. N. and Tietjen, W. H., *J. Ass. Off. Anal. Chem.*, **58** (1975) 620–1.
184. Gertz, C. and Böschemeyer, L., *Z. Lebensmittelunters, u.-Forsch.*, **171** (1980) 335–40.
185. Siriwardana, M. G. and Lafont, P., *J. Dairy Sci.*, **62** (1979) 1145–8.
186. Škrinjar, M. and Žakula, R., *Mljekarstvo*, **35** (1985) 131–7.
187. Scossa-Romano, D. A., Bickel, R. E., Zweifel, U., Reinhardt, C. A., Lüthy, J. W. and Schlatter, C. L., *J. Ass. Off. Anal. Chem.*, **70** (1987) 129–32.
188. Müller, T. and Lepom, P., *Mh. Vet.-Med.*, **40** (1985) 486–9.
189. Ďuračková, Z., Betina, V. and Nemec, P., *J. Chromatogr.*, **116** (1976) 155–61.
190. Prior, M. G., *Can. J. Comp. Med.*, **43** (1979) 352–5.
191. Dickens, J. W. and Whitaker, T. B., *J. Ass. Off. Anal. Chem.*, **66** (1983) 1059–62.
192. Kalinoski, H. T., Udseth, H. R., Wright, B. W. and Smith, R. D., *Anal. Chem.*, **58** (1986) 2421–5.
193. Trenholm, H. L., Warner, R. and Prelusky, D. B., *J. Ass. Off. Anal. Chem.*, **68** (1985) 645–9.
194. Whitaker, T. B., Dickens, J. W. and Giesbrecht, F. G., *J. Ass. Off. Anal. Chem.*, **69** (1986) 508–10.
195. Brumley, W. C., Nesheim, S., Trucksess, M. W., Trucksess, E. W., Dreifuss, P. A., Roach, J. A. G., Andrzejewski, D., Eppley, R. M.,

Pohland, A. E., Thorpe, C. W. and Sphon, J. A., *Anal. Chem.*, **53** (1981) 2003–6.
196. Lee, L. S. and Catalano, E. A., *J. Am. Oil Chem. Soc.*, **58** (1981) 949A–51A.
197. Trucksess, M. W., Brumley, W. C. and Nesheim, S., *J. Ass. Off. Anal. Chem.*, **67** (1984) 973–5.
198. Hurst, W. J., Snyder, K. P. and Martin, R. A., Jr, *Peanut Sci.*, **11** (1984) 21–3.
199. Gimeno, A., *J. Ass. Off. Anal. Chem.*, **67** (1984) 194–6.
200. Wilson, D. M., *Abstr. 100th Ann. Int. Mtg. Ass. Off. Anal. Chem.*, Sept. 15–18, Scottsdale, Arizona, 1986, p. 11.
201. Lansden, J. A., *J. Ass. Off. Anal. Chem.*, **69** (1986) 964–6.
202. Le Tutour, B., Tantaoui-Elaraki, A. and Ihlal, L., *J. Am. Oil Chem. Soc.*, **60** (1983) 835–7.
203. Miller, N., Pretorius, H. E. and Trinder, D. W., *J. Ass. Off. Anal. Chem.*, **68** (1985) 136–7.
204. Destro, O. and Gelosa, L., *Ind. Alim.*, **25** (1986) 868–70.
205. Kamimura, H., Nishijima, M., Yasuda, K., Ushiyama, H., Tabata, S., Matsumoto, S. and Nishima, T., *J. Ass. Off. Anal. Chem.*, **68** (1985) 458–61.
206. Francis, O. J., Jr, Ware, G. M., Carman, A. S. and Kuan, S. S. *J. Ass. Off. Anal. Chem.*, **68** (1985) 643–5.
207. Le Bars, J., *Appl. Environ. Microbiol.*, **38** (1979) 1052–5.
208. Still, P., Eckardt, C. and Leistner, L., *Fleischwirtschaft*, **58** (1978) 876–7.
209. Lafont, P., Siriwardana, M. G., Combemale, I. and Lafont, J., *Fd Cosmet. Toxicol.*, **17** (1979) 147–9.
210. Scott, P. M. and Kennedy, B. P. C., *J. Agric. Fd Chem.*, **24** (1976) 865–8.
211. Scott, P. M. and Lawrence, G. A., *J. Agric. Fd Chem.*, **30** (1982) 445–50.
212. Tarter, E. J., Hanchay, J.-P. and Scott, P. M., *J. Ass. Off. Anal. Chem.*, **67** (1984) 597–600.
213. Goto, T., Manabe, M. and Matsuura, S., *Agric. Biol. Chem.*, **46** (1982) 801–2.
214. Stubblefield, R. D. and Kwolek, W. F., *J. Ass. Off. Anal. Chem.*, **69** (1986) 880–5.
215. Toya, N., *Bull. Kumamoto Women's Univ.*, **37** (1985) 104–10.
216. Lansden, J. A., *J. Ass. Off. Anal. Chem.*, **67** (1984) 728–31.
217. Cohen, H. and Lapointe, M., *J. Ass. Off. Anal. Chem.*, **69** (1986) 957–9.
218. Stubblefield, R. D., Greer, J. I. and Shotwell, O. L., *Abstr. 100th Ann. Int. Mtg. Ass. Off. Anal. Chem.*, September 15–18, Scottsdale, Arizona, 1986, p. 11.
219. Scott, P. M. and Lawrence, G. A., *J. Ass. Off. Anal. Chem.*, **70** (1987), 850–3.
220. Ware, G. M., Thorpe, C. W. and Pohland, A. E., *J. Ass. Off. Anal. Chem.*, **63** (1980) 637–41.
221. Hurst, W. J., Snyder, K. P. and Martin, R. A., Jr, *J. Chromatogr.*, **392** (1987) 389–96.

222. Ibe, A., Nishijima, M., Yasuda, K., Saito, K., Kamimura, H., Nagayama, T., Ushiyama, H., Naoi, Y. and Nishima, T., *J. Fd Hyg. Soc. Jpn*, **25** (1984) 334–41.
223. Mirocha, C. J., Pawlowsky, R. J. and Hewetson, D. W. In: *Diagnosis of mycotoxicoses*, ed. J. L. Richard and J. R. Thurston, Martinus Nijhoff Publishers, Dordrecht, 1986, pp. 305–22.
224. Dantzman, J. and Stoloff, L., *J. Ass. Off. Anal. Chem.*, **55** (1972) 139–41.
225. Romer, T. R., Ghouri, N. and Boling, T. M., *J. Am. Oil Chem. Soc.*, **56** (1979) 795–7.
226. Shotwell, O. L. and Holaday, C. E., *J. Ass. Off. Anal. Chem.*, **64** (1981) 674–77.
227. Metwally, M. M., Dawood, A. E. A. and Aly, A.-E. N. In: *Proc. Int. Symp. Mycotoxins, Cairo, Egypt*, September 6–8, 1981, ed. K. Naguib, M. M. Naguib, D. L. Park and A. E. Pohland, Food and Drug Administration, Washington, DC and National Research Centre, Cairo, 1983, pp. 455–9.
228. Arévalo, G. and Martel, J., *Grasas Aceites*, **34** (1983) 386–91.
229. Pons, W. A., Jr, Cucullu, A. F., Franz, A. O., Jr, Lee, L. S. and Goldblatt, L. A., *J. Ass. Off. Anal. Chem.*, **56** (1973) 803–7.
230. Hald, B. and Krogh, P., *J. Ass. Off. Anal. Chem.*, **58** (1975) 156–8.
231. Holaday, C. E., *J. Am. Oil Chem. Soc.*, **53** (1976) 603–5.
232. Holaday, C. E., *J. Am. Oil Chem. Soc.*, **57** (1980) 491A–2A.
233. Shotwell, O. L., Goulden, M. L. and Hesseltine, C. W., *Cereal Chem.*, **51** (1974) 492–9.
234. Ashworth, L. J., Jr and McMeans, J. L., *Phytopathology*, **56** (1966) 1104–5.
235. Farsaie, A., McClure, W. F. and Monroe, R. J., *J. Fd Sci.*, **43** (1978) 1550–2.
236. McClure, W. F. and Farsaie, A., *Trans. Am. Soc. Agric. Engng.* (1980) 204–7.
237. Schade, J. E. and King, A. D., Jr, *J. Fd. Sci.*, **49** (1984) 493–7.
238. Dickens, J. W. and Whitaker, T. B., *J. Am. Oil Chem. Soc.*, **58** (1981) 973A–5A.
239. Marsh, P. B., Simpson, M. E., Ferretti, R. J., Merola, G. V., Donoso, J., Craig, G. O., Trucksess, M. W. and Work, P. S., *J. Agric. Fd Chem.*, **17** (1969), 468–72.
240. Lillehoj, E. B., Jacks, T. J. and Calvert, O. H., *J. Fd Protect.*, **49** (1986) 623–6.
241. Chu, F. S., Liang, M. Y. C. and Zhang, G. S., *Appl. Environ. Microbiol.*, **48** (1984) 777–80.
242. Dixon, D. E., Warner, R. L., Ram, B. P., Hart, L. P. and Pestka, J. J., *J. Agric. Fd Chem.*, **35** (1987) 122–6.
243. Kemp, H. A., Mills, E. N. C. and Morgan, M. R. A., *J. Sci. Fd Agric.*, **37** (1986) 888–94.
244. Pestka, J. J., Liu, M.-T., Knudson, B. K. and Hogberg, M. G., *J. Fd Protect.*, **48** (1985) 953–7.
245. Wei, R.-D. and Chu, F. S., *Anal. Biochem.*, **160** (1987) 399–408.

246. Woychik, N. A., Hinsdill, R. D. and Chu, F. S., *Appl. Environ. Microbiol.*, **48** (1984) 1096–9.
247. Xu, Y.-C., Zhang, G. S. and Chu, F. S., *J. Ass. Off. Anal. Chem.*, **69** (1986) 967–9.
248. Zhang, G. S., Schubring, S. L. and Chu, F. S., *Appl. Environ. Microbiol.*, **51** (1986) 132–7.
249. Ram, B. P., Hart, L. P., Cole, R. J. and Pestka, J. J., *J. Fd Protect.*, **49** (1986) 792–5.
250. Ram, B. P., Hart, L. P., Shotwell, O. L. and Pestka, J. J., *J. Ass. Off. Anal. Chem.*, **69** (1986) 904–7.
251. Morgan, M. R. A., Kang, A. S. and Chan, H. W.-S., *J. Sci. Fd Agric.*, **37** (1986) 908–14.
252. Märtlbauer, E. and Terplan, G., *Arch. Lebensmittelhyg.*, **36** (1985) 53–5.
253. Warner, R., Ram, B. P., Hart, L. P. and Pestka, J. J., *J. Agric. Fd Chem.*, **34** (1986) 714–17.
254. Pestka, J. J., Lee, S. C., Lau, H. P. and Chu, F. S., *J. Am. Oil Chem. Soc.*, **58** (1981) 940A–4A.
255. Goliński, P. and Grabarkiewicz-Szczęsna, J., *J. Ass. Off. Anal. Chem.*, **67** (1984) 1108–10.
256. Brumley, W. C., Trucksess, M. W., Adler, S. H., Cohen, C. K., White, K. D. and Sphon, J. A., *J. Agric. Fd Chem.*, **33** (1985) 326–30.
257. Ross, P. F. In: *Diagnosis of mycotoxicoses*, ed. J. L. Richard and J. R. Thurston, Martinus Nijhoff Publishers, Dordrecht, 1986, pp. 323–9.
258. Gelderblom, W. C. A., Thiel, P. G., Marasas, W. F. O. and van der Merwe, K. J., *J. Agric. Fd Chem.*, **32** (1984) 1064–7.
259. Anon. In: *Environmental carcinogens—selected methods of analysis. Vol. 5—Some mycotoxins*, ed. H. Egan, L. Stoloff, P. Scott, M. Castegnaro, I. K. O'Neill and H. Bartsch, International Agency for Research on Cancer, Lyon, 1982, pp. 63–84.
260. Pohland, A. E., Schuller, P. L., Steyn, P. S. and Egmond, H. P. van, *Pure Appl. Chem.*, **54** (1982) 2219–84.
261. Pathre, S. V., Mirocha, C. J. and Fenton, S. W., *J. Ass. Off. Anal. Chem.*, **62** (1979) 1268–73.
262. Bennett, G. A. and Shotwell, O. L., *Abstr. 100th Ann. Int. Mtg. Ass. Off. Anal. Chem.*, Sept 15–18, Scottsdale, Arizona, 1986, p. 10.
263. Visconti, A., Mirocha, C. J. and Pawlosky, R. J., *J. Ass. Off. Anal. Chem.*, **70** (1987) 193–6.
264. McKinney, J. D., *J. Ass. Off. Anal. Chem.*, **67** (1984) 25–32.
265. Anon., *Fd Chem. News* (March 16, 1987), 41–2.
266. Whitaker, T. B. and Dickens, J. W., *Peanut Sci.*, **8** (1981) 89–92.
267. Schuller, P. L., Horwitz, W. and Stoloff, L., *J. Ass. Off. Anal. Chem.*, **59** (1976) 1315–43.
268. Friesen, M. In: *Environmental carcinogens—selected methods of analysis Vol. 5—Some mycotoxins*, ed. H. Egan, L. Stoloff, M. Castegnaro, P. Scott, I. K. O'Neill and H. Bartsch, International Agency for Research on Cancer, Lyon, 1982, pp. 85–106.
269. Friesen, M. D., Walker, E. A. and Castegnaro, M., *J. Ass. Off. Anal. Chem.*, **63** (1980) 1057–66.

270. Friesen, M. D. and Garren, L., *J. Ass. Off. Anal. Chem.*, **65** (1982) 855–63.
271. Pons, W. A., Jr, Lee, L. S. and Stoloff, L., *J. Ass. Off. Anal. Chem.*, **63** (1980) 899–906.
272. Shepherd, M. J., Holmes, M. and Gilbert, J., *J. Chromatogr.*, **354** (1986) 305–15.
273. Shotwell, O. L. and Goulden, M. L., *J. Ass. Off. Anal. Chem.*, **60** (1977), 83–8.
274. Egmond, H. P. van and Wagstaffe, P. J., *J. Ass. Off. Anal. Chem.*, **70** (1987) 605–10.
275. Ember, L. R., *Chem. Engng. News* (January 9, 1984) 8–34.
276. Ember, L. R., *Chem. Engng. News* (June 25 1984) 25–8.
277. Reutter, D. J., Hallowell, S. F. and Sarver, E. W., *Abstr. 191st Mtg. Am. Chem. Soc.*, April 13–18, New York, No. ANYL 75, 1986.
278. Trenholm, H. L. and Young, J. C. In: *Mycotoxins: a Canadian perspective*, ed. P. M. Scott, H. L. Trenholm and M. D. Sutton, National Research Council Canada, 1985, pp. 133–6.
279. Castegnaro, M., Hunt, D. C., Sansone, E. B., Schuller, P. L., Siriwardana, M. G., Telling, G. M., Egmond, H. P. van and Walker, E. A. (Eds), *Laboratory contamination and destruction of aflatoxins B_1, B_2, G_1, G_2 in laboratory wastes*, International Agency for Research on Cancer, Lyon, 1980.
280. Thompson, W. L. and Wannemacher, R. W., Jr, *Appl. Environ. Microbiol.*, **48** (1984) 1176–80.
281. Vidal, D., Creach, O., Genton, A., Beaudry, Y. and Fontanges, R., *C.R. Acad. Sci. Paris, Ser. III*, **301** (1985) 183–6.
282. Burrows, E. P. and Szafraniec, L. L., *J. Org. Chem.*, **51** (1986) 1494–7.
283. Burg, W. R., Shotwell, O. L. and Saltzman, B. E., *Am. Ind. Hyg. Ass. J.*, **42** (1981) 1–11.
284. Palmgren, M. S., Lee, L. S., DeLucca, A. J., II and Ciegler, A., *Am. Ind. Hyg. Ass. J.*, **44** (1983) 485–8.
285. Sorenson, W. G., Jones, W., Simpson, J. and Davidson, J. I., *J. Toxicol. Environ. Health*, **14** (1984) 525–33.
286. Baxter, C. S., Wey, H. E. and Burg, W. R., *Fd Cosmet. Toxicol.*, **19** (1981) 765–9.
287. Hayes, R. B., Van Nieuwenhuize, J. P., Raatgever, J. W. and Kate, F. J. W. ten, *Fd Chem. Toxicol.*, **22** (1984) 39–43.
288. Van Nieuwenhuize, J. P., Herber, R. F. M., DeBruin, A., Meyer, I. P. B. and Duba, W. C., *Tijdschr. Soc. Geneesk*, **51** (1973) 754–9.
289. Tosch, D., Waltking, A. E. and Schlesier, J. F., *J. Ass. Off. Anal. Chem.*, **67** (1984) 337–9.
290. Bicking, M. K. L., Knisely, R. N. and Svec, H. J., *Anal. Chem.*, **55** (1983) 200–4.
291. Kostyukovskii, Y. L. and Melamed, D. B., *J. Anal. Chem. USSR*, **39** (1984) 1781–5.
292. Shepherd, M. J. and Gilbert, J., *Fd Addit. Contamin.*, **1** (1984) 325–35.
293. Thiel, P. G., Stockenström, S. and Gathercole, P. S., *J. Liq. Chromatogr.*, **9** (1986) 103–12.

294. Gilbert, J. and Shepherd, M. J., *Fd Addit. Contamin.*, **2** (1985) 171–3.
295. Frisvad, J. C. and Thrane, U., *J. Chromatogr.*, **404** (1987) 195–214.
296. Rajakylä, E., Laasasenaho, K. and Sakkers, P. J. D., *J. Chromatogr.*, **384** (1987) 391–402.
297. Tiebach, R., Blaas, W., Kellert, M., Steinmeyer, S. and Weber, R., *J. Chromatogr.*, **318** (1985) 103–11.
298. Bata, A.,Ványi, A. and Lásztity, R., *J. Ass. Off. Anal. Chem.*, **66** (1983) 577–81.
299. Karppanen, E., Rizzo, A., Berg, S., Lindfors, E. and Aho, R., *J. Agric. Sci. Finland*, **57** (1985) 195–206.
300. Scott, P. M., Kanhere, S. R. and Tarter, E. J., *J. Ass. Off. Anal. Chem.*, **69** (1986) 889–93.
301. Steinmeyer, S., Tiebach, R. and Weber, R., *Z. Lebensmittelunters u.-Forsch.*, **181** (1985) 198–9.
302. Vesonder, R. F. and Rohwedder, W. K. In: *Modern methods in the analysis and structural elucidation of mycotoxins*, ed. R. J. Cole, Academic Press, Orlando, Florida, 1986, pp. 335–51.
303. Scott, P. M., Lau, P.-Y. and Kanhere, S. R., *J. Ass. Off. Anal. Chem.*, **64** (1981) 1364–71.
304. Gilbert, J., Shepherd, M. J. and Startin, J. R., *J. Sci. Fd Agric.*, **34** (1983) 86–92.
305. Salhab, A. S., Russell, G. F., Coughlin, J. R. and Hsieh, D. P. H., *J. Ass. Off. Anal. Chem.*, **59** (1976) 1037–44.
306. Scott, P. M., Panalaks, T., Kanhere, S. and Miles, W. F., *J. Ass. Off. Anal. Chem.*, **61** (1978) 593–600.
307. Scott, P. M. and Kanhere, S. R., *J. Ass. Off. Anal. Chem.*, **63** (1980) 612–21.
308. Mortimer, D. N., Parker, I., Shepherd, M. J. and Gilbert, J., *Fd Addit. Contamin.*, **2** (1985) 165–70.
309. Rosen, R. T., Rosen, J. D. and DiProssimo, V. P., *J. Agric Fd Chem.*, **32** (1984), 276–8.
310. Friedli, F., *J. High Res. Chromatogr. Chromatogr. Commun.*, **4** (1981) 495–9.
311. Brumley, W. C., Andrzejewski, D., Trucksess, E. W., Dreifuss, P. A., Roach, J. A. G., Eppley, R. M., Thomas, F. S., Thorpe, C. W. and Sphon, J. A., *Biomed. Mass Spectrom.*, **9** (1982) 451–8.
312. Miles, W. F. and Gurprasad, N. P., *Biomed. Mass Spectrom.*, **12** (1985) 652–8.
313. Rothberg, J. M., MacDonald, J. L. and Swims, J. C. In: *Xenobiotics in foods and feeds*, ed. J. W. Finley and D. E. Schwass, American Chemical Society, Washington, DC, 1983, pp. 271–81.
314. Black, R. M., Clarke, R. J. and Read, R. W., *J. Chromatogr.*, **388** (1987) 365–78.
315. Tiebach, R., Blaas, W. and Kellert, M., *J. Chromatogr.*, **323** (1985) 121–6.
316. McFadden, W. H., Bradford, D. C., Games, D. E. and Gower, J. L., *Am. Lab.* (Oct., 1977) 55–64.
317. Abramson, D., *J. Chromatogr.*, **391** (1987) 315–20.

318. Eckers, C., Games, D. E., Mallen, D. N. B. and Swann, B. P., *Biomed. Mass Spectrom.*, **9** (1982) 162–73.
319. Voyksner, R. D., Hagler, W. M., Jr, Tyczkowska, K. and Haney, C. A., *J. High Res. Chromatogr. Chromatogr. Commun.*, **8** (1985) 119–25.
320. Park, D. L., DiProssimo, V., Abdel-Malek, E., Trucksess, M. W., Nesheim, S., Brumley, W. C., Sphon, J. A., Barry, T. L. and Petzinger, G., *J. Ass. Off. Anal. Chem.*, **68** (1985) 636–40.
321. Pawlowsky, R. J. and Mirocha, C. J. In: *Diagnosis of mycotoxicoses*, ed. J. L. Richard and J. R. Thurston, Martinus Nijhoff Publishers, Dordrecht, 1986, pp. 299–304.
322. Paré, J. R. J., Greenhalgh, R., Lafontaine, P. and ApSimon, J. W., *Anal. Chem.*, **57** (1985) 1472–4.
323. Smith, R. D., Udseth, H. R. and Wright, B. W., *J. Chromatogr. Sci.*, **23** (1985) 192–9.
324. Plattner, R. D. In: *Modern methods in the analysis and structural elucidation of mycotoxins*, ed. R. J. Cole, Academic Press, Orlando, Florida, 1986, pp. 393–414.
325. Plattner, R. D. and Bennett, G. A., *J. Ass. Off. Anal. Chem.*, **66** (1983) 1470–7.
326. Plattner, R. D., Bennett, G. A. and Stubblefield, R. D., *J. Ass. Off. Anal. Chem.*, **67** (1984) 734–8.
327. Lau, B. P.-Y., Scott, P. M. and Sakuma, T., *Abstr. 32nd Ann. Conf. Mass. Spectrom. Allied Topics*, May 27–June 1, San Antonio, Texas, 1984.
328. Yates, S. G., Plattner, R. D. and Garner, G. B., *J. Agric. Fd Chem.*, **33** (1985) 719–22.
329. Chen, J. T., Mossoba, M. M., Trucksess, M. W., Flood, M. T. and Page, S. W., *Abstr. 99th Ann. Int. Mtg Ass. Off. Anal. Chem.*, Oct. 27–31, Washington, DC, 1985, p. 25.
330. Brockus, C. L. and Prange, C. A., *Abstr. 100th Ann. Int. Mtg Ass. Off. Anal. Chem.* Sept. 15–18, Scottsdale, Arizona, 1986, p. 9.
331. Riesenfeld, G., Kirsch, I. and Weissman, S., *Fd Addit. Contamin.*, **2** (1985) 253–7.
332. Chamkasem, N., Cobb, W. Y. and Phillips, T. D., *Abstr. 100th Ann. Int. Mtg Ass. Off. Anal. Chem.*, Sept. 15–18, Scottsdale, Arizona, 1986, p. 10.
333. Howard, B., Levy, G. B., Berry, V. and Ouchi, G. I., *Am. Lab.* (April 1987) 144–51.

5

Glucosinolates in Seeds and Residues

J. K. Daun

Canadian Grain Commission, Grain Lab. Division, Room 1404–303 Main Street, Winnipeg, Manitoba, Canada R3C 3G8

AND

D. I. McGregor

Agriculture Canada, Research Station, 107 Science Crescent, Saskatoon, Saskatchewan, Canada S7N 0X2

5.1. INTRODUCTION

Glucosinolates are sulphonated oxime thioesters, predominantly of glucose, with the general formula shown in Fig. 5.1. The structure of the side group 'R', which is derived from the amino acids, may be aliphatic, aromatic or heteroaromatic (Table 5.1). Glucosinolates have been found in all hitherto investigated plants of the order Capparales which includes the families: Bataceae, Bretschneideraceae, Capparaceae, Cruciferae, Gyrostemonaceae, Limnanthaceae, Moringaceae, Pentadiplandraceae, Resedaceae, Salvadoraceae, Tovariaceae and Tropaeolaceae.[1] The Cruciferae family supplies many foodstuffs including the common Brassica oilseeds, *Brassica campestris* L. (*Brassica rapa* L.) and *Brassica napus* L., commonly referred to as rapeseed. Varieties of rapeseed with normal levels of glucosinolates but less than 5% (usually less than 2%) erucic acid are called single zero rapeseed or Low Erucic Acid Rapeseed (LEAR), When the glucosinolate content has also been reduced (as in canola), the variety is called a double zero type. Glucosinolates are also found in *Crambe abyssinica* L., grown experimentally for its high level of erucic acid and *Brassica juncea* L., used as a condiment throughout much of the world and as a source of edible oil in the Indian Sub-continent.

A common feature of all glucosinolates is their hydrolysis by the enzyme myrosinase (thioglucoside glucohydrolase, EC 3.2.3.1) which

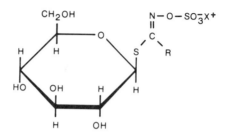

Fig. 5.1. Chemical structure of glucosinolates.

has been found to be present in all glucosinolate-containing plants of the order Capparales[1] (Fig. 5.2). Although intact glucosinolates have generally been considered to be innocuous, the hydrolysis products produce several physiological effects when they are present in large quantities in animal feeds. These include depressed growth related to the goitrogenicity of several hydrolysis products; haemorrhagic livers in poultry possibly related to the presence of epithionitriles, and skeletal abnormalities in poultry. The chemistry and the physiological effects of glucosinolates have been reviewed recently.[2]

The presence of glucosinolates in Cruciferous oilseeds has hindered the use of their meals in animal feed. As a result, in the 1970s plant breeders developed varieties of LEAR which also had lowered levels of glucosinolates. Nutritional studies established that meals from these low glucosinolate varieties could be used to a much greater extent in animal feed rations. In Canada, where low glucosinolate varieties were first introduced, the effect on the oilseed industry was so dramatic that to distinguish this new type of oilseed, low glucosinolate varieties of *B. napus* and *B. campestris* were collectively renamed canola.

The development of low glucosinolate varieties of rapeseed resulted in increased interest in glucosinolate analysis, both from a regulatory and research standpoint. The analysis of glucosinolates has been the subject of several recent reviews and symposia.[3-7] The purpose of this chapter is to present information on the current state of glucosinolate analysis and to recommend those methods which seem best for particular purposes including methods used for the analysis of individual glucosinolates and for the determination of total glucosinolate content of oilseed crops and commodities. Included are methods used for regulation and quality control in the processing industry, methods

TABLE 5.1
(a) Major glucosinolates found in oilseeds

Semi-systematic name of radical group	Trivial name of glucosinolate	Oilseeds
Allyl-	Sinigrin	*Brassica juncea* L.
3-Butenyl-	Gluconapin	*Brassica campestris* L.
		Brassica napus L.
4-Pentenyl-	Glucobrassicanapin	*Brassica campestris* L.
		Brassica napus L.
S-2-Hydroxy-3-butenyl-	Progoitrin	*Brassica campestris* L.
		Brassica napus L.
R-2-Hydroxy-3-butenyl-	*epi*-Progoitrin	*Crambe abyssinica* L.
1-Methoxy-3-indolylmethyl	Neoglucobrassican	*Brassica campestris* L.
		Brassica napus L.

(b) Minor glucosinolates found in oilseeds[a]

Semi-systematic name of radical group	Trivial name of glucosinolate
Methyl-	Glucocapparin
2-Hydroxy-4-pentenyl-	Gluconapoleiferin
4-Hydroxy-3-indolylmethyl-	4-Hydroxyglucobrassican
3-Indolylmethyl-	Glucobrassican
3-Methylsulphinylpropyl-	Glucoiberin
4-Methylsulphinylbutyl-	Glucoraphanin
4-Methylsulphinyl-3-butenyl-	Glucoraphenin
3-Methylthiopropyl-	Glucoiververin
4-Methylthiobutyl-	Glucoerucin
4-Methylthiopentyl-	Glucoberteroin
2-Phenethyl-	Gluconasturtin
p-Hydroxybenzyl-	Sinalbin

[a] Although many of the minor glucosinolates listed are natural constituents of the glucosinolates, others may be present due to contamination with other oilseeds or weeds. For example, the presence of significant amounts of allylglucosinolate (sinigrin) in rapeseed may indicate contamination with mustard (*B. juncea*) or stinkweed (*Thlaspi arvense* L.). Similarly, the presence of *p*-hydroxybenzylglucosinolate (sinalbin) in rapeseed indicates contamination with wild mustard (*Sinapis arvensis* L.) or mustard (*Sinapis alba* L.).

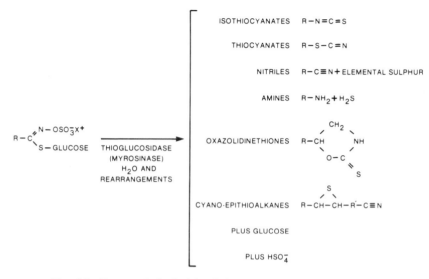

Fig. 5.2. Enzymatic hydrolsis of glucosinolates with myrosinase.

used for biochemical research, and methods used for screening of glucosinolate containing material in breeding programs.

5.2. GLUCOSINOLATES OF CRUCIFEROUS OILSEEDS AND MEALS

5.2.1 Rapeseed and Canola

Rape and canola seeds contain predominantly aliphatic glucosinolates derived from methionine and heterocyclic indole glucosinolates derived from tryptophan.[1] The glucosinolate content of Brassica oilseeds and meals varies widely due to differences between species and cultivars, environmental effects and the effects of processing. Even the method of analysis can affect reported contents.

B. napus rapeseed varies between 140 and 170 μM/g oil-extracted air-dried meal while *B. campestris* varies between 80 and 110 μM/g oil-extracted air-dried meal[8] (Table 5.2). Canola varieties of both species have been bred for lower glucosinolate content. By definition, they have less than 30 μM/g oil-extracted air-dried meal of the aliphatic glucosinolates, 3-butenylglucosinolate, 4-pentenylglucosinolate, 2-

TABLE 5.2

Glucosinolates in different varieties of rapeseed and conola seed, analysis by HPLC

Species	Glucosinolate Content (μM/g, oil-free, 8·5% moisture)					
	Variety	Bu	Pe	BuOH	Ind	Oth	Total
B. napus							
(canola)	Westar	3·0	0·5	5·5	6·0	4·7	19·7
	Glacier	6·8	0·8	14·6	5·1	4·8	32·1
(rape)	Tandem	7·5	2·4	17·3	4·3	3·8	35·2
	Turret	26·7	6·1	66·6	7·2	15·2	121·7
	Jet Neuf	35·4	5·2	67·8	5·8	10·9	125·2
B. campestris							
(canola)	Tobin	6·1	3·3	10·0	5·9	3·3	28·7
(rape)	Torch	18·1	14·0	12·3	5·9	9·8	60·1
(Sarson)	R-500	119·5	1·3	0·0	4·4	8·8	133·9

Bu, 3-butenylglucosinolate; Pe, 4-pentenylglucosinolate; BuOH, 4-hydroxy-3-butenylglucosinolate; Ind, 4-hydroxy-3-indolylglucosinolate.
Others include traces of sinigrin, 5-hydroxy-4-pentenylglucosinolate, 2-phenethylglucosinolate, methylthio- and methylsulphinylglucosinolates.

hydroxy-3-butenylglucosinolate and 2-hydroxy-4-pentenylglucosinolate.[9]

Environment and in particular sulphur fertility[10,11] can have a large effect on glucosinolate content. Where the availability of sulphur is limited, either due to lack in the soil or the inability of the plant to incorporate sulphur as in drought, glucosinolate levels are substantially reduced due to preferential incorporation of available sulphur into protein.[12] In Australia, late sown seed was found to have increased glucosinolate contents.[13] An example of the environmental variability of glucosinolates in canola is shown in Table 5.3. In this case, the variation is thought to be due to a combination of sulphur availability and water stress at the different sites.

Processing conditions commonly used to extract oil from rapeseed and canola can result in a 40–60% reduction in glucosinolate content[4,14–16] (Table 5.4). This reduction has been generally assumed to be a thermal decomposition which takes place during desolventizing and toasting of the extracted meal. The composition of residual

TABLE 5·3
Variation of glucosinolates with growing environment. Samples from the 1987
Western Canadian Cooperative Test

| | Glucosinolates (μM/g oil-free) | |
	B. napus cv. Westar	B. campestris cv. Tobin
In seed sown	9	16
In seed harvested at		
Winnipeg	10	19
Saskatoon	17	26
Melfort	12	26
Edmonton	11	23
Beaverlodge	11	29

Sum of four aliphatic glucosinolates.

TABLE 5·4

(a) Effect of commercial processing on the glucosinolates content (μM/g, oil-free, dry basis) of B. napus cv. Tower[14]

| | Hydrolysis products | | | | | | Intact glucosinolates | | | |
| | CHB | | OZT | | SCN | | SCN precursor | | Others | |
Sampling point	Sple 1	Sple 2	Sple 1	Sple 2	Sple 1	Sple 2	Sple 1	Sple 2	Sple 1	Sple 2
Seed	tr.	tr.	0·7	0·5	2·6	3·2	11·6	17·4	14·7	10·9
Pre-expeller	0·1	0·2	nd	nd	2·4	3·0	14·0	19·4	24·4	17·8
Pre-solvent extractor	0·1	0·2	nd	nd	2·1	3·2	15·3	20·3	20·5	17·7
Pre-toasting	0·1	0·2	nd	nd	2·1	2·8	14·8	19·8	23·2	18·6
Meal	1·2	1·1	nd	nd	4·7	4·2	6·8	9·5	17·0	13·9

CHB, 1-cyano-2-hydroxybutene; OZT, oxazolidinethione; SCN, thiocyanate ion.
nd, not detected.
tr., trace.

(b) Loss of intact glucosinolates on toasting (%)

Sple	SCN precursors	Others
1	54·1	26·7
2	52·0	25·3

TABLE 5.5

Intact glucosinolates and glucosinolate hydrolysis products in commercial rapeseed and canola meals[14]

Glucosinolates (μM/g sample) Constituent	Canola meals	Rapeseed meals
1-cyano-2-hydroxy-3-butene	1·1–5·6	8·6–14·4
5-vinyl-2-oxazolidinethione	Trace	Trace–0·3
Thiocyanate ion	1·9–5·8	4·1–6·9
Intact glucosinolates	10·3 20·5	36·2–105·4

decomposition products in the meal is not well understood[14,17] but the major components found have been nitriles and thiocyanate ion (Table 5.5).

Collaborative and methodology studies have pointed out analytical difficulties in obtaining good agreement between methods and laboratories.[18–22] The reasons for these differences have not been fully elucidated. They do, however, indicate the need for standardization of methodology.

5.2.2 Crambe

Crambe abyssinica L. seed has been grown on an experimental basis in the USA, as a potential source of erucic acid for industrial and chemical uses. Crambe seed contains substantial amounts of (S)-2-hydroxy-3-butenylglucosinolate [epi-progoitrin] which limits the use of crambe meal as a high protein animal feed ingredient. In a commercial scale processing study,[23] the amount of intact glucosinolates in crambe seed (105–164 μM/g, whole seed, defatted) was found to decrease substantially during desolventization (to 20–80 μM/g). The nitrile formed during hydrolysis of the glucosinolate, 1-cyano-2-hydroxy-3-butene was found to be present in the desolventized meal in amounts of 15–50 μM/g.

5.2.3 Mustard

There are two major mustard seeds in commerce, *Brassica juncea* (L.) Coss. (Oriental Mustard) and *Sinapis alba* L. (yellow or white mustard). In the Indian subcontinent, *B. juncea* is grown as a source of edible oil. In North America and Europe, however, mustards are grown for use as a condiment, the desirable flavor of which is due to hydrolysis products of the glucosinolates present in the seed. *Sinapis* alba contains upwards of 200 μm/g oil-extracted air-dried meal of 4-hydroxybenzylglucosinolate (sinalbin). *Brassica juncea* of European or North American origin contains 150–200 μm/g oil-extracted air-dried meal of allylglucosinolate (sinigrin). *B. juncea* from the Indian subcontinent contains variable amounts of allylglucosinolate and 3-butenylglucosinolate. Recently, plant breeders have been attempting to develop lines of *B. juncea* which are low in both erucic acid and glucosinolates as sources of edible oil. While low erucic acid seed stocks were identified in the late 1970s, a low glucosinolate *B. juncea* line has only recently been identified.

5.2.4 Common Cruciferous Weeds

The most common cruciferous weed seeds associated with rapeseed or canola in Canada are *Sinapis arvensis* L. (wild mustard) and *Thlaspi arvense* L. (stinkweed). Analyses in the authors' laboratories of *S. arvensis* seed have shown up to 200 μm/g oil-extracted air-dried meal of 4-hydroxybenzylglucosinolate while the seed *Thlaspi arvense* was found to have similar contents of allylglucosinolate.

5.3. CHEMISTRY OF GLUCOSINOLATE ANALYSIS

The chemical features of glucosinolates which have major roles to play in their analytical chemistry are:

1. their ionic nature;
2. hydrolysis by myrosinase or sulphatase;
3. the nature of the 'R' group;
4. the presence of glucose;
5. the presence of a bisulphate ion;
6. the presence of a sulphur atom;
7. the presence of an oxime group.

The anionic nature of glucosinolates permits isolation and concentration using ion exchange chromatography. Because of their relatively weak acidity, most glucosinolates have been purified using basic ion exchange media with active sties in the form of diethylaminoethyl (DEAE sephadex)[24] or tertiary amino (Ecteola cellulose)[25] groups. Once bound on the ion exchanger, impurities, particularly carbohydrates, may be removed by washing with water. The glucosinolates may then be eluted with solutions of stronger ionic strength such as potassium sulphate[26] or pyridine.[27] Alternatively, glucosinolates may be hydrolysed directly on the ion exchange medium with either the sulphatase[28] or myrosinase[29] enzyme and the products removed with water for further analysis.

The sulphatase enzyme (aryl sulphatase, EC 3.1.6.1. from *Helix pomatia*) removes the bisulphate group (Fig. 5.3). The corresponding desulphoglucosinolates may be determined by spectrometry as a group or individually by gas liquid chromatographic techniques.[30,31]

Under uncontrolled hydrolysis, or autolysis, the products of myrosinase hydrolysis are diverse, consisting of nitriles and epithionitriles as well as isothiocyanates (Fig. 5.2). The exact composition of the autolysis products depends on the pH of the solution and the presence of cofactors such as metal ions and product-specifier proteins.[32] By buffering to pH 7, most aliphatic glucosinolates, but not all, hydrolyse quantitatively to form stable isothiocyanates which may be determined by gas chromatography[33] or spectrophotometrically after conversion to thioureas.[34]

Isothiocyanates from some glucosinolates, such as 2-hydroxy-3-butenylglucosinolate, the major aliphatic glucosinolate in rapeseed, cyclize to form oxazolidinethiones which are insufficiently volatile for accurate determination by gas chromatography. Other glucosinolates, such as 4-hydroxybenzylglucosinolate, the major glucosinolate of the seed of *S. alba* and the indolyl glucosinolates, found in the seed and other plant parts of Brassicas including rapeseed, form unstable isothiocyanates on hydrolysis which rapidly hydrolyse to form thiocyanate ion. Both oxazolidinethione[35] and thiocyanate ion[14] may be determined spectrophotometrically.

Glucose and bisulphate ion, both products of myrosinase hydrolysis, have been used to give estimates of total glucosinolate content.

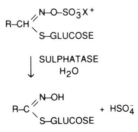

Fig. 5.3. Enzymatic hydrolysis of glucosinolates with sulphatase.

Glucose may be determined spectrophotometrically using a glucose oxidase/peroxidase system or a hexokinase system.[36,37] The glucose oxidase system has been used in the form of glucose test paper for rapid screening of glucosinolates.[38-40] Bisulphate has been determined gravimetrically,[41] by acid–base titration[42] or indirectly by emission spectrometry[43] although these methods are generally tedious, time-consuming and lack precision. The presence of sulphur atoms has been used as a basis for an X-ray fluorescence procedure.[44]

The glucosinolate oxime was found to complex with palladium ions and this reaction has been used as the basis of a spectrophotometric procedure.[45]

5.4. EXPRESSION OF RESULTS OF GLUCOSINOLATE ANALYSIS

Scientists at a symposium on the analysis of rapeseed and its products in 1980 agreed that the unit for expression of results of glucosinolate analysis should be micromoles per gram ($\mu M/g$).[46] Previously, results had been expressed in several different forms, often depending on the analytical method employed which could lead to confusion or inability to compare results. Frequently, when glucosinolate composition was unknown, or the method failed to separate and determine individual glucosinolates, results were expressed in equivalents of one glucosinolate known to be present or used as an internal standard. Methods relying on mass measurements (e.g. '%' or 'mg/g') were often misleading because of failure to report whether calculations were based on the glucosinolates, their hydrolysis products or derivatives thereof. Expression of results in micromoles per gram accounts for the number of reactive centres regardless of the molecular weight of compounds being measured.

The major discrepancies in today's literature appear to be caused by the 'basis' rather than the units upon which the results are expressed and upon which glucosinolates are included in the quoted values. The 1980 accord recommended that for oilseeds, and for rapeseed in particular, results be reported on an oil-free, air-dry meal basis.[46] For rapeseed it was agreed that analyses for total glucosinolates by gas chromatography include the sum of allylglucosinolate, 3-butenylglucosinolate, 4-pentenylglucosinolate, 2-hydroxy-3-butenylglucosinolate and 4-hydroxy-benzylglucosinolate unless otherwise stated. In Canada, this recommendation (without allyl- and hydroxybenzylglucosinolates) was incorporated into the definition of canola[9] and eventually was incorporated into the Canadian General Standard.[47]

One problem with the recommendation of 1980 is the definition of 'air-dry'. In one of the authors' laboratories, air-dried meal samples ranged in moisture content from 6% to 12% depending on the relative humidity of the laboratory air. It would seem preferable to either report results on a constant moisture basis or to indicate the moisture content of the 'air-dried' meal. Furthermore, it is not specified if the seed should be cleaned or *telle quelle*.

The second problem arose when the European Economic Community (EEC) decided to report glucosinolates on a whole seed basis.[48,49] Although this can facilitate the reporting of glucosinolate content by eliminating the need to know the oil content or to extract the oil prior to analysis, the presence of both forms of reporting in the literature may be confusing unless the basis of reporting is clearly stated.

The number and nature of glucosinolates included in the 'total' result also differs between Canada and Europe. In Canada and other countries using the canola definition,[9] only the aliphatic glucosinolates are reported. In Europe, the EEC method specifies the reporting of all glucosinolates measured.[48,49] The difference can be expected to result in a 10–12 μM/g oil-extracted air-dried meal higher value (due mostly to indolyl glucosinolates) for glucosinolate content reported according to the European as opposed to the Canadian standard (Table 5.2).

Differences in the approach to expressing glucosinolate content have resulted from the relatively recent discovery of the presence of indolyl glucosinolates in rapeseed.[50] In Canada, it was decided to base the canola definition only on those glucosinolates which could be accurately measured by gas chromatography of the trimethylsilyl derivatives, the accepted method of analysis at the time. In addition, there was reason to believe that the indole glucosinolates may not be as nutritionally important as they do not form stable isothiocyanates upon myrosinase hydrolysis but break down further to release free thiocyanate ion. The chemistry and nutritional significance of these glucosinolates has been reviewed recently.[51]

5.5. METHODS OF GLUCOSINOLATE ANALYSIS

5.5.1 Introduction

Methods of glucosinolate analysis may be divided into two types, methods which involve measurement of individual glucosinolates and methods which measure total glucosinolate content. Methods which determine individual glucosinolates involve some sort of chromatographic separation and thus are somewhat more time consuming. These methods, however, have been preferred as primary standard procedures because they provide both qualitative and quantitative information. Methods which determine total glucosinolate content, although sometimes less precise, are generally faster and thus are more suited for quality control work in the process industry or for screening in plant breeding programs, situations which require analyses of large numbers of samples.

5.5.2 Analysis of Individual Glucosinolates

5.5.2.1 Indirect measurement

Indirect determination of total glucosinolates in rapeseed by measurement of one or more of the hydrolysis products from the R-group was prevalent in the 1960s and early 1970s before methods for direct measurement of glucosinolates became available. Problems with reproducibility of these methods were traced to the source of myrosinase used in the method, the ratio of water to seed during myrosinase deactivation and to the order of addition of reagents.[52]

In recent years, determination of glucosinolate content based on estimation of isothiocyanates formed on myrosinase hydrolysis is mostly used in nutritional studies when information on the amounts and nature of hydrolysis products formed during processing may be required. Gas–liquid chromatography is still the method of choice for determination of isothiocyanates, nitriles and epithionitriles in commercial meals[14,17,53] while spectrophotometry is used for the determination of oxazolidinethiones[14] and thiocyanate ion.[54] Episulphides[55] and nitriles[56] also may be determined by spectrophotometry. High performance liquid chromatography (HPLC) methods for the determination of glucosinolate hydrolysis products have also been proposed.[57,58] All of these methods suffer from the lack of a suitable means of concentrating the small amounts of volatile hydrolysis products found in commerical meals after processing or in animal tissues after consumption of glucosinolate-containing feed.

5.5.2.2   Direct measurement

GLC.   Glucosinolates may be derivatized to form trimethylsilyl ethers which are sufficiently volatile to separate by GLC.[59] Derivatization is accompanied by removal of sulphate so it is actually the trimethylsilyl derivatives of the desulphoglucosinolates which are separated. Ion exchange chromatography coupled with enzymatic desulphation has been used to advantage for isolating and purifying glucosinolates prior to derivatization.[30] This elegant procedure involves binding the glucosinolates, which have been extracted from the sample using water or a water and alcohol mixture, to an ion exchange column of DEAE-Sephadex. Impurities, particularly carbohydrates, are eluted with water and the glucosinolates are then desulphated on column using a commercially available sulphatase enzyme extracted from snails (*Helix pomatia*). The desulphated glucosinolates are removed from the column with water, dried, derivatized and separated by GLC on either a packed glass column or a silica or glass open tubular column. Temperature programming has permitted separation of all the major glucosinolates of rapeseed and mustard including the indole glucosinolates,[60] although the latter cannot be quantitated with precision. Methylsulphinyl glucosinolates are also difficult to derivatize quantitatively.

The gas chromatographic procedure was adopted in Canada as the official method for determining aliphatic glucosinolates in canola seed[9] and has been the subject of several 'round-robin' studies conducted under the auspices of the International Standards Organization (ISO).[22] Recent publications have dealt with optimization of the conditions of analysis. In one study, prolonged wet heat treatment during the extraction of the glucosinolates resulted in a substantial reduction in the indolyl glucosinolate content.[61] A dry heat treatment (95°C) for 15 min followed by 3 min extraction with hot (95°C) water was found to be optimum. In another study, replacement of pyridine by methylimidazole was found to allow reduction of silylation temperature to 80°C without interfering with the yield of indolyl-glucosinolates.[62] Yet another study has highlighted the importance of the condition of the analytical column and the cleanliness of the injection port in maintaining reproducible results.[63] There has been some uncertainty over the requirement to remove oil from the seed prior to the analysis. In a study between the Oilseeds Laboratory of the Canadian Grain Commission (CGC) and Centre Technique

Interprofessionnel des Oléagineux Métropolitains (CETIOM) analytical laboratory in France, no differences were found between oil free
and full fat samples ground using a number of grinders (Table 5.6).
Analysis without prior removal of oil from seed samples is becoming a
more common practice.[64,35]

HPLC. Two approaches have been taken to the HPLC separation and
quantitation of glucosinolates, HPLC of desulphoglucosinolates and
HPLC of intact glucosinolates.[1] HPLC of intact glucosinolates has the
advantage of being able to detect minor glucosinolates which have
various groups attached to the glucose moiety. HPLC of desulphoglucosinolates has been more widely applied, however, because
glucosinolates may be readily purified by on column desulphation in a
manner similar to that used in sample preparation for GLC of
trimethylsilyl derivatives. Also, the solvent system used for HPLC of
desulphoglucosinolates, water and acetonitrile, is simpler than the
solvent system employing ion paired reagents which is required for
HPLC of intact glucosinolates. Detection has mostly been with
ultraviolet absorption, usually in the region of 225–230 nm. Response
factors may be determined by collection of separated glucosinolates

TABLE 5.6

Comparison of glucosinolate content of seed samples analysed by GLC of
pertrimethylsilylated desulphoglucosinolates with and without extraction of oil
before extraction of glucosinolates

Sample	Oil-free samples glucosinolate content (μM/g, oil-free, dry basis)			Full-fat samples glucosinolate content (μM/g, oil-free, dry basis)		
	CETIOM[a]	CGC[b]	Mean	CETIOM	CGC	Mean
Jet Neuf	146	160	153	168	160	164
Tandem	34	39	37	33	40	36
Tobin	30	31	31	34	32	33
Westar	13	16	15	16	17	16
Torch	85	79	82	91	83	87

Analysis of variance showed no significant differences between seed treatment
or laboratories.
[a] Centre Technique Interprofessionnel des Oléagineux Métropolitains,
Orleans, France.
[b] Canadian Grain Commission, Winnipeg, Canada.

followed by determination with thymol reagent.[65,66] Quantitation may be effected by addition of an internal standard during the extraction of the glucosinolates or to the eluate from the DEAE-Sephadex column. Allylglucosinolate or benzylglucosinolate have been used as internal standards when not present in the material under examination.[67] Alternatively, sulphanilic acid and o-nitrophenyl-b-D-galactopyranoside have been used.[65] In the latter case, care must be taken to heat the DEAE-Sephadex eluate prior to the addition of standard as commercial sulphatase has been observed to contain an active enzyme that will degrade o-nitrophenyl-b-D-glactopyranoside.

For HPLC of intact glucosinolates, water, or water/methanol extracts have been purified by ion exchange chromatography using Ecteola cellulose.[1] Glucosinolates were eluted from the column with either sodium bicarbonate or phosphate solution. Following filtration, the glucosinolates have been separated by reversed phase ion-pairing chromatography using a phosphate buffer and methanol[25] or more recently, acetonitrile[1] as a modifier and tetraheptylammonium bromide as a counter ion. The nature and concentration of the modifier, counter ion and pH were observed to have a substantial effect on resolution. Use of a counter ion has been avoided by using aqueous ammonium acetate and acetonitrile as the chromatographic solvents.[68]

5.5.3 Methods Used for Total Glucosinolates

5.5.3.1 Spectrophotometric

Glucosinolates and desulphoglucosinolates have ultraviolet absorption maxima in the region of 225 nm. Although the ultraviolet absorption of glucosinolates was used in developmental studies for the TMS–GLC method,[30] no indications of general use, except in HPLC, have appeared in the literature. This may be due to differences in specific absorptivities for different glucosinolates as well as to the large number of possible interfering components which have strong absorbances in this region of the UV spectrum.

Palladium. The discovery of complex formation between some transition elements and desulphoglucosinolates or glucosinolate ions[69] led to the development of a spectrophotometric procedure based on the absorption characteristics of a palladium glucosinolate complex.[45] Although this procedure has been found to be useful as a screening procedure, and has been adapted for flow injection analysis[70] it has performed rather poorly in collaborative studies[22] and would not be useful as a reference procedure.

Glucose determination. Two enzymatic approaches have been used to measure glucose released by myrosinase hydrolysis, one involving hexokinase and glucose-6-phosphate dehydrogenase, the other glucose oxidase and peroxidase.[38]

Glucose oxidase/peroxidase impregnated test paper has been used for years to screen for glucosinolate content in plant breeding programmes. Glucose released by endogenous myrosinase when two or three seeds are crushed is reacted with glucose oxidase impregnated in the paper to produce gluconic acid and hydrogen peroxide. The hydrogen peroxide then reacts with peroxidase to change the colour of a chromagen also impregnated in the paper.[38,40,71] The method is effective because mature seeds have a low content of free glucose. A limitation of the method is the apparent inhibition of peroxidase by phenolic compounds present in the crude extracts. This has been overcome by purification with ion exchange chromatography.[72] Charcoal has also been used to selectively adsorb phenolics.[73] Addition of charcoal can enhance sensitivity and reproducibility to the point where canola seed may be identified. A test based on the addition of charcoal was used in the mid-1970s to monitor the conversion of commercial production of rapeseed to the new low glucosinolate canola.[39] More recently this approach has been used in Europe[40] and attempts to remove subjectivity from the approach by using a reflectometer to measure the colour of the glucose test paper[74] have met with some success.[75]

In recent years, the glucose oxidase/peroxidase approach has been gaining popularity as a quantitative method for total glucosinolates.[18,36,76–79] Recent methods for removing the interfering components have involved chromatography on DEAE-Sephadex[29,30] or precipitating along with proteins using sodium tungstate.[80] A safer chromophore, 4-aminophenazone, has also been described.[80]

The hexokinase–dehydrogenase system is more sensitive than the oxidase–peroxidase system but this system involves measurement in the ultraviolet and for greatest accuracy, a kinetic measurement must be used.[35] This system has been adapted for use in flow injection analysis.[81]

Glucose released from hydrolysis of glucosinolates has also been determined by polarography,[82] amperometrically[83] and even by gas–liquid chromatography.[84]

Acid hydrolysis and spectrophotometric determination of the resulting thioglucose with phenol or thymol has recently gained interest as an analytical procedure.[26,85,86] In this method, the glucosinolates are eluted from the column with salt solution eliminating the time consuming enzymatic hydrolysis step.

5.5.3.2 Physical

Near infrared reflectance spectroscopy (NIRS). NIRS has been investigated as a rapid, simple and non-destructive method for determination of glucosinolate content.[87,88] the potential of this technique for analysis of the chemical constituents of agricultural commodities was realized when near infrared spectrophotometry was combined with reflectance spectrometry and correlation transform spectrometry.[89] Reflectance spectroscopy permitted rapid measurement with minimal sample preparation. Correlation transform spectroscopy permitted accurate quantitative analysis of complex mixtures, typified by agricultural commodites and products. Speed is facilitated by the fact that several constituents may be analysed simultaneously in only a few seconds. Usually the seed samples are simply ground, but in some instances whole seed may be used. Non-consumption of sample, also called non-destructive testing, allows re-analysis of the material, analysis by another procedure or, in the case of plant breeding, growing of the seed to produce subsequent generations.

When applied to the rapeseed range of glucosinolate content (10–160 μM/g) results suitable for screening purposes have been obtained.[90,91] However when restricted to the canola range (10–30 μM/g oil-extracted air-dried meal) using either whole or ground seed, accuracy was not adequate even for screening purposes.[92]

X-ray fluorescence. The estimation of glucosinolates in seeds and meal by X-ray fluorescence determination of total sulphur in the sample has received considerable interest in Europe recently.[44,93–96] For glucosinolate contents of 20–130 μM/g oil-extracted air-dried meal, glucosinolate sulphur amounts to 40–260 μM/g while protein sulphur amounts to 200 μM/g of protein or about 50 μM/g of oil-extracted air-dried meal. Thus glucosinolate sulphur can range from about 45% of the total sulphur for low glucosinolate seed to 75% of total sulphur for high glucosinolate seed. Suitability of the procedure, even for screening purposes, would seem to depend on the precision, which in turn would depend largely on the variability in protein content in different seed lots. Although the environmental variation in protein content in European rapeseed may be only ±1·5%, the variation in protein content of samples of Canadian seed has been found to be much larger, ranging from about 29% to about 47% (oil-free, 8·5% moisture). This range would correspond to a maximum error in glucosinolate estimation of about ±25 μM/g (oil-free) or ±12·5 μM/g on a seed basis. This technique is extremely rapid, however, and has gained acceptance in Europe where the variation in protein is small. It has been approved by several member states of the EEC for intervention purposes.

5.5.3.3 Other methods

Methods for glucosinolate determination have been proposed including spectrophotometric determination with 3,5-dinitrosalicylic acid,[97] a turbidimetric procedure based on the release of sulphate,[98] an extremely sensitive method using infrared spectroscopy,[99] a method for TLC detection based on N,2,6-trichloro-p-benzoquinoemine[100] and an enzyme linked immunosorbent assay (ELISA).[101]

5.5.4 Methods Used for Regulation

Regulatory and other official agencies have adopted several different methods for the analysis of glucosinolates. The ISO originally had a method ISO 5504[102] for the determination of isothiocyanates (ITCs) and vinyl oxazolidinethiones (OZTs) released in oilseed residues. Due to poor results in interlaboratory studies and the failure of this method to detect all glucosinolates, the ISO has been in the process of replacing this method. HPLC; glucose by oxidase/peroxidase, thymol and hexokinase/dehydrogenase; NIRS and X-ray fluorescence have been considered and the HPLC method has advanced to the Draft International Stage. Gas–liquid chromatography of trimethylsilyl derivatives has been abandoned by ISO because of poor results in two interlaboratory collaborative studies while the glucose release methods have been dropped although several are being used as National Standards.[22]

FOSFA[103] and the AOAC[104] still (at the time of writing) carried a method for rapeseed and mustard seed which measures isothiocyanates following hydrolysis with myrosinase by steam distillation and argentimetric titration. Results are expressed as allyl isothiocyanate which, although the main constituent in *B. juncea* (oriental mustard), is not a major constituent in rapeseed. The method is not applicable to yellow mustard *B. hirta*, as *p*-hydroxybenzylglucosinolate, the major glucosinolate in this species, yields an unstable isothiocyanate on hydrolysis which hydrolyses further to inorganic thiocyanate ion. The recommended methods of these agencies require replacement or, at least clarification as to the species for which they are applicable.

The EEC originally adopted the GLC–TMS method as its official method[49] but since this method was subsequently shown to give erratic results for indolyl glucosinolates, it will likely adopt an HPLC method as the official method in the near future.* There are at least three methods based on glucose release, as well as the X-ray fluorescence method which are being used as National Standard Procedures acceptable for intervention in the EEC.

In Canada, the Canola Council of Canada adopted the GLC–TMS method as the official method for determining glucosinolates in canola.[105] This method was later incorporated into official standards. Since the development of Canadian low glucosinolate cultivars preceded the discovery of the presence of indolylglucosinolates in brassica seeds,[65] the Canadian method only specifies the determination of aliphatic glucosinolates. As witnessed by an ongoing Canadian collaborative study,[4] good interlaboratory precision has been achieved using this method for measurement of the aliphatic glucosinolates. It is likely, however, that the official reference methods used in Canada will be changed to HPLC in the near future in order to include the indolyl glucosinolates and thereby give a truer value for total glucosinolate content. In Japan, there is no official method for glucosinolate analysis but processors have tested imported seed using a method which measures ITCs and OZTs formed upon myrosinase hydrolysis.[33] Australia also has yet to adopt an official method for glucosinolates but a method based on determination of glucose released on hydrolysis using glucose oxidase/peroxidase is used by industry.[106]

FOSFA specifies EEC methods for EEC rapeseed, ISO 5504 (1983) for high glucosinolate rapeseed and the CGC method for other low glucosinolate rapeseed.

---

* HPLC method adopted by EEC on 26 June 1990 (Official Journal L 170/27 of 3 July 1990, but contains minor errors. A better description of same method is in ISO/DIS/9167-1 (1991).

5.6. STANDARDS

One of the major difficulties facing any analyst who wishes to determine glucosinolates has been lack of pure reference glucosinolates. The only glucosinolate regularly available from commerical sources has been allylglucosinolate and this glucosinolate has been preferred in many cases as an internal standard even though the presence of small amounts of allylglucosinolate in rapeseed, especially as a result of contamination from weed seeds, has complicated the analysis. Benzylglucosinolate, isolated from *Tropaolium majus* L.[3] has been available from the Canola Council of Canada for use as an internal standard.

The original isolation of glucosinolates from brassica seeds carried out by Gmelin[107] relied on the use of alumina columns as weak ion exchange columns. The availability of weak ion exchange resins has renewed interest in the isolation of glucosinolates and several publications[27,108–112] have described techniques which allow the isolation of glucosinolates from plant sources. Unfortunately, glucosinolates may be easily isolated in a high degree of purity only from plant sources wherein they are the major glucosinolate present. In some cases, only glucosinolate mixtures are isolated and these must be further purified by preparative scale HPLC.

Another aid to analysts is a standard reference sample. A standard seed sample containing 25 μM/g total glucosinolates on a whole seed basis has been prepared by the EEC's Community Bureau of Reference in Brussels.[113] Further samples, with lower levels of glucosinolates, are being prepared.

5.7. RECOMMENDED METHODS

The method chosen for glucosinolate analysis will depend somewhat on the aim of the analyst. Plant breeders, for example, require methods which are rapid and give an accurate estimation of the approximate level of glucosinolates for initial screening. Methods giving individual glucosinolates are required for confirmatory plant breeding work. Regulatory methods must be accurate and precise. Where the regulations do not define individual glucosinolates, a method which determines total glucosinolates may be suitable. For payment of subsidies, etc., as in the EEC intervention system, it is only necessary to known if the concentration is above or below a stipulated value. When the actual concentration is well below this level rough screening methods can be acceptable, although recourse to more accurate methods will be necessary in samples shown to be borderline. In this instance it is not always necessary for regulatory methods to be accurate and precise.

For screening samples in plant breeding studies, the authors use the glucose oxidase/peroxidase test paper method[39] and gas–liquid chromatography utilizing capillary columns[114] for testing samples from cooperative tests and packed columns[105] for monitoring commercial crop samples. The gas–liquid chromatography procedure will shortly be replaced by an HPLC procedure, probably based on the method which may be adopted by the EEC. A method which determines total glucose utilizing thymol[86] has been found useful for the accurate determination of total glucosinolates but methods based on glucose release may be equally accurate, and in some cases, more convenient. The three detailed methods in the Appendix have been used successfully in the authors' laboratories.

At present there seems to be a consensus that HPLC is the method of choice for determination of individual glucosinolates. There is some discussion about whether HPLC of intact glucosinolates or HPLC of desulphoglucosinolates is preferable but most analysts prefer to analyse desulphoglucosinolates as this procedure does not require the use of a counter ion. Analysis of intact glucosinolates may be necessary for samples of plant material where it is desired to determine complex glucosinolates such as cinnamoyl glucosinolates which are difficult to determine in the desulphated form. Individual glucosinolate hydrolysis products in meals are still best analysed by gas–liquid chromatography and spectrophotometric determination of thiocyanate ion although there is considerable room for development of improved methodology in this area.

Several methods are available which are suitable for screening of breeders material or for rough screening of samples submitted for intervention pricing. Of these methods, the ones based on rapid determination of glucose released on myrosinase hydrolysis using glucose sensitive paper (with visual or instrumental estimation of the colour intensity) seem to offer the most economical solution. NIR and X-ray fluorescence have also been used as more expensive alternatives.

For accurate determination of total glucosinolates, it is first necessary to purify the glucosinolates using ion exchange chromatography. The glucosinolates can then be determined by analysis of the glucose released on hydrolysis with myrosinase or by the formation of a thymol–thioglucose complex on hydrolysis with sulphuric acid. These methods are probably equivalent in their accuracy and precision and their adoption should be determined on the local requirements and budgetary restraints.

In future, it is likely that the interest in glucosinolate analysis generated by the development of low glucosinolate varieties will decrease. The HPLC methods will continue to be refined in order to analyse the glucosinolates present in very low ($<2\,\mu\text{M}/\text{g}$) varieties which are currently being developed. It is likely that methods based on the release of glucose on hydrolysis will serve well into the future. The main area of challenge will be in the development of improved methods to determine residual glucosinolate hydrolysis products in meals and protein concentrates and in the continued development of inexpensive rapid methods for use in screening.

REFERENCES

1. Bjerg, B. and Sorensen, H. In: *World crops: production, utilization, description. Vol. 13, Glucosinolates in rapeseeds: analytical aspects*, ed. J. P. Wathelat, Martinus Nijhoff Publishers, Dordrecht, 1987, pp. 125–50.
2. Fenwick, G. R., Heaney, R. K. and Mawson, R. In: *Toxicants of plant origin, Vol. II: Glycosides*, ed. P. R. Cheeke, CRC Press, Boca Raton, 1989.
3. McGregor, D. I., Mullin, W. J. and Fenwick, G. R., *J. Ass. Off. Analyt. Chem.*, **66** (1983) 825–49.
4. Daun, J. K., *J. Japn Oil Chem. Soc. (Yukagaku)*, **35** (1986) 426–34.
5. Heaney, R. K., Spinks, E. A., Hanley, A. B. and Fenwick, G. R., *Technical Bulletin: Analysis of glucosinolates in rapeseed*, AFRC Food Research Institute, Norwich, UK, 1986.
6. Wathelet, J. P., Marlier, M., Severin, M. and Biston, R., *Rev. Agric.*, **39** (1986) 1047–59.
7. Wathelet, J. P. (ed.), *World crops: production, utilization, description, Vol. 13. Glucosinolates in rapeseeds: analytical aspects*, Martinus Nijhoff Publishers, Dordrecht, 1987.
8. Sang, J. P. and Salisbury, P. A., *J. Sci. Food Agric.*, **45** (1988) 255–61.
9. Canadian certification mark registration No. 234139. Consumer and Corporate Affairs, Trademarks, Ottawa, April 18, 1980, Amended August 18, 1982.

10. Arnaud, F., *Bull. CETIOM,* **96** (1988) 16–17.
11. Schnug, E., In: *Double low rapeseed forum proceedings,* National Institute of Agricultural Botany, Cambridge, 1988, pp. 28–30.
12. Mailer, R. J. and Wratten, N., *Aust. J. Exp. Agric.,* **25** (1985) 932–8.
13. Sang, J. P., Bluett, C. A., Elliott, B. R. and Truscott, R. J. W., *Aust. J. Exp. Agric.,* **26** (1986) 607–11.
14. Campbell, L. D. and Cansfield, P. In: *7th Progress Report. Research on Canola meal, oil and seed,* ed. E. McGregor, Canola Council of Canada, Winnipeg, 1983, pp. 50–4.
15. Velisek, J., Pokorny, J., Davidek, J., Simicova, Z., Brenner, V., Lachoutova, D., Hrdlicka, J., Cmolik, J. and Janicek, G., *Scientific Papers of the Prague Inst. of Chem. Tech. Food,* **56** (1983) 153–75.
16. Dem'yanchuk, G. T., *Visn. Sil's'kogospod. Nauk.,* **10** (1987) 48–50; *Chem. Abstr.,* **108,** 149032t.
17. Gardrat, C., Coustille, J. L., Gauchet, C. and Prevot, A., *Rev. Franc. Corps Gras,* **35** (1988) 99–104.
18. Feibeg, H.-J. and Kallweit, P., *Fat Sci. Tech.,* **89** (1987) 135–9.
19. Wathelet, J. P., Cwikowki, M., Marlier, M. and Severin, M., *Rev. Franc. Corps Gras,* **35** (1988) 177–9.
20. Schung, E. and Haneklaus, S., *Fat Sci. Technol.,* **92** (1990) 101–6.
21. Marquard, R. and Schlesinger, V., *Fette, Seifen, Anstrichm.,* **87** (1985) 471–6.
22. Nouat, E. and Ribaillier, D., Determination of glucosinolates. Results of the second international collaborative test (November 1986–January 1987), ISO/TC 34/SC 2 N 360 E, May 1987, AFNOR, Paris.
23. Carlson, K. D., Baker, E. C. and Mustakas, G. C., *J. Am. Oil. Chem. Soc.,* **62** (1985) 897–905.
24. Thies, W. In: *Proc. 4th Int. Rapeseed Conf.,* Geissen, GFR, June 4–8, 1974, pp. 275–82.
25. Helboe, P., Olsen, O. and Sorensen, H., *J. Chromatogr.,* **197** (1980) 199–205.
26. Brzezinski, W. and Mendelewski, P., *Z. Pflanzenzucht.,* **93** (1984) 177–83.
27. Olsen, O. and Sorensen, H., *Phytochemistry,* **18** (1979) 1547–52.
28. Thies, W., *Naturwissenschaften,* **66** (1979) 364–5.
29. Heaney, R. K. anf Fenwick, G. R., *Z. Pflanzenzucht.,* **87** (1981) 89–95.
30. Thies, W. In: *Proc. 5th Int. Rapeseed Conf.,* Malmo, Sweden, June 12–16, 1978, Vol. 1, pp. 136–9.
31. Sorensen, H., *GCIRC Bulletin,* **4** (1988) 17–19.
32. Kondo, H., Yamauchi, M., Nakamura, T. and Nozake, H., *Agric. Biol. Chem.,* **49** (1985) 3587–9.
33. Youngs, C. G. and L. R. Wetter, *J. Am. Oil Chem. soc.,* **44** (1967) 551–4.
34. Wetter, L. R. and Youngs, C. G., *J. Am. Oil Chem. Soc.,* **53** (1976) 162–4.
35. Gardrat, C. and Prevot, A., *Rev. Franc. Corps Gras,* **34** (1987) 457–61.
36. Smith, C. A. and Dacombe, C., *J. Sci. Food Agric.,* **38** (1987) 141–50.
37. Heaney, R. K., Spinks, E. A. and Fenwick, G. R., *Analyst,* **113** (1988) 1515–18.

38. Lein, K. A., *Z. Pflanzenzucht.*, **63** (1970) 137–54.
39. McGregor, D. I. and Downey, R. K., *Can. J. Plant Sci.*, **55** (1975) 191–6.
40. Ribaillier, D. and Quinsac, A., *Bull. CETIOM*, **97** (1988) 11–13.
41. Changling, Y., Bing, F., Yuqin, J. and Zhijing, H., *Fat Sci. Technol.*, **89** (1987) 342–5.
42. Croft, A. G., *J. Sci. Food Agric.*, **30** (1979) 417–23.
43. Schnug, E., *Fat Sci. Technol.*, **89** (1987) 438–42.
44. Schnug, E. and Haneklaus, S., *Fresenius Z. Anal. Chem.*, **326** (1987) 441–5.
45. Thies, W., *Fette Seifen Anstrichm.*, **84** (1982) 347–50.
46. Recommendation of Canada/Sweden exchange on rapeseed research symposium on the analytical chemistry of rapeseed and its products. In: *Analytical chemistry of rapeseed and its products—a symposium*, ed. J. K. Daun, D. I. McGregor and E. E. McGregor, Canola Council of Canada, Winnipeg, 1980, p. 99.
47. Canadian General Standards Board, National Standard of Canada, Canola Meal, CAN/CGSB-32. 301-M87, Minister of Supplies and Services, Ottawa, 1987.
48. Annex VII (to Regulation (EEC) No. 1470/68) Colza and rapeseed determination of the content of glucosinolates, ISO/TC 34/SC 2N 348, AFNOR, Paris, May 1985.
49. Determination de la teneur en glucosinolates (Methode communautaire), Annex 1. Bull. Oleagin., June, 1986, Assoc. Gen. Prod. Oleagin. and Fed. Franc. Coop. Oleagin. Proteagin.
50. McGregor, D. I., *Can. J. Plant Sci.*, **58** (1978) 795–800.
51. McDanell, R., McLean, A. E. M., Hanley, A. B., Heaney, R. K. and Fenwick, G. R., *Food Chem. Toxic.*, **26** (1986) 59–70.
52. Radwan, M. N. and Lu, R. C.-Y., *J. Am. Oil Chem. Soc.*, **63** (1986) 1442–3.
53. Daxembichler, M. E., Spencer, G. F., Kleiman, R., VanEtten, C. H. and Wolff, I. A., *Analyt. Biochem.*, **38** (1970) 373–82.
54. Srivatava, V. K. and Hill, D. C., *Can. J. Biochem.*, **53** (1975) 630–3.
55. Petroski, R. J., *J. Ass. Off. Anal. Chem.*, **66** (1983) 309–11.
56. Kononova, R. V., *Zhavad. Lab.*, **52** (1986), 14–15; *Chem. Abstr.*, **105** (1986) 633 (Abstract No. 105:224729u).
57. Maheshwari, P. N., Stanley, D. W., Gray, J. I. and Van de Voort, F. R., *J. Am. Oil Chem. Soc.*, **56** (1979) 837–41.
58. Harris, J. R., Hopkins, R. G. and Baker, P. G., *Analyst*, **104** (1979) 457–61.
59. Underhill, E. W. and Kirkland, D. F., *J. Chromatogr.*, **57** (1971) 47–54.
60. Heaney, R. K. and Fenwick, G. R., *J. Sci. Food Agric.*, **31** (1980) 593–59.
61. Slominski, B. A. and Campbell, L. D.,, *J. Sci. Food Agric.*, **40** (1987) 131–43.
62. Landerouin, A., Quinsac, A. and Ribaillier, D. In: *World crops: production, utilization, description. Vol. 13, Glucosinolates in rapeseeds: analytical aspects*, ed. J. P. Wathelat, Martinus Nijhoff Publishers, Dordrecht, 1987, pp. 26–37.

63. Hase, A., Johansson, M.-L. and Viljava, T.-R., *J. Am. Oil Chem. Soc.*, **65** (1988) 647–51.
64. Wathelet, J. P., Marlier, M., Severin, M. and Biston, R. In: *World crops: production, utilization, description, Volume 13, Glucosinolates in rapeseeds: analytical aspects*, ed. J. P. Wathelat, Martinus Nijhoff Publishers, Dordrecht, 1987, pp. 109–24.
65. McGregor, D. I., *Eucarpia-Crucifierae Newsletter No. 10* (1985) 132–6.
66. Buchner, R. In: *World crops: production, utilization, description. Vol. 13, Glucosinolates in rapeseeds: analytical aspects*, ed. J. P. Wathelat, Martinus Nijhoff Publishers, Dordrecht, 1987, pp. 50–8.
67. Heaney, R. K. and Fenwick, G. R. In: *World crops: production, utilization, description. Vol. 13, Glucosinolates in rapeseeds: analytical aspects*, ed. J. P. Wathelat, Martinus Nijhoff Publishers, Dordrecht, 1987, pp. 177–91.
68. Bjorqvist, B. and Hase, H., *J. Chromatogr.*, **435** (1988) 501–7.
69. Cansfield, P. and Campbell, L. D. In: *Analytical chemistry of rapeseed and its products—a symposium*, ed. J. K. Daun, D. I. McGregor and E. E. McGregor, Canola Council of Canada, Winnipeg, 1980, pp. 91–8.
70. Moller, P., Olsen, O., Ploger, A., Rasmussen, K. W. and Sorenson, H. In: *Advances in the production and utilization of cruciferous crops*, ed. H. Sorensen, Martinus Nijhoff/Dr W. Junk, Dordrecht, 1985, pp. 111–26.
71. Pec, K., Dabrowski, K. and Krygier, K., *Tluscze Jadalne*, **23** (1985) 14–22.
72. Bjorkman, R., *Acta Chem. Scand.*, **26** (1972) 1111–16.
73. Van Etten, C. H., McGrew, C. E. and Daxembichler, M. E., *J. Agric. Food Chem.*, **22** (1974) 483–7.
74. Thies, W., *Fette Seif. Anstrich.*, **87** (1985) 347–50.
75. Robbelen, G., In: *World crops: production, utilization, description. Vol. 13, Glucosinolates in rapeseeds: analytical aspects*, ed. J. P. Wathelat, Martinus Nijhoff Publishers, Dordrecht, 1987, pp. 26–37.
76. Craig, E. A. and Morgan, A. G. In: *Analytical chemistry of rapeseed and its products—a symposium*, ed. J. K. Daun, D. I. McGregor and E. E. McGregor, Canola Council of Canada, Winnipeg, 1980, pp. 81–5.
77. Mailer, R. J. In: *ARAB V. Australian Rapeseed Agronomist and Breeders 5th Research Workshop*, University of Western Australia, 1985.
78. Smith, D. B., Parsons, D. G. and Starr, C., *J. Agric. Sci., Camb.*, **105** (1985) 597–603.
79. Heaney, R. K., Spinks, E. A. and Fenwick, G. R., *Analyst*, **113** (1988) 1593–612.
80. Saini, H. S. and Wratten, N., *J. Ass. Off. Anal. Chem.*, **70** (1987) 141–5.
81. Rugraff, L., Chemin-Douaud, S. and Karleskind, A., *Rev. Franc. Corps Gras*, **33** (1986) 207–15.
82. Iori, R., Leoni, O. and Palmieri, S., *Analyt. Biochem.*, **134** (1983) 195–8.
83. Kuan, S. S., Ngeh-Ngainbi, J. and Guibault, G. G., *Analyt. Lett.*, **19** (1986) 887–99.
84. Olsson, K., Theander, O. and Aman, P., *Swedish J. Agric. Res.*, **6** (1976) 225–9.

85. McGregor, D. I., *Cruc. News* (1986) 132–3.
86. DeClercq, D. R. and Daun, J. K., *J. Am. Oil Chem. Soc.*, **66** (1989) 788–91.
87. Tkachuk, R., *J. Am. Oil Chem.. Soc.*, **59** (1981) 819–22.
88. Starr, C., Suttle, J., Morgan, A. G. and Smith, D. B., *J. Agric Sci. Camb.*, **104** (1985) 317–23.
89. Norris, K. and Hart, J. R. In: *Humidity and moisture*, ed. P. N. Win, Van Nostrand-Reinhold, Princeton, N.J., 1965, pp. 19–25.
90. Starr, C., Morgan, A. G. and Smith, D. B., *J. Agric Sci. Camb.*, **97** (1981) 107–18.
91. Biston, R., Dardenne, P., Cwikowski, M., Marlier, M., Severin, M. and Wathelet, J.-P., *J. Am. Oil Chem. Soc.*, **65** (1988) 1599–600.
92. McGregor, D. I., New trends in the analytical chemistry of Brassica seed. Presentation to *The Symposium on new trends and developments in canola/rapeseed research*, June 18, Third Chemical Congress of North America, Toronto.
93. Schnug, E. and Haneklaus, S., *J. Sci. Food Agric.*, **45** (1988) 243–54.
94. Schnug, E. and Kallwit, P., *Fat Sci. Technol.*, **89** (1987) 377–81.
95. Schnug, E. and Haneklaus, S., *Fat Sci. Technol.*, **89** (1987), 32–7.
96. Schung, E. and Evans, E., *Oilseeds*, **6** (1988) 9–10.
97. Yin, T. and Zheng, H., *Zhongguo Nongye Kexue (Beijing)* (1984) 49–54.
98. Kononova, R. V., Levitskii, A. P., Chaika, I. K. and Lukashenok, E. V., *Zavod, Lab.*, **51** (1985) 15–16.
99. Yang, Z.-H., Xiu, J.-H. and Zhu, Y.-M., *Analyst*, **113** (1988) 355–7.
100. Yiu, S. H., Collins, F. W., Fulcher, R. G. and Altosaar, I., *Can. J. Plant Sci.*, **64** (1984) 869–78.
101. Hassan, F., Rothnie, N. E., Yeung, S. P. and Palmer, M. V., *J. Agric. Food Chem.*, **36** (1988) 398–403.
102. Oilseeds and oilseed residues—determinations of isothiocyanates and vinyl thiooxazolidone, International Standard ISO 5504, 1983.
103. FOSFA Official Method Determination of volatile oil in mustard seed and rapeseed, FOSFA Standard Contractual Methods List, FOSFA, 1986, London, pp. 292–3.
104. Volatile oil in mustard seed, AOAC 30.026, *Official methods of analysis of the Association of Official Analytical Chemists*, ed. W. Harrowitz, AOAC, Washington, DC, 1980, p. 499.
105. Daun, J. K. and McGregor, D. I., Glucosinolate analysis of rapeseed (canola) method of the Canadian Grain Commission Grain Research Laboratory, Canadian Grain Commission, Winnipeg, 1983.
106. Glucosinolate content of rapeseed and meal, glucose method. NSW Department of Agriculture (May 1987), Wagga Wagga, NSW, Method, OP-49.
107. Gmelin, R., *Praparative und analytische Versuche uber Senfolglucoside* (1954) Dissertation, Tubingen.
108. Hanley, A. B., Heaney, R. K. and Fenwick, G. R., *J. Sci. Food Agric.*, **34** (1983) 869–73.
109. Truscott, R. J. W., Minchinton, I. and Sang, J., *J. Sci. Food Agric.*, **34** (1983) 247–54.

110. Lange, R. and Linow, F., *Nahrung,* **31** (1987) 837–41.
111. Peterka, S. and Fenwick, G. R., *Fat Sci. Technol.,* **90** (1988) 61–4.
112. Thies, W., *Fette Seif. Anstrich.,* **90** (1988) 311–14.
113. Wathelet, J. P., Wagstaffe, P. J., Biston, R., Marlier, M. and Severin, M., Rapeseed reference materials for glucosinolate analysis. Study of stability of glucosinolates in rapeseed extracts. Presentation to the 19th World Congress of the International Society for Fat Research (ISF) and the 27th Annual Meeting of the Japan Oil Chemists' Society (JOCS), Tokyo, September 26–30, 1988.
114. Sosulski, F. W. and Dabrowski, K. J., *J. Agric. Food Chem.,* **32** (1984) 1172–75.

APPENDIX: DETAILS OF METHODS

Most of the methods for glucosinolate analysis requre a prior aqueous extraction and purification of the glucosinolates on ion exchange media prior to their determination by GLC, HPLC or spectrophotometry. In order to simplify the presentation of the three methods, the method for determination by GLC will be given in its entirety and the other methods will be given referencing similar portions of the GLC method.

A. Determination of Glucosinolates by Gas–Liquid Chromatography of the Trimethylsilylethers

A1. Reagents

A1.1 Pyridine acetate. (Reagent grade pyridine and reagent grade acetic acid.)

A1.1.1 Pyridine acetate, 0·5 M. Place 930 ml water in a 1 litre flask and add 30 ml glacial acetic acid and 40 ml pyridine. [Do not add glacial acetic acid directly to the pyridine as a violent reaction may occur.]

A1.1.2 Pyridine acetate, 0·02 M. Place 4 ml of 0·5 M pyridine acetate into a 100 ml flask and dilute to 100 ml.

A1.2 Sodium acetate, 1 M. Weigh 20·5 g sodium acetate into a 250 ml flask and dilute to 250 ml with water.

A1.3 Barium acetate and lead acetate, 0·5 M solution. Weigh 63·9 g of barium acetate and 94·9 g of lead acetate in a 500 ml flask and dilute to 500 ml with water.

A1.4 Sodium hydroxide, 0·5 N solution. Weight 2 g of sodium hydroxide into a 100 ml flask and dilute to 100 ml with water.

A1.5   Internal standards

A1.5.1   Allylglucosinolate (sinigrin) (monohydrate potassium salt), 1 mM. Weigh 0·415 g potassium allylglucosinolate monohydrate into a 100 ml flask and dilute to 100 ml with water.

A1.5.2   Benzylglucosinolate tetramethylammonium salt, 1 mM. Weigh 0·241 g tetramethylammonium benzylglucosinolate into a 100 ml flask and dilute to 100 ml with water.

A1.6   DEAE Sephadex A-25 and SP Sephadex C-25.

A1.6.1   DEAE Sephadex A-25. Weigh 25 mg DEAE Sephadex A-25 into a micro ion exchange column (A2.8). Add 10 ml water and allow to swell overnight. Pass 5 ml 0·5 N sodium hydroxide (A1.4) through the column and then wash with 5 ml water to remove excess sodium hydroxide. (Check to ensure the pH of the effluent is neutral.) Pass 5 ml of 1 M sodium acetate (A1.1) through the column followed by 7 ml water. (Check to ensure the pH of the effluent is neutral.)

A1.6.2   SP Sephadex C-25. Weigh 25 mg SP Sephadex C-25 into a micro ion exchange column (A2.8). Add 10 ml water and allow to swell overnight.

A1.7   Sulphatase (aryl sulphatase, Helix Pomatia, type H1).

A1.7.1   Purification of sulphatase. Weigh about 70 mg of sulphatase into a 16 × 150 mm test tube. Add 3 ml water to dissolve the sulphatase and dilute with an equal volume of ethanol. Centrifuge for 10 min at 2000g. Decant the supernatant fluid into a second test tube and discard the precipitate. Add 9 ml of ethanol to the supernatant fluid and centrifuge again at 2000g. Discard the supernatant fluid and dissolve the precipitate in 2 ml of water. Pass the enzyme solution through the DEAE Sephadex A-25 column and then through the SP Sephadex C-25 column (for convenience, the DEAE column may elute directly into the SP column). Wash the columns with 10 ml water and dilute the combined eluted enzyme solution to 35 ml with water. (It may be more convenient to carry out this procedure in triplicate and dilute the combined eluted enzyme solutions to 100 ml). The purified enzyme should be stored at −20°C and thawed immediately before use. It is stable for about 2 weeks if stored in a refrigerator.

A1.8   Pyridine (Silylation Grade)

A1.9   MSTFA (or MSHFBA). N-Methyl-N-trimethylsilyl-trifluoro-acetamide (or N-Methyl-N-trimethylsilyl-heptafluorobutyramide). MSHFBA, although more expensive, does not produce deposits in the combustion chamber of the flame ionization detector.

A1.10   TMCS Trimethylchlorosilane, Silylation Grade.

A2. Apparatus

Usual laboratory apparatus, and in particular:

A2.1   Micro-grinder, e.g. coffee mill.

A2.2   Desiccator.

A2.3   Analytical balance.

A2.4   Water bath, boiling, or block-type heater at 100°C and at 95 ± 2°C.

A2.5   Oven, at 120 ± 2°C.

A2.6   Centrifuge, reaching a centrifugal acceleration of 5000g.

A2.7   Stirrer, Vortex type, for text tubes.

A2.8   Micro ion exchange columns. May be purchased commercially or prepared from pasteur pipettes or disposable pipette tips. The details given here are for a column containing about 100 mg (dry weight) of DEAE-Sephadex A-25 ion exchange resin which has been swelled with water, swirled to remove air bubbles and allowed to settle giving an ion exchange bed of 15 mm × 8 mm (diameter). Smaller columns or columns of different dimensions may be prepared but these should be tested for capacity, for the amount of enyzme required, and for the amount of solvent required to completely elute the glucosino-lates or desulphoglucosinolates.

A2.9   Tube, borosilicate glass, 8 ml capacity, with PTFE lined caps.

A2.10   Tube, borosilicate glass, 2–4 ml capacity, with PTFE lined caps which may be screw-type or crimp-type for use with autosamplers.

A2.11   Gas chromatograph. With temperature programming capability and flame ionization detector. If a packed column instrument, injector and detector connections should be glass lined.

Packed column instruments should be equipped with a 1·2 m × 6 mm o.d., 2 mm i.d. glass column containing 2% OV-7 on Chromosorb AW DMCS 100/120 mesh. The column should be conditioned by attaching to the injection port of the chromatograph and then, with a small flow of helium, programming at 1°C/min from 100°C to 280°C and holding overnight. Capillary column instruments may be equipped with a 25 m × 0·20 mm i.d. fused silica column coated with cross linked methyl silicone (0·1 μm). The column should be conditioned in a similar manner to the packed column (above). A precolumn (1 cm) packed with 1% silicone gum rubber (SE 30) on Chromosorb W high performance 80/100 mesh may be used. After attaching to the detector end, initial operating conditions are:

	Packed column	Capillary column
Injector temperature (°C)	280	220
Detector temperature (°C)	280	300
Initial column temperature (°C)	180	200
Initial time (min)	5	0
Program rate (°C/min)	5	4
Final temperature (°C)	280	280
Final time (min)	0	0
Carrier gas (helium)	30 ml/min	30 cm/s
(hydrogen is an acceptable alternative)		
Air (FID) (ml/min)	500	500
Hydrogen (FID) (ml/min)	50	50
Detector range	1	1
Detector attenuation	64	16
Split ratio	N/A	100:1

Settings may vary somewhat from instrument to instrument.

A3. Procedure

A3.1 Preparation of the test sample.

This method assumes that the sample has been selected according to appropriate sampling procedures. Grind the sample using the microgrinder (A2.1). If the oilseeds have a water content greater than 10% (m/m), they must be dried beforehand to c. 8% moisture with an air flow at about 45 °C.

Note: for the analysis of seeds treated with pesticide, wash with dichloromethane prior to grinding.

A3.2 Determination of moisture and volatile matter content.

Determine the moisture and volatile matter content of a portion of the ground test sample in accordance with the appropriate standard procedure (see Chapter 2). Dry a 5 g sample at 103 ± 2°C to constant weight.

A3.3 Determination of oil content (if the results are desired on an oil-free basis). Determine the oil content in accordance with the appropriate standard procedure (see Chapter 2).

A3.4 Test portion

Weigh into a tube (A2.9) to the nearest 0·1 mg 200 mg of the ground test sample of oilseeds or 100 mg of the test sample of ground meal.

A3.5 Extraction of glucosinolates

A3.5.1 Extraction. Place the tube containing the test portion into the water bath (A2.4) at 95°C and leave for 15 min. Then add 1 ml of boiling water. Wait 2 min and withdraw the tube, add 1 ml internal standard (A1.5.1 or A1.5.2) and stir with the stirrer (A2.7). Cap the tube with the lined cap and replace the tube in the water bath at 95°C for a further 1 min. Remove from the water bath and leave to cool to room temperature. Add 100 μl of the lead barium acetate solution (A1.3), then centrifuge for 5 min at an acceleration of 5000g.

A3.5.2 Calibration mixture. For GLC with benzyl glucosinolate as the internal standard, prepare one or more samples containing 1 ml of internal standard (A1.5.2) and 1 ml of sinigrin solution (A1.5.1) and carry this mixture through the extraction procedure (A3.5.1).

A3.6 Preparation of ion exchange columns. Pass 5 ml of 0·5 N sodium hydroxide (A1.4) through the hydrated column (A2.8). Pass 10 ml distilled water through the column and continue washing with water until the eluant is neutral to test paper. Pass 5 ml of the 0·5 M pyridine acetate (A1.1) through the column followed by 10 ml of water. Leave sufficient water to form a meniscus for ease of sample application.

A3.7 Isolation of glucosinolates. Transfer 1 ml of the extracted glucosinolates containing internal standard (A3.5.1) to the ion exchange column (A3.6). When the solution has run onto the column, wash with 3 ml of 0·02 M pyridine acetate (A1.1.1). Add 0·5 ml of the purified sulphatase solution (A1.7.1) to the column. When the enzyme solution has completely entered the column, stop the flow and cover the column to reduce evaporation. Allow to stand at room temperature overnight (a minimum of 4 h at 20°C). Elute the desulphoglucosinolates with 1·5 ml water into a small tube (A2.10.1).

A3.8 Preparation of pertrimethylsilyl esters. Evaporate the moisture from the desulphoglucosinolates under a stream of nitrogen at 60°C. To the dried desulphoglucosinolates (A3.7) add 100 μl pyridine (A1.8), 100 μl MSTFA or MSHFBA (A1.9) and 10 μl TMCS (A1.10) and securely cap the tube. Autosampler vials may be capped after addition of pyridine and the MSFTA and TMCS added through the septum

(Note: It is essential that all reagents be dry. It is advisable to add the reagents to one sample at a time to minimize exposure to the atmosphere.) Heat at 120°C for 20 min.

A3.9 Separation and determination by GLC. Inject about 2 μl of the pertrimethylsilylated desulphoglucosinolates (A3.8) onto the OV-7 column (A2.11). Run the temperature program according to A2.11. If the internal standard used is benzylglucosinolate, the first sample run should be the calibration mixture containing equimolar benzyl and allyl gluosinolates. Typical chromatograms showing the relative retention times for major glucosinolates are shown in Fig. 5A1.

(Note: Samples silylated with MSTFA produce a small peak with a retention time similar to silylated 2-hydroxy-4-pentenyl desulphoglucosinolate. If this peak is inseparable, it may be necessary to adjust results for this glucosinolate downwards. When samples are silylated with MSHFBA, the peak occurs between 4-hydroxybutenyl- and 2-hydroxy-4-pentenyl desulphoglucosinolate and is separated.)

(Note: When analysing samples with a wide range of glucosinolate content, some 'memory' of the glucosinolate content of the preceding sample may occur (for example, some residue may remain in the injection syringe). For example, when analysing with and without allylglucosinolate internal standard, analysis of a sample which does not contain allylglucosinolate immediately after a sample which does contain allylglucosinolate may result in an overestimate of the allyl-glucosinolate in the second sample. The reverse situation may also occur. Repeated analysis of the same sample until stable results are obtained may be necessary. It is also good practice to carry out several injections of the first sample prior to beginning an analytical session when the system has had a period of inactivity (e.g. overnight).

A4. Quantification of results

A4.1 Collect the peak areas for:
  3-butenylglucosinolate
  4-pentenylglucosinolate
  2-hydroxy-4-butenylglucosinolate
  2-hydroxy-4-pentenylglucosinolate
  allylglucosinolate (if present)
  4-hydroxybenzylglucosinolate (if present)
  indolyl-3-methylglucosinolate
  4-hydroxy-indolyl-3-methylglucosinolate
  benzylglucosinolate

A4.2 Internal standards
A4.2.1 For GLC with benzylglucosinolate Internal Standard. Since aromatic glucosinolates give different responses in the flame ionization

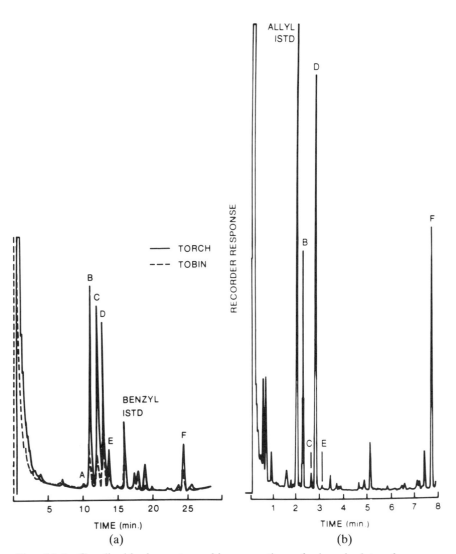

Fig. 5A.1. Gas–liquid chromatographic separation of glucosinolates from rapeseed. Glucosinolates identified: A = allyl-; B = 3-butenyl-; C = 4-pentenyl-; D = 2-hydroxy-3-butenyl-; E = 2-hydroxy-4-pentenyl-; F = 1-hydroxy-D-indolyl-methyl-; benzyl- or allyl- internal standards (ISTD). (a) packed column and (b) capilliary column.

detector to aliphatic glucosinolates, it is necessary to correct the benzylglucosinolate response factor. The theoretical value for benzylglucosinolate relative to allylglucosinolate is 0·86 but because of day to day variation in columns and equipment, the equimolar mixture of allyl- and benzylglucosinolates (A3.5.2) is used for this purpose. Divide the area obtained for benzylglucosinolate by the area obtained for allylglucosinolate to obtain the relative response factor R_r for the particular conditions of analysis.

A4.2.2 For GLC with allylglucosinolate Internal standard. Small amounts of allylglucosinolate may be present in the sample either as a minor component of the rapeseed glucosinolates or as a result of contamination from weed seeds (especially Thlaspi arvense (stinkweed)). In order to correct for naturally occurring allylglucosinolate, samples must be run with and without allylglucosinolate internal standard. The true peak area (A) of allylglucosinolate in the internal standard may then be calculated as:

A(true) = A(found) − A(contam) where

A(true) is the area due to added allylglucosinolate (internal standard)

A(found) is the total area for allylglucosinolate in the chromatogram with added internal standard

A(contam) is the area due to natural contamination and is calculated as

A(contam) = A(allyl−) × A(2-OH-3-butenyl+)/A(2-OH-3-butenyl−) where

A(allyl−) is the area of the allylglucosinolate peak in the chromatogram without added internal standard

A(2-OH-3-butenyl+) is the area of the 1-OH-3-butenylglucosinolate peak in the chromatogram with added internal standard

A(2-OH-3-butenyl−) is the area of the 2-OH-3-butenylglucosinolate peak in the chromatogram without added internal standard.

A4.3   Calculation of glucosinolate content.

The content of each individual glucosinolate may be calculated from the appropriate one of the following equations: For GLC with benzylglucosinolate internal standard,

μM Glucosinolate/g sample = $A_{gluc}/A_{IS} \times R_f/R_r \times \mu M_{IS}/g$(sample)

For GLC with allylglucosinolate internal standard,

μM Glucosinolate/g sample = $A_{gluc}/A_{IS} \times R_f \times \mu M_{IS}/g$(sample) where:

g(sample) = grams of sample extracted

A_{gluc} = peak area for glucosinolate in consideration

A_{IS} = peak area of internal standard (may be corrected if allylglucosinolate)

R_f = response factor from table below

R_r = determined response factor for benzyl- relative to allylglucosinolate (4.2.1)

μM_{IS} = micromoles of internal standard added

	GLC Response Glucosinolate
Factor (R_f)	
Allyl-	1·0000
3-butenyl-	0·9615
4-pentenyl-	0·9259
2-Hydroxy-3-butenyl-	0·8621
2-Hydroxy-4-pentenyl-	0·8333
Benzyl-	R_r (A4.2.1)
4-Hydroxybenzyl-	0·7813
Indolyl-3-methyl-	1·55
4-Hydroxy-indolyl-3-methyl-	1·55

A5. Expression of results

Total glucosinolates may be reported as the sum of all the individual glucosinolates found. The moisture, oil and seed purity of the reporting basis should be stated in the report. For canola, the basis for reporting should be oil-free with 8·5% moisture in the flour unless the analysis is carried out on oil-free flour in which case actual moisture basis of the oil-free flour is determined. For canola, the sum of only the four aliphatic glucosinolates (3-butenyl-, 4-pentenyl-, 2-hydroxy-3-butenyl- and 2-hydroxy-4-pentenylglucosinolates) are reported.

If so desired, calculate the concentration of the glucosinolate in the oil-free sample from the formula

($\mu M/g$ wf) $\times$ 100/(100-OC) = $\mu M/g$ oil-free where

$\mu M/g$ wf = glucosinolate content in whole fat material

OC = oil content of material

Similarly, expression of results on different moisture bases may be calculated from:

($\mu M/g$ dry) $\times$ 100/(100-MB) = $\mu M/g$ MB where

$\mu M/g$ dry = glucosinolate content on dry basis

MB = desired moisture basis.

A6. Test report
The test report shall show the method used and the result obtained. It shall also mention all operating details not specified regarded as optional, as well as any incidents which may have influenced the result.
The test report shall include all details required for the complete identification of the sample.

B. Determination of Glucosinolates by High Performance Liquid Chromatography

B1. Reagents
Reagents as for the GLC Method from A1.1 to A1.9.
B1.1   Mobile phases for HPLC.
B1.1.1   Eluant A. Deionized water purified by passing through an activated carbon system.
B1.1.2   Eluant B. Acetonitrile (HPLC GRADE), 25% (v/v) solution in purified water.

B2. Apparatus
Apparatus as for the GLC Method from A2.1 to A2.9.
B2.10   Filtration apparatus for filtering small samples through 2–5 μM filters.
B2.11   Small vials or autosample vials for HPLC.
B2.12   High performance liquid chromatograph, with capability for gradient elution and temperature control of the column at 30°C connected to a spectrophotometric detector permitting measurements at 229 nm. The detector should be connected to a recording and integrating system.
B2.13   Analytical HPLC column packed with reversed phase C18 packing. Typical columns will be 150–250 mm in length and 4–4·6 mm in diameter. Packings should be 4–6 μM silica based. Columns should be checked for performance, preferably against a mixture of desulpho-glucosinolates isolated from rapeseed. The column should not cause degradation of 4-hydroxyglucosinolate. New columns should be operated (including the injection of samples) for some time before analysis in order to obtain stable results.

B3. Procedure
Follow the procedure for the GLC method from A3.1 to A3.7.
B3.8 Filter the collected eluant containing the desulphogluco-sinolates (A3.7) through the micro-filtration apparatus (B2.10) and collect the eluant in the sample vial or autosample vial (B2.11).
B3.9 Chromatography.
B3.9.1 Chromatographic conditions.

Flow rate of mobile phase	1.5 ml/min
Column temperature	30°C
Detection wavelength	229 nm

B3.9.2 Elution gradient (all gradients linear).

Time (min)	% Eluant A	% Eluant B
0	100	0
1	100	0
19	0	100
22	0	100
25	100	0
30	100	0

The gradient profiles can be modified to give optimum separation according to the column used.
B3.9.3 Inject 20 μl of the desulphoglucosinolate solution obtained in B3.8 and carry out the separation using the gradient in B3.9.2.

B4. Quantification of results
Typical chromatograms showing the relative retention times for major glucosinolates are shown in Fig. 5A2.
B4.1 Collect the peak areas for:
  3-butenylglucosinolate
  4-pentenylglucosinolate
  2-hydroxy-4-butenylglucosinolate
  2-hydroxy-4-pentenylglucosinolate
  allylglucosinolate (if present)
  4-hydroxybenzylglucosinolate (if present)
  indolyl-3-methylglucosinolate
  4-hydroxy-indolyl-3-methylglucosinolate
  benzylglucosinolate

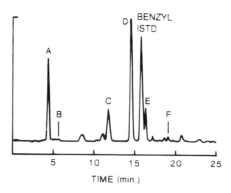

Fig. 5A.2. High performance liquid chromatographic separation of glucosinolates from rapeseed. Glucosinolates identified: A = 2-hydroxy-3-butenyl-; B = 2-hydroxy-4-pentenyl-; C = 3-butenyl-; D = 4-hydroxy-3-indolylmethyl-; E = 4-pentenyl-; F = 3-indolylmethyl- benzyl internal standard (ISTD).

B4.2   Calculation of glucosinolate content

The content of each individual glucosinolate may be calculated from the following equation:

μM glucosinolate/g sample = $A_{gluc}/A_{IS} \times R_f \times \mu$M(IS)/g(sample)

where:

g(sample) = grams of sample extracted

A_{gluc} = peak area for glucosinolate in consideration

A_{IS} = peak area of internal standard (may be corrected if allylglucosinolate)

R_f = response factor from table below

μM(IS) = micromoles of internal standard added

HPLC Response

Glucosinolate	Factor (R)
Allyl-	1·00
3-Butenyl-	1·00
4-Pentenyl-	1·00
2-Hydroxy-3-butenyl-	1·00
2-Hydroxy-4-pentenyl-	1·00
Benzyl-	0·95
4-Hydroxybenzyl-	0·95
Indolyl-3-methyl-	0·25
4-Hydroxy-indolyl-3-methyl-	0·25
4-Methoxy-indolyl-3-methyl-	0·25

B5. Expression of results
Same as GLC method.

B6. Test report
Same as GLC method.

C. Determination of Total Glucosinolates by Thymol Sulphuric Acid

C1. Reagents
Reagents as for the GLC Method from A1.1 to A1.4 as well as:
C1.5   DEAE Sephadex A-25.
C1.6   Potassium sulphate, 0·3 M. Weight 52·3 g anhydrous potassium sulphate into a 1 litre flask and dilute to 1 litre with water.
C1.7   Formic acid (30% v/v). Add 150 ml formic acid to 350 ml water in a 600 ml beaker. Acetic acid may be substituted and may help to avoid formation of bubbles on addition of sulphuric acid.
C1.8   Sulphuric acid, 80% v/v. Place 40 ml water in a 500 ml beaker and add 160 ml concentrated sulphuric acid gradually with stirring. (The dissolution is extremely exothermic and precautions (e.g. ice baths) should be used to avoid overheating the solution while adding the sulphuric acid.)
C1.9   Thymol (5-methyl-2-{1-methylethyl}phenol), 1% w/v in ethanol. Place 1 g thymol in a 100 ml flask, dissolve in ethanol and dilute to 100 ml with ethanol.
C1.10   Sinigrin (allylglucosinolate), 0·3 mmol/litre solution. Place 12·46 mg of monohydrated potassium allylglucosinolate in a 100 ml flask, dissolve with water and dilute to 100 ml with 0·3 M potassium sulphate solution (C1.6).

C2. Apparatus
Apparatus as for the GLC Method from A2.1 to A2.9 as well as:
C2.10   Flask, volumetric, 10-ml, borosilicate glasses.
C2.11   Tube, borosilicate glass, 12 ml capacity, with PTFE lined caps.
C2.12   Spectrophotometer, capable of holding test tubes (C2.11) and of displaying absorbance readings at 505 nm.

C3. Method
Follow the procedure for the GLC method from A3.1 to A3.4.
C3.5   Extraction. Place the tube containing the text portion into the water bath (A2.4) at 95°C and leave for 2 min. Then add 1 ml of

boiling water. Wait 2 min, withdraw the tube and stir with the stirrer (A2.7). Cap the tube with the lined cap and replace the tube in the water bath at 95°C for a further 15 min. Remove from the water bath and leave to cool to room temperature.

C3.6   Add 100 μl of lead and barium acetates (A1.3) and centrifuge for 5 min at an acceleration of 5000g. Decant the supernatant into the 10 ml graduated tube or flask (C2.11). Extract the pellet with a further 4 ml portion of distilled water, heating in the water bath for 5 min before centrifuging and decanting into the same 10 ml tube. Make the combined extracts in the tube up to 10 ml with distilled water.

C3.7   Add 3.0 ml of the combined extracts (C3.5) to the ion exchange columns. Wash the columns with 2×2 ml distilled water followed by 2×2 ml 30% formic acid solution (C1.7). Elute the glucosinolates into clean 10 ml volumetric flasks (C2.10) with $2 \times 4\cdot5$ ml $0\cdot3$ M potassium sulphate solution (C1.6). (Note: It is extremely important that the flasks be scrupulously clean and free from dust as small amounts of organic matter will generate large errors in the procedure.) Fill the flask to the 10 ml mark with distilled water.

C4. Quantification of results

C4.1   Spectrophotometric determination with thymol.

C4.1.1   Preparation of thymol–thioglucose complex. Transfer aliquots of 1 ml of the isolated glucosinolates (C3.7) to each of 2 clean glass tubes (C2.11). To each add 7 ml of 80% sulphuric acid (C1.8) followed by 1 ml of 1% thymol solution (C1.9). Cap the tubes, mix gently, and heat at 100°C for 60 min.

C4.1.2   Spectrophotometry. Cool the tubes under tap water and read the absorbance of the solution at 505 nM against a blank prepared by carrying distilled water through steps C3.7 and C4.1.

C4.1.2.1   Determination of the micromolar extinction coefficient. With each set of samples prepare a set of 4 blanks consisting of 1 ml of $0\cdot3$ M potassium sulphate, 7 ml 80% sulphuric acid and 1 ml 1% thymol solution. Also prepare 4 sinigrin standards with 1 ml $0\cdot3$ mM/litre sinigrin (C1.10) 7 ml 80% sulphuric acid and 1 ml 1% thymol solution. Heat, cool and measure the standards and blanks as in C4.1.2. Calculate the micromolar extinction coefficient k as

$k = (A - B)/0\cdot3$ where

A = the mean absorption of the standards and

B = the mean absorption of the blanks

C4.1.2.2   Calculation of results for unknown samples
Calculate the concentration of the glucosinolate in the sample from the formula:
$A/k \times DF/W \times 100/(100 - m)$ where
A = optical density of the sample at 505 nm (corrected)
k = micromolar extinction coefficient (C4.1.2.1).
DF = dilution factor (in this case 30)
W = weight of sample in grams
m = moisture content of the sample.

C5. Expression of results
Same as GLC method.

C6. Test report
Same as GLC method.

6

Extraction of Fats from Fatty Foods and Determination of Fat Content

I. D. LUMLEY AND R. K. COLWELL

Laboratory of the Government Chemist, Queens Road, Teddington, Middlesex TW11 0LY, UK

6.1. INTRODUCTION—WHAT IS FAT AND LIPID

The determination of the fat content of a sample is one of the most common analyses performed in a food or feeding stuffs laboratory, however the quantitative extraction of fat from a sample is far from straightforward. We first have to define what we mean by 'fat', not always an easy task, then select an appropriate extraction procedure which is compatible with our definition and with the sample on hand. What is meant by 'fat' and how does this relate to the term 'lipid'? Typical dictionary definitions are:

> Fat—glycerides of fatty acids
> Lipid—any of the various substances that are soluble in organic solvents including fats, waxes, phosphatides, cerebrosides and related and derived compounds.

From the above definition of fat it could be argued that the analyst would find it almost impossible to determine just 'glycerides of fatty acids' in the majority of foodstuffs, and the definition of lipid could include organic solvent soluble substances such as carbohydrate and protein fractions and other food additives which the analyst would certainly not class as lipid. These points are discussed later.

Other terms commonly used are 'simple' and 'complex' lipids. Simple lipids are esters of fatty acids and alcohols, usually glycerol, although sterol and wax esters fall into this category. This description is similar to, but more comprehensive than the dictionary definition above. Fat and simple lipids are also often called 'neutral' lipids because of their solubility in non-polar solvents but this classification on perceived solubility is not sufficiently rigorous for routine use.

Complex lipids include phospholipids, glycolipids and lipoproteins which yield fatty acid, glycerol and other molecular species such as phosphate, carbohydrate and protein on hydrolysis. The term 'polar' lipids is used to describe complex lipids because of the polar nature of some of their functional groups and their resulting preferential solubility in polar organic solvents. Thus the analyst has a range of definitions and classifications for fat and lipid but do these usefully relate to the material actually extracted from a sample when we determine 'fat'? When using a standard procedure for the determination of fat in a particular product or product type analysts may not need to know what they are extracting and determining because the method itself defines fat content under the conditions of the test (assuming that the field of application clause has been adhered to). Many methods for the determination of fat show good precision for specific foods or types of food, but with the ever increasing range of processed, composite and novel foods available, the analyst faces an increasingly difficult task when selecting appropriate methods for fat extraction and determination. If a relevant standard method is not available the analyst must decide if free (crude) or total fat, or lipid is to be determined. Free or crude fat may be determined by a simple extraction with a non-polar solvent whereas the determination of total fat or lipid may involve sample digestion and the use of a cocktail of solvents with different polarities for subsequent extraction. Whichever method is used the analyst will almost certainly extract lipid and non-lipid components of the sample which do not match the definitions of fat or lipid given above.

Hannant[1] has described three entities;

Fat—a mixture of glycerides of fatty acids
Fat, fats—commercial oils and fats
Fatty matter—fatty chemicals obtained in an analytical extract.

He qualifies the third definition by stressing that the analytical extract will contain co-extractants which are patently not of natural fat or lipid origin, e.g. emulsifiers, stabilisers, caffeine and lactose. He states that a stricter definition of a specific fat could therefore be; 'anhydrous fatty matter extracted from a single source, consisting mainly of fatty acid triglycerides with relatively small proportions of co-extracted fat soluble substances, excluding any additives or further ingredients, but including such a fat after refining and/or hydrogenation'. A useful definition but again, how can analysts be sure that the fat they are determining complies with this definition without considerable additional analytical effort? As stated above, analytical methods involve the determination of fatty matter and not fat, although standard methods generally refer to free fat, crude fat and total fat. The AOAC states that 'fat is triglyceride and other ether extractable substances', however several methods for the determination of lipids are prescribed (see Appendix).

Hannant points out that the use of prescribed or standard methods to define fat leads to a multitude of definitions because each method has a limited application and he states that a means of avoiding this is to require any extract to satisfy a single property before it can be called fatty matter. One criterion often used is solubility in a relatively small volume of light petroleum and relative non-volatility at 100°C. This leads Hannant to propose a definition of total fatty matter as:

the total amount of substances contained in the sample which, with no chemical change other than acid/base effects, are soluble in a relatively small volume of warm light petroleum spirit and are relatively non-volatile at 100°C.

A detailed treatment of the definition of lipids was first carried out by Bloor[2,3] over 50 years ago. In essence, he described lipids as a major class of biological substances including fatty acids, their naturally occurring compounds, and other substances related to them chemically, or found naturally in association with them. Bloor further characterised lipids as follows: 'Lipids are insoluble in water but soluble in fat solvents such as diethyl ether, chloroform, benzene and boiling ethanol'. These were their most distinguishing properties in contrast with the other main groups of biological substances. These properties were recognised as not being absolute, i.e. some lipids such as lecithins form dispersions in water that approach true solutions. On the other hand, some lipids were known to be relatively insoluble in some fat solvents, e.g. lecithins in acetone, cephalins in alcohol and sphingomyelins and glycolipids in diethyl ether.

'They are related to fatty acids as esters, either actually or potentially and they are utilised by living organisms'.

The above characterisation does contain certain flaws, as there is some uncertainty as to the lipid character of the fat soluble vitamins, because although they are considered, and their association with fats emphasised, they are not specifically included in Bloor's classification. Also, sterols are included, but steroids, although then known, were not considered to be lipids until later (nowadays sterols are considered to be steroids).

Since Bloor's pioneering attempt at classification and characterisation a number of new lipids and lipid complexes have been discovered including for example dipolar lipids, gangliosides, proteolipids and lipoproteins. Although a surprising number of these can be accommodated in Bloor's definition it is inevitable that broader approaches are needed to encompass many of the new compounds that for one reason or another should logically be regarded as lipids.

For the purposes of this chapter the term fat is used as defined in standard methods under discussion or as fatty matter extracted under the analytical conditions prescribed. Lipid is used where this term is included in a cited paper or where attempts have been made to extract or determine total fatty matter.

6.2. THE DETERMINATION OF FAT CONTENT—THE USE OF STANDARD METHODS

6.2.1 A Description of Analytical Procedures Commonly Used

As previously stated the analytical procedure used will chemically define the fat or lipid content because no one method will give an absolute value and this is illustrated several times in this chapter.

A wide range of procedures appear in the literature for the determination of fat, but we have taken the view that the majority of analysts would prefer, and should use, standard methods where they are available and applicable. The appendix lists standard methods for the determination of fat in a range of human foodstuffs and animal feeds as prescribed by national and international standards organisations. The methods can be used for the analysis of samples outside the stated field of application of the standard but the analyst must be aware of the possible empirical nature of the results obtained.

Standard methods for the determination of fat content can generally be classified within one of six common basic analytical procedures known by the name of their original proposers. The first automatic solvent extraction apparatus for example was designed by Franz von Soxhlet (1939) and his name is most commonly associated with solvent extraction systems although others are used (e.g. Goldfisch, Butt,

Bolton, Bailey-Walker). The Soxhlet procedure involves the extraction of fat from the sample by an organic solvent (usually hexane or ether) using a Soxhlet or similar extraction apparatus, and subsequent gravimetric determination. Samples with an appreciable moisture content may have to be dried before extraction. Methods for the determination of 'total' fat usually involve a digestion step prior to solvent extraction. The Weibull–Berntrop (WB) procedure involves boiling the sample with hydrochloric acid, passing the digest through a filter paper which retains the released fat, drying the filter, then extraction of the fat in a Soxhlet type apparatus followed by gravimetric determination. This procedure is useful for products with mixed ingredients, e.g. fruit, vegetables, starch and high levels of carbohydrate. The disadvantages of the method are the number of manipulative steps involved and the overall time taken to perform the determination. Acid digestion can produce non-fat ether extractable compounds particularly if sugars are present, but the filtration stage in the WB procedure is designed to remove this source of possible overestimation of fat content.

The Werner–Schmid (WS) and Schmid–Bondzynski–Ratzlaff (SBR) methods are similar and again involve acid digestion (hydrochloric acid) but fat is solvent extracted directly from the digest using Mojonnier type extraction flasks or extraction tubes fitted with a syphon or wash bottle fitting. These procedures are relatively rapid to perform but they are not always suitable for samples which contain high levels of carbohydrate (because of formation and subsequent extraction of non-lipid ether extractable compounds), or for samples which are relatively inhomogeneous, as small sample weights (1–3 g) are taken. A WB type procedure should be used if these factors pose a problem.

The Rose–Gottlieb (RG) method is used mainly for dairy products and involves precipitation and solubilisation of protein by ethanol and ammonia respectively, and subsequent extraction of fat with a mixture of diethyl ether and light petroleum. This method cannot be used for most cheeses because the protein is not soluble in ethanolic ammonia and partially hydrolysed triglycerides and free fatty acids are not extracted from the ammoniacal phase by the diethyl ether–light petroleum mixture.

Volumetric methods for dairy products used in Europe are generally based on the Gerber procedure (Babcock method in USA) which involves the direct measurement of the volume of liquid fat in a

butyrometer after acid digestion of the sample. The procedure is
described later in this text. Van Gulik butyrometers are also described. The Babcock method prescribed in the USA for some
applications is technically similar.

The names associated with these six basic methods are used in the
Appendix to identify the type of analytical procedure used in the
corresponding Standard, however several other named methods used
in the Appendix are discussed later in this chapter.

6.2.2 Discussion of Standard Methods

6.2.2.1 Meat and meat products

ISO 1443 prescribes a reference method for the determination of the
total fat content of meat and meat products, and involves the boiling
of the test portion with dilute hydrochloric acid to free the occluded
and bound lipid fractions. The hot acid digest is then passed through a
filter paper, which retains the freed lipid fractions, the paper is dried,
and the lipid then extracted with hexane or light petroleum in a
Soxhlet or related extraction apparatus. After evaporation of the
solvent the extracted fat thus obtained is determined gravimetrically.
This method is similar in principle to those described by Weibull[4,5] and
modified by Berntrop.[6] ISO 1443 is technically the same as BS 4401:
Part 4: 1970, Method A. Total fat in these two standards is defined as
'the fat extracted under the operating conditions described'. The WB
type procedure is widely used in laboratories in Europe and is gaining
in popularity in the UK as the reference procedure, however current
preference in the UK is probably still for Method B in BS 4401: Part 4.
This involves boiling the test portion with hydrochloric acid in an
extraction tube. The liberated fat is extracted by shaking with portions
of diethyl ether and the extract transferred to an evaporation flask via
a wash bottle fitting. After evaporation of the solvent and weighing,
the fat is removed from the flask with light petroleum and any
remaining insoluble matter is weighed and subtracted from the
calculated fat content. This procedure is commonly referred to as the
Werner–Schmid process.

ISO 1444 (BS 4401: Part 5: 1970 is identical) describes a procedure
for the determination of free fat content and involves the direct
extraction of fat from a dried test portion using hexane and a Soxhlet
or similar extraction apparatus. The AOAC procedure (method
24·005) specifies the use of anhydrous ether for extraction. Anhydrous

conditions are achieved using sodium hydroxide and maintained by storage of the ether over sodium.

6.2.2.2 Milk and dairy products

ISO 2446 (1976) prescribes a routine method for the determination of the fat content of liquid milk using a Gerber butyrometer. Sulphuric acid is added to an appropriate Gerber butyrometer and the required volume of milk is then added using a milk pipette so that it forms a layer on top of the acid. Amyl alcohol is added, the butyrometer stoppered, and then shaken until all the protein has completely dissolved. The butyrometer is then centrifuged, placed in a water bath (65°C) for at least 3 min and the volume of fat then read from the graduated scale. The scale provides the result as percentage fat (m/m) in the milk sample. Slightly modified procedures are given for skimmed and homogenised milk. ISO 2446 makes reference to ISO 488 which specifies the characteristics of butyrometers with graduations designed to measure fat over the 0–0·5%, 0–5%, 0–8% and 0–10% (m/m) ranges.

BS 696 Parts 1 and 2, 1989, supersede Part 1 of 1955 and Part 2 of 1969, and provide a technically similar procedure to ISO 2446. Part 1 of the standard provides specifications for butyrometers, pipettes, a centrifuge and other apparatus, and takes into account those aspects of ISO 488 which are applicable to practice in the UK. Part 2 provides details of routine methods for the determination of fat in raw, pasteurised, homogenised sterilised and UHT milk, partly skimmed, skimmed, buttermilk and whey, cream, dried milk, reconstituted dried milk and cheese. For milks containing about 2·5%–4·5% butterfat, the Gerber method gives results which are in close agreement with those obtained by the more accurate RG procedure described in BS 1741: Part 3.

ISO 3433 gives details of the Van Gulik method for the determination of the fat content of cheese whilst ISO/DIS 3432 specifies characteristics of a 0–40% butyrometer for use in ISO 3433. AOAC methods are similar but make reference to the Babcock procedure.

International Dairy Federation (IDF) standard 1C: 1987 (a revision of IDF 1B: 1983) specifies a reference method for the determination of the fat content of raw and processed liquid milk and skimmed milk using a method based on the RG principle. Reference is given to the use of Mojonnier-type fat extraction flasks and the use of a centrifuge to separate solvent layers, however an Appendix gives details of a procedure using fat extraction tubes fitted with syphon or wash-bottle fittings. A technically similar procedure is ISO 1211: 1984 and BS 1741: Part 3. When greater accuracy is required for skimmed milk, whey and buttermilk IDF 22B: 1987 should be used. Increased accuracy is achieved by extraction of two test portions and combination of the extracts. The corresponding ISO text is ISO 7208: 1984.

Similar reference methods based on the RG procedure are available for dried milk (IDF 9C: 1987, ISO 1736: 1985), evaporated and condensed milk (IDF 13C: 1987, ISO 1737: 1985), cream (IDF 16C: 1987, ISO 2450: 1985) and skimmed milk, whey and buttermilk (IDF 22B: 1987, ISO 7208: 1984).

The protein in cheese and processed cheese products does not readily dissolve in ammonia and ripened products contain free fatty acids and mono- and di-glycerides which are not easily extracted from an aqueous ammonia solution by ethers; a RG type procedure is therefore not applicable for the determination of fat content. A suitable method is based on the SBR procedure (IDF 5B: 1986, ISO 1735: 1975, BS 770: Part 3: 1976). Cheeses usually contain low levels of sugars (e.g. lactose less than 5% m/m) therefore acid digestion is applicable, but care should be taken particularly if analysing cheeses with added fruit, honey or sugars. If this is the case a WB procedure should be used (IDF 126: 1985, ISO/DIS 8362/3). The WB procedure is also used to determine the fat content of fresh cheeses such as cottage cheese, because larger test portions can be taken and sample inhomogeneity thus overcome. Since whey cheese contains a high percentage of lactose, acid digestion is not appropriate, but whey cheese usually dissolves in ammonia so that a RG method can be used (IDF 59: 1970, ISO 1854: 1972).

Standard methods for other dairy products are given in the Appendix. A procedure for the determination of the fat content of whole milk using an infra-red spectrometer is described in IDF 141: 1988 (ISO/DP 9622) and is based on the absorption of the carbonyl groups of the ester bonds of the glycerides ($5 \cdot 73 \, \mu$m) or the absorption by the CH groups ($3 \cdot 48 \, \mu$m). The method of calibration of the instrument will vary according to manufacturer, but the standard prescribes linearity and calibration checks using cream/skimmed milk mixtures and milk samples respectively. Calibration is compared with results obtained using a standard gravimetric reference procedure (RG method) although Gerber methods can be used if care is taken.

6.2.2.3   Oilseeds and residues

The methods in the standards are based on the use of an extraction thimble in a Soxhlet or similar extractor; ISO 659 (oleaginous seeds) and 734 (oilseed residues) prescribe the use of hexane or light petroleum as solvent, however ISO 736 (oilseed residue) prescribes the use of diethyl ether. This difference in solvent choice has arisen because the oil content of oleaginous seeds (ISO 659) is obtained using *n*-hexane or light petroleum. Therefore to enable the oil industry to control manufacture the determination of the oil content of oilseed residues should be carried out in the same way. The major users of oilseed residues however are animal feed manufacturers who have traditionally determined oil content by extraction with diethyl ether and have thus accumulated much data from this method. Extraction with hexane does not necessarily give the same result as with diethyl ether therefore it was considered necessary to establish ISO 736. These applications are discussed in more detail in Chapters 2 and 3.

6.2.2.4   Animal feeding stuffs

Two methods for the determination of crude oils and fats in animal feeding stuffs (not oil seeds and oleaginous fruits) are given in the *Official Journal of the European Communities* (Directive 84/4/EEC, Annex 1 No L15/29 and L15/30 18.1.84) as a Commission Directive amending Directives 71/393/EEC, 72/199/EEC and 78/633/EEC, establishing Community methods of analysis for the official control of feeding stuffs. Method A involves a Soxhlet type extraction with light petroleum whilst method B involves acid hydrolysis, filtration and Soxhlet extraction of the dried filter paper residues (WB type procedure). The Standard describes sample types to be analysed by methods A and B. ISO DP6492 describes technically similar procedures but this standard refers to 'determination of hexane extract', i.e. hexane rather than light petroleum is used as the extraction solvent.

6.2.2.5   Cereal products

ISO 7302 (ICC Standard No. 136) for the determination of total fat in cereal and cereal products (including baked products and pasta) employs an interesting and slightly different digestion procedure. The dry sample is dispersed in ethanol then formic and hydrochloric acids are added and the mixture placed in a 75°C waterbath under reflux. After hydrolysis and cooling, ethanol and hexane are added and the mixture stirred vigorously, then the phases are allowed to separate.

The hexane lyaer is poured off and the residue extracted three more times with hexane. The hydrolysis and extraction are performed in a flask fitted with a side-tube of sufficient volume to retain the aqueous phase and a small portion of the hexane phase when decanting the majority of the hexane layer. The formic acid assists digestion and the ethyl formate formed on heating acts as a lipid solvent.

6.2.2.6 Miscellaneous standard methods

A range of AOAC and AACC methods prescribed for the determination of fat in miscellaneous foodstuffs are listed in the Appendix including a reference to a rapid method for the volumetric determination of fat in seafood using the Babcock procedure (AOAC, 18.045).

6.3. NON-STANDARD PROCEDURES

6.3.1 Volumetric Methods

The volumetric determination of fat in dairy products by the Gerber, and to a lesser extent the Babcock and van Gulik butyrometers, has been described earlier in this chapter, however there are numerous non-standard methods described in the literature and several comparisons of results obtained by volumetric and gravimetric procedures. The Gerber method can be applied to ice-cream using a milk, cheese or cream butyrometer[7] whilst perhaps surprisingly Rosenthal et al.[8] describe the use of the Gerber procedure for the determination of fat in foods such as avocado, olives, peanuts, pecans and soyabeans. The authors provide experimental conditions such as digestion time and temperature found suitable for the foodstuffs examined, and regression analysis of results obtained by petroleum ether Soxhlet extraction. The described Gerber procedure showed no significant difference between results.

Pearson[9] describes a volumetric method for the determination of fat in meat products using a van Gulk butyrometer for use with 5 g of meat product where Salwin acid reagent (1:1 v/v glacial acetic acid: perchloric acid) is used instead of sulphuric acid for sample digestion. Salwin reagent is also used in AOAC method 18.045 for seafood.

The sulphuric acid used in the Gerber procedure is extremely corrosive and alternative reagent mixtures have been proposed.[10] A mixture of trisodium citrate, sodium salicylate, EDTA, Tween, *n*-butanol and methylated spirit as proposed by Macdonald[11] can be used as a direct replacement for sulphuric acid.

Many studies have compared the Babcock and ether extraction methods for the determination of the fat content of milk and most have found the Babcock procedure to give higher results than the various ether extraction procedures.[12,13] Babcock and ether extraction procedures are used to calibrate infra-red instruments used for fat determination; several workers have therefore attempted to evaluate the apparent difference in results. Barbano et al.[14] report the results of a collaborative study designed to compare and evaluate modified Babcock and Mojonnier ether extraction methods. The ether extraction procedure demonstrated consistently better within- and between-laboratory agreement and the Babcock overall mean test value was significantly higher (0·021% fat) than that for the ether extraction. The difference in results from the two methods was shown to vary for milk from different farms, and the mean difference between percentage fat determined by both methods varied with the month of sampling. Also there was no correlation in the difference between the methods for samples with a fat content of 2·7–5·6%. The methods as modified and collaboratively tested were recommended for adoption as official first action in the AOAC.

Kleyn et al.[15] report the comparison of a Monjonnier ether extraction procedure and a Gerber method by collaborative study.

6.3.2 Solvent Extraction of Fat and Lipid

It is clear from the Appendix that the most common solvents for lipid extraction specified in standard methods are diethyl ether, light petroleum and hexane. What solvents should the analyst use with non-standard methods? Light petroleum and hexane are non-polar whereas diethyl ether is of higher polarity, however all three are good solvents for fat and lipids which contain few polar groups. These three solvents therefore are used to extract triglycerides and other simple (non-polar) lipids from samples in order to determine what is commonly termed 'free' fat content. The lipid components extracted as free fat often occur in the sample matrix in storage tissue and can easily be extracted, but samples must be dry and well ground to a small particle size to enable efficient solvent penetration. In biological samples such as meat and meat products and dairy products some simple lipid may be physically entrained within the structure of the sample matrix, and complex lipids may be physically and chemically bound to proteins, polysaccharides and other cellular components by ionic and hydrogen bonding and Van der Waals' forces. If we wish to determine 'total' fat in such samples a digestion step is incorporated to release entrained and bound lipid prior to extraction with hexane, light petroleum or diethyl ether. Lipids containing very polar functional groups, although now 'free' after digestion may be insoluble or only partially soluble in these relatively non-polar solvents, therefore although the analyst may be determining total fat he may not be determining total lipid. (In the context of the previous sentence hexane, light petroleum and diethyl ether are all 'relatively non-polar' in comparison to chloroform or methanol, which may be required to solubilise very polar lipids.) As stated earlier, standard methods define the fat content and so this may not be important, but if non-standard methods are to be used analysts must decide what they wish to determine, i.e. free fat, total fat or total lipid.

All the three commonly used solvents mentioned above have advantages and disadvantages; all have the advantage of volatility, an important factor for gravimetric analysis. Light petroleum and hexane extract neither starches nor protein and do not absorb water, however diethyl ether is a better solvent for fat, i.e. triglyceride. Diethyl ether has the disadvantage that it will absorb water and then extract some sugars and a range of other non-fat constituents. Several standard methods take advantage of the better solubility of fat in diethyl ether by extracting initially with this solvent, then re-extracting the fat into light petroleum with subsequent gravimetric fat determination. Any non-fat material extracted by the diethyl ether remains as a residue after this re-extraction. Mixtures of diethyl ether and light petroleum can also be used to reduce the co-extraction of non-fat material.

6.3.2.1 Extraction of total lipid

Although we are of the view that analysts should use Standard methods this is clearly not always feasible or desirable. Analysts may not only wish to quantify all the various lipid components in a sample but also examine these components further by chemical and chromatographic procedures. This means that often we must attempt to extract all lipid components using a procedure that does not significantly alter structural or chemical composition. Over the years analysts have used a range of solvents and solvent mixtures with polarities spanning the eluotropic series in attempts to penetrate biological tissue, overcome hydrogen and ionic bonds and Van der Waals forces to free bound, simple and complex lipids and solubilise all lipid components.

Bloor[16] first showed that if a water miscible solvent such as ethanol was added to diethyl ether the mixture would penetrate more easily into biological tissue containing water, and thus effect more efficient extraction of lipid. A range of procedures based on the extraction of 'total' lipids from biological materials (including foodstuffs) using chloroform–methanol mixtures have been subsequently published and extensively used over the years.

The chloroform–methanol procedure proposed by Folch *et al.* for the extraction of brain lipids[17] and animal tissues[18] is the most widely known. This procedure involves shaking the sample with a 2:1 (v/v) chloroform–methanol mixture and washing with saline solution. After phase separation the lower organic phase contains the lipid material. Several extraction, washing and filtration stages are involved and the ratio of chloroform–methanol–water must be carefully controlled (8:4:3 by volume) if losses of lipid material are to be avoided.

Many variations of the Folch procedure have been published and analysts need to select a procedure which is suitable for their analytical purpose and which is compatible with desired sample throughput and staff expertise.

The method of Bligh and Dyer[19] as modified by Hanson and Olley[20] provides a thorough and relatively rapid method for the extraction of lipid from tissues and foodstuffs which contain significant amounts of water. The method is based on the homogenisation of the sample with chloroform, methanol and water in such proportions as to form a single phase miscible with the water in the sample. On addition of further quantities of chloroform and water, two separate phases are formed. Lipid material is contained in the lower chloroform phase whilst non-lipid material is contained in the aqueous methanol phase. Lipid can be extracted from about 2 g of 'dry' sample and up to about 20 g of a 'wet' sample, e.g. fish or meat. The water content of the sample is adjusted (if necessary) to 16 ml; for example, 16 ml of water would be added to a 'dry' sample whilst 10 g of fish would require the addition of about 8 ml of water to achieve a total water content of 16 ml. Chloroform and methanol are then added (20 ml and 40 ml respectively) and the mixture blended for 2 min. Water (20 ml) is added and the mixture again blended. Addition of a further aliquot of chloroform (20 ml) is followed by phase separation (centrifugations may be necessary). Aliquots of the lower layer can be taken for lipid analysis or a known volume taken for gravimetric fat determination. Again maintenance of the correct chloroform–methanol–water ratio

(2:2:1·8 in final mixture) is essential if phase separation and quantitative lipid extraction is to be achieved and co-extraction of unwanted material avoided. The advantage of this procedure is that filtration and washing stages are eliminated. The method has been used satisfactorily for many years in the authors' laboratory on a wide range of foodstuffs and feeds prior to chemical and chromatographic analysis of the extracted lipid. However it is rarely used for the quantitative determination of fat. It has been shown that the method can give falsely high results for dry cereal products which contain a high percentage of non-lipid extractable material. Marinetti[21] and Winter[22] have also proposed methods based on chloroform–methanol extraction.

Southgate[23] developed a quantitative procedure for the extraction of fat from foods using a 2:1 v/v mixture of chloroform–methanol. The sample is boiled in the solvent mixture and immediately filtered; this extraction is performed three times. The solvent is distilled from the filtrate, and the fat re-dissolved in light petroleum. Extraction into light petroleum avoids interference caused by non-lipid sample components which may be extracted from samples with appreciable water content. The procedure provides results which are comparable to those obtained by methods involving acid or alkaline pre-treatment of the sample, and extracted fat can be used for chemical or physical investigation. The method is also simple and relatively rapid and has been widely used.

Lento and Daugherty[24] modified the chloroform–methanol procedure of Bligh and Dyer and showed that it was applicable to a wide variety of food types for fat determination. Samples are digested with the enzyme preparation Clarase 40,000 (Miles Laboratories) prior to blending with chloroform–methanol and gravimetric determination of fat content. The procedure has been collaboratively tested by 15 laboratories on a range of processed foods and statistical analysis of the results indicated that the method was adequate for determining the fat content for nutritional labelling.[25] The study report states that the ratio of chloroform to methanol to water is critical for the quantitative extraction of fat and the moisture content of the sample may have to be determined in order to optimise the amount of water (as enzyme solution) added to the system. The method is now official first action in the 14th edition of the AOAC *Official Methods of Analysis* (method 42.275–277).

The Bligh and Dyer technique is also very useful when it is necessary to determine the peroxide value of the extracted fat, as the chloroform solution can be used directly for this determination. Use of the Bligh and Dyer technique, and direct use of the chloroform solution avoids heating the sample or the extract, a step which can modify the peroxide value. An aliquot of the extract is also evaporated in order to determine the fat content as this is used in the calculation of the peroxide value of the extracted fat.

Sahasrabudhe and Smallbone[26] used the extraction methods and solvents in Table 6.1 'to determine the amount of neutral and polar lipids in beef', i.e. simple and complex lipids. Samples of the extracted lipid were fractionated by column chromatography (on silica) into: (i) triglycerides, (ii) free fatty acids, mono- and di-glycerides, and sterols, (iii) polar lipids. Identity of the three fractions was confirmed by TLC. The AOAC Soxhlet procedure extracted less than 75% of total lipid, 89% of triglycerides and 15% of polar lipids from lean beef as compared to the other methods. Methods 5 and 6 using Soxhlet extraction with chloroform–methanol and dichloromethane–methanol respectively gave results for all lipid fractions which were comparable to those obtained by the Folch, and Bligh and Dyer procedures, whilst method 4 involving boiling under reflux with ethanol and diethyl ether (Bloor's solvent mixture) extracted more polar lipid from high fat samples compared to the other procedures. The fatty acid composition of triglycerides and polar lipids extracted were generally within experimental error when methods of extraction were compared. The authors conclude that the AOAC Soxhlet extraction procedure using ether is not satisfactory for the determination of the lipid composition of meat.

A somewhat similar comparison of lipid extraction methods was

TABLE 6.1

Method	Solvents	Reference
1 Folch *et al.*	Chloroform–methanol	18
2 Bligh and Dyer	Chloroform–methanol	19
3 Hara and Radin	*n*-hexane-2-propanol	27
4 Sheppard	Ethanol–diethyl ether	28
5 Sahasrabudhe	Chloroform–methanol	29
6 Sahasrabudhe and Smallbone	Dichloromethane–methanol	26
7 AOAC	Light petroleum or diethyl ether	30

performed by de Koning et al.[31] using fish meals as the analyte. Boiling under reflux with hexane, hexane Soxhlet, hot Sandler (boiling under reflux with chloroform and methanol), modified Bligh and Dyer,[20] and an EEC hexane Soxhlet[32] procedure were used for this comparison. The results showed that boiling under reflux with hexane and the hexane Soxhlet methods give essentially the same result but both procedures extracted less material than the others. The modified Bligh and Dyer procedure extracted most material, more even than the modified Sandler extraction despite the fact that chloroform and methanol are used for each method. One explanation is that the modified Bligh and Dyer method extracts more lipid because of the vigorous homogenisation step involved, the other is that the ratio of sample weight to volume of chloroform, methanol and water is more ideal for lipid extraction in the Bligh and Dyer procedure. Phosphorus and nitrogen were determined in the lipid extracted by each method and the known ratio of phosphorus to nitrogen in the fish phospholipids was used to calculate the 'extra' nitrogen from proteinaceous material solubilised in the extracting solvent. The amount of protein could then be subtracted to give a corrected percentage of lipid extracted. The authors state that the modified Bligh and Dyer procedure was the most effective and convenient extraction technique of those examined.

Khor and Chan[33] investigated the use of alternative solvents to chloroform for extraction of lipids from soyabeans. Their study was performed in an attempt to find a less toxic solvent than chloroform which gave good recovery of lipid material. Lipids were extracted using chloroform–methanol according to Folch et al.,[18] dichloromethane–methanol[34] and hexane-2-propanol.[35] Total lipid was fractionated by column chromatography (Florisil) into neutral, glyco- and phospho- lipid fractions. Dichloromethane–methanol proved to be as efficient as chloroform–methanol for extraction of the lipid classes however the hexane–isopropanol procedure gave low results for all classes of lipid. The fatty acid and sterol composition of extracted lipid was the same (within experimental error) irrespective of the extraction solvents used.

From previous statements in this chapter it is clear that diethyl ether or hexane extract less lipid material from samples which contain polar lipids compared to solvents such as chloroform–methanol mixtures. Firth et al.[36] compared the efficiency of ether (diethyl ether and light petroleum mentioned in paper) and chloroform for the Soxhlet extraction of lipid from freeze-dried animal tissues. Their reason for the comparison was an attempt to find a non-flammable extraction solvent. The authors took the view that chloroform was acceptable as an extractant despite its toxicity if good laboratory practice was followed. Twenty three samples of each of four tissue groups (carcass, non-carcass, muscle, adipose tissue) were extracted in duplicate in each solvent; fat content covered the range 10% to more than 95%. Regression analysis showed that the intercepts and gradients were not 0 and 1 respectively, thus showing that the solvents did not give identical results. After considering paired *t*-test data the authors state that chloroform and ether values cannot replace each other but that for any given study conclusions reached about treatment differences should not be affected by the solvent used; chloroform was therefore considered a satisfactory replacement for ether. Care must be exercised if this view is widely taken as data obtained by several laboratories may not be compared absolutely. The contrary argument, supported by the present authors, is that the Firth *et al.* data provide an argument for the use of a standard approach and the use of standard methods.

A study performed at the Laboratory of the Government Chemist (LGC)[37] compared the Bligh and Dyer (as modified by Hansen and Olley), Southgate and standard WS procedures for the determination of fat in offal products. Results from the three methods for the majority of products agreed well, however the Southgate procedure gave significantly higher results for lambs liver and heart perhaps because of the high levels of phospholipid present. Results are shown in Table 6.2.

A subsequent study at the LGC compared six methods of fat determination on a range of foods with differing fat, carbohydrate and phospholipid contents. The results obtained are shown in Table 6.3. As expected the Soxhlet extraction with diethyl ether generally produced the lowest results, and the RG method gave low results for the non-dairy type products. The WS procedures gave the highest results for mayonnaise and the instant dessert topping, with the 9M acid tending to give higher results than the 6M. On examination the extracted fat was shown to contain sugars and some protein as a result of the acid digestion thus explaining the falsely high results for fat content using the WS methods. The modified Bligh and Dyer procedure gave low results for mayonnaise and it was thought that partial glycerides were partitioning between the chloroform and

TABLE 6.2
Determination of the fat content of offal products: comparison of fat
extraction procedures

Sample type	Fat content (g/100 g)		
	Bligh and Dyer	Southgate procedure	WS procedure
Liver sausage	53·0	54·3	55·2
	54·8	54·8	
Brawn	38·6	40·4	39·3
	38·0	39·8	
Faggots	31·6	33·7	34·0
	32·0	33·3	
Saveloy	44·0	45·5	46·8
	44·8	44·9	
Polony	42·8	41·7	43·3
	42·9	41·8	
Black pudding[a]	42·0	40·3	38·9
	41·0	—	
White pudding[a]	39·5	37·2	40·0
	—	38·4	
Haggis	39·4	36·7	40·2
	39·2	37·1	
Lamb liver	24·1	29·9	23·1
	24·4	30·2	
Lamb heart	22·6	26·1	23·5
	23·2	26·3	

[a] Heterogeneous samples.
WS, Werner–Schmid procedure.

aqueous phases thus producing a low result. The Southgate method tended to give higher results than Bligh and Dyer for samples which contained high levels of phospholipid (phospholipid: fat ratio equal to or greater that about 0·3), however results from both methods generally compared well. For batch extractions and general applicability analysts tended to favour the modified Bligh and Dyer procedure. An extension of these studies included comparison of solvents of varying polarity in a Soxhlet extraction system, a modified Bligh and Dyer method (as modified by Hanson and Olley) and a WS method on a range of food types.

The results in Table 6.4 for rapeseed, turkey and herring samples show an increased efficiency of Soxhlet fat extraction as the polarity of

TABLE 6.3

Comparison of six commonly used fat extraction procedures

Sample		WS 6M	WS 9M	Soxhlet DEE	Bligh and Dyer[a]	Southgate	RG
		Fat content (g/100 g)					
Mayonnaise	X	82·9	83·1	81·4	79·8	81·8	82·8
	SD	0·3	0·5	0·7	0·6	1·0	0·3
Instant	X	48·2	49·2	32·0	47·4	44·8	47·9
dessert	SD	0·6	0·4	0·4	0·5	2·7	0·5
topping							
Milk	X	30·7	30·4	29·4	30·4	29·7	29·9
chocolate	SD	0·3	0·3	0·2	0·3	0·3	0·2
Whole	X	40·8	38·8	30·8	41·8	42·2	38·2
egg	SD	1·4	1·1	0·8	0·8	0·8	1·0
dried							
Herring	X	12·9	11·2	9·2	12·2	13·1	9·5
meal	SD	1·1	0·2	0·03	0·2	0·9	0·4
Chicken liver	X	19·3	19·2	16·3	17·8	18·6	12·0
	SD	0·7	0·4	0·3	0·5	0·6	1·2

X, mean value; SD, standard deviation; WS, Werner–Schmid; 6M and 9M refer to the strength of hydrochloric acid used for digestion; RG, Rose–Gottlieb; DEE = diethyl ether.
[a] Bligh and Dyer as modified by Hansen and Olley.[20]

the solvent used increases. This is due to the increased ability of polar solvents to overcome the forces previously mentioned which bind lipids within the sample matrix. The results for cheese and sausage show the opposite trend and this has not yet been fully explained. The modified Bligh and Dyer procedure produced the highest result in all cases studied except for the pork and beef sausage. This low result was due to the extraction of water soluble components during blending of the sample with the chloroform–methanol–water mixture which inhibited phase separation on centrifugation. The WS method produced reasonable results for all samples except rapeseed but this is to be expected as this procedure is not recommended for vegetables and oilseeds. Soxhlet extraction using chloroform and methanol extracted large amounts of non-fat material (10–20% by weight of fat extract) which was water soluble and had the appearance of a gum. This material was insoluble in light petroleum and so fat could be

TABLE 6.4

Comparison of Soxhlet (various solvents), Bligh and Dyer and Werner–Schmid fat extraction procedures

Sample		Fat content (g/100 g)					
		Soxhlet extraction				Bligh and Dyer[a]	Werner–Schmid
		Light petroleum	n-Hexane	Chloroform	Methanol–chloroform	Methanol–chloroform	Petroleum + diethyl ether
Rapeseed	X	37·7	38·6	42·5	40·4	42·6	33·6
	SD	0·4	0·3	0·5	0·1	0·5	0·3
Turkey	X	17·3	17·2	19·4	21·7	21·9	17·8
	SD	0·03	0·3	0·3	0·3	0·1	0·2
Cheese	X	30·6	23·3	22·5	24·9	34·5	30·8
	SD	0·1	0·3	0·1	0·5	0·7	0·1
Sausage	X	17·2	16·2	14·3	14·8	13·4	17·0
	SD	0·4	0·5	0·8	0·6	0·7	0·5
Herring	X	7·6	10·9	10·3	16·0	16·4	14·1
	SD	0·1	0·8	0·5	0·8	0·3	0·5

X, mean value; SD, standard deviation.

The turkey sample was freeze-dried and no correction has been made to the results above to allow for removal of water.

[a] The Bligh and Dyer procedure referred to in this table was the modified version as described by Hansen and Olley.[20]

determined after re-extraction into this solvent. Examination of the rapeseed lipid from each extraction procedure showed the absence of diglycerides, monoglycerides and free fatty acids indicating that extraction conditions were not causing detectable breakdown of triglycerides.

Fatty acid isomer analysis by capillary GLC showed differences in the profiles of fatty acids from the various lipid extracts reflecting the different composition of lipid extracted according to solvent polarity and solvent mixture used. Chloroform–methanol Soxhlet extracts, or extracts from the Bligh and Dyer or WS methods (Table 6.3) produced the most consistent and reproducible fatty acid isomer profiles in this study.

Determination of sterols in the lipid extracts showed that lower results were generally obtained when polar solvents were used for sample extraction (including WS method). Thus if the analyst needs to determine the amount of lipid in a sample and then determine the fatty acid isomer profile and sterol content, careful thought needs to be given to the extraction procedures used. An easy way to avoid problems is to saponify the sample direct for these analyses which avoids errors associated with possible incomplete and unrepresentative solvent extraction.

6.3.2.2 Column fat extraction procedures

Maxwell *et al.*[38] developed a dry column procedure for the extraction of total lipid from meat tissues. In outline, the ground meat sample is mixed in a mortar with anhydrous sodium sulphate and Celite 545. The mixture is then packed into a glass column and the fat extracted by passing a mixture of dichloromethane–methanol (9:1 v/v) through the packed column. Solvent is removed from the eluant by evaporation and the fat determined gravimetrically. A determination takes 2·5 h or less and has been shown to give higher results than ether extraction particularly for samples containing complex lipids. The total fat content of meat samples and the phospholipid content (via phosphorus determination) of the extracted fat compared very well with results obtained from a modified Folch procedure.

The dry column procedure was modified to enable the quantitative extraction and simultaneous class separation of lipids from muscle tissue.[39] Various elution mixtures were evaluated to minimise non-lipid co-elution and optimise the separation of neutral lipid from polar lipid. Sequential elutions using first dichloromethane to elute neutral lipids followed by dichloromethane–methanol (9:1 v/v) to elute polar lipids is the resulting procedure.

Maxwell and colleagues[40] later used the dry column method to isolate peanut lipids and separate polar from neutral ('crude') lipids. The efficacy of the method for lipid quantitation and separation was evaluated and results compared with those obtained by traditional methods. Total lipid was obtained from the peanut/sodium sulphate/Celite column by isocratic elution with dichloromethane–methanol (9:1 v/v). Neutral lipids were obtained by elution with dichloromethane only, then the 9:1 solvent mixture could be used sequentially to elute the polar lipids. Total lipid results obtained by the column procedure were slightly higher than those obtained by a standard Soxhlet procedure (AOCS method[41]) because of more complete extraction of polar lipids. The gravimetrically determined sum of polar and neutral lipids by the sequential column procedure was in good agreement with the non-sequentially obtained total lipid values (e.g. 47·45% versus 47·51% for whole peanuts). TLC analysis confirmed that pressed peanut oil had almost an identical composition to the neutral lipid fraction obtained by the column procedure. Phosphorus determination indicated that 2·7% of the total lipid obtained from the dry column procedure was phospholipid whereas the Folch extraction contained only 0·58%. Maxwell *et al.* conclude that this is due to the fact that the Folch procedure was designed for brain tissue which has a higher moisture content, and that the method is not suitable for the isolation of total lipid from peanuts.

Maxwell later described the use of the dry column procedure for the extraction of total fat in canned petfoods.[42] The point is made that the AOAC Soxhlet procedure (method 7·056 13th edn) only determines 'crude' fat content or ether-extractable fat which may be misleading as an indicator of fat content because the potentially large amount of phospholipid present in such products is not accounted for. Again Maxwell describes the determination of neutral ('crude') fat and total fat and states that no non-lipid artifacts were found in the extracted total fat. Reference is made to the fact that small but inconsistent amounts of phospholipid are co-extracted with neutral (crude) fat when diethyl ether is used in the Soxhlet procedure although this does not occur if light petroleum is used.[43] Total fat as determined by the column procedure gave significantly higher results than the AOAC Soxhlet method because of complete extraction of polar lipids.

Zubillaga and Maerker[44] used the dry column procedure of Marmer and Maxwell[39] to extract total lipid and lipid sub-classes from cured meats and were then performing experiments on the extracted fat to gain information on the mechanism by which sodium nitrite inhibits oxidation of meat fats during storage. They found that the polar lipid fractions obtained from the column procedure contained polar and pigmented contaminants that interfered with subsequent experiments; they therefore developed three modifications to the Marmer and Maxwell procedure to overcome this problem. Several additives to the column trap were investigated for their ability to enhance purity retention. Aluminium oxide and Florisil were unacceptable because they retained some polar lipids (particularly phosphatidyl ethanolamine) but activated carbon and magnesium oxide were both satisfactory. The latter was preferred because it permits visual observation of the progress of pigments through the column. A second modification involved the insertion of a mixture of anhydrous sodium sulphate and

Celite 545 (9:1) between the sample and the standard column trap. This was designed to absorb any residual moisture remaining in the eluting polar lipid and to decrease the rate of elution of impurities. A third modification involved a change in the composition of eluting solvent. TLC was used to examine the column eluates and evaluate each modification. All three modifications were effective but the use of the magnesium oxide trap was favoured because of convenience, and lipid fractions thus obtained appeared to be the most pure.

More recently Maxwell has reviewed the use of the dry-column procedure[45] and makes reference to its application to the extraction of lipid from legumes and milk.

6.3.2.3 Supercritical fluid extraction of fat

Since the advent of the use of supercritical fluids there has been an interest in the use of supercritical carbon dioxide for the extraction of fats. It has been reported by several workers[46–48] that the extraction of triglyceride mixtures is easily accomplished, but much of the published material relates to non-analytical applications such as the removal of oil from vegetable seeds, spices and animal muscles to aid food processing. Carbon dioxide is the most commonly used medium for this method of extraction as it is readily transformed into a supercritical state at relatively low pressures. At the critical temperature of a substance, the vapour and liquid phases have identical densities. A gas cannot be liquefied when it is above its critical temperature, irrespective of the pressure it is subjected to. Beyond the critical temperature and pressure the substance is said to be a supercritical fluid. Supercritical fluids possess certain physical characteristics that are intermediate between those of the gas and liquid states for example, the solubility of an analyte is related to solvent density. Thus, any given analyte has increased solubility in a supercritical fluid relative to the gas phase at the same temperature and similar to the liquid phase at the same temperature.

King *et al.*[49] recently reported an analytical method for the extraction of fat tissue from meat products using supercritical carbon dioxide. They conducted experiments at 37–70 MPa and 80°C on samples with fat contents from 2 to 35% by weight and reported that greater than 96% of the theoretical fat content could be removed. They concluded that, although in its infancy as an analytical extraction medium, supercritical carbon dioxide is an effective agent for the quantitative removal of fat from meat matrices.

Two advantages of supercritical fluids for solvent extraction are the ease with which the solvent is removed after extraction and the inert nature of the extraction medium. It is expected that the use of supercritical fluid extraction will attract much attention in the analytical community in the near future.

6.4. PHYSICAL METHODS FOR THE DETERMINATION OF FAT

6.4.1 Near Infra-Red Reflectance (NIR)

The near infra-red analysis of foodstuffs has been developed over the past 20–25 years into a routine, rapid and non-destructive technique for determining many constituents including fat, moisture, carbohydrate and protein. The technique is particularly suitable for process control in food manufacture. The near infra-red region of the electromagnetic spectrum is generally regarded as being from 750 to 2500 nm and spectroscopic analysis of absorptions within this area are correlated to an analytical factor of the foodstuff.

Most absorptions in the NIR region are thought to be overtones of absorption bands in the fundamental infra-red region of the electromagnetic spectrum, i.e. from 2500 to 25 000 nm. These bands are brought about by the vibration of atoms within molecules whilst those in the NIR region are 'echoes' or overtones of these fundamental absorptions arising from anharmonicity.

The resultant NIR spectrum of a foodstuff tends to be very complex with each absorption band being made up of combinations of symmetric and asymmetric stretching fundamentals rather than the first overtone of a single fundamental. In light of this most NIR spectroscopists use assignments of absorptions tentatively. Some NIR absorptions important to the food analyst are listed in Table 6.5 along with tentative assignments.

NIR determines the composition of a foodstuff by the measurement of diffuse reflectance (DR). DR can be described as radiation that has been tramsitted through a portion of the sample and, due to internal scattering, emerges from the illuminated surface over a wide angle. Therefore, DR is dependent on the absorption and light scattering properties of the sample.

The instrumentation used in NIR spectroscopy can vary in complexity according to type of wavelength selection, detection system and degree of sophistication of data analysis. The simplest instrument type

TABLE 6.5
NIR Absorptions and their tentative assignments

Wavelength (nm)	Constituent	Assignment
1200	Lipid	C—H
1440–1450	Water, carbohydrate	O—H
1720–1730	Lipid, protein	C—H
1780	Lipid	C—H
1940	Water	O—H
1980	Protein, water	N—H, O—H
2080–2100	Carbohydrates	O—H
2180–2190	Protein	C=O, N—H
2310–2320	Lipid	C—H
2340–2350	Lipid	C—H

uses infra-red emitting diodes and employs photodiode detection whereas others use tungsten–quartz–halogen sources and lead sulphide detectors.

The use of NIR spectroscopic data was pioneered by K. H. Norris at the US Department of Agriculture in the late 1960s and one of the first applications to fat analysis was by Ben-Gara and Norris[50] in 1968. They measured 2 mm thick samples of meat emulsions and interpreted the spectra in terms of absorption of C—H stretching vibrations combined with scatter losses. The difference in absorption between 1725 nm and 1650 nm gave high correlation with fat content. Their technique predicted fat contents within a standard error of ±2·1%.

Kruggel et al.[51] used a Technicon Infra Alyser instrument to analyse 79 samples of emulsified beef and 65 samples of ground lamb and compared fat content data with that obtained by Goldfisch extraction. They found that the standard error percentage ranged from 1·8–2·04 for beef to 2·41–2·58 for lamb. The lower correlation coefficients and higher standard error for ground samples were attributed to the effect that particle size has on reflectance measurements coupled with the inherent relative inhomogeneity of ground samples.

Kaffka and co-workers[52] analysed 60 cocoa powder samples from various sources with a fat content range of 8·92–17·2% by mass. They found good between-sample correlation and low standard error although they took no account of particle size. They stressed that as NIR is a technique based on correlations the accuracy of analytical data from calibrants are critically important and that better reproducibility and repeatability may be obtained using a wider range of data treatments.

Nadai's work[53] has proved to be very important in the development of NIR as a reliable analytical tool as he set out to measure the effect of sample thickness, sample position, temperature and homogeneity on data obtained for meat used in sausage production. He concluded that wide variations in spectral quality are obtained for thin samples highlighting the differences between a 1 mm and 2 mm slice. The optimum thickness seemed to be from 8 to 10 mm. The effects of position and temperature were less severe if controlled and maintained constant. The greatest variation was found, not unsurprisingly, in the degree of sample preparation.

Many other types of foodstuffs have been analysed for their fat content by NIR[54,55] proving its usefulness as an efficient, rapid technique in process control. Indeed, today more applications are being developed for on-line use.

6.4.2 Wide Line Nuclear Magnetic Resonance

Wide line or broad band NMR has been used to evaluate the fat content of mainly whole seeds or beans since the early 1960s when Conway[56] determined the fat content of bulk samples of corn, as well as single kernels. This technique has been applied to oilseeds in both continuous and pulse wave forms[57–60] and is discussed in detail in Chapter 2.

6.4.3 Determination of Fat Content by Titration

Bosch Serrat *et al.*[61] developed a titrimetric method for the determination of fat in liquid and solid foodstuffs. Triglycerides were hydrolysed by saponification and carboxylate anions thus produced were precipitated using a standard solution of barium chloride. Titration of the excess barium(II) with the magnesium potassium dihydrate salt of EDTA using Eriochrome Black as an indicator was used to quantify the fat content. Calculation is based on the average relative molecular mass of the triglycerides in the fat of a sample (calculated from fatty acid composition) and the indirect stoichiometric ratio between the triglyceride and the barium(II) which acts as a precipitating agent. As barium does not precipitate short chain fatty acids present in milk (i.e. butyric, caproic, caprylic) a correction factor is used in the calculation. The fat content of milk samples as determined by titration agreed well with a Majonnier type prcocedure but titrimetric results on some solid foods (almonds, peanuts, hazlenuts, sausages) were slightly lower than those obtained by a diethyl ether Soxhlet extraction. The authors state that the Soxhlet results are perhaps high due to extraction of material other than triglcyeride; relative standard deviation for the titrimetric procedure was in the range 1·04–1·53. It is clear that this type of procedure has the advantages of relative practical simplicity, flammable and toxic solvents are not required and some common interferences can be avoided, but there are a range of disadvantages. Knowledge of the fatty acid composition of each sample is required, short chain fatty acids are not precipitated and so corrections have to be made according to the proportion of short chain fatty acids present, and lipid components other than triglycerides will not be fully accounted for.

6.5. CONCLUSION

In conclusion we have tried to give the analyst an overview of the standard methods available for the extraction and determination of fat and also highlight some of the non-standard approaches. With the increasing range of foodstuffs reaching the marketplace and the increasing complexity and range of ingredients it is almost certain that no one procedure for the extraction and determination of fat will become available and universally acceptable. For these reasons new approaches and techniques for fat determination will continue to be developed and published. New food products and methods of analysis will continue to pose challenges to the analyst for the foreseeable future.

REFERENCES

1. Hannant, G., *J. Ass. Public Anal.*, **20** (1982) 117.
2. Bloor, W. R., *Proc. Soc. Exp. Biol. Med.*, **17** (1920) 138.
3. Bloor, W. R., *Chem. Rev.*, **2** (1925) 243.
4. Weibull, M., *Z. angew. Chemie* (1892) 450.
5. Weibull, M., *Z. angew. Chemie* (1894) 199.
6. Berntrop, J. C., *Z. angew. Chemie* (1902) 11.
7. Hyde, K. A. and Rothwell, J., *Ice Cream*, Churchill Livingstone, Edinburgh, 1973.
8. Rosenthal, I., Merin, U., Popel, G. and Bernstein, S., *J. Ass. Off. Anal. Chem.*, **68**(6) (1985) 1226.
9. Pearson, D., *Laboratory techniques in food analysis*, Butterworths, London, 1973.
10. Houston, J., *J. Soc. Dairy Technol.*, **8** (1955) 47.

254 *I. D. Lumley and R. K. Colwell*

11. Macdonald, F. J., *Analyst*, **84** (1959) 287, 747.
12. Marshall, R. T., *J. Dairy Sci.*, **68** (1985) 1642.
13. Packard, V., Ginn, R. E., Gulden, D. and Arnold, E., *Dairy Food Sanit.*, **6** (1986) 332.
14. Barbano, D. M., Clark, J. L. and Dunham, C., *J. Ass. Off. Anal. Chem.*, **71**(5) (1988) 898.
15. Kleyn, D. H., Trout, R. J. and Weber, M., *J. Ass. Off. Anal. Chem.*, **71**(4) (1988) 851.
16. Bloor, W. R., *J. Biol. Chem.*, **17** (1914) 377.
17. Folch, J., Ascoli, I., Lees, M., Meaths, J. A. and Le Baron, F. N., *J. Biol. Chem.*, **191** (1951) 833.
18. Folch, J., Lees, M. and Sloane Stanley, G. H., *J. Biol. Chem.*, **226** (1957) 497.
19. Bligh, E. G. and Dyer, W. J., *Can. J. Biochem. Physiol.*, **37** (1959) 911.
20. Hanson, S. W. F. and Olley, J., *Biochem. J.*, **89** (1963) 101.
21. Marinetti, G. V., *J. Lipid Res.*, **3** (1962) 1.
22. Winter, E. Z., *Z. Lebensmittelunters. u-Forsch*, **123** (1963) 205.
23. Southgate, D. A. T., *J. Sci. Food Agric.*, **22** (1971) 590.
24. Lento, H. G. and Daugherty, C. E., 94th Annual Meeting of the AOAC, Washington, DC, 1980, Abstract 80.
25. Daugherty, C. E. and Lento, H. G., *J. Ass. Off. Anal. Chem.*, **66**(4) (1983) 927.
26. Sahasrabudhe, M. R. and Smallbone, B. W., *J. Am. Oil Chem. Soc.*, **60**(4) (1983) 801.
27. Hara, A. and Radin, N. S., *Anal. Biochem.*, **90** (1978) 420.
28. Sheppard, A. J., *J. Am. Oil Chem. Soc.*, **40** (1963) 545.
29. Sahasrabudhe, M. R., *J. Am. Oil Chem. Soc.*, **56** (1979) 80.
30. *Official Methods of Analysis of the Association of Official Analytical Chemists*, 13th edn, AOAC, Washington, DC, 1980.
31. de Koning, A. J., Evans, A. A., Heydenrych, C., de Purcell, C. J. and Wessels, P. H., *J. Sci. Food Agric.*, **36** (1985) 177.
32. *Official Journal of the European Communities*, Part 4 (1971) 55–57, 71/393/EEC.
33. Khor, H. T. and Chan, S. L., *J. Am. Oil Chem. Soc.*, **62**(1) (1985) 98–9.
34. Chen, I. S., Shen, C. S. J. and Sheppard, A. J., *J. Am. Oil Chem. Soc.*, **58** (1981) 599.
35. Hara, A. and Radin, N. S., *Anal. Biochem.*, **90** (1978) 420.
36. Firth, N. L., Ross, D. A. and Thonney, M. L., *J. Ass. Off. Anal. Chem.*, **68**(6) (1985) 1228.
37. Christie, A. A., *Report of the Government Chemist*, HMSO, London, 1974, p. 28.
38. Maxwell, R. J., Marmer, W. N., Zubillaga, M. P. and Dalickas, G. A., *J. Ass. Off. Anal. Chem.*, **63**(3) (1980) 600.
39. Marmer, W. N. and Maxwell, R. J., *Lipids*, **16**(5) (1981) 365.
40. Adnan, M., Argoudelis, C. J., Tobias, J., Marmer, W. N. and Maxwell, R. J., *J. Am. Oil Chem. Soc.*, **58** (1981) 550.
41. *American Oil Chemists Society, Official and Tentative Methods*, Ab 3–49 (corrected 1984).

42. Maxwell, R. J., *J. Ass. Off. Anal. Chem.*, **67**(5) (1984) 878.
43. Hagan, S. N., Murphy, E. W. and Shelley, L. M., *J. Ass. Off. Anal. Chem.*, **50** (1967) 250.
44. Zubillaga, M. P. and Maerker, G., *J. Food Sci.*, **49** (1984) 107.
45. Maxwell, R. J., *J Ass. Off. Anal. Chem.*, **70**(1) (1987) 74.
46. Peter, S. and Brummer, G., *Angew. Chem. Int. Ed. Engl.*, **16** (1978) 746.
47. Quinn, K. W., *Fette Seifen Anstrichm.*, **84** (1982) 460.
48. Falton, M., Bulley, N. R. and Meisen, A., *J. Agric. Food Chem.*, **35** (1987) 739.
49. King, K. W., Johnson, J. H. and Friedrich, J. P., *J. Agric. Food Chem.*, **37** (1989) 951.
50. Ben-Gara, I. and Norris, K. H., *J. Food Sci.*, **33** (1968) 64.
51. Kruggel, W. G., Field, R. A., Riley, M. L., Radlof, H. D. and Norton, K. M., *J. Ass. Off. Anal. Chem.*, **64**(3) (1981) 692.
52. Kaffka, K. J., Norris, K. H., Kulcsar, F. and Draskovits, I., *Acta Alimentaria*, **11**(3) (1982) 271.
53. Nadai, B. T., *Acta Alimentaria*, **12**(2) (1983) 119.
54. Biggs, D. A. and McKenna, D., *J. Ass. Off. Anal. Chem.*, **72**(5) (1989) 724.
55. Osborne, B. G., Fearn, T. and Randall, P. G., *J. Food Technol.*, **18** (1983) 651.
56. Conway, T. F., *Proceedings of Symposium on High Oil Corn*, College of Agriculture, University of Illinois–Urbana, 1961, 29–32.
57. Conway, T. F. and Earle, F. R., *J. Am. Oil Chem. Soc.*, **40** (1963) 265.
58. Collins, F. I., Alexander, D. E., Crodgers, R. and Silvera, L., *J. Am. Oil Chem. Soc.*, **44** (1967) 708.
59. Robertson, W. J. A. and Morrison, H., *J. Am. Oil Chem. Soc.*, **56** (1979) 961.
60. Robertson, J. A. and Windham, W. R., *J. Am. Oil Chem. Soc.*, **58** (1981) 993.
61. Bosch Serrat, F., Zuriaga Cosin, M. F. and Chamorro Pascual, P. R., *Analyst*, **114**(4) (1989) 485.

APPENDIX: STANDARD METHODS PRESCRIBED FOR THE EXTRACTION AND DETERMINATION OF FAT IN FOODS

Standard and issuing body	Digestion/ solubil- isation medium	Extracting solvent	Method types
Foods			
AOAC 43.275 Lipid in foods	Enzyme	Chloroform +methanol	Bligh and Dyer
Meat and meat products			
Total fat			
ISO 1443 1973	HCl	Hexane	WB
BS 4401: Part 4 Method A2:1970		or LP	
BS 4401:Part 4 Method B:1970	HCl	Hexane or LP	WS
Free fat			
ISO 1444 1973	None	Hexane	Soxhlet
BS 4401:Part 5:1970		or LP	
AOAC 24.005	None	DEE	Soxhlet
AOAC 24.006 Meat, rapid method	None	Tetra- chloro- ethylene	Foss–Let fat analyser
Milk and dairy products			
ISO 2446 1976, Liquid-milk (routine method)	H_2SO_4	None	Gerber
BS 696:Part 1 and Part 2:1989 Milk and milk products, Gerber method	H_2SO_4	None	Gerber
ISO 3433-1975 Cheese (routine method)	H_2SO_4	None	Van Gulik butyrometer
AOAC 16.065, Liquid milk, Babcock method	H_2SO_4	None	Babcock
AOAC 16.177, Cream, Babcock method	H_2SO_4	None	Babcock
ISO 1211 1984, Liquid milk Gravimetric reference method	Ammonia Ethanol	DEE + LP	Rose–Gottlieb
IDF 1C:1987 Milk	Ammonia Ethanol	DEE + LP	Rose–Gottlieb
BS 1741:Part 3 Liquid milk and cream	Ammonia Ethanol	DEE +	Rose–Gottlieb
ISO 1736 1985, Dried milk Reference method	Ammonia Ethanol	DEE + LP	Rose–Gottlieb
IDF 9C:1987 Dried milk, whey, buttermilk and dried butter serum, reference method	Ammonia Ethanol	DEE + LP	Rose–Gottlieb
AACC 30–16 Dry milk products reference method	Ammonia Ethanol	DEE + LP	Rose–Gottlieb
ISO 1737–1985 Evaporated Milk and sweetened condensed milk, reference method	Ammonia Ethanol	DEE + LP	Rose–Gottlieb
IDF 13C:1987	Ethanol	DEE + LP	Rose–Gottlieb
AOAC 16.192 Evaporated unsweetened			
AOAC 16.205 Evaporated sweetened			
ISO 7208–1984 Skimmed milk whey and buttermilk, reference method	Ammonia Ethanol	DEE + LP	Rose–Gottlieb

Standard and issuing body	Digestion/solubilisation medium	Extracting solvent	Method types
IDF 22B:1987	Ethanol	DEE + LP	Rose–Gottlieb
ISO 2450–1972 Cream, reference method	Ammonia	DEE +	Rose–Gottlieb
IDF 16C:1987	Ammonia	DEE +	Rose–Gottlieb
AOAC 16.176	Ammonia	DEE +	Rose–Gottlieb
ISO 1735–1975 Cheese and processed cheese products, reference method	HCl	DEE +	SBR
BS 770:Part 3:1976	HCl	DEE +	SBR
IDF 5B:1986	HCl	DEE +	SBR
AOAC 16.284	HCl	DEE +	SBR
ISO 1854–1972 Whey Cheese reference method	Ammonia Ethanol	DEE + LP	Rose–Gottlieb
IDF 59:1970	Ethanol	LP	Rose–Gottlieb
AOAC 16.285	Ethanol	LP	Rose–Gottlieb
ISO 7328–1984 Milk based edible ice and ice mixes	Ammonia Ethanol	DEE + LP	Rose–Gottlieb
AOAC 16.316	Ethanol	LP	Rose–Gottlieb
ISO 8262/1 1987 Infant Food IDF 124:1985	HCl	Hexane or LP	WB
ISO 8262/2 1987 Edible ices and ice mixes IDF 125:1985	HCl	Hexane or LP	WB
ISO 8262/3 1987 Milk and milk products, special cases IDF 126:1985	HCl	Hexane or LP	WB
ISO/5543 1986 Caseins and caseinates	HCl	LP	WB
IDF 127:1985	HCl Ethanol	DEE + LP	SBR
BS 6248:Part 10:1986	Ethanol Ethanol	LP LP	SBR SBR
ISO/DP 9622 Whole milk, mid-infra-red instruments	None	None	Physical-IR absorption
IDF 141:1988	None	None	IR absorption
AOAC 16.083	None	None	IR absorption
ISO 8381 1987 Milk based infant foods, very low starch content	Ammonia	DEE + LP	Rose–Gottlieb
IDF 123:1985	Ammonia	LP	Rose–Gottlieb
AOAC 16.233 Butter Indirect method % fat = 100 − (% water + % residue)			
AOAC 16.233 Butter Solvent extraction			

Oil seeds and feeds

ISO 0659–1979 Oleaginous seeds, det" of oil content BS 4289:Part 4:1982 IUPAC 1.122	None	Hexane or LP	Soxhlet
ISO 0734 1979 Oilseed residues, oil content BS 4325:Part 4:1982	None	Hexane or LP	Soxhlet
ISO 0736 1977 Oilseed residue detn of diethylether extract	None	DEE	Soxhlet

Standard and issuing body		Digestion/ solubil- isation medium	Extracting solvent	Method types
ISO 5511 1984 Oilseeds oil content by low resolution NMR IUPAC 1.123		None	None	Physical- NMR
ISO/DP 6492 Feedingstuffs (not oilseed or oilseed residues)				
	Method A	None	Hexane	Soxhlet
	Method B	HCl	Hexane	WB
EC Directive 84/4/EEC				
	Method A	None	LP	Soxhlet
	Method B	HCl	LP	WB
AOAC 7.060 Animal feed, grain, stock feeds		None	DEE	Soxhlet
AACC 30–20		None	DEE	Soxhlet
AOAC 14.066 Wheat, rye, oats, corn, rice, buckwheat, barley, soybeans and products refers to AOAC 7.062		None	DEE	Soxhlet
Cereal and cereal products				
ISO DIS 7302 1982 Cereals and cereal products ICC Standard No 136		Formic and hydro- chloric acids	Ethanol + hexane	See text
AOAC 14.019 Flour		HCl	DEE	SBR
AOAC 14.104 Bread		HCl	DEE	SBR
AACC 30–10 Flour, bread and baked cereal products		HCl	DEE	SBR
Miscellaneous AOAC and AACC standards				
AOAC 13.031 Cacao products		None	LP	Knorr tube
AOAC 13.032–033 Cacao products (chocolate)		None	LP	WB
AACC 30–12A Cocoa		None	LP	WB
AOAC 14.126 Fig bars, raisin crackers		HCl	LP	SBR
AOAC Macaroni products		HCl	LP	SBR
AOAC 7.063 Baked pet foods		HCl	LP	SBR
AACC 30–14 Baked dog food		HCl	DEE	SBR
AOAC 17.012 Liquid/dried eggs		HCl	DEE/LP	SBR
AACC 30–18 Egg yolk/dried egg		HCl	DEE/LP	SBR
AOAC 27.006 Nuts and nut products (refers to AOAC 7.062)		None	DEE	Soxhlet
AOAC 18.043 Seafood		HCl	LP	SBR
AOAC 18.045 Seafood, rapid method		Acetic/ perchloric	None	Babcock
14.028 Lipid in flour		None	Ethanol + DEE, re-extract in chloro- form	
14.139 Lipid in macaroni products		None	None	
17.014 Lipid in eggs		None	Chloroform + ethanol	

Standard organisations
ISO	International Organization for Standardization
BSI	British Standards Institution (UK)
AOAC	Association of Official Analytical Chemists (USA)
IDF	International Dairy Federation
AACC	American Association of Cereal Chemists
IUPAC	International Union of Pure and Applied Chemistry
ICC	International Association of Cereal Scientists and Technologists

Method types: abbreviations
SBR	Schmid–Bondzynski–Ratzlaff
WS	Werner–Schmid
WB =	Weibull–Berntrop

Extracting solvents: abbreviations
DEE	Diethyl ether
LP	Light petroleum

Notes
1. Grouped methods are technically similar but not necessarily exactly the same in all practical detail.
2. The 'field of application' clause of the Standard procedures should be carefully read to ensure suitability for a particular application.
3. All references to AOAC methods refer to the *Official Methods of Analysis of the Association of Official Analytical Chemists*, 14th edn, 1984.

7

Vegetable Oils and Fats

J. B. ROSSELL

Oils and Fats Section, Leatherhead Food R. A., Randalls Road, Leatherhead, Surrey KT22 7RY, UK

7.1. INTRODUCTION

The contrasts between the compositions of vegetable oils and animal fats, mammalian butters and fish oils have been highlighted in recent years by emphasis on the role of fats in nutrition and disease. Thus, in 1982, S. Bergstrom and B. I. Samuelson were two of the three recipients of the Nobel Prize for medicine, in recognition of their work in elucidating the structure of prostaglandin and its metabolic formation from arachidonic acid. Arachidonic acid is in turn a metabolite of linoleic acid, which is now termed an essential fatty acid (EFA). The richest common sources of linoleic acid are the vegetable oils derived from sunflower seeds, soyabeans, maize germs and safflower seeds. The third of the three 1982 Nobel Prize laureates, J. R. Vane, underscored this work on EFA metabolism by relating prostaglandins to pain and disease. In consequence of the nutritional importance of EFAs, the '*n-*' or EFA nomenclature has been introduced, in which it is presumed that, unless stated otherwise, the double bonds are situated along the chain in the methylene interrupted or *cis,cis*-1,4-diene structure found in most vegetable oils, and the location of the double bonds is counted from the methyl end of the chain. This is contrary to the conventional IUPAC system, but enables metabolic relationships to be more clearly demonstrated.[1] Thus, normal linoleic acid is *cis,cis*-9,12-octadecadienoic acid, or C18:2-9,12-cc, under the IUPAC system, but C18:2*n*-6 in the EFA nomenclature. This latter system will be used most frequently in this chapter, in which analyses for oil quality and oil composition, including fatty acid composition, will be discussed.

Vegetable oils also contrast with animal carcass fats, mammalian butters and fish oils, in that they have fewer component acids and a simpler triglyceride composition. Thus vegetable oils seldom contain either branched chains or odd-numbered fatty acids, or unsaturated fatty acids with fewer than 16 or more than 20 carbon atoms, but animal fats and fish oils frequently include these acids. Likewise, the triglyceride compositions of vegetable oils and fats generally follow a pattern in which the fatty acids at the central or 2-position of the glyceride molecule are unsaturated, with linoleic acid (C18:2n-6) being favoured more than oleic (C18:1n-9) or linolenic (C18:3n-3) acids. Saturated acids are found at the 2-position only when there is a very high overall saturated fatty acid concentration in the fat. Some hard tropical fats with sharp melting points, the prime example of which is cocoa butter, have very high levels of saturated acids in the overall fatty acid composition, but very little at the triglyceride 2-position. This gives the fats very simple triglyceride structures, and is the chemical reason for the physical phenomenon of their sharp melting behaviour. This pattern gives simpler triglyceride compositions to vegetable oils and fats, in contrast to animal fats, which often contain appreciable concentrations of saturated acids at the 2-position.

It is perhaps relevant to remark here that most fatty acids derive their names from plant species; thus it is easy to remember the major acids in the oils from palm (palmitic acid), olive (oleic), linseed (linoleic and linolenic), castor *Ricinus communis* bean (ricinoleic), parsley *Petroselinium sativum* seed (petroselenic), and nutmeg *Myristica officinalis* (myristic). In other cases acids have been named after the vegetable oil in which they were first found, e.g. *Sterculia foetida* (sterculic).

Another contrast of animal and vegetable oils and fats is in the presence of minor components such as tocopherols and sterols. In general, animal fats are almost free from tocopherols, whilst many vegetable oils contain up to 1000 mg/kg. Animal fats contain high concentrations of the zoo-sterol cholesterol, namely up to 1000 and 7000 mg/kg, or more, for animal fats and fish oils, respectively. Vegetable oils contain at most 100 mg/kg, and usually less than 25 mg/kg cholesterol, but instead contain higher concentrations of phyto-sterols like sitosterol.

Although vegetable oils may have relatively straightforward compositions, they vary so much from one to another that the group as a whole encompasses a vast range of chemical and physical properties. Not surprisingly, it is those that have commercial importance, especially for food use, that have been cultivated and exploited, and are consequently the best known.

Coconut and palm kernel oils are highly saturated fats, solid at cool ambient temperatures, and ideally suited for food manufacture; palm oil is of intermediate composition, whilst olive and groundnut oils are sufficiently unsaturated to remain fluid, but not so unsaturated that they develop rancid off-flavours and have short shelf-lives. Cottonseed, maize germ, rapeseed, safflowerseed, soyabean and sunflower seed oils are all liquid at room temperature, and all contain high levels of polyunsaturated 18 carbon EFAs. Most of the polyunsaturated oils have limited shelf-lives in manufactured foods, and they are, in consequence, frequently partially hydrogenated.[2] This process converts the oil into a semi-solid fat, the physical properties depending on the nature and extent of hydrogenation. Hydrogenation also improves the resistance to oxidation, flavour stability and shelf-life of the oil.

Other vegetable oils are often used for industrial purposes. Linseed oil has a very high degree of unsaturation, which enables the oil to combine with atmospheric oxygen and form a film of oxidised and polymerised material ideally suited for application in the paint, varnish and linoleum industries. Castor bean oil with its high content of ricinoleic (12-hydroxy-octadeca-9-enoic) acid, and tung oil, which contains 65–80% of eleostearic (octadeca-9,11,13-trienoic) acid are also useful in these industrial fields.

Vegetable oils and fats are also produced in largest quantity, as shown in Table 7.1, where it can be seen that in 1988 a total of over 54 million tonnes (Mt) of vegetable oils were produced in comparison with about 18 Mt of animal fats and 1.5 Mt of fish oils.[3] Table 7.1 also illustrates the fact that the largest tonnage vegetable oil is soyabean, at over 15 Mt/year, but palm oil is growing the most rapidly, followed by rapeseed oil. Production of sunflower seed oil is also increasing, although not as rapidly as previously, while that of olive oil, and the two industrial oils, linseed and tung, has fallen.

For most uses the suitability of an oil depends as much on its quality as on its composition, and the first section of this chapter will therefore review analytical tests for quality. Oil type and purity are normally assessed by reference to the fatty acid composition (FAC), which also can often identify the relative proportions of different oils in a blend.

TABLE 7.1
World production—types of oils and fats ('000 tonnes; in terms of oil or fat)

	1974/75	1981/82	1982/83	1983/84	1984/85	1985/86	1987[c]	1988[c]
Liquid oils								
Cottonseed	2 930	3 280	3 025	3 040	3 810	3 365	3 213	3 645
Groundnut	3 130	3 455	2 900	3 190	3 395	3 350	3 460	3 603
Soyabean	8 630	13 885	15 255	13 540	15 110	15 675	15 473	15 387
Sunflower	3 840	5 250	6 000	5 530	6 540	7 190	7 022	7 586
Olive	1 545	1 495	2 025	1 575	1 720	1 595	1 738	1 877
Sesame	680	730	620	700	670	730	603	612
Rapeseed	2 385	4 040	4 905	4 700	5 750	6 370	7 434	7 822
Maize	365	695	745	785	895	930	1 248	1 293
Safflower	230	260	245	260	245	270	—	—
Total	23 735	33 090	35 720	33 320	38 135	39 475	40 191	41 825
Palm-type oils								
Coconut	2 485	2 990	2 860	2 435	2 555	3 325	3 080	2 587
Palm Kernel	480	700	730	760	905	1 060	1 020	1 095
Palm	2 620	5 410	5 410	5 645	6 505	7 825	7 883	8 734
Total	5 585	9 100	9 000	8 840	9 965	12 210	11 983	12 416
Industrial-type oils								
Linseed	745	670	815	670	715	780	802	701
Castor bean	425	350	350	375	400	435	365	334
Tung	115	95	100	95	90	95	—	—
Total	1 285	1 115	1 265	1 140	1 205	1 310	1 167	1 035
Animal fats								
Butter/ghee	5 240	5 705	6 200	6 255	6 110	6 260	6 146	6 114
Lard	4 235	5 050	5 130	5 185	5 350	5 430	5 251	5 445
Tallow	5 415	6 200	6 100	6 110	6 255	6 270	6 437	6 735
Total	14 890	16 955	17 430	17 550	17 715	17 960	17 834	18 294
Fish oil	1 175[a]	1 300	1 090	1 265	1 460	1 520	1 459	1 466
Minor/ unspecified oils[b]	725	945	980	1 030	1 025	1 050	—	—
World total	47 495	62 505	65 485	63 145	69 505	73 525	72 633	75 037

[a] Includes 130 000 t whale (including sperm whale) oil.
[b] Includes rice bran, mustard, babassu, oiticica, mowrah, hempseed, perilla seed and other minor oils.
[c] Taken on a different basis (viz. from Oil World) to figures for 1974–1986, giving slightly lower values in several cases.
Source: International Association of Seed Crushers[3].

In some cases, however, determination of the FAC alone is not sufficient, and recourse to more sophisticated tests such as analysis of the acids at the 2-position of the triglyceride molecule, or measurements of the tocopherol and sterol compositions of the oil may be appropriate.

When analysing fatty foods it is usual first to measure the fat content of the food, especially when it is necessary to comply with the requirements of fat labelling legislation, and the methods discussed in Chapter 6 are then relevant. This will often involve separation of the oil or fat, after which its properties can be determined by the techniques discussed below.

7.2. TESTS FOR OIL QUALITY (AND IDENTITY)

High quality unrefined, or 'virgin', vegetable oils, produced from fresh raw materials, can have an attractive flavour and be immediately used as food without further processing. This is the case, for instance, with pure primed pressed (PPP) cocoa butter, virgin olive oil, or some cold pressed seed oils. These oils are costly however, as they are obtained from selected raw materials and in relatively low yield in comparison with yields from modern industrial procedures. Most commerical vegetable oils, apart from soya, which is usually all extracted, are therefore obtained by processes of first crushing the seed to remove a proportion of the oil (expelled oil) and then extracting the remaining oil by application of volatile solvents (usually industrial hexane). The solvent is removed by distillation to give the remainder of the oil as extracted oil. The expelled oil is often obtained in higher yield, and is often of higher quality than the extracted oil, for example, it has a lower acidity, phospholipid and pigment content. With some seeds there can be additional differences and with undecorticated sunflower seed for example, the expelled oil also has a lower wax content. Expelled and extracted oils are nevertheless blended in the same proportions as yielded by the separate processes to give normal crude oil. However, blends containing higher than normal proportions of expelled oil may be selected for a specific purpose, the surplus extracted oil being sold to unsuspecting third parties. Furthermore, unsatisfactory post-harvest husbandry of the seeds, and poor or extended storage of the crude oil, can lead to a deterioration in oil quality. On the other hand, the increased use of physical or steam refining, in contrast to the more traditional caustic refining, calls for crude oils of better and more uniform quality.[4,5] A variety of tests are

TABLE 7.2
Typical analytical properties of some crude vegetable oils and fats

Oil or fat	Refractive index[a] n_D^{40}	Saponification value	Unsaponifiable matter (%)	Titre (°C)	Melting point (°C)	Iodine value	Codex Alimentarius Standard
Aceituno	1·4595[b]	190–198	ca. 0·5	—	ca. 32	51–60	—
Almond (sweet) kernel	1·4620–1·4656	188–196	0·4–1·0	—	–10[b]	95–100	—
Apricot kernel	1·4646[b]	192–198	—	—	—	107[b]	—
Babassu kernel	1·448–1·451[c]	245–256[c]	0·2–0·9	22–23	24–26	10–18[c]	128–1981
Blackcurrant seed	1·479–1·481 (20°C)	185–195	0–1·0	—	—	173–182	—
Borneo tallow (Illipe)	1·4560–1·4572	189–200	0·7–2·0	51–53	37–39	31–38	—
Brazil nut	1·4580–1·4620	192–202	below 1·0	27–34	–2[b]	97–106	—
Cashew nut	1·4613–1·4626	193–196	0·4–1·5	28–30	—	79–84	—
Castor bean	1·466–1·473	176–187	0–1·0	—	–12 to –10	81–91	—
Chinese vegetable tallow	1·455–1·457	200–218	0·5–1·5	45–55	42–45	16–30	—
Cocoa butter	1·456–1·459[c]	188–198[c]	Not more than 0·5[c]	45–50	31–35[c]	33–42	86–1976
Coconut	1·448–1·450[c]	248–265[c]	0–0·5	20–24	23–26	6–11[c]	124–1981
Corn (maize)	1·465–1·463[c]	187–195[c]	0·5–2·8	14–20	–12 to –10	103–128[c]	25–1981
Cottonseed	1·458–1·465[c]	189–198[c]	0·5–1·5	30–37	–2 to 2	99–119[c]	22–1981
Crambe seed[b]	1·472	—	c. 0·5	—	—	93	—
Cuphea sp.	—	—	—	—	—	5–20	—
Dhupa fat	1·4576–1·4590	187–192	1·2–2·5	—	30–40	36–41	—
Evening primrose seed[b]	1·4791 (20°C)	192·8	1·5–2·0	—	—	155	—
Grapeseed	1·473–1·477[c]	188–194[c]	0·2–2·0	–17	–10	130–138[c]	127–1981
Groundnut	1·460–1·465[c]	187–196[c]	0·2–0·8	26–32	–2[b]	80–106[c]	21–1981
Hazelnut	1·4559–1·4633	189–198	—	19–20	—	81–92	—
Illipe[d] (see borneo tallow)							
Illipe bassia- or Indian-illipe:	1·459–1·462	186–200	1·4–2·3	—	25–29	50–60	—

Jojoba[b]	1·4590	92[b]	51[h]	—	6·8–7·0	83·3	—
Kapok seed	1·460–1·466	189–195	0·5–1·0	27–32	c. 30	90–110	—
Kokum butter[b]	1·456	192	c. 2·3	60	40–42	33–36	—
Linseed	1·474–1·475	188–196	0·1–1·7	19–21	−20 to −24	155–205	—
Lupin seed (lupinus alba)[b]	—	—	2·4	—	—	106	—
(lupinus mutabilis)[b]	1·4670	188	1·0	—	—	114	—
Mango kernel[b]	1·4598–1·4602	188–195	0·5–2·9	42–50	25	39–48	—
Mowrah	1·458–1·461	188–192	1·0–3·0	—	25–31	50–70	—
Mustard seed	1·461–1·469[c]	170–184[c]	0·7–1·5	6–8	−16	92–125[c]	34–1981
Neem[b]	1·4615 (15°C)	194·5	—	26–30	−3	71	—
Niger seed	1·4659–1·4688	188–192	0·5–1·0	—	—	128–134	—
Nutmeg butter[f]	1·4659–1·4704	170–190	—	34–45	40–50	48–85	—
Olive[e]	1·4677–1·4705[e]	184–196[c]	0·7–2·5[c] (up to 1·5[e])	17–26	−3 to 0	75–94[c]	33–1981[e]
Palm (fruit)	1·449–1·455 (50°C)[c]	190–209[c]	0·3–1·2	40–47	33–40	50–55[c]	125–1981
Palm (kernel)	1·448–1·452[c]	230–254[c]	0·2–0·8	20–28	24–26	14·5–19[g]	126–1981
Pecan	1·4742–1·4760	189–193	1·0–1·5	—	−5[b]	193–205	—
Poppy seed	1·4678–1·4688	192–196	0·4–0·6	15–19	−20 to −15	130–138	—
Pumpkin seed	1·466–1·469	185–198	0·6–1·5	11–15	c. −15	117–130	—
Rapeseed							
High erucic	1·465–1·469[c]	168–181[c]	0·2–2·0	20–22	−9	94–120[c]	24–1981
Low erucic	1·465–1·467[c]	188–193[c]	0·2–1·8	—	−20	110–126[c]	123–1981
Rice bran	1·466–1·471	179–195	3–7	25[b]	0–8	85–109	—
Safflower seed	1·467–1·470	186–198[c]	0·3–1·5	15–18	−18 to −13	135–150[c]	27–1981
Sal fat	1·456–1·457	186–194	—	51	30–35	38–44	—
Sesame seed	1·465–1·469[c]	187–195[c]	0·9–2·0	20–25	−4 to 0	104–120[c]	26–1981
Shea nut	1·463–1·467	178–190	4–8	49–54	32–42	52–67	—
Soya bean	1·467–1·470[c]	188–195[c]	0·5–1·6	20–21	−23 to −20	120–143[c]	20–1981
Stillingia (kernel)	1·482–1·484 (25°C)	200–212	0·5–3·0	4–12	—	169–190	—

J. B. Rossell

TABLE 7.2—contd.

Oil or fat	Refractive index n_D^{40} [a]	Saponification value	Unsaponifiable matter (%)	Titre (°C)	Melting point (°C)	Iodine value	Codex Alimentarius Standard
Sunflower seed	1·467–1·469[c]	188–194[c]	0·3–1·3	16–20	−18 to −16	110–143[c]	23–1981
Tea seed	1·462–1·464	190–195	0·1–1·0	13–14·5	−9 to −5	80–89	—
Tobacco ssed	1·4678–1·4717	187–200	below 1·5	—	−14 to −16	119–146	—
Tomato seed	1·4665–1·4682	180–192	below 1·2	—	−9 to −12	110–125	—
Tucum kernel	1·4495–1·4506	240–246	c. 0·8	26·2	30–33	12–14	—
Tung	1·503–1·515	189–195	0·4–1·0	36–37	4[b]	170–187	—
Walnut	1·469–1·471	189–198	0·5–1·0	14–16	−16 to −12	138–148	—
Wheat germ	1·470–1·480	185–192	2–5	—	c. 0°C	115–125	—
Winged bean	1·4625–1·4635	186–188	0·3–1·1	—	—	80–91	—

[a] Data at 40°C unless otherwise stated.
[b] Typical specimen.
[c] Figures given in the Codex Alimentarius Standard cited. Other values are from laboratory records and general literature.
[d] Bassia or Indian Illipe is known in botanical circles as 'true illipe'. Borneo Tallow Illipe, from the shorea species, is the illipe of commerce, used in cocoa butter type fats.
[e] Codex figures for virgin olive oil.
[f] Nutmeg butter has a variable composition depending on method of preparation. Expressed fat has saponification value of about 200 and iodine value of about 35·7.
[g] Codex figures as amended February 1987 (Alinorm 87/17, Appendix X).
[h] Jojoba 'oil' is a chemical wax, and the unsaponifiables include the long chain alcohols liberated by saponification.

therefore used to measure crude oil quality. Some of these are traditional tests of longstanding, many of which have been recently reviewed.[6] These traditional tests may also be used as identity or purity criteria, as in the Codex Alimentarius Oils and Fats Standards, data from which, along with other values from laboratory records and the scientific literature,[7–12] are presented in Table 7.2. These fairly broad ranges are normal for agricultural products, and are inter-related and/or dependent on the fatty acid composition. Thus, the refractive index varies in line with the iodine value, and the proportion of unsaturated fatty acids in the oil. An estimate of the iodine value can in fact be calculated from the fatty acid composition by the AOCS method, Tz 1c-85. The saponification value, on the other hand, is related to the mean molecular weight of the constituent fatty acids.[6] The titre, in quantifying the melting point of the fatty acids derived from the fat, bears a relationship to the melting point of the parent fat. Despite these inter-relationships each criterion has its own value, and all are specified in standards and trade contracts.

The resistance of a fat to oxidation, and its oxidative condition at the time of production and use, are also very important. Tests relating to these aspects have been reviewed.[13–15] The importance of these tests with refined oils is emphasised in Chapter 10. Nevertheless, the demands of modern processing mentioned above require stringent quality assessments, and a number of trading rules and guidelines have been developed to assist the trade in judging which oils are of Good Merchantable Quality (GMQ). Notwithstanding the importance and usefulness of these trading rules, they are inadequately discussed in the scientific literature. A number of typical examples are therefore presented in the following tables.

Table 7.3 summarises the trading rules of the Netherlands Oils, Fats and Oilseeds Trade Association (NOFOTA), while Tables 7.4 and 7.5 give synopses of Brazilian and North American trading rules, the former refering to FOSFA International Contracts. Table 7.6 presents a synopsis of the National Institute of Oilseed Products (NIOP) guideline specification for oils and fats, which reflects current in-dustrial practice in the USA. Table 7.7 presents data from Malaysian National Standards for palm oil, stearin and olein; and palm kernel oils. FOSFA International includes certain criteria in its contracts, and has published guideline specifications[16] that relate mainly to the authenticity of crude oil, as discussed later.

Comparison of Table 7.2 with Tables 7.3 and 7.7, illustrates the fact

TABLE 7.3
Synopsis of NOFOTA trading rules (the actual trading rules should be consulted for trading purposes)

	Crude degummed soyabean oil	Crude degummed rapeseed oil	Crude rapeseed oil
FFA (oleic) (%)	0·75 basis 1·25 max.	1·75 max.	2·0 max.
Moisture and volatiles (%)	0·20 basis 0·25 max.	} 0·4 max.[a]	} 0·5 max.[a]
Impurities (%)	0·10 basis 0·125 max.		
Phosphorus (%)	0·02 basis 0·025 max.	0·03 max.	0·075 max.
Gardner Break Test, sediment (%)[b]	0·10 max.		
Colour, Lovibond 1-inch cell	50 Y max. 5·0 R max.		
Flash point (°F) (min.)	250	250	250
(°C) (min.)	121	121	121
Erucic acid (%)	—	5·0 max.	5·0 max.

[a] Moisture, volatile matter and impurities combined.
[b] Sediment (Gardner Break Test) by AOCS Ca. 10–40.

that Codex identity characteristics relate mainly to oil identity and purity, while trade criteria relate much more closely to oil quality. A number of the official bodies, federations and associations that issue contracts, trading rules, guidelines or standards are listed in the Appendix. Several of the analytical tests specified in Tables 7.3–7.6 relate to the quality control of refined oils, and are discussed in Chapter 10. Others are described fully in the reviews mentioned above,[2,6,8,9,13] but some other, less common, tests deserve more detailed consideration here.

7.2.1 Lecithin (Phosphorus)

Crude oils contain phospholipids that were extracted with the oil from the protoplasm of the animal or plant cells in the raw material. Some important phospholipids are phosphatidylcholine (true lecithin), phosphatidylethanolamine (cephalin), phosphatidylinositol and phosphatidylserin. The most familiar of these is lecithin, and the mixture of phospholipids that separate out as so-called 'gum' when

TABLE 7.4

Synopsis of quality clauses in Brazilian trading contracts (the actual contracts should be consulted for trading purposes)

ANEC[a] Contract No.	81	110	111	112
Description of Brazilian oil	Degummed soyabean oil	Crude groundnut oil	Semi-refined cottonseed oil	Crude sunflower seed oil
FFA { (basis)[b] (%)	1·0 (oleic)	2·0 (oleic)	0·25 (oleic)	2·0
FFA { (max.) (%)	1·25 (oleic)	3·0 (oleic)		3·0
M and V (max.) (%)	0·2	}0·5[c]	}0·25[c]	}0·5[c]
Imps. (max.) (%)	0·1			
Phosphorus { (basis) (%)	0·02			
Phosphorus { (max.) (%)	0·025			
Flash point (min.) (°C)	121	121	121	121
Unsap. (max.) (%)	1·5			
Colour { (max.) R	5(1″)		12(5·25″)	
Colour { (max.) Y	50(1″)		35(5·25″)	
Sediment (max.) (%)	0·1[e]			0·3
Wax (max.) (%)				1·0

[a] ANEC = Associacao Nacional dos Exportadores de Cereais.
[b] Where (basis) and (max.) are given, allowances on contract price are specified between the two values.
[c] Moisture, other volatile matter, and impurities insoluble in petroleum ether combined.
[d] Lovibond colour by AOCS Cc 13b–45. Cell length as shown.
[e] Sediment (Gardner Break Test) AOCS Ca. 10–40.

crude oils are treated with water in the de-gumming operation[4] is usually called 'lecithin'. Crude rapeseed and soyabean oils contain the highest concentrations of lecithin, with up to 2·5 and 3·2%, respectively, while oil from maize germ contains up to 2·0%, sunflower seed 1·5%, cottonseed 1·0% and groundnut 0·4%. Sesame seed oil contains only 0·1% lecithin, and palm kernel oil almost none.

Methods for the determination of 'lecithin' in processed oils are reviewed in Chapter 10. However, the analytical difficulties are greater with crude oils and there are occasional trade disputes. It is, therefore, appropriate also to consider the methods for 'lecithin' and phosphorus in crude oils.

When crude oils are allowed to become damp the 'lecithin' hydrates and settles out on standing as 'sediment' or 'foots'. An early method of

TABLE 7.5
Synopsis of the American trading rules for grades and quality (the actual

			12·90-1	12·90-3		12·75-2
NIOP Rule[a]						
AFOA Rule	6	8, 11 and 12			14	15
NSPA Rule	103 (A)	103 (C)				
Description of oil	Crude degummed soyabean	Fully refined soyabean and cotton	Crude rape (low erucic)	Super-degummed rape (low erucic)	Crude sunflower	Crude safflower (export)
FFa, basis %[b]	0·75 (oleic)					
FFA max %	1·25 (oleic)	0·05 (oleic)	1·0 (oleic)	1·0 (oleic)	2·0 (oleic)	2·0 (oleic)
M & V. basis % ⎱[e] Imps. basis % ⎰						1·0[e]
M & V, max % ⎱ Imps, max % ⎰	}0·3[e]	0·10	}0·5[e]	}0·3[e]	0·5 / 0·3	0·8 / 0·3
Unsap., max %	1·5	1·5			1·3	1·5
I.V.						140–148
Colour.,[c] R(		2·0 basis[g] 2·5 max (see notes[g,i])	1·5 (max)[k] 15 (max)[k]	1·5 (max)[k] 15 (max)[k]	2·5 (max)[k]	
Colour.,[c] Y(						
Oleic acid[d]						
Linoleic acid[d]						72% (min)
Linolenic acid[d]					1·0% (max)	
Erucic acid			2·0% (max)	2·0% (max)		
Flash point °C	121 (min)		150 (min)	150 (min)	121 (min)	121 (min)
Phosphorus, basis,[b]	0·020%					
Phosphorus, max.[b]	0·025%			0·005%		
Notes	a	g,h,i	j	j	l,m	m

[a] NIOP = National Institute of Oilseed Products; AFOA = American Fats and Oils Trading Association; NSPA = National Soya bean Producers Association.

[b] Where 'basis' and 'max' are given, allowances on contract price are specified between the two values.

[c] Red (R) and Yellow (Y) colour by AOCS Lovibond method Cc13b–45, Cell length $5\frac{1}{4}''$.

[d] As mass percentage of total fatty acids.

[e] Moisture, other volatile matter, and impurities combined.

[f] Negative insoluble bromide test for fish oil (AOAC method 28.132 p 526 in 14th edn, 1984).

[g] NSPA rule 103C specifies for refined soya oil, colour 2.OR/20Y (max), a negative fish oil test by AOAC method (note[f]), bland flavour, PV 2·0 m, eq/kg (max), cold test by AOCS Cd 8–53, $5\frac{1}{2}$ h (max), and a PV of 35 m. eq/kg (max) after 8 h AOM aeration by AOCS method Cd 12–57.

[h] AFOA rule 8 specifies that 0·005% citric acid or 0·006% monoisopropyl citrate be added to oils during cooling after deodorisation.

[i] AFOA rules 11 and 12 specify cottonseed oil and cottonseed cooking oil colours as 4·0R/35Y

trading rules must be consulted for trade purposes)

12·77	12·80		12·60	12·61	12·65	12·66	12·70	12·97
	19	13						
Crude high oleic safflower	Crude corn (maize)	Crude peanut or groundnut	Crude (Manila type) coconut	Refined coconut	Crude palm	Neutral unbleached palm	Crude palm kernel	Crude sesame
	3·0 (oleic)	2·0 (oleic)	6·0 (oleic)				5·0 (lauric)^s	3·0 (oleic)
	5·0 (oleic)	3·0 (oleic)	12·0 (oleic)	0·1^q	5·0 (p'tic)^q	0·25 (p'tic)^q	6·0 (lauric)^q	5·0 (oleic)
	0·5^e		1·0^e	1·0	1·0	1·0		
0·5 0·3 1·3	}1·0^e	}0·5^e				}0·1^{e,q}	}1·0^{e,q}	}0·5^e
78–88			10·5 (max)^q			50–55	19·0 (max)	
2·5 (max)^n			15 (max)^q	1 (max)^q				
25 (max)^n			100 (max)^q	10 (max)^q				
77% (min)								
121 (min)			121 (min)	121 (min)				
	o	p						

(max). National Cottonseed Producers Association show these colours as 'basis' in rule E-8, with PV = 1·0 (basis), 2·0 (max) (see note^q), and additives as in Note^h.

^{j} Chlorophyll 30 ppm (max), sulphur 10 ppm (max), Canadian Standard CAN-CGSB-32-300-M87 also specifies the data shown.

^{k} Refined, bleached colour.

^{l} Saponification Value 188–194.

^{m} Negative Halphen reaction.

^{n} Colour after bleaching.

^{o} Refractive Index at 25°C = 1·467–1·469.

^{p} See NIOP Guideline Specification for corn oil.

^{q} At time of shipment, colour for crude coconut oil must be no more than 18R on arrival.

^{r} Melting point 33–39°C.

^{s} On arrival.

TABLE 7.6
Synopsis of guideline specifications published by the NIOP,

Description of oil	Once refined corn (maize)	Refined bleached winterised deodorised corn	Refined bleached deodorised low erucic rape[g]	Crude high oleic sunflower	Edible high oleic sunflower
FFA (max) (%)	0·15 (oleic)	0·05 (oleic)	0·05 (oleic)		0·5
M and V (max) (%)	0·5	0·1	}0·05[e]	0·5	0·1
Impurities (max) (%)				0·3	
Unsap. (max) (%)	2·0	2·0	(see note[g])	1·3	1·3
PV (m.eq/kg) max		0·5	2·0		
Colour (max)[a] R	14	3·5	1·5	2·5[f]	1·5
Colour (max)[a] Y			15	25[f]	15
Melting point (°C)[c]					
Oleic acid (%)[c]				77 (min)	77 (min)
Linolenic acid (%)[c]	1·5 (max)	1·5 (max)			
Erucic acid (%)[c]			2·0 (max)		
IV (Wijs)	122–128	122–128	(see note[g])	78–88	78–88
Refractive Index, 25°C			(see note[g])	1·467–1·469	1·467–1·469
Flash point (min) (°C)				121	121
Smoke point (min) (°C)			232		232
Phosphorus mg/kg (max)	10		2		
Soap (mg/kg) (max)	150 (max)	below 18			
Cold test (h) (min)		20	12		
Cloud point (max) (°C)[d]					
Sap. value			(see note[g])		

[a] AOCS Method Cc 13b–45 ($5\frac{1}{4}''$ cell).
[b] AOCS Method Cc 3–25, note special stabilisation at 10°C for palm products.
[c] As mass percentage of fatty acids.
[d] AOCS method Cc 6–25.
[e] Moisture, other volatiles, and impurities combined.
[f] Colour, as in note[a], after bleaching.
[g] Canadian Standard CAN/CGSB-32-300-M87 also specifies data shown, and in addition specifies: unsaponifiable matter at 1·5% (max), saponification value as 182–193, refractive index as 1·465–1·467, iodine value as 110–126, Crismer value as 67–70, and relative density at 20°C against water at 20°C as 0·914–0·920.
[h] Wiley method.

assessment was therefore to allow the oil to stand for 96 h at a temperature of between 15° and 20°C and then measure the volume of the sediment. This procedure is standardised in Paragraph 11 of the BS specification for linseed oil (BS 6900 (1987), ISO 150 (1980)). A development from this approach standardised in Paragraph 12 of this reflecting current industrial practice

Refined bleached deodorised coconut	Refined bleached deodorised palm	Crude palm olein	Refined bleached deodorised palm olein	Crude palm stearin	Refined bleached deodorised palm stearin	Refined bleached deodorised palm kernel
0·1 (oleic)	0·1 (palmitic)	5·0 (palmitic)	0·1 (palmitic)	5·0 (palmitic)	0·1 (palmitic)	0·1 (lauric)
}0·1^e	}0·1^e	}0·5^e	}0·1^e	}0·5^e	}0·1^e	}0·1^e
2·0	3·0 (max)		3·0 (max)		3·0 (max)	2·0
20						20
75–79°F^h	33–39	24 (max)	24 (max)	44–50	44–50	
11 (max)	51–55	57 (min)	57 (min)	43–48	43–48	14–19
		10	10			
		10	10			
250–264						245–255

British Standard, and also in IUPAC method 2.422, is the phosphoric acid test (PAT). This is also for linseed oil and involves treating a test portion of the oil with 85% phosphoric acid. The 'lecithin' precipitates or 'breaks' from the oil, is separated by centrifuging, washed free of oil with acetone, dried and weighed. The precipitate is not lecithin in the accepted sense of hydrated phospholipid, but is a complex mixutre, which has not been characterised. The equivalence of 0·25% m/m PAT value to 1% V/V foots by BS sedimentation for 96 h was well established for raw linseed oil on more than 500 test results. However, efforts to extend this to groundnut oil were unsuccessful owing to poor reproducibility.

The early development of methods to determine the phosphatide content of drying oils such as linseed was a consequence of its deleterious effect in paints and varnishes. When these oils are

TABLE 7.7
Malaysian palm oil, stearin and olein, and palm kernel oil, criteria

Characteristic[a]	Palm oil	Palm stearin	Palm olein	Palm kernel oil Grade 1	Grade 2
Relative density, 50°C/water at 25°C	0.8919–0.8932	0.8816–0.8915[b]	0.9001–0.9028[b]	—	
Refractive index, $n_D^{50°C}$ (AOCS Cc 7-25)	1.4546–1.4560	1.4472–1.4511[b]	1.4586–1.4592[b]	1.4500–1.4518[c,k]	
Saponification value (mg KOH/g oil)	190.1–201.7	193–205	194–202	243–249[k]	
Unsaponifiable matter (%) (AOCS Ca 6a-40)	0.15–0.99	0.3–0.9	0.3–1.3	0.1–0.8[k]	
IV, Wijs	50.6–56.1	21.6–46.0	56.5–60.6	16.2–19.2[k]	
Slip M/Pt (°C) (AOCS Cc 3-25)	30.8–37.6	46–56[d]	19.0–23.0[e]	25.9–28.0[k]	
Total carotenoids[f] (as β-carotene) mg/kg	500–1 000	300–600	500–1 200	4.3–11.8[k]	
FFA (see note[g]), crude	5.0% (max)	5.0% (max)	5.0% (max)	3.0% (max)[l]	6.0% (max)[l]
FFA (see note[g]), special quality	2.5% (max)	—			
FFA (see note[g]), neutralised	0.25% (max)	0.25% (max)	0.25% (max)		
FFA (see note[g]), refined, bleached, deodorised	0.10% (max)	0.20% (max)	0.10% (max)		
Moisture and impurities, crude	0.25% (max)	0.25% (max)	0.25% (max)	0.35% (max)[l]	0.5% (max)[l]
Moisture and impurities, special quality	0.25% (max)				
Moisture and impurities, neutralised	0.10% (max)	0.15% (max)	0.10% (max)		
Moisture and impurities, refined, bleached deodorised	0.10% (max)	0.15% (max)	0.10% (max)		

Peroxide value, special quality meq/kg	3 (max)	
Peroxide value, refined, bleached, deodorised	2 (max)	
Anisidine value, special quality	4 (max)	
Anisidine value, refined, bleached, deodorised	4 (max)	
Colour		
(refined, bleached and deodorised	6 Red units (max) [i]	6 Red units (max) [j]
product) 5¼ inch cell, Lovibond	[k]	[l]
Reference notes	[h]	

[a] Commodities are crude unless stated otherwise.
[b] At 60°C; density relative to water at 25°C.
[c] At 40°C; density relative to water at 25°C.
[d] Given as 48°C min for final product specification.
[e] Given as 24°C max for final product specification.
[f] Carotenoids from absorption at 446 nm.
[g] As palmitic (m wt 256) for palm oil, palm stearin and palm olein, as lauric (m wt = 200) for palm kernel oil.
[h] Malaysian Standard MS 814 (1983).
[i] Malaysian Standard MS 815 (1983).
[j] Malaysian Standard MS 816 (1983).
[k] PORIM Technology No. 6, September 1981.
[l] Malaysian Standard 3.10 (1973).

heat-bodied to improve performance, any phosphatides 'break' from the oil when the temperature reaches about 300°C, depositing flocculent particles of phosphatide, together with any impurities held in suspension by the emulsifying action of the former. The surface active properties of the 'lecithin' also reduce the resistance of surface coatings to water and, in addition, as phosphatides act as antioxidant synergists,[17,18] they retard the drying of the oil by oxidation. For these reasons, phosphatides must be removed from drying oils by steam or water washing.

A development used with soyabean oil was the Gardner Break Test, standardised by the American Oil Chemists Society (AOCS) in method Ca10–40. In this, three drops of concentrated hydrochloric acid are added to 25 g of oil, which has been previously heated to 75°C for 5 min. The oil is then further heated to 289°C (550°F), and cooled to 25°C, without stirring. Carbon tetrachloride is added to dissolve the oil, which is then allowed to stand, with stirring every 15 min, for 1 h. The precipitate is quantitatively filtered, washed with carbon tetrachloride and weighed. The 'break' is calculated as four times the weight of the recovered precipitate. Although this test is specified in NOFOTA trading rules and ANEC contracts (Tables 7.3 and 7.4), it has been declared surplus by the AOCS and will not be printed in further copies of their manual.

There are various other physical and chemical test for phosphorus. In AOCS Recommended Practice Ca19-86, a nephelometric method based on work by Sinram[19] is described. The oil sample is first heated to 50°C, then filtered, and an amount of between 0·3 and 8·3 g is weighed into a 50 ml volumetric flask, which is then made up to the mark with acetone. The solution is poured into a nephelometric cell and after 5 min the turbidity is measured, in a previously calibrated turbidimeter, in nephelometric turbidity units. A correction is then made for an acetone blank, and the phosphorus level calculated by means of one of a series of equations depending on the oil type and stage of processing. This procedure is clearly suited for factory quality control where a large number of related samples need routine evaluation.

The chemical determination of phosphorus may be accomplished by the related AOCS or IUPAC methods Ca12-55 or 2.421 respectively, or by the AOCS Recommended Practice Ca12b-87.

In the IUPAC method, 0·1–10 g of sample is weighed into a previously ignited crucible containing 0·1 g magnesium oxide. The fat is carefully burnt off, and the crucible heated to yield a white ash at 800–900°C. This magnesium-containing ash is then dissolved in 5·0 ml of 6N nitric acid solution, treated with 20·0 ml of a 50/50 mixture of 5% ammonium molybdate and vanadate solutions, and allowed to stand for 20 min. The extinction is then measured at 460 nm against a blank. The amount of phosphorus is determined by reference to a previously prepared calibration curve. The AOCS method Ca12-55 is basically similar except that the sample is ashed in the presence of zinc oxide and the phosphorus measured as molybdenum blue at 650 nm.

AOCS Recommended Practice Ca12-87 is for an atomic absorption method previously calibrated with samples well characterised by the colorimetric method Ca12-55.

When the 'lecithin' content is needed it is usual to calculate this from the phosphorus content by use of the multiplication factor 30 as specified in AOCS method Ca12-55. However, various other factors have been proposed,[20,21] depending on the composition of the phosphatide mixture.

While the above methods are generally satisfactory for recently extracted or expelled crude oils, and for degummed or processed oils, they are often unsatisfactory for crude oils that have been transported or held in storage for some time. In the author's view, this is due to the fact that in the presence of moisture the phosphatides hydrate and precipitate out. No satisfactory guidance is given in any of the methods about the sampling of the heterogeneous mixture and the incorporation of the precipitate. Indeed, some methods call for prior filtration or use of a clear oil. Obviously, filtration or decantation to obtain a clear oil will leave behind the 'lecithin' present in the sediment. While this will not invalidate the result for the clear oil, the result will not relate to the whole cargo. Furthermore, determination on clear oil samples taken from the cargo after different periods of storage may well give different answers if 'lecithin' has, in the meantime, precipitated out as the hydrated form. This aspect, which relates to sampling and sub-sampling rather than analysis, is probably responsible for the majority of the trade disputes about the 'lecithin' content of a cargo. In the author's view, unless it has been agreed otherwise, original samples should be fully representative of the total cargo, including sediment, which should be re-incorporated into the laboratory sample before the analysis is undertaken. As insoluble sediments are more easily handled in the ashing procedure used in the colorimetric methods, these are the author's preferred techniques for crude oils.

The AA approach, on the other hand, is much faster and ideal for routine analyses of filtered or part-processed oils having no sediment. Several experimental techniques have appeared in the literature, e.g. determination of phospholipids by the Iatroscan Chromorod method,[22] by conventional thin layer chromatography (TLC),[23] and by high performance liquid chromatography (HPLC),[23,24] but here again none of the papers describes the analysis of non-homogeneous crude oil samples.

Phosphatides are good emulsifiers and are used in various foods. Several informative reviews of their properties and uses have been published.[16–18,25,26]

7.2.2 Cold Test

The cold test described in AOCS method Cc11-53 is often specified with regard to the suitability of oil for domestic table or salad oil use, or in the manufacture of mayonnaise. In the former case it is important that the oil should not deposit an unsightly 'sludge' of high melting triglycerides during refrigerator storage. In mayonnaise the formation of stearin crystals 'breaks' the emulsion, presumably because the film of emulsifier surrounding the spherical oil droplet cannot 'stretch' to surround the stearin crystal, which being non-spherical will have increased surface area. The 'cold' test was developed as a check of stearin removal in winterisation of cottonseed oil, or lightly hydrogenated soyabean oil. The liquid oil sample is first filtered to remove dust, etc., heated to 130°C to melt any crystal nuclei completely, and then a 4 oz (115 ml) bottle is completely filled and stoppered at 25°C. The bottle is held at 0°C in a constant temperature bath, usually of ice/water, and the time taken for cloudiness to develop is noted. In AOCS test Cc11-53, the oil is held to have passed the test if it is still clear and bright after 5·5 h at 0°C, although longer periods are sometimes specified (Table 7.6). Small, finely dispersed air bubbles should, however, not be confused with fat crystals. Slight dampness of the oil can also confuse the test, as the moisture may separate out during the low temperature storage, causing a clouding of the oil, which should therefore be free of moisture before starting the test.

In more recent years there has been an increase in the use of sunflower seed oil, on account of its high linoleic (essential) fatty acid content coupled with a low linolenic acid content. Sunflower seed oil, when extracted from whole seeds including husks, contains up to 0·2%

wax, which crystallises very slowly from the oil. A sunflower seed oil may therefore pass the cold test, but still become cloudy, or be associated with oiling out of a mayonnaise, unless it has been de-waxed. This is seen as a major problem in the application of the cold test. The determination of wax is described below.

7.2.3 Wax

Sunflower seed hulls, or husks, contain a vegetable wax,[27-29] which comprises an ester of long chain acids with long chain alcohols. The wax is extracted from the seed hull along with the oil in the kernel, and is normally present at a concentration of about 0·2% in the oil. Extracted oil contains more wax than expelled oil, and if a batch of sunflower seed oil from a conventional expelling/extraction plant contains a higher than normal proportion of the extracted oil, then it may contain more than the normal level of 0·2% wax. Safflower, maize and rice bran oils also contain similar waxes, the removal of which is described in Chapter 10. The wax is sparingly soluble in the oil, but its melting point is above 60°C, and it slowly crystallises from cold oil. However, crystallisation is inhibited by any phosphatides present, as discussed by Rivarola *et al.*[30] Un-dewaxed sunflower seed oil may have a satisfactory cold test, but later show turbidity; the threshold of the cold test has been shown to be about 80 mg wax/kg oil.[31] Waxes may also separate from a crude oil some time after it has been de-gummed, or following hydration and precipitation of the phosphatides during storage.

Although the wax content of sunflower seed oil is sometimes specified in trading contracts (see for example, Table 7.4), there is at present no universally accepted method for its determination.[32] Most trade problems relating to wax concern that in sunflower seed oil rather than safflower, maize or rice bran oils.

Methods currently mentioned in the literature fall into four categories:

(a) precipitation from oil/solvent mixtures at low temperatures;
(b) gas–liquid chromatography (GLC) determination of wax components such as esterified long chain alcohols after separation from the oil;
(c) HPLC; and
(d) exotic techniques such as laser spectroscopy or ultrasound-assisted crystallisation.

A favoured method for the precipitation, recovery and determination of wax by low temperature crystallisation from an oil/solvent mixture is described by Gracien and Arevalo.[33] In their method, 50 g of homogeneous sample are dissolved in 100 litre of dry acetone and cooled to −5°C. The waxes precipitate out overnight (16 h) and are filtered off, washed, dried and weighed. As mentioned above,[30] precipitation of waxes may be inhibited by the presence of phospholipids, especially at low wax concentrations. In addition, the accuracy of this approach may be influenced by slight variations in the temperature of crystallisation, the oil/solvent ratio, the moisture content, and the varying solubilities of the individual chemical compounds that constitute 'wax'. Thus precipitation techniques will give a satisfactory quality control (QC) answer with crude oils only when levels much below 150 mg/kg are not encountered.

Various other related precipitation techniques have been recommended, e.g. the IASC method[34] stipulates that a 50/50 mixture of oil and hexane is held at 0°C for 24 h, Cancalon[35] recommends holding a 1:10 mixture of oil: petroleum ether at 4°C for 36 h, and Ostric–Matijasevic and Turkulov[27] recommended crystallisation from hexane at 1–3°C over a period of 16 h, but they do not specify the oil: solvent ratio.

Related techniques that may give a quicker answer have been sought, and papers by Brimsberg and Wretensjo,[36] Morrison,[37] and Moulton,[38] describe rapid techniques in which the microcrystalline waxes are determined by nephelometry. However, attempts by the AOCS to ring test a method based on this approach were abandoned, owing to the poor reproducibility of the method.[39] A convenient 'in-house' procedure[40] is to bleach 200 g of crude oil with 2% active earth at 80°C for 20 min, then filter; 100 g of this oil are diluted with 100 ml of hexane, cooled, and held at 0°C for 24 h. The solution is filtered through a weighed paper (Schleicher and Schull 589 or equivalent) in a cooled Buchner funnel. The precipitate is washed four times with 5 ml portions of hexane at 0°C, which has been previously saturated with wax having an IV less than 5, having been 'worked-up' from plant material. After the hexane wash the filter paper and deposited wax are dried to constant weight at 50–60°C. The method gives the amount of wax capable of being removed by de-waxing, rather than the absolute quantity of wax present.

Several methods[31,35,41,42] have been reported in which the wax esters are first saponified, the long chain fatty alcohols liberated and determined by GLC. Unfortunately, the esters are very resistant to saponification,[35,43,44] and none of the proposed methods includes a satisfactory check for complete saponification. Furthermore, it is held[45] that not all of the waxes in the mixture crystallise out on standing, so this method tends to over-estimate the amount of wax that may separate.

Vigneron *et al.*[46] proposed an HPLC method in which a small quantity of homogeneous sample is injected onto an HPLC column of silver nitrate impregnated Partisil and eluted with methylene chloride. The waxes elute first, followed by methyl esters and triglycerides. This technique appears to give quantitative results in a few minutes, and although not yet applied in the author's laboratory, it does appear to be the best of the techniques so far described in the literature. It may, however, suffer from the disadvantage mentioned above, in that it will give a value for the total amount of wax rather than the amount of wax that will crystallise on standing.

7.2.4 Insoluble Bromide Test for Fish Oils

Some trading contracts specify (Table 7.5) that liquid oils should give a negative insoluble bromide test. The procedure standardised in AOAC method 28.132,[47,48] for detection of fish oils in the absence of metallic salts, involves dissolving 30 drops of oil in 8 ml chloroform, and adding 10 ml Wij's iodine solution. A 1:3 mixture of bromine and glacial acetic acid is added to the stirred solution from a burette, until there is a slight excess, as indicated by the colour. The well-mixed solution is then poured into a flat-bottomed tube and allowed to stand until a clear supernatant layer is obtained. The amount of precipitate is then compared with a series of standards containing known amounts of fish oil, including zero.[49] While it is accepted that the test is sufficiently sensitive to detect as little as 1% fish oil in a liquid vegetable oil, the proportionality is disputed,[50] and it is therefore wise to prepare standards reasonably close to the expected fish oil content. Ghose and Pal[51] have claimed that a variation of the test can be used to detect partially hydrogenated fish oils.

Notwithstanding the claimed sensitivity of the above methods, the present author prefers the analysis of the derived fatty acid methyl esters (FAME) by gas–liquid chromatography with a capillary column, as described in the next section, the FACs of fish oils being sufficiently different from those of vegetable oils to permit ready identification of all but the lowest levels of contamination.

7.3   FATTY ACID COMPOSITION

The fatty acid composition (FAC) of an oil is its most useful chemical feature. Many of the chemical tests for oil identity or purity can be related to the fatty acid composition, as discussed elsewhere.[52-54] Edible oils and fats do, of course, comprise triglycerides, and while the composition of the triglyceride mixture is itself interesting, knowledge of the amounts and types of the fatty acids in the triglyceride is widely regarded as more useful. This is particularly so from a nutritional standpoint, as the triglycerides are in any case broken down into the constituent fatty acids during digestion. The FAC of oils is also a very convenient measure of their properties and purities, and has been widely used in establishing oil authenticities.[7,53,55-56] This section will therefore cover methods for the preparation of fatty acid methyl esters (FAME) as well as other esters, a necessary first step in the determination of FAC, followed by analyses of FAME by GLC. In many cases, however, more sophisticated analyses are necessary, and methods for determination of 2-position acids, *trans* acids, essential fatty acids and erucic acid are also described.

7.3.1   Preparation of Fatty Acid Methyl Esters

Although FACs may be determined by classical chemical tests involving separation of silver or lead salts of the fatty acids,[6,8] such techniques have been almost entirely replaced by GLC. The fatty acids are first liberated, and then converted to simple ester derivatives, as this increases the volatility and improves chromatographic behaviour and recovery. Alternatively, the esters may be prepared directly from the triglyceride by a transesterification reaction. It is usual to prepare the methyl esters, but as the FAME of the very short chain acids (C_4–C_{10}) are very volatile, losses can occur, leading to inaccuracies. Esters of longer-chain alcohols, such as butyric, may therefore be prepared in these cases.

Although the initial preparation of FAME is an important part of the analytical procedure, and probably the most common chemical reaction carried out by lipid chemists, it is often poorly understood.[57] Only a few reviews have appeared on this subject.[57-61] It is necessary to consider several techniques, each procedure being best suited to particular circumstances. The International Standards Organisation (ISO) specifies four methods in ISO 5509 (1978). This is dual numbered with BS 684, Section 2.34 (1980), while IUPAC method 2.301 is technically equivalent. The general method, which is the preferred technique, involves saponification of the fatty acids with refluxing methanolic sodium hydroxide solution, and conversion of the fatty acids to methyl esters with methanolic boron trifluoride. This aspect of the ISO/BSI/IUPAC tests is technically equivalent to AOCS method Ce2-66, but while the former authorities specify additional methods for use when the general BF_3 method is unsuitable, the AOCS text is silent on this issue.

A small quantity (about 350 mg) of clear dry oil is weighed into a 50 ml flask fitted with a reflux condenser, and 6 ml of approximately 0·5 N methanolic sodium hydroxide solution added by pipette. The air in the flask is removed by flushing with oxygen-free nitrogen to prevent oxidation, and the mixture is then boiled under reflux for 5–10 min, and in any case until droplets of fat disappear and the mixture becomes homogeneous. Methanolic boron trifluoride solution (7 ml) of 12–15% (m/m) are added through the reflux condenser and boiling is continued for a further 2 min. About 3 ml of chromatographic quality heptane are then added, again through the reflux condenser, and boiling continued for a further 1 min. The reaction mixture is cooled, and the flask is swirled several times with a small amount of saturated aqueous sodium chloride solution. Additional sodium chloride solution is next added until the flask is full to the neck. About 1 ml of the upper heptane layer is removed, dried with anhydrous sodium sulphate crystals, and used directly for GLC analysis. It should contain about 100 mg/ml of methyl esters. Evaporation of the solvent (heptane) should be avoided as this can lead to losses of the more volatile FAMEs.

This technique is generally applicable, but can lead to errors if the sample has any fatty acids with secondary oxygen groupings (e.g. hydroxyl- as with ricinoleic in castor oil, hydroperoxy-, keto-, or epoxy-), cyclopropane or cyclopropene (e.g. sterculic as in kapok or cottonseed oils), or conjugated polyunsaturated or acetylenic compounds. The method is applicable to free fatty acid mixtures and in this case the initial saponification with methanolic sodium hydroxide may be omitted. However, the BF_3 methanol reagent is toxic, and does not keep for long periods unless refrigerated; it can also lead to the formation of artefacts[59] if old or too concentrated.[60–61]

An alternative procedure described in ISO 5509 relates to the transesterification (methanolysis) of a substantially neutral oil (FFA less than 1%) in methanolic potassium hydroxide solution, a procedure that avoids many of the potential errors cited for the general procedure. Oils having an initial FFA of over 1% should be first neutralised with dry methanolic sodium methylate solution, followed by alkaline methanolysis under reflux, and then acid catalysed methanolysis of the originally present free fatty acids by addition of an excess of methanolic hydrogen chloride and further refluxing. Sodium chloride frequently precipitates out in this reaction, and can be filtered off to prevent its causing 'bumping'. After reaction, the methanolic solution is diluted with water and the FAMEs are extracted with heptane. The heptane solution may be used directly for GLC after drying.

ISO 5509 also describes a cold procedure intended for neutral oils and fats, such as butter, which have short chain acids with, for example, four carbon atoms. The test sample (1 g) is dissolved in chromatographic grade heptane and treated with dry methanolic potassium hydroxide solution in a stoppered tube. The tube is shaken vigorously to mix the contents. It should become clear, showing that reaction is complete, after which it again grows cloudy as free glycerol separates. The upper layer containing the methyl esters is separated by decantation. Readers wishing to use this section of ISO 5509 (BS 684, Section 2.34) should take note of amendment No. 5838, which describes the preparation of reference solution II by weighing, to the nearest 0·1 gm, 200 mg methyl pentanoate and diluting to volume in a 100 ml volumetric flask with heptane. This reference solution can be used as an internal standard for accurate measurement of the butyric acid content, when this is less than 1%.

Another formerly popular method for the formation of FAME from free fatty acids is by the use of diazomethane.[60] This reagent is conveniently prepared in small quantities for immediate use, but as it is extremely toxic and explosive, and forms potent carcinogens, it has been recommended that it should be used only when no other reagent is available.[60]

An alcoholic solution of 1–2% (v/v) concentrated sulphuric acid may also be used to transesterify fatty acids, in much the same way as with methanolic hydrogen chloride.[60] Toluene should be added to assist solubilisation of the non-polar glycerides, giving a reagent that is relatively quick to prepare and easy to store. If it is intended for use in the preparation of FAME the proportions to use are sulphuric acid/toluene/methanol (1/10/20 by volume). The test sample (50 mg) is refluxed with 5 ml of reagent for 1 h, the solution diluted with water and the FAMEs extracted with heptane, dried and analysed by GLC. Use of alternative alcohols, e.g. propionic or butyric, accompanied by appropriate adjustment of the reagent blend and reaction conditions, enables preparation of fatty acid propionic or butyric esters. These can be especially useful in the study of short-chain fatty acids as they are less volatile, and thus less prone to losses; they also have longer GLC retention times, giving better separation from the solvent peak and other background in the GLC analysis.[62]

The analysis of FAMEs should be undertaken as soon as possible after they have been prepared. Unsaponifiable matter is not removed in the procedures described, and if present in large quantities may interfere with subsequent analysis of the FAMEs. If this is suspected, the alkaline solution of fatty acids should be diluted with water and the unsaponifiable matter extracted by diethyl ether or light petroleum. The aqueous solution is then acidified, the fatty acids extracted, and converted to FAMEs by one of the foregoing methods.

While the methods described are applicable to the majority of oils and fats, each method being best suited to a specific class of raw material as indicated above, they may not be suited to other lipids such as phospholipids, wax esters, sterol esters or amide bound fatty acids such as sphingolipids, and in these cases reference to more exhaustive texts is recommended.[57,58,60]

7.3.2 Analysis of Fatty Acid Methyl Esters by Gas–Liquid Chromatography (GLC)

Until recently, the majority of FAME analyses were carried out by packed-column GLC. The technique of capillary column gas chromatography (CCGC) has been known for many years,[63] but was judged to be too temperamental for routine application, perhaps because of the lack of a sufficiently stable detector. Standard texts therefore specify packed-column analyses, and most reference tables of FACs relate to analyses carried out with packed columns.[7,9,10,53,55–56] this is the case with the related GLC procedures specified in ISO 5508 (1978), BS 684 Section 2.35, IUPAC method 2.302 and AOCS method Ce1-62, texts to which reference should be made for greater detail on the analytical conditions.

In essence, the methods specify that a chromatographic column, 1–3 m long, packed with an inert material supporting a non-volatile stationary phase, which is liquid at the operating temperature, is housed in a thermostatically controlled oven capable of holding a uniform temperature of up to 220°C. The column is fitted with an injection port, and a suitable detector, preferably a Flame Ionisation Detector (FID), both of which may be held at a thermostatically controlled temperature above that of the column. An inert oxygen-free gas such as nitrogen, argon, helium or hydrogen is passed through the column and into the detector, which, in the case of an FID, is also provided with the auxiliary gases hydrogen and oxygen (or air) to maintain the flame. Samples of FAME are injected into the column with a long needled syringe, separated into the basic components by the action of the column and burnt in the flame of the FID. The changes in the ionisation of the flame as the components are burnt are detected in the FID, which gives a linear response according to the weight of FAME eluted. This is converted into an electrical signal, which is then amplified and displayed on a strip chart recorder. Many instruments are also fitted with electronic integrators which enable easy calculation of the mass of each component eluting into the FID.

The identification of the peaks and, if necessary, calibration of the apparatus, is accomplished by analysis of standard mixtures of methyl esters. AOCS text Ce1-62 lists several companies supplying such cocktails. Unlike the situation with the preparation of methyl esters, there has been a large number of comprehensive reviews of GLC procedures and operating conditions.[9,46,60,61,63-67] It is generally agreed that the liquid phase is the principal factor governing the separations of FAME, and the standard texts allow a wide range, including polyesters such as polyethylene glycol adipate (PEGA) and (poly)-diethylene glycol succinate (DEGS), and cyanosilicone phases. However, some work carried out in the author's laboratory on behalf of FOSFA International[68] has shown that the cyanosilicone stationary phases are far superior, as analysts using these phases in a ring test consistently returned more uniform results with fewer outliers or stragglers than analysts using polyester phases.

The ISO and related texts allow for the calculation of correction, or response, factors, especially in the presence of FAME of fatty acids with fewer than 8 carbon atoms, with acids having secondary groups, when using thermal conductivity detectors, or with FIDs when the highest accuracy is needed. These are calculated by analysis of a reference mixture of FAME, the composition of which is accurately known. The practice has been challenged by Craske and Bannon,[66] who argue that this approach is fundamentally unsound as it may obscure faulty practice. Rather, they argue, technique should be optimised so that theoretically valid results are obtained from a primary FAME standard. Craske and Bannon[66] identified eight facets of the analysis that need to be adressed for optimum accuracy. These are: use of a computing integrator; use of only the theoretical FID relative response factors to correct raw peak areas; use of a primary standard of saturated FAME to optimise chromatographic procedures; use of a primary standard of saturated triglycerides to optimise overall procedure; optimisation of FAME preparation; optimisation of FID linearity; optimisation of injection technique; and use of 'grade' of analysis to identify and correct errors.

This list of important features contains the often overlooked feature discussed by Christie,[60] namely that of methyl ester preparation. Many laboratories optimise their GLC equipment by analysis of standard mixtures of FAME, and may be unaware of errors in the preparation of FAME. However, the recommendations of Craske and Bannon fail to take into account possible losses of polyunsaturated fatty acids as a result of oxidation. In order to overcome such aspects the Bureau of Community Reference (BCR), a department of the European Economic Community, has prepared a series of ampouled standard oil blends which are of accurately known composition and which can be used for standardisation of the full analytical procedure. Packs of two such ampouled reference oil blends, which come provided with an analytical certificate,[69,70] are shown in Fig. 7.1.

Table 7.8 presents ranges and means of FACs of eleven commer-

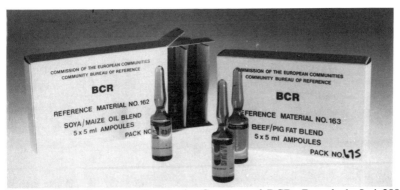

Fig. 7.1. BCR Reference Standards. Courtesy of BCR, Rue de la Loi 200, B-1049 Brussels, Belgium.

TABLE 7.8

Ranges (means) of fatty acid compositions of commercially important vegetable oils (% m/m)

Fatty acid	Oils analysed										
	Palm kernel	Coconut	Cotton-seed[a]	Soyabean	Maize	Groundnut	Palm	Sunflower seed	High erucic rapeseed	Low erucic rapeseed	Safflower seed
C6	ND–0.8 (0.3)	0.4–0.6 (0.5)	—	—	—	—	—	—	—	—	—
C8	2.5–4.7 (3.3)	6.9–9.4 (7.8)	—	—	—	—	—	—	—	—	—
C10	2.8–4.5 (3.5)	6.2–7.8 (6.7)	—	—	—	—	—	—	—	—	—
C12	43.6–51.4 (47.5)	45.9–50.3 (47.5)	tr–0.1 (tr)	tr (tr)	tr–0.3 (0.1)	—	ND–0.2 (0.1)	—	—	—	—
C14	15.3–17.2 (16.4)	16.8–19.0 (18.1)	0.7–1.0 (0.8)	tr–0.2 (0.1)	tr–0.3 (0.1)	tr–0.1	0.8–1.3 (1.0)	tr–0.1 (tr)	0.1	tr–0.2 (0.1)	tr–0.2 (0.1)
C16	7.2–10.0 (8.5)	7.7–9.7 (8.8)	21.4–26.4 (23.9)	9.9–12.2 (11.2)	10.7–13.6 (12.0)	9.2–13.9 (11.6)	43.1–46.3 (44.3)	5.6–7.4 (6.4)	2.8–5.1	3.4–6.0 (4.9)	5.3–8.0 (6.8)
C16:1	—	—	0.3–1.1 (0.8)	tr–0.2 (0.1)	tr–0.4 (0.2)	tr–0.1 (tr)	tr–0.3 (0.15)	tr–0.1	0.2–0.5	0.2–0.6 (0.4)	tr–0.2 (0.1)
C18	1.9–3.0 (2.4)	2.3–3.2 (2.65)	2.3–3.2 (2.7)	2.6–5.4 (4.1)	1.8–3.3 (2.4)	2.2–4.4 (3.4)	4.0–5.5 (4.6)	3.0–6.3 (4.6)	0.7–1.3	1.1–2.5 (1.5)	12.1–2.9 (2.4)

C18:1	11·9–18·5 (15·3)	5·4–7·4 (6·2)	14·7–21·4 (19·0)	17·7–25·5 (21·7)	24·6–42·2 (32·2)	36·6–65·3 (44·9)	36·7–40·8 (38·7)	14·0–34·0 (20·4)	9·8–49·8	52·0–65·7 (58·2)	8·4–21·3 (12·0)
C18:2	1·4–3·3 (2·4)	1·3–2·1 (1·6)	46·7–57·7 (52·3)	50·5–56·8 (53·9)	39·4–60·4 (50·8)	15·6–40·7 (32·6)	9·4–11·9 (10·5)	55·5–73·9 (67·1)	13·0–22·9	16·9–24·8 (20·8)	67·7–83·2 (77·3)
C18:3	tr–0·7 (0·7)	tr–0·2 (tr)	0·1–0·2 (0·1)	5·5–9·5 (7·5)	0·7–1·3 (0·9)	tr–0·1 (tr)	0·1–0·4 (0·3)	tr–0·1 (tr)	7·0–10·3	6·5–14·1 (10·3)	tr–0·1 (0·1)
C20	0·1–0·3 (0·1)	tr–0·2 (tr)	0·2–0·4 (0·3)	0·2–0·6 (0·4)	0·3–0·6 (0·5)	1·1–1·7 (·5)	0·1–0·4 (0·3)	0·2–0·3 (0·3)	0·2–1·0	0·2–0·8 (0·3)	0·2–0·4 (0·3)
C20:1	ND–0·5 (0·1)	tr–0·2 (tr)	0·1 (0·1)	0·2–0·3 (0·2)	0·2–0·4 (0·3)	0·8–(1·7) (1·1)	ND–0·3 (tr)	0·1–0·2 (0·1)	2·6–9·4	1·2–3·4 (1·7)	0·1–0·2 (0·2)
C22	—	—	0·2	0·3–0·7 (0·5)	0·1–0·5 (0·2)	2·1–4·4 (3·3)	—	0·6–1·0 (0·8)	0·3–0·9	0·1–0·5 (0·3)	0·2–0·8 (0·3)
C22:1	—	—	—	ND–0·3 (0·1)	—	tr–0·3 (0·1)	—	ND–0·2 (tr)	6·5–51·6	tr–4·7 (0·9)	tr–1·0 (0·2)
C24	—	—	0·1	ND–0·4 (0·2)	0·1–0·4 (0·2)	1·2–2·2 (1·5)	—	0·2–0·3 (0·2)	0·1–0·3	0·1–0·2 (0·1)	0·1 (0·1)
C24:1	—	—	—	—	—	—	—	—	0·3–1·1	0·1–0·4 (0·2)	0·1–0·2 (0·1)

[a] Cottonseed oil also contains small amounts of cyclopropenoid fatty acids. These normally decompose during conventional GLC of the methyl esters.
Mean values not calculated for high erucic acid rapeseed oil due to wide variations in these samples.
Means are shown below ranges.
ND = not detected; tr = trace.
Where single values are shown, all samples had same concentration within experimental error.

cially important vegetable oils. These were determined by packed-column GLC of almost 600 authentic vegetable oil samples at the Leatherhead Food RA in a project jointly funded by the (UK) Ministry of Agriculture, Fisheries and Food, FOSFA International, the Leatherhead Food RA and the Egyptian Government.[55,56,71,72] The Gudelines Specifications published by FOSFA International[16] give more detailed information on this topic, as the fatty acid composition of each of the eleven major oil types is categorised according to the geographical origin of the commodity. Values for a selection of several less important vegetable oils are presented in Table 7.9, while those for several lauric fats, and oils rich in gamma linolenic acid (GLA) are shown in Table 7.10.

Capillary column gas chromatography is now becoming widely used in analytical laboratories, especially since the introduction of reliably sensitive, linear, flame ionisation detectors (FID), and fused silica capillary columns with chemically bonded stationary phases. The essential feature of a capillary column, and one that distinguishes it from a packed column, is that it is open tubular, with no packing, the stationary phase being on the inside wall of the column. Capillary columns are often 25 metres long, and 0·2–0·8 mm internal diameter. Other features are essentially the same as for packed-column, except that optimisation of the flow rate of the carrier gas is more important. Furthermore the pressure drops along the column cause expansion of the carrier gas, and increased linear flow towards the end of the column. To minimise this distortion, gases of low viscosity should be used, hydrogen being best, followed by helium. Nitrogen is seldom used in capillary column work. Several dedicated CCGC instruments are now on the market, a typical instrument being shown in Fig. 7.2. Indeed, it is now becoming difficult to purchase a conventional packed-column instrument. A number of excellent, modern reviews of capillary column gas chromatography have appeared,[9,60,63,67,73,74] and provided the instruments are correctly calibrated, both packed and capillary column methods should give equivalent FACs, although small differences may accrue when there is a wide spread of fatty acid chain lengths.[75] The capillary column technique is better at resolving isomers of fatty acids, as shown in Fig. 7.3, where the chromatogram of hydrogenated vegetable biscuit dough fat blend is illustrated. The packed-column technique seldom separates *cis* and *trans* isomers, an aspect that can confuse a comparison of results by the two procedures. If a capillary column analysis of an unhdyrogenated oil is to be

TABLE 7.9

Fatty acid compositions of vegetable oils and fats, typical values (% m/m)[a]

Fatty acid	Aceituno oil	Allanblackia oil	Almond kernel oil	Avocado (fruit coat) oil	Borneo illipe butter	Cashew nut oil	Castor bean oil	Chinese vegetable tallow (fruit coat)	Citrus seed oil	Cocoa butter	Grape seed oil	Hazel nut oil	Hemp seed oil
C12	tr	tr			0·2				tr	tr			
C14	tr	tr	tr		0·1	tr		tr	0·5	0·1	0·1	tr	
C16	10·0	1·2	5·6	20	16·0	11	1	64·7	28	26·0	7·1	5·6	5·6
C16:1	tr		0·1	8		0·5		0·2	0·5	0·1	0·4	0·1	0·3
C18	25·5	52·5	1·6	0·5	46·0	8	1	1·5	3·5	33·0	5·1	1·6	2·6
C18:1	56·5	43·8	75·8	57·5	36·0	61	3·5	32·1	23	35·0	15·8	75·8	10·6
C18:2	6·5	0·9	16·1	13	0·5	19	5	1·4	37·5	4·2	70·6	16·1	59·4
C18:3		1·1	0·2	0·5		0·3	0·5	0·1	6		0·3	0·2	19·4
C20	1·5	0·4	0·1	0·5	1·2	0·3	tr	tr	1	1·0	0·3	0·1	1·9
C20:1		0·2	0·2			tr	0·5				0·2	0·1	tr
Others							87·5[b]						

[a] See Table 7.8 for palm kernel and coconut oils, Table 7.10 for other lauric and gamma linolenic acid (GLA) rich vegetable oils and fats.
[b] Ricinoleic acid (C18:1-OH).
tr = trace.

Table 7.9—*contd.*

Fatty acid	High oleic safflower seed oil	Kapok seed oil[c]	Kokum butter	Linseed (flaxseed) oil	Mango kernel fat	Mowrah fat	Oat germ oil	Olive oil (fruit coat)	Olive kernel oil	Pentad-esma fat	Perilla seed oil	Poppy seed oil	Rice bran oil
C12			tr					tr					
C14			tr				0·2	tr				tr	0·5
C16	5·5	10	2·0	6·5	6·9	20	17·1	12·0		2·4	7	10·0	16
C16:1	tr			tr	0·2		0·5	0·5		0·1			0·5
C18	2	9	56·3	4·5	41·4	22	1·4	2·5	11	50·3	2	2·0	1·5
C18:1	80·5	46	41·0	19·5	44	44	33·4	75·5	67	46·3	13	11·0	42·5
C18:2	12	34	0·5	16·5	4·6	11	44·8	8·5	19	0·6	14	76·0	35·5
C18:3				53·0	0·3			0·5		0·1	64	1·0	1·0
C20		1	0·2		2·5	3	0·2	0·5	3	0·2		tr	0·5
C20:1	tr							tr					0·5
C22													0·5
C24													0·5

[c] Kapok seed oil also contains up to 15% cycloprcpenoid fatty acids.

Table 7.9—*contd.*

Fatty acid	Rubber seed oil	Sal seed fat	Sesame seed oil	Shea nut oil[d]	Sugar cane oil[d]	Tall oil[d]	Teaseed oil	Tobacco seed oil	Tomato seed oil	Tung oil[d]	Walnut oil	Wheat germ oil
C12		tr		0·4	4				0·5			
C14		tr	tr	0·3	1		0·5	0·5	0·5		tr	
C16	8	5·5	10·0	4·6	25	5	8	13·5	15	4	7·3	19
C16:1					1				0·5	1	0·2	tr
C18	9	45·0	5·5	40·4	4	3	1	tr	6·5	8	2·2	1
C18:1	30	40·5	39·5	44·9	14	46	78	3·5	23	4	17·5	17
C18:2	33	2·8	43·0	7·5	36	41	12	14·5	50·5	3	59·4	56
C18:3	20				13	3		67	2		12·8	6
C20			1·0	0·4	1	2		1	0·5		0·1	0·3
C20:1		6·2	1·0	1·6			0·5				0·2	0·3
C22											0·1	
C22:1												0·3

---

[d] Shea nut oil contains up to 8% unsaponifiable matter. Tall oil contains 45% fatty acids, 42% rosin acids and 13% terpenes. Tung oil contains about 80% eleostearic acid (C18:3-*cis* 9-*trans* 11-*trans* 13).

TABLE 7.10
Fatty acid compositions of lauric and gamma linolenic acid (GLA), C18:3 n-6, rich vegetable oils and fats
(typical values % m/m)

Fatty acid	Babassu kernel oil	Coconut parings oil	Cohune nut oil	Mumururu tallow	Ouri-curi tallow	Tucum butter	Ucuhuba tallow	Oil of evening primrose	Boraginaceae oils	Blackcurrant seed oil
C6			0.3	0.1						
C8	5.5	5.9	8.7	1.2	10	2	0.3			
C10	5.5	4.3	7.2	1.5	9	2	0.8			
C12	43	40.9	47.4	46.2	46	47	16.6			
C14	16	19.0	16.2	32.4	9	23	72.1			0.1
C16	9	11.8	7.7	5.6	8	7	4.3	7.0	6–12	6–7
C18	3.5	2.0	3.2	2.2	2	3	0.7	2.0	2–6	1–2
C18:1	15	11.2	8.3	8.9	13	13	3.7	9.0	13–39	9–10
C18:2	2.6	4.9	1.0	1.5	3	3	0.6	72.0	13–38	45–50
C18:3 n-3							0.3		tr–45	12–14
C18:3 n-6								10.0	4–15	15–19
C18:4									0–12	2.5–4
C20				0.2			0.1		0–1.6	0.5
C20:1									0–6	0.5

See Table 7.8 for palm kernel and coconut oils.
tr = trace.

Fig. 7.2. A modern capillary column gas–liquid chromatograph fitted with autosample attachment (courtesy: Carlo Erba).

compared with values shown in Tables 7.8, 7.9 and 7.10, the sum totals of the contributions of the relevant isomers (e.g. C18:1, *n*-7c and C18:1, *n*-9c) must be used.

At present no full standard or official texts specifying capillary column analysis of FAME have been issued by IUPAC, ISO, BSI or AOCS, but groups are working on the finalisation of these. Draft International Standard 5508 (1989) has recently been published on this topic, but it may be a year or so before the full ISO standard is approved.

One problem with capillary column analysis can be the allocation of peaks to the correct isomers, especially with hydrogenated products such as fish oils, where there can be a large number of isomers, differing not only in unsaturation but also double bond position and *cis, trans* configuration. A useful check is to calculate the iodine value from the FAC by AOCS method Tz-1c-85,[54] and compare this with the

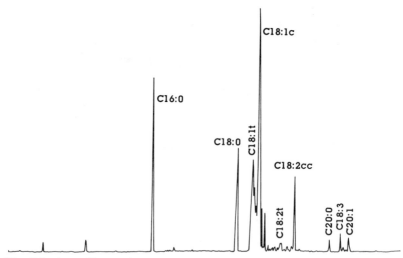

Fig. 7.3. A capillary column chromatogram of a vegetable biscuit dough fat blend showing isomer separation. 50 mm × 0·25 mm CP Sil88 fused silica column, programmed from 110°C to 170°C at 1·2°C/min, then 170°C to 200°C at 3·0°C/min. Split injection ratio 1:90, carrier gas hydrogen flow rate 2 ml/min.

Wijs titrated IV. Sebedio and Ackman used this approach to demonstrate errors in peak identification during analysis of hydrogenated menhaden oil by CCGC.[76]

Another commonly overlooked factor is that the result obtained by GLC of FAME is the composition of the constituent fatty acid methyl esters. The triglycerides in the oil contain glycerol, and conversion to FAME involves a molecular re-arrangement, the consequences of which, in analytical terms, have been discussed by Hammond.[61] In addition, an oil or fat may contain unsaponifiable matter (Table 7.2). For these reasons, the concentration of a particular acid, say linoleic, in an oil, is not equal to its percentage contribution to the fatty acid composition as determined by GLC of the FAME. It will always be lower than this value. When it is necessary to know the actual concentration of an acid in an oil, say for nutritional or labelling purposes, then an internal standard such as margaric (C17:0) acid should be added, and its concentration in the final FAC used to calculate the actual concentrations of the individual acids in the original sample.

In view of the underlying uncertainties with some samples, and bearing in mind the fact that many CCGCs are used for tasks other than just analysis of FAME, it is the author's view that many instruments will in the future be fitted with small, sophisticated mass spectrometry (MS) detectors. The GC/MS technique has been known for many years,[77,78] but it has been too expensive and specialised a technique for routine application. With continued improvements and automation, however, it is highly likely that it will become routinely applied to a variety of GC analyses in the future, enabling more confident identification of components in complicated mixtures of FAME.

7.3.3 Analysis of Fatty Acids at the Triglyceride 2-Position

Pancreatic lipase enzyme liberates fatty acids at the triglyceride 1-, and 3-positions in preference to the fatty acid at the triglyceride 2-position.[9] Although there is some migration of fatty acids between the 1-, 2- and 3-positions in monoglycerides, partial hydrolysis of a triglyceride with pancreatic lipase, under controlled conditions, followed by recovery and analysis of the monoglyceride rather than the liberated fatty acids, enables an accurate determination of the composition of the acids located at the centre, or 2-position of the triglycerides in a fat. The procedure is standardised in IUPAC method 2.210. The triglycerides in the fat are first isolated by neutralisation of free fatty acids (if present), and removal of any partial glycerides initially present by column chromatography. The triglycerides are melted (if necessary) and cooled to a temperature of at most 42°C. Pancreatic lipase and a buffer solution of pH 8 are added, and the mixture gently shaken, after which sodium cholate and calcium chloride solutions are added. The solution is shaken carefully for 1 min and then vigorously for 2 min, at a temperature of about 40°C, during which liberation of the fatty acids at the 1-, and 3- positions takes place. The mixture is then cooled, acidified with hydrochloric acid, and treated with a small amount of diethyl ether to extract the monoglycerides. The organic layer is removed from the surface of the mixture and separated into glyceride fractions by TCL over silica gel using a developing solvent of *n*-hexane:diethyl ether:formic acid (70:30:1). The developed plate is dried and sprayed with 2′,7′-dichlorofluorescein solution to identify the monoglyceride band. This is then scraped from the plate and the monoglycerides are converted to FAME and analysed in the usual way.

The IUPAC method recommends that hard fats with melting points above 40°C should be dissolved in a small amount of hexane, but it has been found in the author's laboratory that this can cause errors,[55,79] and it is therefore recommended that addition of hexane should be avoided whenever possible. It has instead been found that it is usually possible to maintain hard fats in a liquid condition by first melting, cooling to 42°C and then combining with reagents already warmed to 40°C. Most fats, except the very hardest, will remain in a supercooled state at 42°C for long enough for the reaction to liberate sufficient monoglyceride for a reliable determination of the 2-position acids. Of course, the hydrolysis itself will reduce the melting point of the fat in many cases, while fats having a melting point of less than 42°C cause no problem. Nevertheless, the enzyme should not be exposed to temperatures of over 42°C as this will deactivate it.

Oils and fats having appreciable concentrations of fatty acids with 14 or fewer carbon atoms, such as palm kernel, coconut or dairy butter, should not be investigated by this technique as it has been shown[80,81] that the short-chain acids are preferentially hydrolysed, giving misleading results. The IUPAC method is not applicable to materials containing highly unsaturated fatty acids with four or more double bonds and over 20 carbon atoms (such as fish oils) or oxygenated groups other than the acid group. It is also claimed[82] that gamma linolenic acid (GLA) is resistant to lipase hydrolysis, in contrast to alpha, and that other techniques must be used in its presence.

The composition of the 2-position acids can be a useful quality criterion. The joint Codex Alimentarius/International Olive Oil Council standard for olive oils specifies[83] that the concentration of fatty acids at the triglyceride 2-position should be not more than 1·5% for virgin olive oil; not more than 1·8% for refined or blended oil; not more than 2·2% for refined olive pomace oil; and not more than 2·0% for blends of refined olive pomace oil with virgin olive oil.

Mathematical manipulation[84-86] enables calculation of the concentration of the acids at the glyceride 1- and 3-positions. Application of the 1-, 3-random-2-random distribution law then enables convenient and reasonably accurate calculation of the triglyceride compositions of most natural fats,[84-86] although care must be taken not to apply the theory to fat fractions or blends.[87]

In a range of samples corresponding to a specific oil type, the concentration of an acid may rise and fall according to botanical variation within the species, climatic influences, etc. It has been found that the concentration of an acid at the triglyceride 2-position bears a constant ratio to its concentration in the fat as a whole, and furthermore that the ratio of these two concentrations varies with the oil type,[55,56] a feature that can be important in establishing oil purity. The concentration of an acid at the 2-position also has important dietary consequences, and it has been shown that infants thrive on feeds with an enriched content of palmitic acid at the triglyceride 2-position.[88,89] On the other hand, the distribution of saturated and unsaturated fatty acids between the 1-, 3- and 2-positions of different varieties of groundnut oils, before and after processing, has been implicated in a variation of atherogenic potential.[90–93] In high erucic acid rapeseed oil (HEAR) the erucic acid is located exclusively at the 1- and 3-positions. The detrimental nutritional effects of feeding HEAR are reduced by interesterification, which relocates some of the erucic to the 2-position.[94] Furthermore, the distribution of gamma linolenic acid (GLA) between the 1-, 3- or 2-positions has been linked to the bioavailability or useful rate[95] of GLA and the efficiency of the different GLA rich oils in alleviating essential fatty acid (EFA) deficiency.[82,95]

7.3.4 Determination of Essential Fatty Acids

Essential fatty acids (EFA) are those having a *cis,cis*-1,4-diene structure, useful in alleviating EFA deficiency syndrome and in preventing coronary heart disease.[1,9,72,94,95–97] The most common and useful of these is *cis,cis*-9,12-linoleic acid, commonly abbreviated (C18:2,*n*-6). There are, however, others such as alpha (C18:3,*n*-3) and gamma (C18:3,*n*-6) linolenic acids (GLA), arachidonic acid (C20:4,*n*-6), eicosapentaenoic acid or EPA (C20:5,*n*-3), and docosahexaenoic acid or DHA (C22:6,*n*-3). The last three acids occur mainly in fish oils, while linoleic acid is a major constituent in sunflower seed and soyabean oils (Tables 7.8 and 7.9). The GLA contents of several oils are given in Table 7.10. In unprocessed, edible, vegetable oils there is seldom a problem in identifying EFA, as these polyunsaturated fatty acids (PUFA) naturally have the *cis,cis*-1,4-diene structure and are therefore EFAs. This is also the case with fresh unprocessed fish oils. Problems can arise, however, when processed, especially partially hydrogenated oils, are present, as in these there will have been some double bond migration and/or isomerisation to the *trans* form. Capillary column GLC analysis is frequently employed to separate the peaks corresponding to EFA from those corresponding to the other, non-essential PUFA.[1,9,63,74,76,82,98] In cases of doubtful allocation of peaks to corresponding fatty acid the IV can be calculated from the FAC and compared with the titrated value, as described earlier.[54,77]

An alternative procedure is the lipoxygenase assay originally developed by MacGee[99] and others,[100,101] and subsequently standardised by the AOAC,[102] IUPAC (method 2.209[103]), and in BS684, Section 2.43 (1988) and ISO 7847 (1987). The method involves saponification of the sample to liberate the fatty acids, followed by enzymic oxidation of the acids having a cis,cis-1,4-diene structure using lipoxydase enzyme. The oxidation causes conjugation of the double bonds, the concentration of which is then measured by UV absorption at the peak maximum near to 235 nm. Correction for the optical absorption in a blank sample treated in the same way but with inactivated enzyme enables calculation of the concentrations of EFA.

The test is not applicable to samples having acids with n-8 or n-9 unsaturation, or branched chain acids. Undue mixing of the solutions should be avoided, as mixing in excess of that specified in the method causes an increase in optical density, of both blank and measuring solutions. The reasons for this are not understood,[103] but it is of little consequence if sample handling is standardised and applied equally to both sample and blank solutions. It should also be noted that the nature of the results will depend upon the method of calibration. This is normally against the conventional FAC of a pure unprocessed vegetable oil, high in PUFA which is assumed to be all EFA. In this case the results will be expressed as a percentage of the fatty acids. If the results are needed as a m/m % of the total oil, then the EFA of the calibrant oil should be determined against an internal standard.[103]

Collaborative testing of the AOAC[102] and IUPAC[104,105] versions of the method has shown reasonable accuracy at EFA contents in the 10–60% range, but increased error at concentrations of less than 5%. While some workers may, on these grounds, prefer to employ CCGC analysis, it must be remarked that there has been no corresponding collaborative test of CCGC analysis. The present author therefore prefers to use the IUPAC (or ISO/BSI) lipoxygenase method in cases where there may be confusing isomers, as this is a fully standardised, ring tested, procedure which gives a single unambiguous result not open to individual interpretation.

7.3.5 Determination of Fatty Acids with *Trans* Double Bonds

Unsaturated fatty acids contain double carbon–carbon bonds, which are usually in the *cis* form. In this condition, like groups, in this case the —CH_2— alkyl chains, are on the same side of the double bond, about which there can be no rotation. The —CH_2— groups are more bulky than the hydrogen atoms on the other side of the double bond: this causes steric strain, and the *cis* double bond is energetically less stable than its *trans* isomer. During oil processing, especially hydrogenation, a free radical, usually a hydrogen atom, may momentarily become attached to one of the carbons of the double bond, temporarily converting it to a single bond about which there may be rotation. For this reason, processed oils, especially those that have been partially hydrogenated, contain some *trans* double bonds. If the isomerisation is allowed to proceed to equilibrium, about 70% of the original *cis* double bonds will isomerise to the *trans* form.

Fatty acids with *trans* double bonds have higher melting points than their *cis* isomers, and as a consequence their formation is encouraged in the production of speciality fats, especially those for the confectionery industry.[2,9,72,106–109] Biological hydrogenation, such as in the rumen, also causes *cis* to *trans* isomerisation and for this reason the milk and depot fats of ruminating animals contain up to about 8% of *trans* fatty acids. Although *trans* fatty acids have been shown to present no health hazards,[110] they have been linked with saturated fatty acids.[111] Polyunsaturated fatty acids (PUFA) with one or more *trans* double bonds should, therefore, not be described as PUFA or EFA in any nutritional claims.[111]

For these reasons, it is important to measure the *trans* fatty acid content of relevant fats. There are three general approaches; capillary column GLC, infra-red (IR) absorption spectroscopy, and argentation TLC.[112] In the IR absorption method, the fat or its methyl ester is dissolved in a suitably transparent solvent such as carbon disulphide or bromoform and the absorption of a peak occurring at around 970 cm^{-1} is measured. In IUPAC method 2.207, a solution of methyl esters of the fat is evaluated against a corresponding solution of methyl stearate. This eliminates a background absorption due to the methyl ester group and enables greater accuracy at low *trans* values. Calibration is against standard mixtures of methyl stearate and methyl elaidate of purity over 99%, and results are calculated as equivalent methyl elaidate content. With tests of this type, construction of the baseline to the absorption peak is always a problem, and the IUPAC method specifies that it should be drawn between the absorption minima at around 1000 and 925 cm^{-1}. AOCS method Cd 14–61 is basically similar to the IUPAC procedure except that the spectrum is recorded against the pure solvent, triglycerides are permitted when *trans* contents are above 15%, and the baseline is drawn between specific wavelengths (10·02–10·59 μm for methyl esters).

The IR method has been in use for many years and its advantages and limitations are well known;[112,114] for instance, it is only applicable to isolated *trans* double bonds and gives misleading results when conjugated bonds are present. Furthermore, the spectrum obtained in the presence of ricinoleic acid can be difficult to interpret. The method gives an overall result of the *trans* content of the sample, with no detail about the individual fatty acids. When information is needed about these, recourse must be made to GLC. One important problem can be the lower (85%) absorption found with a *cis trans* diene, and the higher (166%) absorption with a *trans trans* diene,[115] in comparison with a *trans* monoene, effects that to some extent compensate but cause lack of accuracy and confusion in fats with an appreciable diene content. Recently Belton *et al.*[116] and Sleeter and Matlock[117] have described improved Fourier Transform IR (FTIR) methods. In one of these[116] an attenuated total reflectance (ATR) cell was used. Use of the ATR cell dispensed with the need to use toxic inflammable solvents, as the neat fat sample is applied directly to the ATR cell, and even when the anticipated *trans* content is below 15% there is no need to convert to methyl esters. In addition, the use of the dedicated computer of the FTIR system is claimed to give greater reliability and accuracy.[116–118]

Gas chromatographic methods are now more applicable with the general availability of capillary columns. Figure 3 shows the chromatogram of the vegetable biscuit dough fat blend in which the *trans* isomers are clearly separated. Reasonable separation can also be achieved on packed columns provided a sufficiently polar stationary phase and columns of 6 m in length are used.[112] This is the basis of AOCS method Cd17-85 which is generally applicable to the hydrogenated vegetable oil shortenings available in the USA,[119] but is not applicable to the multiplicity of European blends containing hydrogenated fish oils.[112] Capillary column GC is more appropriate to these hydrogenated fish oil blends but here again there is no standard method, and the variety of stationary phases, columns and conditions used confuse inter-laboratory comparisons.

A resolution of this confusion is to carry out a preliminary separation of the methyl esters by TLC over silver nitrate impregnated silica gel as specified in IUPAC method 2.208. The band corresponding to the *trans* isomers is quantitatively removed from the plate in the presence of an internal standard, such as methyl heptadecanoate or myristate, and analysed by GLC. This is a very reliable method with hydrogenated vegetable oils, and is claimed to give reliable results even at low *trans* contents. The method is also an improvement on other approaches when hydrogenated fish oils are present, although some band overlap can occur.[112]

Various additional methods occur such as acetoxy-mercuric-adduct formation or ozonolysis, techniques that were used to measure unequivocally the *trans trans* diene contents of several Canadian margarines,[120] present at concentrations around 1%. This was considered important as *trans*-9-*trans*-12-C18:2 has been shown to interfere with lipid metabolism,[121] and its level in dietary fats should therefore be restricted to not more than 1%. In this work[120] it was observed that the *cis* mono-unsaturated positional isomers were the major source of error in the interpretation of results from chromatograms of the total fatty acids. These sophisticated techniques are, however, only necessary when there is a need to measure concentrations of specific *trans* isomers present at these low levels of around 1%.

A comprehensive review of the various techniques for measuring *trans* isomers has been published by Strocchi.[122]

7.3.6 Erucic Acid

Prior to 1968 almost all rapeseed varieties contained an oil with a high concentration, of 45% or so, of erucic acid, a mono-unsaturated 22 carbon acid (C22:1,*n*-9, *cis*; or *cis*-13-docosenoic). The fatty acid composition of the current varieties of high erucic acid rapeseed (HEAR) oil are shown in Table 7.8. However, erucic acid was alleged to cause nutrition problems[1,94,97,123] and low erucic acid rapeseed (LEAR) strains were introduced by plant breeding, initially in Canada. Although the nutritional benefits of the LEAR strains are not universally accepted,[94,124] it appeared prudent to recommend the production of edible oils with a reduced erucic acid content. The EEC therefore proposed[125] in 1967 that the concentration of erucic acid in the fatty acids of oils and fats for human consumption should not exceed 10%, with the level reducing to 5% in 1979. This was followed by the Erucic Acid in Foods Regulations 1977 in the UK,[126] which restricted the erucic acid content of fat in food initially to a maximum of 10%, and then to 5% for food manufactured after 1st July 1979. Related measures were introduced in many other countries. As the regulations relate only to true erucic acid, i.e. C22:1,n-9-cis, it became necessary to distinguish erucic acid from the various other C22:1 isomers, in particular cetoleic acid C22:1,n-11 cis, a component of fish oils.[124,127] Furthermore, it was known that partial hydrogenation of rapeseed or fish oils would produce other isomers, some of which would have $trans$ double bonds.[127]

Simple GLC of FAME with packed or capillary columns, as described earlier, provides a rapid screening test, by which it can be established that most oils are free from any C22:1 isomers, and are thus not in contravention of the regulations. In cases where this rapid screening test shows that the total content of C22:1 is above 5%, and when it is known or suspected that a proportion of the total C22:1 content will be isomers of erucic acid, it becomes necessary to distinguish erucic acid itself from the other C22:1 fatty acids. Two methods have been proposed; a CCGC method using Silar-5-CP stationary phase,[127] and a low temperature ($-25°$C) TLC method in which the silica stationary phase is impregnated with silver nitrate.[128] Although the CCGC method gave good reproducibility in a ring test,[127] the collaborative work did not establish that the CCGC procedure would give an accurate, true, answer in the presence of other C22:1 isomers. The low-temperature TLC method was collaboratively evaluated on the same set of samples as used for the CCGC work.[128] Ackman[129] compared the results of the two tests. Notwithstanding the fact that some experts favoured the CCGC approach,[129] the low temperature argentation TLC method was adopted by IUPAC (method 2.311), ISO (ISO 8209-1986) and BSI (BS 684 Section 2.41-1987) and became the official regulatory method in the EEC.[130]

In this method the oil is first converted to FAMEs by the standard methods (e.g. ISO 5509), and if appropriate the FAMEs are first analysed by GLC. If the total C22:1 is below the appropriate limit, e.g. 5% for the Erucic Acid in Food Regulations, or if the C22:1 acids content is well above some contracted limit and there is no reason to suspect the presence of other isomers, e.g. in unprocessed but suspect rapeseed oil, then there is no reason to progress further. In some cases, for example, with blends of rapeseed oil and hydrogenated fish oil, other C22:1 acids will be present, and in these cases the FAMEs are further analysed by low-temperature TLC over silver nitrate impregnated silica.

A short 50-mm 'streak' of 50 μl of a solution of the FAME in hexane is applied to the prepared and activated $AgNO_3/SiO_2$ plate, parallel to and about 10 mm from the bottom of the plate. A separate streak of 100 μl of a 50/50 mixture of the FAME and a prepared solution of methyl erucate is then applied, also 10 mm from the bottom of the plate. If desired, a streak of 50 μl of a prepared solution of methyl erucate may also be applied to assist identification of the erucate band on the TLC Plate after development. The plate is then dipped in diethyl ether, which is allowed to rise until the diethyl ether is 15 mm above the bottom of the plate. The plate is then removed from the ether and allowed to dry. This latest step concentrates the FAME into a sharp narrow band, and overcomes any 'spread' caused during the initial application.

A mixture of 10% hexane/90% toluene (v/v) is added to a TLC developing tank until about 5 mm deep, a lid is fitted, and the tank is placed in a freezer at between $-20°$ and $-25°C$. After 2 h, the TLC plate, which may optionally also be cooled, is placed in the tank, and the solvent allowed to rise until the solvent front has risen 50–66% up the plate. The plate is then removed, allowed to dry, and then redeveloped in the same tank until the solvent reaches the top of the plate. The plate is then removed, again dried, and sprayed with a solution of 2',7'-dichlorofluorescein.

The plate is irradiated with UV light and the band corresponding to the erucic acid esters is marked. The erucate and non-erucate bands are then quantitatively removed from the plate, taking care to avoid cross-contamination, and transferred to separate 50 ml beakers. An internal standard of methyl tetracosanoate and 10 ml of diethyl ether are added to each beaker, and the contents stirred to redissolve the FAME. The two solutions are then each transferred to short silica gel chromatography columns from which the FAME are eluted with diethyl ether. The eluates are collected, evaporated and redissolved in hexane. The resulting solutions are then analysed by GLC according to, for example, ISO 5508. The amount of erucic acid in the sample is then obtained by comparison of the relevant peak areas.

If the initial sample contains any tetrasosanoic acid, a correction must be made, using, for example, the intrinsic palmitic acid content as reference in the analyses. However, the method does not allow separation from cis-15-docosenoic acid, which if present will be reported as erucic. Duplicate determinations carried out simultaneously or in rapid succession by the same analyst should not differ by more than 10% of the mean value. Thus for a sample containing 5% erucic acid, the duplicates should not differ from one another by more than 0·5% (absolute). It is often found that analyses carried out in different laboratories at different times show a 'spread' 3 or 4 times the 'within laboratory' spread. Thus the 'spread' of results obtained by different laboratories analysing the same sample at different times may be up to 2%, an aspect worth bearing in mind in any discussions.

7.4. STEROLS AND TOCOPHEROLS

Sterols and tocopherols are the most important of the minor components of oils and fats. Other minor components include trace metals, tannins, flavanoids, phenolic compounds, free fatty acids, mono- and di-glycerides, phospholipids, glycolipids, e.g. monogalactosylglycerides, pigments such as chlorophyll, wax esters and aliphatic alcohols. Hydrocarbons, both paraffinic and in the form, for example, of squalene and carotene, are also present in many oils. Several of these minor components, such as the pigments, have been linked with oil quality for some time and analytical methods for their determination have been described in other parts of this book or elsewhere.[6,8,9,18,50,103] Other minor components, such as trace metals may be described as contaminants or impurities, and need be determined only if their presence is suspected. Trace metals, especially Fe and Cu, reduce an oil's resistance to oxidation considerably.[15] The range of substances that can contaminate an oil or fat is vast and each problem must be considered on its own merits. In some cases, as with hydrocarbons, it is difficult to determine whether a component is naturally present or is a contaminant, since analytical results on the naturally occurring components sometimes conflict. In the case of soyabean oil, for instance, the natural hydrocarbon content has been variously reported as 3800 ppm,[131] 900 ppm[132] and 60 ppm.[133] Fortunately the situation with sterols and tocopherols is much better understood, and reliable methods have been published by, for example, ISO, BSI and IUPAC.

7.4.1 Sterols

Sterols are 27–29 carbon compounds, which occur at low levels in vegetable oils and fats. In addition to the conventional or des-methyl sterols, there are also 4-mono-methyl, and 4,4-dimethyl sterols, but as these are usually found at much lower concentrations than the des-methyl sterols, their analytical determination will not be discussed here. Kochhar[134] reports the structures of the various des-methyl, mono-methyl and dimethyl sterols, all of which may be present in the free form, or esterified with a fatty acid.[135] The composition of the sterols in an oil enables one to distinguish vegetable oils from animal fats or fish oils, as vegetable products contain only very low levels of the sterol cholesterol. At one time it was thought that vegetable oils contained no cholesterol, but it is now known[136,137] that this is not the case as concentrations of 10–20 mg/kg are frequently found, while some vegetable oils such as rape or maize may contain up to 100 mg/kg. In contrast animal fats and fish oils contain up to 1000 and 7000 mg/kg, or more, respectively, and dairy butter contains 2000–3000 mg/kg cholesterol. While some dietary experts worry about the different cholesterol contents of animal and vegetable fats, it is worth remembering that egg yolk contains about 12 500 mg/kg cholesterol. In view of the higher concentrations in animal products cholesterol is termed a zoo-sterol, while the sterols like sitosterol found in vegetable oils are called phytosterols.

The method for the separation and analysis of the sterol mixture in an oil or fat specified in BS 684 Section 2.38 (1983) and ISO 6799 (1983) respectively, closely resembles IUPAC method 2.403 and the AOAC[138] procedure.

About 5 g of the sample are first saponified by boiling with 50 ml of freshly prepared, molar, ethanolic potassium hydroxide solution. Water (100 ml) is added to the cooled solution through the top of the reflux condenser, and the unsaponifiable matter, which contains the sterols, is quantitatively extracted from the solution with a total of 300 ml diethyl ether. The ethereal extract is washed with 40 ml water, and evaporated to about 1 ml at a temperature not exceeding 50°C. The solution is then streaked on to a previously cleaned and activated thin-layer chromatography plate having a silica layer 0·25 mm thick. The plate is developed with chloroform as eluting solvent, dried and sprayed with an indicator solution such as a 0·05% solution of rhodamine 6G in 95% ethanol. The plate is then viewed under UV light and the position of the sterol band marked. If the position of the sterol band cannot be ascertained from previous experience it is recommended that spots of a cholesterol solution are applied to the plate on either side of the original streak. The sterol band is then quantitatively removed from the plate, and the sterols extracted from the silica by refluxing three times each with 5 ml diethyl ether. The extracts are filtered, combined and evaporated under a stream of nitrogen, and the residue redissolved in a minimum quantity of diethyl ether.

The silylether derivatives of the sterols are next prepared by reaction with hexamethyldisilazane and trimethylchlorosilane in the presence of pyridine. Two layers form, the upper layer being used for the final stage of GLC analysis. ISO 6799 (1983) specifies a packed column, but at the time of writing a new Draft International Standard (DIS 6799–1989) updates the method and permits use of a capillary column instrument. In both cases a column with a stationary phase of methylphenyl polysiloxanes (OV17) or alternatively methylpolysiloxanes (SE30) is used. Peaks corresponding to the separate sterols are identified by reference to the analysis of a standard mixture, or the chromatogram of the sterols from a well characterised oil, e.g. a BCR reference sample. Retention times according to the two recommended stationary phases (OV17 and SE30) are also given in the ISO and BSI standards. Calibration factors should be used to convert the peak areas of the individual sterols into the mass percentage contributions to the total sterol content. Unfortunately, calibration factors differ according to the type of detector used, and standard mixtures of pure sterols are difficult to obtain for calibration purposes. It is hoped that the growing collection of BCR reference samples will help to alleviate this problem. In the meantime, many results are expressed as a percentage of the peak areas or assume that calibration factors do not differ from one sterol to another. This is unlikely to be the case, in fact, if only because the molecular weights vary (cholesterol = 387; sitosterol = 415). When the quantitative content of the sterols in the oil is required it is convenient to add an internal standard such as betulin or 5-α-cholestane to the sample. Alternatively the total sterol content may be separately determined, as described in IUPAC method 2.404, by enzymic oxidation of the separated sterols to sterones using cholesterol oxidase, which gives quantitative formation of hydrogen peroxide. The hydrogen peroxide is used to convert methanol quantitatively to formaldehyde, which in turn converts acetylacetone to lutidine, which is determined spectrophotometrically. The collaborative evaluation of this method has been reported by Naudet and Hautfenne.[139]

The method specified is quite tedious, having several stages, all of which are prone to experimental error. During the saponification stage some of the sterols might remain unsaponified, especially if they are more resistant to saponification than the triglycerides. Sterols still in the form of steryl esters will of course separate differently on the TLC plate, and not be recovered, giving variable results both in terms of total recovery and possibly compositional estimate.[133,134] Other sources of error are extraction of the unsaponifiable material, and quantitative recovery of the sterols from the TLC Plate. It is also possible that oxidation of the sterols could take place while they are on the TLC plate as they are then spread in a thin layer giving maximum access to atmospheric oxygen. This problem is most likely to occur if solvent is allowed to evaporate from the plate over an extended period at an increased temperature or while exposed to bright light, e.g. sunlight. It is reported[113,134] that silylated sterol derivatives can decompose on storage, even after only a few days at 4°C, giving rise to doubts about some of the widely varying sterol distributions in early literature.[133,134]

In an effort to avoid some of these problems, some workers[140] have eliminated the TLC stage, instead forming the steryl derivatives directly from the extract of unsaponifiable matter. This mixture is then immediately subjected to analysis by capillary column GLC, and it is claimed[140] that the increased resolution of the capillary column enables identification of the appropriate sterol peaks. It is also claimed[140] that the tocopherols in most oils can be determined at the same time, although it is admitted that in some cases the preliminary purification is still necessary. In the present author's laboratory, this abbreviated method was found wanting, as the α-tocopherol peak overlapped with that of the small amount of cholesterol masking its presence. In fact, the proponents of this method do not report the known presence of cholesterol, and its varying but unresolved contribution may explain their greater variation for α-tocopherol (CV = 1·4%) in comparison with the other sterol compounds (CVs of 0·2–0·6%). Homberg and Bielefeld[141] studied the influence of the minor components on the determination of sterols and found that failure to carry out the purification stage caused errors in both the relative composition of the sterols and in their overall concentrations. This was attributed evenly to the interference of 4-methyl sterols and triterpenes present in the unsaponifiable material. They report that the errors introduced depend on the amount and composition of the 4-methyl sterols and triterpenes in the oil in comparison with those of the conventional sterols.

An alternative procedure is specified in IUPAC method 2.402. The oil or fat is first saponified with ethanolic potassium hydroxide. Water is added, and for a qualitative examination the fatty acids are precipitated by addition of hydrochloric acid and filtered off on a previously moistened filter paper. The fatty acids are melted and treated with an ethanolic digitonin solution which precipitates the steryl digitonides. The digitonides are filtered off, washed, dried and converted to acetates by treatment with refluxing acetic anlydride. After recrystallisation two or three times from hot ethanol the melting point of the mixture of the steryl acetates may be determined. The free sterols may be prepared by saponification of the acetates with ethanolic potassium hydroxide, and recrystallised on a microscope slide from ethanol to enable examination of the crystallinity. The melting point range of plant sterols (phytosterols) is 126–127°C, and of animal sterols (zoo-sterols—mainly cholesterol) 113–115°C. Under a microscope phytosterols are rectangular and show parallel extinction under polarised light, while zoo-sterols have broader crystals with oblique extinction.

If quantitative determination of the sterols is needed, the digitonin solution is added directly to the homogeneous solution obtained from the initial saponification of the fat. The solution is cooled and allowed to stand for several hours, preferably overnight. The steryl digitonides crystallise out and may be filtered off and weighed. The content (% m/m) S of sterols in the fat is then given by the formula

$$S = 24a/m$$

where a = mass of the digitonides, and m = mass of the test portion.

Digitonin is of course a very hazardous heart stimulant. It should be handled with gloves and, as it is usually in the form of a fine powder efforts must be taken to avoid inhalation, for example, use of a fume cupboard and early preparation of stock solutions.

Homberg and Bielefeld[142,143] have reviewed sterol analyses by a number of alternative methods, while in an earlier series of papers[144-7] they reported on the contents of free and bound sterols, including sterol esters and acylated sterol glucosides, and the influence of refining on these. Kochhar[134] reviewed the literature on the influence of refining on the composition of sterols in edible vegetable oils, while Worthington and Hitchcock[148] developed a method using column chromatography, TLC and GLC for the separation and quantification of free sterols and sterol esters in groundnut and maize germ oils. The analysis of sterols in olive oil has been extensively studied in Mediterranean countries; Leone *et al.*[149] presented work on the sterol, methyl sterol and dimethyl sterol compositions of olive oils at various stages of processing, and in comparison with those of several other vegetable oils. Extensive work on sterol compositions has also been reported by Castang[150] and Itoh *et al.*[151]

Work carried out at the Leatherhead Food RA on the authenticity of oils and fats,[55,56] funded by FOSFA, MAFF, the Food RA and the Egyptian Government, led to the determination of sterol compositions for a wide range of samples covering nine common oil types.[152] The results of this work, which was carried out by M. J. Downes, are shown in Table 7.11. The FOSFA Guideline specifications,[16] and several Food RA reports,[152] contain additional information about this work.

7.4.2 Tocopherols

Most vegetable oils contain tocopherols, the more unsaturated oils containing higher concentrations—up to 1000 mg/kg or more—while the more saturated vegetable oils like coconut or palm kernel contain almost none. There are several types of tocopherol, the alpha-, beta-, gamma- and delta-forms differing from one another in the position and number of methyl groups on the phenol ring. There is also a series of corresponding tocotrienols in which the 16 carbon side chain is unsaturated. The eight compounds, collectively known as tocols, are sometimes called tocopherol *isomers,* but this is not strictly accurate as the compounds have different molecular formulae.

Tocols are natural antioxidants, the higher levels in unsaturated vegetable oils thus offering greater protection. Animal fats, in contrast, have no natural tocopherol content, whilst fish oils may contain small amounts.[153] The tocotrienols have greater antioxidant potential than the tocopherols, while delta tocopherol is more potent than alpha tocopherol. Tocols also have vitamin E potency, and in contrast with their antioxidant potential, alpha tocopherol has the greatest activity. It is therefore prepared synthetically and often added to foods such as domestic margarine, usually in the form of tocopheryl acetate. This

TABLE 7.11

Ranges (means) of sterol compositions in vegetable oils (percentage of sterol fraction; totals in mg/kg) determined on an OV-17 column in GLC stage of analysis

Sterol	Oils analysed								
	Palm kernel	Coconut	Cottonseed	Soyabean	Maize	Groundnut	Palm	Sunflower seed	Rapeseed
Cholesterol	1·0–3·7 (1·7)	0·6–3·0 (1·7)	0·7–2·3 (1·0)	0·6–1·4 (0·9)	0·2–0·6 (0·4)	0·6–3·8 (1·5)	2·7–4·9 (3·5)	0·2–1·3 (0·5)	0·4–1·3 (0·7)
Brassicasterol	ND–0·3 (0·1)	ND–0·9 (0·5)	0·1–0·9 (0·3)	ND–0·3 (0·1)	ND–0·2 (0·05)	ND–0·2 (0·0)	ND (ND)	ND–0·2 (<1)	5·0–13·0 (9·6)
Campesterol	8·4–12·7 (10·0)	7·5–10·2 (8·7)	7·2–8·4 (7·9)	15·8–24·2 (19·5)	18·6–24·1 (21·3)	12·3–19·8 (17·0)	20·6–24·2 (23·0)	7·4–12·9 (9·5)	18·2–38·6 (34·1)
Stigmasterol	12·3–16·1 (13·7)	11·4–13·7 (12·5)	1·2–1·8 (1·4)	15·9–19·1 (17·5)	4·3–7·7 (5·5)	5·4–13·3 (8·7)	11·4–11·8 (11·7)	8·6–10·8 (9·4)	ND–0·7 (tr)
Sitosterol	62·6–70·4 (67·0)	42·0–52·7 (46·7)	80·8–85·1 (83·2)	51·7–57·6 (54·3)	54·8–66·6 (63·4)	48·0–64·7 (58·5)	56·7–58·4 (57·7)	56·2–62·8 (59·9)	45·1–57·9 (49·9)
Δ^5-Avenasterol	4·0–9·0 (6·2)	20·4–35·7 (26·6)	1·9–3·8 (2·6)	1·9–3·7 (2·4)	4·2–8·2 (5·7)	8·3–18·8 (12·3)	2·1–2·7 (2·5)	1·9–6·9 (3·4)	ND–6·6 (4·5)
Δ^7-Stigmastenol	ND–2·1 (0·6)	NS–3·0 (2·4)	0·7–1·4 (1·0)	1·4–5·2 (2·9)	1·0–4·2 (1·8)	ND–5·2 (2·1)	0·4–0·8 (0·7)	7·0–13·4 (10·4)	ND–1·3 (0·5)
Δ^7-Avenasterol	ND–1·4 (0·1)	0·6–3·0 (1·1)	1·4–3·3 (0·7)	1·0–4·6 (0·6)	0·7–2·7 (0·5)	ND–5·5 (1·2)	0·4–2·5 (1·0)	3·1–6·5 (4·8)	ND–0·8 (0·4)
Others	ND–2·7 (0·7)	ND–3·6 (1·1)	ND–1·5 (0·7)	ND–1·8 (0·6)	ND–2·4 (0·5)	ND–1·4 (0·5)		ND–5·3 (2·0)	ND–4·2 (0·5)
Total (mg/kg)	792–1 187 (1 025)	470–1 110 (807)	2 690–5 915 (4 490)	1 837–4 089 (3 199)	7 931–22 137 (13 776)	901–2 854 (1 575)	389–481 (446)	2 437–4 545 (3 387)	4 824–11 276 (7 516)

ND = not detected; NS = not separated; tr = trace.

has good storage properties, but no antioxidant activity. It is hydrolysed to free tocopherol during digestion.

In addition to their vitamin activity, tocopherols and especially tocotrienols have been claimed to give protection against heart attack.[154] This has in turn led to claims that oils containing tocotrienols, such as palm oil, and oils obtained from cereals, such as wheat, oats or barley, may be beneficial in this respect.

The concentrations of the different tocol compounds vary from oil to oil, perhaps in conformity with the different requirements of the corresponding plants. In addition to providing a useful nutritional guide to the vitamin E content of the oil, a knowledge of its tocol composition may also be a useful purity criterion. As tocopherols are antioxidants, their concentrations will fall as an oil ages or becomes oxidised. Processing also reduces their levels, and in view of these factors the tocol composition must be used with caution in any purity assessment. However, the ratios of the concentrations remain reasonably constant. Above all, most oils contain an almost zero concentration of at least one of the tocols, while other oils may contain high levels of this same compound. An elevated level of a normally absent tocol is therefore clear proof of impurity or adulteration.[55,56,71,152]

In contrast to the reduction in tocopherol levels caused by most forms of oil processing or storage, high temperature or hydrogenation can caused an apparent increase.[155] This is because mild oxidation of the oil can cause the formation of an ether by reaction of the tocopherol with the hydroperoxide of, say, linoleic acid. Heating, and especially hydrogenation, can cause rupture of the ether linkage with liberation of the tocopherol and an apparent increase in concentration. This little known reaction has led to several analytical disputes about the tocopherol contents of oils. It must also be remembered that tocopheryl acetates are added to some foods. Methods that determine total tocopherol will, in these cases, clearly give a higher answer than methods that respond only to the free tocopherol.

The method preferred at the Leatherhead Food RA for the determination of free tocols in vegetable oils was developed initially by Hatina and Thompson,[156] and subsequently employed by Taylor and Barnes.[157] This is an HPLC method, which involves direct injection of a solution of about 0·5 g of oil in 10 ml heptane on to an HPLC instrument fitted with a 50 mm × 50 mm internal diameter (i.d.) guard column with He-Pellosil packing, and a 250 mm × 4·9 mm analytical column packed with 5 μm Partisil 5, or LiChrosorb SI 605. A solvent mixture of heptane: damp heptane: isopropanol (49·55; 49·55; 0·9) is pumped through the column at a flow rate of 1 ml/min, and detection is by fluorescence with an excitation wavelength of 290 nm and an emission wavelength of 330 nm. When tocol esters are present it is necessary to saponify the fat to liberate the free tocols, but extreme care should be taken to avoid oxidation, as the free tocols are much more reactive in alkaline solution. The samples may therefore be protected by pyrogallol, and kept at −20°C prior to analysis. Frozen oils and solid fats should be warmed as gently as possible during melting to avoid degradation by overheating, and they should be protected from incident UV light (e.g. sunlight). If a composite food is being analysed, the oil should be extracted by a method unlikely to damage the tocols, as described in Chapter 6. The preferred Food RA procedure is described in full by Slack.[158]

A closely related method using fluorescence or UV abosrbance detection appears in IUPAC method 2.432. This test is also being considered for ISO 9936, still in draft form at the time of writing. This method has been collaboratively tested and developed over a number of years until it is now at a good stage of reliability. The paper by Pocklington and Dieffenbacher[159] describing the series of collaborative trials mentions a number of factors found to influence the accuracy of this method, in particular the low purities of some commercial tocopherol standards, and the advantage of fluoresence over UV detection. Provision is made for the differing UV absorbances of the separate tocopherols during preparation of the reference samples, methanolic solutions of 1 µg/ml of alpha, beta, gamma and delta-tocopherols having absorbances of 0·0076 at 292 nm, 0·0089 at 296 nm, 0·0091 at 298 nm and 0·0087 at 298 nm, respectively. The tocotrienols have very similar UV absorbances to the corresponding tocopherols. Identification of the separate tocols can be made by reference to the relative retention times (RRTs) shown in Table 7.12,[160] provided the conditions mentioned above are used. Calibration factors and RRTs for a number of other HPLC systems have been published by Gertz and Herrman.[161]

Speek et al.[162] used an HPLC method with fluorescence detection, similar to that described above, for the determination of tocols in a number of seed oils, finding no significant difference between hot and cold pressed oils from the same origin, although they did find a variation depending on the origin of the seed. In this work the identity of the separated tocol compounds was confirmed by mass spectro-

TABLE 7.12
Ranges (means) of tocopherol and tocotrienol levels in vegetable oils

Tocol	Relative retention times	Oils analysed									
		Palm kernel[a,b]	Coconut[a]	Cotton-seed	Soyabean	Maize	Ground-nut	Palm	Sun-flower seed	High erucic rapeseed	Low erucic rapeseed
αT	1·0	ND–44	ND–17	136–543 (338)	9–352 (99·5)	23–573 (282)	49–304 (178)	4–185 (89)	403–855 (670)	39–305	100–320 (202)
βT	1·5	ND–248	ND–11	ND–29 (16·9)	ND–36 (7·7)	ND–356 (54)	0–41 (8·8)	—	9–45 (27)	24–158	16–140 (65)
γT	1·8	ND–257	ND–14	158–594 (429)	409–2 397 (1 021)	268–2 468 (1 034)	99–389 (213)	6–36 (18)	ND–34 (11)	230–500	287–753 (490)
δT	2·7	—	ND–2	ND–17 (3·3)	154–932 (421)	23–75 (54)	3–22 (7·6)	—	ND–7 (0·6)	5–14	4–22 (9)
αT_3	1·1	ND–tr	ND–5	—	—	ND–239 (49)	—	4–336 (128)	—	—	—
βT_3	1·7	—	—	—	—	ND–52 (8)	—	—	—	—	—
γT_3	1·95	ND–60	ND–1	—	—	ND–450 (161)	—	42–710 (323)	—	—	—
δT_3	3·2	—	—	—	—	ND–20 (6)	—	tr–148 (72)	—	—	—
Total		ND–257	tr–31	410–1 169 (788)	575–3 320 (1 549)	331–3 402 (1 647)	176–696 (407·4)	98–1 327 (630)	447–900 (709)	312–928	424–1 054 (766)

[a] Means are negligible.
[b] High values may be due to migration of some palm oil into the palm kernels before separation.
Means are shown below ranges. T = tocopherol; T_3 = tocotrienol.

scopy, establishing that the tocols elute in the order alpha, beta, gamma, delta, as shown in Table 7.12, RRTs given elsewhere[159] being incorrect.

Fourie and Basson[163] also used a related HPLC method to study changes in the tocopherol contents of several types of nut during 16 months storage at 30°C. Almond kernels, which had the highest initial tocopherol levels, had the best shelf-life, the total tocopherol contents of all the samples falling during storage. This was clearly due to the interaction of the tocopherols in the oxidation process through their antioxidant activity.

An alternative approach is described in IUPAC method 2.411 (7th edn, method 2.404 in 6th edn). The fat is first saponified with potassium hydroxide solution, in the presence of pyrogallol to inhibit oxidation, the unsaponified matter is extracted from the aqueous soap solution with diethyl ether, and the resulting solution is evaporated to dryness. The residue is redissolved in hexane, again evaporated to dryness and redissolved in heptane prior to immediate separation into components by TLC. Steps to avoid oxidation of the tocopherols during the extraction and thin layer chromatography are specified, and the separate components are then identified by spraying reference areas of the plate with ferric chloride and 2,2'-dipyridyl solutions. The appropriate parts of the silica layer are then quantitatively removed and the tocopherols are extracted from the silica with solvent. After evaporation of the solvent the tocopherols are determined colorimetically by treatment with solutions of 5,5'-dipyridyl and ferric chloride, the absorbances being measured at 520 nm against a blank. Alternatively the separated tocopherols can be derivatised by treatment with hexamethyldisilazane and trimethylchlorosilane in the presence of pyridine, and analysed by GLC. Relative retention times of the tocols under the GLC conditions used are specified in the method. The AOAC method 43.129 describes a related technique of extraction, saponification, purification by TLC and estimation in a colour reaction with ferric chloride.

These long, involved colorimetric methods are not favoured by the present author, as there are several stages in which the tocols can be lost by oxidation. In particular, tocols are especially prone to oxidation in alkaline solution, as in the initial saponification stage. Also, there is danger of oxidation during the TLC operation, as the tocopherol is exposed as a thin film on the TLC plate. The technique has also been criticised by several other authors.[157,159,161] The method does, of course, recover total tocopherols including esters, a factor that should be kept in mind, in comparison with HPLC results. A colorimetric method involving direct reaction of oil in toluene solution with bipyridine and ferric chloride was nevertheless used to good effect by Wong *et al.*[164] to study changes in the concentrations of tocols in palm oil during processing, a simple clean up and direct reaction being sufficient for the comparative objectives of their work.

The polarographic, or voltametric, technique has found several champions.[164,165] In this method carbon electrodes are used, as the dropping mercury electrode oxidises at +0·4 V while potentials of about +2·0 V are required for assessment of tocopherols, easily achievable with a glassy carbon electrode. The tocopherols are directly extracted into an organic solvent suitable for the subsequent voltammetry. A good review of the various methods for determining tocopherols appears in the book by Ball.[166]

The previously mentioned work into edible oil purity at the Leatherhead Food RA, and which involved the analysis of a large number of different, authentic edible vegetable oil samples, led to the ranges and means presented in Table 7.12. Additional details are available in Leatherhead Food RA reports and papers,[55,56,71,79,152,160] and in the Guideline Specifiations published by FOSFA International.[16]

ACKNOWLEDGEMENTS

The author wishes to thank the IASC, NOFOTA, ANEC, NIOP, AFOA, NSPA NCPA, BGCC, and PORIM for permission to reproduce data in Tables 7.1–7.7, the BCR for Fig. 7.1 and Carlo Erba for Fig. 7.2. Full copies of trading specifications and BSI or ISO Standards can be obtained form the appropriate body as detailed in the Appendix.

Thanks are also due to Mrs G. Wheatley and Miss S. Goddard for help with typing the manuscript, and Mrs A. Pernet for editorial assistance.

REFERENCES

1. Gurr, M. I. In: *The role of fats in human nutrition*, ed. F. B. Padley and J. Podmore, Ellis Horwood, Chichester, 1985, p. 23.

2. Patterson, H. B. W., (Ed.) *Hydrogenation of fats and oils,* Applied Science Publishers, London and New York, 1983.
3. International Association of Seed Crushers, *Congress Proceedings,* 1985–1988, Rome, New Delhi, India, Vancouver.
4. Young, F. V. K. In: *Fats and oils, chemistry and technology,* ed. R. J. Hamilton and A. Bhati, Applied Science Publishers, London and New York, 1980, p. 135.
5. Various authors, *Refining of soy oil—caustic or physical?* Proceedings *Second American Soybean Association Symposium,* Antwerp, American Soybean Assocation, 1000 Brussels, Belgium, 1981.
6. Rossell, J. B. In: *Analysis of oils and fats,* ed. R. J. Hamilton and J. B. Rossell, Elsevier Applied Science Publishers, London and New York, 1986, p. 1.
7. Codex Alimentarius Commission, *Standards for edible fats and oils,* 1982 (Revised 1987, Codex Alinorm 87/17). Issued by the joint FAO/WHO Food Standards Programme, via delle Terme di Caracalla 00100, Rome.
8. Williams, K. A., (ed.) *Oils, fats and fatty foods,* 4th ed., J & A Churchill, London, 1986.
9. Gunstone, F. D., Harwood, J. L. and Padley, F. B., *The lipid handbook,* Chapman and Hall, London and New York, 1986.
10. Thomas, A. In *Ullmans encyclopaedia of industrial chemistry,* V. C. H. Verlagsgesellschaft mbH, Weinheim, West Germany, 1987.
11. Hudson, B. J. F., Personal communication, 1988.
12. Morris, A. J., Personal communication, 1988.
13. Rossell, J. B. In *Rancidity in foods,* (2nd edn) J. C. Allen and R. J. Hamilton, Elsevier Applied Science, London and New York, 1989.
14. Laubli, N. W. and Bruttel, P. A., *J. Am. Oil Chem. Soc.,* **63** (1986) 792.
15. Robards, K., Kerr, A. F. and Patsalides, E., *Analyst,* **113** (1988) 213.
16. Farrer, K. T. H. (ed.) *The Shipment of Edible Oils.* IBC Technical Services, London, 1990.
17. Hudson, B. J. F., *Lebensm-Wiss und Technol.,* **17** (1984) 191.
18. Patterson, H. B. W., *Handling and storage of oilseeds, oils, fats and meal,* Elsevier Applied Science, London and New York, 1989.
19. Sinram, R. D., *J. Am. Oil Chem. Soc.,* **63** (1986) 667.
20. Pardun, H., *Fette Seifen Anstrichmittel,* **83** (1981) 240.
21. Chapman, G. W., *J. Am. Oil Chem. Soc.,* **57** (1980) 229.
22. Du Plessis, L. M. and Pretorius, H. E., *J. Am. Oil Chem. Soc.,* **60** (1983) 1261.
23. Hamilton, R. J. In: *Analysis of oils and fats,* eds R. J. Hamilton and J. B. Rossell, Elsevier Applied Science, London, 1986, p. 254.
24. Sotirhos, N., Ho, C. T. and Chang, S. S., *Fette Seifen Anstrichm.,* **88** (1986) 6.
25. Szuhaj, B. F. and List, G. R. (eds) *Lecithins,* American Oil Chemists Society, Champaign, Illinois, 1985.
26. Various authors, Papers from a symposium, *J. Am. Oil Chem. Soc.,* **58** (1981) 886.
27. Ostric-Matijasevic, B. and Turkulov, J., *Rev. Franc. Corps, Gras.,* **20** (1973) 5.

28. Morrison, W. H., Akin, D. E. and Robertson, J. A., *J. Am. Oil Chem. Soc.*, **58** (1981) 969.
29. Pritchard, J. R. L., *J. Am. Oil Chem. Soc.*, **60** (1983) 328.
30. Rivarola, G., Anon, M. C. and Calvelo, A., *J. Am. Oil Chem. Soc.*, **65** (1988) 1771.
31. Turkulov, J., Dimic, E., Karlovic, Dj and Vuska, V., *J. Am. Oil Soc.*, **63** (1986) 1360.
32. Downes, M. J. (1989) Scientific and Technical Survey No. 167, Leatherhead Food RA.
33. Gracien, J. and Arevalo, G., *Gracas y Aceitas*, **32** (1981) 1.
34. International Association of Seed Crushers, *Sub-Committee handbook, analysis of oilseeds, oils and fats,* 3rd edn, IASC, 1980, p. 75.
35. Cancalon, P., *J. Am. Oil Chem. Soc.*, **48** (1971) 629.
36. Brimsberg, U. C. and Wretensjo, I. C., *J. Am. Oil Chem. Soc.*, **56** (1979) 857.
37. Morrison, W. H., *J. Am. Oil Chem. Soc.*, **59** (1982) 284.
38. Moulton, K. J., *J. Am. Oil Chem. Soc.*, **65** (1988) 367.
39. Berner, D., American Oil Chemists Society. Technical Director, Personal communication, 1988.
40. McGinley, L., Personal communication, 1989.
41. Henon, G., *Rev. Franc. Corps, Gras.*, **33** (1986) 475.
42. Lawrence, J., Iyengar, J. R., Page, B. D. and Conacher, H. B. S., *J. Chromatogr.*, **236** (1982) 403.
43. Kleiman, R., Earle, F. R. and Wolff, I. A., *J. Am. Oil Chem. Soc.*, **46** (1969) 505.
44. Guillaumin, R. and Drouhin, N., *Rev. Franc. Corps, Gras.*, **13** (1966) 21.
45. Thomas, A., Personal communication, 1989.
46. Vigneron, P. Y., Henon, G., Monseigny, A., Levacq, M., Stochlin, B. and Delvoye, P., *Rev. Franc. Corps. Gras.*, **29** (1982) 423.
47. Association of Official Analytical Chemists, *Official methods of analysis,* 14*th edn,* Detection of Fish and Marine Animal Oils, Test 28.132, 1984, p. 526.
48. Lauro, M. F., *J. Am. Oil Chem. Soc.*, **47** (1970) 234.
49. Schwein, W. G., *J. Assoc. Off. Anal. Chem.*, **57** (1974) 1005.
50. Mehlenbacher, V. C., *The analysis of fats and oils,* (1960), Garrard Press, Champaign, Illinois, 1960, p. 277.
51. Ghose, M. N. and Pal, H. K., *Analyst,* **60** (1935) 240.
52. Koman, V. and Kotuc, J., *J. Am. Oil Chem. Soc.*, **53** (1953) 563.
53. Spencer, G. F., Kowlek, W. F. and Princen, L. H., *J. Am. Oil Chem. Soc.*, **56** (1979) 972.
54. American Oil Chemists Society, Recommended practice, Tz 1c-85, Calculated iodine value, 1985.
55. Rossell, J. B., King, B. and Downes, M. J., *J. Am. Oil Chem. Soc.*, **60** (1983) 333.
56. Rossell, J. B., King, B. and Downes, M. J., *J. Am. Oil Chem. Soc.*, **62** (1985) 221.
57. Christie, W. W., *Analyst,* **97** (1972) 221.

58. Darbre, A., *Handbook of derivatives for chromatography*, ed. K. Blan and G. S. King, Heyden & Son, London, 1978, p. 36.
59. Sheppard, A. J. and Iverson, J. L., *J. Chromatogr. Sci.*, **13** (1975) 448.
60. Christie, W. W., *Lipid analysis*, 2nd edn, Pergamon Press, Oxford, New York, Sydney, 1982, p. 52.
61. Hammond, E. W., *Analysis of fats and oils*, ed. R. J. Hamilton and J. B. Rossell, Elsevier Applied Science, London, 1986, p. 122.
62. Iverson, J. L. and Sheppard, A. J., *J. Ass. Off. Anal. Chem.*, **60** (1977) 284.
63. Ackman, R. G. In: *Analysis of fats and oils*, ed. R. J. Hamilton and J. B. Rossell, Elsevier Applied Science, London, 1986, p. 137.
64. Hitchcock, C. and Hammond, E. W. In: *Developments in Food Analysis Techniques—2*, Ed. R. D. King, (1980). Elsevier Applied Science, London, 185.
65. Christie, W. W. In: *Fats and oils—chemistry and technology*, ed. R. J. Hamilton and A. Bhati, Applied Science Publishers, London, 1980, p. 1.
66. Craske, J. D. and Bannon, C. D., *J. Am. Oil Chem. Soc.*, **64** (1987) 1413.
67. Smith, R. M., *Gas and liquid chromatography in analytical chemistry*, John Wiley, Chichester, New York, Brisbane, 1988.
68. Cooke, M. V. and Gillatt, P. N., Report to FOSFA International on Collaborative Test P13, Personal communication, 1984.
69. Wagstaffe, P. J. In: *Biological reference materials*, ed. W. R. Wolf, John Wiley, New York, Chichester, Toronto, 1985, p. 63.
70. Pocklington, D. In: *Analysis of oils and fats*, ed. R. J. Hamilton and J. B. Rossell, Elsevier Applied Science, London, New York, 1986, p. 91.
71. Rossell, J. B. and Turrell, J. A., *Food Flavourings Ingredients Processing Packaging*, **8**(3) (1986) 33.
72. Rossell, J. B., In: *Food industries manual*, ed. M. D. Ranken, Blackie & Sons, Glasgow, 1988, p. 168.
73. Traiter, H., *Prog. Lipid Res.*, **26** (1987) 257.
74. Ackman, R. G., *J. Am. Oil Chem. Soc.*, **66** (1989) 293.
75. Jordan, M. A., James, M. J. and Muddeman, C. C. A., Leatherhead Food RA Research Report No. 628, Members only, 1988.
76. Sebedio, J. L. and Ackman, R. G., *J. Am. Oil Chem. Soc.*, **60** (1983) 1986.
77. Chapman, J. R., *Practical organic mass spectrometry*, John Wiley, New York, Chichester, Toronto, 1985.
78. Fjeldsted, J. and Truche, J., *Int. Lab.*, **19** (1989) 28.
79. King, B. and Sibley, I., Leatherhead Food RA Research Report No. 462, Members only, 1984.
80. Jenson, R. G., Sampugna, J. and Pereira, R. L., *J. Dairy Sci.*, **47** (1964) 727.
81. Sampugna, J., Quinn, J. G., Pitas, R. E., Carpenter, D. L. and Jensen, R. G., *Lipids*, **2** (1967) 397.
82. Lawson, L. D. and Hughes, B. G., *Lipids*, **23** (1988) 313.
83. Codex Alimentarius Commission, Codex Standard 33–1981 updated at Feb. 1987 meeting, Codex Alinorm 87/17, Appendix VIII, 1987.

84. Van der Waal, R., *Advances in lipid research, Vol. 2*, Academic Press, New York, 1967, p. 1.
85. Coleman, M. H. and Fulton, W. C., *Enzymes of lipid metabolism*, Pergamon Press, New York, 1961, p. 127.
86. Litchfield, C., *Analysis of triglycerides*, Academic Press, New York, London, 1972, p. 250.
87. Rossell, J. B., Russell, J. and Chidley, J. E., *J. Am. Oil Chem. Soc.*, **55** (1978) 902.
88. Tomarelli, R. M. and Bernhart, F. W., US Patent 3542560, 1970.
89. Droese, W., Page, E. and Strolley, H., *Eur. J. Pediatr.*, **123** (1976) 277.
90. Kritchevsky, D., Tepper, S. A., Vesselinovitch, D. and Wissler, R. W., *Atherosclerosis*, **14** (1971) 53.
91. Kritchevsky, D., Tepper, S. A., Vesselinovtich, D. and Wissler, R. W., *Atherosclerosis*, **17** (1973) 225.
92. Myher, J. J., Marai, L., Kuksis, A. and Kritchevsky, D., *Lipids*, **12** (1977) 775.
93. Manganaro, F., Myher, J. J., Kuksis, A. and Kritchevsky, D., *Lipids*, **16** (1981) 508.
94. Food and Agriculture Organisation, *Dietary fats and oils in human nutrition—Report of an expert commission*, FAO, UNO, Rome, 1980.
95. Muderhwa, J. M., Dhuique-Mayer, C., Pina, M., Galzy, P., Grignac, P. and Graille, J., *Oléagineux*, **42** (1987) 207.
96. Holman, R. T., *Prog. Lipid Res.* (1981) 907.
97. Padley, F. B. and Podmore, J., *Role of fats in human nutrition*, Ellis Horwood, Chichester, 1985.
98. Lee, T. W., *J. Assc. Off. Anal. Chem.*, **70** (1987) 702.
99. MacGee, J., *Anal. Chem.*, **31** (1959) 298.
100. Prosser, A. R., Sheppard, A. J. and Hubbard, W. D., *J. Assc. Off. Anal. Chem.*, **60** (1977) 895.
101. Sheppard, A. J., Waltking, A. E., Zmachinski, H. and Jones, S. T., *J. Assc. Off. Anal. Chem.*, **61** (1978) 1419.
102. Madison, B. L. and Hughes, W. J., *J. Assc. Off. Anal. Chem.*, **66** (1983) 81.
103. IUPAC Method 2.209 Note 6, *IUPAC standard methods for the analysis of oils, fats and derivatives*, 7th Revised and Enlarged edn, ed. C. Paquot and A. Hautfenne, Blackwell., Oxford, London, Melbourne, 1987, p. 110.
104. IUPAC Working Report 1978–1979 (issued May 1979), Appendix 9–79.
105. Levin, O., *Rev. Franc. Corps, Gras*, **27** (1980) 571.
106. Caverly, B. L., Meertens, G. J. H. and Rossell, J. B., British Patent No. 1214321, 1970.
107. Caverly, B. L. and Rossell, J. B., British Patent No. 1228139, 1971.
108. Cottier, D. C. and Rossell, J. B., British Patent No. 1521884, 1978.
109. Cochran, W. M., US Patent No. 2972 541; British Patent No. 867615, 1958.
110. Senti, F. R., *Health aspects of dietary trans fatty acids*, Special Report prepared for the Centre for Food Safety and Applied Nutrition, Food and Drug Administration, Department of Health and Human Services,

Washington, DC, under contract no. FDA 223-83-2020, by the Federation of American Societies for Experimental Biology (FASEB), 1985.
111. Department of Health and Social Security, Committee on Medical Aspects of Food Policy (COMA). Report of the Panel on Diet in Relation to Cardiovascular Disease, Report on Health and Social Subjects No. 28, HMSO, London, 1984.
112. Jordan, M. A., *Nutrition Bulletin,* **12** (1987) 156, British Nutrition Foundation, London.
113. Kochhar, S. P. and Matsui, T., *Food Chem.,* **13** (1984) 85.
114. Kochhar, S. P. and Rossell, J. B., *Int. Analyst,* **1** (1987) 23.
115. Scholfield, C. R. In: *Geometric and positional fatty acid isomers,* ed. E. A. Emken and M. J. Dutton, American Oil Chemists Society, Champaign, Illinois, 1979, p. 17.
116. Belton, P. S., Wilson, R. H., Sadeghi-Jorabchi, H. and Peers, K. E., *Lebensm. Wiss u. Technol.,* **21** (1988) 153.
117. Sleeter, R. T. and Matlock, M. G., *J. Am. Oil Chem. Soc.,* **66** (1989) 121.
118. Lanser, A. C. and Emken, E. A., *J. Am. Oil Chem. Soc.,* **65** (1988) 1483.
119. Ottenstein, D. M., Witting, L. A., Hometchko, D. J. and Pelick, N., *J. Am. Oil Chem. Soc.,* **61** (1984) 390.
120. Marchand, C. M. and Beare-Rogers, J. L., *Can. Inst. Food Sci. Technol. J.,* **15** (1982) 54.
121. Privett, O. S., Phillips, F., Shimasaki, H., Nozawa, T. and Nickell, E. C., *Am. J. Clin, Nutr.,* **30** (1977) 1009.
122. Strocchi, A., *La Riv. delle Sos. Grasse,* **63** (1986) 99.
123. Gurr, M. I., *Role of fats in food and nutrition.* Elsevier Applied Science Publishers, London, New York, 1984.
124. Barlow, S. M. and Stansby, M. E., *Nutritional evaluation of long-chain fatty acid in fish oil,* Academic Press, London, New York, 1982.
125. Commission directive of 20 July 1976, *Official Journal of the European Communities,* No. L202/35, 28-7-76 (76/621/EEC).
126. The Erucic Acid in Food Regulations 1977, Statutory Instrument No. 691, 1977.
127. Ackman, R. G., Barlow, S. M. and Duthie, I. F., *J. Chromatogr. Sci.,* **15** (1977) 290.
128. Player, R. B. and Wood, R., *J. Assc. Publ. Anal.,* **18** (1980) 77.
129. Ackman, R. G., Practical methods for the determination of erucic acid in oil/fat mixtures. In: *Nutritional evaluation of long-chain fatty acids in fish oil,* ed. S. M. Barlow and M. E. Stansby, Academic Press, London, New York, 1982.
130. Commission Directive of 25 July 1980, *Official Journal of the European Communities,* No. L254/35, 27-9-80 (80/891/EEC).
131. Bastic, M., Bastic, Lj., Jovanovic, J. A. and Spiteller, G., *J. Am. Oil Chem. Soc.,* **55** (1978) 886.
132. Itoh, T., Tamura, T. and Matsumoto, T., *J. Am. Oil Chem. Soc.,* **50** (1973) 122.
133. Trost, V. W., *J. Am. Oil Chem. Soc.,* **66** (1989) 325.

134. Kochhar, S. P. K., *Progr. Lipid Res.*, **22** (1983) 161.
135. Worthington, R. E. and Hitchcock, H. L., *J. Am. Oil Chem. Soc.*, **61** (1984) 1085.
136. Brumley, W. C., Sheppard, A. J., Rudolf, T. S., Shen, C. H. J., Yasaei, P. and Spon, J. A., *J. Assc. Off. Anal. Chem.*, **68** (1985) 701.
137. Seher, A., *Dtsch. Lebensm. Rundschau*, **82** (1986) 349.
138. Official Methods of the Association of Official Analytical Chemists, 14th edn, Method 28.110, AOAC, 1984, p. 522.
139. Naudet, M. and Hautfenne, A., *Rev. Franc. Corps. Gras.*, **33** (1986) 167.
140. Slover, H. T., Thompson, R. H. and Merola, G. V., *J. Am. Oil Chem. Soc.*, **60** (1983) 1524.
141. Homberg, E. and Bielefeld, B., *Fett. Wiss. Technol.*, **91** (1989) 105.
142. Homberg, E. and Bielefeld, B., *Fett. Wiss. Technol.*, **89** (1987) 255.
143. Homberg, E. and Bielefeld, B., *Fett. Wiss. Technol.*, **89** (1987) 353.
144. Homberg, E. and Bielefeld, B., *Fett. Wiss. Technol.*, **84** (1982) 141.
145. Homberg, E. and Bielefeld, B., *Fett. Wiss. Technol.*, **86** (1984) 135.
146. Homberg, E. and Bielefeld, B., *Fett. Wiss. Technol.*, **87** (1985) 61.
147. Homberg, E. and Bielefeld, B., *Fett. Wiss. Technol.*, **87** (1985) 333.
148. Worthington, R. E. and Hitchcock, H. L., *J. Am. Oil Chem. Soc.*, **61** (1984) 1085.
149. Leone, A. M., Liuzzi, V., La Notte, E. and Santaro, M., *Riv. Ital. Sost. Grasse*, **61** (1984) 69.
150. Castang, J., *Annls Falsif. Expert. Chim.*, **74** (1981) 697.
151. Itoh, T., Tamura, T. and Matsumoto, T., *J. Am. Oil Chem. Soc.*, **50** (1973) 122.
152. King, B., Turrell, J. A. and Zilke, S. A., Leatherhead Food RA Research Reports No. 563, 1986.
153. Young, F. V. K. and Rossell, J. B., Leatherhead Food RA Symposium Proceedings No. 33, 1986, p. 58.
154. Querishi, A. A., Burger, W. C., Peterson, D. M. and Elson, L. E., *J. Biol. Chem.*, **261** (1986) 10544.
155. Koch, G. K. Han, R. K. W., Hoogeboom, J. J. L., Mutter, M. and Van Tilborg, H., *Chem. Phys. Lipids*, **17** (1976) 85–7.
156. Hatina, G. and Thompson, J. N., *J. Liquid Chromatogr.*, **2** (1979) 327.
157. Taylor, P. and Barnes, P. *Chemy Ind.* (20) (1981) 722.
158. Slack, P. T., *Analytical methods manual*, Leatherhead Food RA, 1987.
159. Pocklington, W. D. and Diffenbacher, A., *J. Jpn Oil Chem. Soc. (Yugkagaku)*, **37** (1988) 1169.
160. King, B., Food RA Research Reports No. 438 (Members only), 1983.
161. Gertz, C. and Herrmann, K. Z. *Lebensm. Unters und Forsch.*, **174** (1982) 390.
165. Speek, A. J., Schrijver, J. and Schreurs, W. H. P., *J. Food Sci.*, **50** (1985) 121.
163. Fourie, P. C. and Basson, D. S., *J. Am. Oil Chem. Soc.*, **66** (1989) 1113.
164. Wong, M. L., Timms, R. E. and Goh, E. M., *J. Am. Oil Chem. Soc.*, **65** (1988) 258.
165. Hart, J. P., *Trends Anal. Chem.*, **5** (1986) 20.

166. Ball, G. F. M., *Fat soluble vitamin assays in food analysis—a comprehensive review*, Elsevier Applied Science, London, New York, 1988.

APPENDIX: OFFICIAL BODIES, FEDERATIONS AND ASSOCIATIONS THAT ISSUE CONTRACTS, TRADING RULES, GUIDELINE SPECIFICATIONS OR STANDARDS

American Oils and Fats Association Inc. (AFOA), Western Union International Plaza, New York, New York 10004, USA

Associanco Nacional Dos Exportadores De Cereais (ANEC), Av. Senador Queiroz, 611-Sao Paulo, Brazil

Board of Grain Commissioners for Canada, Winnipeg, Manitoba, Canada

National Cottonseed Producers Association, 2400 Poplar Avenue, Memphis, Tennessee 38112, USA

National Soyabean Processers Association (NSPA), 1255 Twenty-Third Street NW, Washington DC 20037, USA

Netherlands Oils, Fats and Oilseeds Trade Association (NOFOTA), Oppert 34, Rotterdam 3001, The Netherlands

Syndicat General des Fabricants d'Huiles et de Torteaux de France (SGFHT), 10 rue de la Paix, Paris, France

GROFOR Deutscher Verband des Grosshandels mit Oelen Fetten und Oelrohstoffen e.v., Gotenstrasse 21, D-2000 Hamburg 1, West Germany

Kobenhavns Bedommelses og Voldgiftudvalg for Korn og Foderstof-handelen, Copenhagen, Denmark

Office International du Cacao et du Chocolat (OICC), Verlag Max Glätti, Postfach 6934, Bioggio, Switzerland

International Olive Oil Council (IOOC), Juan Bravo 10 2°–3°, 28006 Madrid, Spain

American Soya Association, Centre International Rogier, Box 521, 1000 Brussels, Belgium

Codex Alimentarius Commission (of the Joint FAO/WHO Food Standards Programme), Via delle Terme di Caracalla 00100, Rome, Italy

Federation of Oils, Seeds and Fats Associations Ltd (FOSFA International), 19–21 Bury St, London, EC3A 5AU

National Institute of Oilseed Products (NIOP), 2600 Garden Road, 208 Monterey, California 93940, USA

Palm Oil Research Institute of Malaysia (PORIM), No. 6 Persiarian Instituti, Bandor Baru Bangi, 4300 Kajang, Selangor, Malaysia

NB: Standards are also available from National Standards Organisations such as the British Standards Institution (BSI).

8

Animal Carcass Fats and Fish Oils

M. ENSER

Department of Meat Animal Science, University of Bristol Veterinary School, Churchill Building, Langford, Bristol BS18 7DY UK

8.1. ANIMAL CARCASS FATS

8.1.1 Introduction

In this chapter the intention is to deal in some detail with the composition of animal fats, including sheep, beef, pig and fish oils. A description of the procedures used in extracting the fats is included as these can influence the chemical purity of the end product. More particularly the fatty acid and triacylglycerol compositions are discussed. It is shown that these vary with the depot site, fatness, age and sex of the animal and can moreover be regulated to some extent by the composition of fat inclusion in the diet. Non-fatty acid-containing compounds are also described, including steroids, with particular emphasis on cholesterol levels, bearing in mind the link between the cholesterol level of the diet and human health. Specific fatty acids are referred to by their chemical or common name or by the number of carbon atoms followed by the number of double bonds, e.g. *cis*-9-octadecenoic, oleic acid, 9c-18:1. When referring to unresolved mixtures of isomers, e.g. from packed-column GLC, 18:1 includes all positional and geometrical isomers having eighteen carbon atoms and one double bond.

8.1.2 Extraction Procedures

Commercially the rendering of animal fatty tissues consists of dispersion of the tissue by mincing followed by heating, either in water, in partially purified lipid from the previous cycle, or by dry heat or steam injection. The fat may then be recovered by separation, centrifugation or pressing. Fats produced by other processes may be clarified by centrifugation to remove solids and water. If the product is coloured it may be treated with bleaching earth, and high concentrations of free fatty acids, arising from poor quality tissue or the rendering process, may be removed by alkali refining. The chemical purity of the product will depend upon the processes used. Although the refined products may have come from lower quality starting materials using processes designed to obtain maximum yields, they may be whiter and more bland in flavour than the traditional lard or premier jus (top quality beef tallow) which retain their colour and flavour.

While the composition of the rendered fat is important for trading purposes and in nutrition, it is unlikely to be identical with that of the tissue in the living animal. Storage of the tissue before rendering may result in enzymic hydrolysis of some of the lipid, as will exposure to hot water. Some oxidation will be initiated at the high temperature in the presence of oxygen, especially as there will be release of iron from the haems present in the tissue. However, natural antioxidants in the meat, mainly α-tocopherol, will prevent significant oxidation occurring. Specific extraction processes have been developed for analytical purposes and these are discussed in depth in Chapter 6. All the lipid in fatty tissues is present in the fat cells or adipocytes. The lipid droplet present in the cells is surrounded by the aqueous phase of the cell and the cell membrane which, despite its lipid centre, is hydrophilic. To extract the intracellular triacylglycerols and the structural lipids of the membrane requires the destruction of the hydrophilic surfaces. This may be achieved through removal of water by freeze-drying or by the use of polar solvent mixtures which are miscible with a limited amount of water. These both remove the water from the aqueous phase of the cell and also denature the proteins of the cell membranes and other cell structures so that the lipid is readily extracted into the solvent.

Dry tissues are usually extracted with diethyl ether, petroleum spirit of boiling range 40–60°C or hexane, and the use of these solvents forms the basis of standard procedures for determining the fat content of meat. They have two disadvantages: they give low recoveries of triacylglycerols when the lipid content of the sample is low, for example muscle from very lean animals, and they extract only a portion of the tissue phospholipids,[1] the amount increasing as the lipid content of the sample increases.[2] Petroleum spirit, being less polar than diethyl ether, extracts approximately half as much phospholipid. Complete extraction of the polar and non-polar lipids is usually achieved with a chloroform:methanol mixture by the procedure of Folch *et al.*[3] or the more rapid procedure of Bligh and Dyer[4] and it is against these that other procedures are judged.[5,6] Diethyl ether:ethanol,[7] heptane:isopropanol,[8] hexane:isopropanol[9] and methylene chloride:methanol[10-12] have also been used to extract total lipid. In a comparison of these solvent systems Sahasrabudhe and Smallbone[2] found that extraction of freeze-dried meat with chloroform:methanol (2:1, v/v) or methylene chloride:methanol (2:1, v/v) using a soxhlet apparatus was as effective in extracting the total lipid as the Folch or Bligh and Dyer methods. The composition of the refluxing solvents was 12·5% methanol and 87·5% chloroform in the constant minimum boiling mixture. The ethanol:ether procedure gave consistently higher values for the polar lipid content of the meat than the other procedures. Despite differences in the efficiency of extraction, the fatty acid composition of the neutral and polar lipids were reported to be similar for all the procedures. However, further work is necessary to confirm that there is no loss of polyunsaturated fatty acids through oxidation in those procedures which involve hot solvents over periods of several hours. The dry column extraction procedure of Maxwell *et al.*[10] is reported to give values for total fat which are comparable with those obtained by the Folch procedure. A modification[13] allowed the neutral lipids and phospholipids to be eluted sequentially, thus avoiding the need for subsequent separation by other procedures. However, while the neutral lipid is recovered in a pure form, some of it is retained by the celite used in the separation and is subsequently eluted with the phospholipids. Whilst this will not affect quantitation based on lipid phosphorus it will bias gravimetric determinations and fatty acid compositions, particularly when they are markedly different, as in ruminants.

8.2. SHEEP LIPIDS

The fatty acid composition of sheep adipose tissue triacylglycerols is complex, with odd-chain, branched-chain and *trans* unsaturated fatty acids, as well as the more common even-chain and *cis* unsaturated monoenoic and polyenoic fatty acids present. However, there are no reports of the total composition which include *trans* isomers and those with branched chains since most of the latter are present in small quantities and their complexity makes resolution difficult. Although modern gas–liquid chromatography (GLC) capillary columns coated with polar silicones such as SP2340, silar 10C and CP sil 88 give good resolution of the *trans* isomers and many of the branched-chain fatty acids, identification is difficult. Since standards are not available for most of these minor components, gas chromatography mass spectrometry (GCMS) is the most reliable method but the loss of resolution creates problems with the branched-chain fatty acids.

8.2.1 Depot Site and Fatty Acid Composition

The fatty acid composition of two internal depots and four subcutaneous sites from mature sheep is shown in Table 8.1. The major fatty acids are palmitic, stearic and oleic. The internal perirenal (surrounding the kidneys) and mesenteric (between the intestines) depots contain more stearic acid and less 18:1 than the subcutaneous depots. The proportion of palmitic acid is similar in the two internal depots and the chest and rump subcutaneous sites, although less is present in the subcutaneous fat of the leg and ear. As the degree of exposure to the environment increases the fat becomes softer, mainly through a decrease in the proportion of stearic acid and an increase in the proportion of oleic acid. Margaric acid and br-17:0 acid behave like stearic acid whilst 16:1, 17:1, linoleic acid and linolenic acid follow oleic acid. The total *trans* fatty acid content determined by infra-red absorption ranged from 1% to 12% but they were not present in the fats from the leg and ear. In young lambs the differences in composition between the subcutaneous (loin) fat and the perinephric fat are reported to be much less.[15] In the subcutaneous depot the stearic acid (23%) was higher and the 18:1 (38.1%) lower whereas in the perirenal depot the stearic acid (27·1%) was lower and the 18:1 (37·9%) higher than in mature sheep. In both studies there were no marked differences in the proportions of linoleic acid between the internal and external sites. In the sheep linoleic acid was marginally higher in the external depots whereas in the lambs it was higher in the perirenal depot (7·6% compared with 7·2%). It must be pointed out that both of these values are rather high for lamb fat.

8.2.2 Effects of Fatness, Age and Sex on Fatty Acid Composition

Since fat animals are likely to be the main source of rendered lipid, and as fatter animals are usually older, the effect of age on fatty acid composition is of relevance. The sex of lambs is also important because of its effect on carcass fatness. Up to 1 year of age the fat of

TABLE 8.1

Fatty acid composition of adipose tissue triacylglycerols from different sites in sheep carcasses

Depot site	14:0	br-15:0	16:0	16:1	br-17:0	17:0	17:1	18:0	18:1	18:2	18:3
Perirenal	2·4 ± 0·22[a]	0·73 ± 0·12	25·6 ± 1·2	1·4 ± 0·06	1·4 ± 0·04	1·7 ± 0·15	0·5 ± 0·02	33·9 ± 1·20	29·5 ± 1·49	0·9 ± 0·14	0·7 ± 0·16
Mesenteric	2·0 ± 0·21	0·55 ± 0·08	24·2 ± 1·3	1·3 ± 0·11	1·6 ± 0·08	1·7 ± 0·24	0·3[b]	33·3 ± 1·79	32·5 ± 2·02	1·1 ± 0·13	0·7 ± 0·14
Chest	2·3 ± 0·18	0·53 ± 0·06	26·2 ± 1·2	1·6 ± 0·11	1·5 ± 0·09	1·5 ± 0·16	0·5 ± 0·07	20·7 ± 1·84	42·6 ± 3·21	1·0 ± 0·04	1·0 ± 0·25
Rump	2·2 ± 0·20	0·52 ± 0·06	23·4 ± 1·3	3·0 ± 0·21	1·3 ± 0·13	1·0 ± 0·14	1·0 ± 0·11	11·1 ± 0·50	53·2 ± 1·91	1·2 ± 0·12	1·4 ± 0·43
Leg	0·9 ± 0·11	ND	17·0 ± 0·9	5·8 ± 0·53	0·6 ± 0·10	0·3 ± 0·06	1·2 ± 0·17	3·0 ± 0·54	69·2 ± 1·05	1·2 ± 0·17	0·9 ± 0·20
Ear	1·3 ± 0·06	ND	17·1 ± 1·0	7·8 ± 1·01	0·6 ± 0·27	0·4 ± 0·09	1·2 ± 0·14	3·1 ± 0·55	63·6 ± 1·01	2·0 ± 0·32	1·9 ± 0·20

[a] Results are weight % ± SEM (standard error of the mean) for 4–6 sheep age 4–5 years.

[b] Only detected in 2 animals.

ND not detected. Based on Duncan and Garton.[14]

lambs becomes harder with an increase in stearic acid and a decrease in oleic acid.[15,16] Subsequently there is a decrease in the proportion of stearic acid and an increase in oleic acid,[17] which the authors considered to be a fatness effect since covariant analysis to the same body weight gave a similar composition to the body fat at 15 and 27 months of age. Linoleic acid is very low at birth, increases initially and then slowly decreases. In lambs growing from 32 to 52 kg it fell from 7·0 to 5·4% in the subcutaneous fat whilst stearic acid increased from 17·2 to 26·3%, and similar changes occurred in the perinephric fat.

Whereas there were very small differences between the fatty acid composition of fat from ewes and wethers (castrated males),[15] other work has revealed marked differences between the fatty acid composition of subcutaneous fat from rams and wethers[18] (Table 8.2). These differences probably stem, at least in part, from differences in fatness: ewe lambs and wethers are equally fat but ram lambs are much leaner at the same age. The major differences were lower proportions of palmitic acid, stearic acid and 18:1 in the rams, mainly caused by an increase in odd-numbered and branched-chain fatty acids. Concentrations of linoleic acid were similar in both groups and at the usual level for lamb fat. In a study in which branched-chain fatty acids were at a much lower level in the fat, the wethers also had higher proportions of

TABLE 8.2
Fatty acid composition of subcutaneous fat from ram
and wether lambs

Fatty acid	Fatty acids (% by weight)	
	Rams	Wethers
14:0	1·7	1·7
15:0	1·4	0·9[a]
16:0	13·0	15·2[a]
16:1	1·5	1·0[a]
17:0	2·5	2·7[a]
18:0	5·2	9·0[a]
18:1	32·8	34·9[a]
18:2	2·7	2·5
Branched	6·2	3·7[a]

Results are means for 32 animals in each group.
[a] Means are significantly different, $P < 0·05$.
Taken from Busboom *et al.*[18]

stearic acid and 18:1 but the proportions of linoleic and linolenic acids were lower in the subcutaneous fat.[19] Surprisingly these differences occurred despite the absence of significant differences in fatness.

8.2.3 Effects of Dietary Fat on Fatty Acid Composition

The effects of dietary fat on the fatty acid composition of sheep depot fats are relatively small because of the partial or complete hydrogenation of the unsaturated fatty acids in the rumen. During the reduction of polyunsaturated fatty acids new positional or geometrical isomers are produced such as *trans* vaccenic acid, although only small amounts of conjugated or *trans* di-unsaturated fatty acids occur. The degree of hydrogenation depends upon the nature of the rest of the diet. Diets high in roughage cause maximum hydrogenation whereas it is somewhat less when diets are high in available carbohydrate and low in roughage. When relatively high concentrations of linoleic acid are fed under these conditions the level in the depot fats may rise to 7–8% (Table 8.3).[15,19,20]

It is possible to treat dietary lipid to prevent its metabolism in the rumen. Essentially this converts the system to that in monogastric

TABLE 8.3

Sheep fed control and sunflower oil-supplemented diets: dietary and depot fatty acid composition

Fatty acid	Dietary fatty acids (g/kg)		Depot fatty acids (%)			
			Subcutaneous		Perirenal	
Diet	C	S	C	S	C	S
14:0	—	—	2·2	2·4	2·3	2·3
16:0	5·0	8·9	27·1	24·4	24·9	21·2
16:1	—	—	2·6	2·2	1·4	1·6
18:0	0·4	1·8	16·1	15·7	32·9	27·2
18:1	4·5	16·5	44·0	44·0	33·2	36·9
18:2	11·8	44·1	2·6	7·7	2·1	7·0
18:3	0·6	0·4	0·1	0·6	0·1	0·3
15:0 + 17:0	—	—	2·3	1·6	1·6	1·5
br-15:0 + br-17:0	—	—	1·7	0·9	1·0	0·9

C, control diet; S, sunflower oil-supplemented diet.
Data taken from Gibney and L'Estrange.[20]

animals, with the result that very high concentrations of linoleic acid are deposited in adipose tissue. The process involves microencapsulation of the oil by producing an oil-in-water emulsion using added protein or that in the oilseed, and then cross-linking the protein with formaldehyde.[21,22] The cross-linked protein is stable in the rumen but is hydrolysed in the duodenum and the fat is released, hydrolysed and absorbed. The concentration of linoleic acid in the adipose tissue can reach 20–30% within 6 weeks of starting to feed protected oils. Although developed as a way to improve the nutritional acceptability of lamb fat by making it less unsaturated and hence potentially less atherogenic, the process is little used because of the cost and problems of oxidative stability of the fatty tissues. Nonetheless, it means that the presence of high concentrations of linoleic acid in sheep tallow is not necessarily a sign of adulteration.

Dietary components, other than fatty acids, can have significant effects on the type of fatty acids found in the depot fat of animals as a result of their effects on the *de novo* synthesis of fatty acids. In lambs the energy density of the diet is particularly important. Forage diets have a low energy density and restrict the rate of animal growth, particularly fat deposition. Cereal grains, however, contain large amounts of readily digested starch which, when added to a forage diet, markedly increases its energy density. In the rumen the starch is degraded to volatile fatty acids, mainly acetic and propionic, which are readily absorbed and used as the substrates for fatty acid synthesis. Acetic acid is the normal substrate for fatty acid synthesis, yielding even-chain-length saturated fatty acids which may be desaturated to monoenoic acids. Propionic acid can replace acetic acid to a limited extent, either as the primer, in which case an odd-chain–length fatty acid is produced, or as methyl malonyl-Coenzyme A (CoA) in place of malonyl-CoA, which gives a methyl branch on an even-numbered carbon atom along the chain. However, since acetic acid is the preferred substrate only small quantities of odd-chain and branched-chain fatty acids are usually produced and most propionate is converted to succinyl-CoA via methylmalonyl-CoA. When cereal, or concentrate diets as they are called, are given to sheep there may be a marked increase in the deposition of branched-chain fatty acids. The factors controlling this process are not fully understood but some breeds are affected more than others and rams more than castrates or females. These fatty acids are deposited to a greater extent in the subcutaneous fat of the back rather than other sites.[23]

The composition of the branched-chain fatty acids of lamb fats has been investigated using GLC[24] and mass spectrometry (MS).[25] It is extremely complex: Smith *et al.*[26] reported the presence of 50 different monomethyl-, 65 dimethyl-, 30 trimethyl- and 9 tetramethyl-substituted acids in a concentrate of branched-chain fatty acids obtained from sheep subcutaneous fat. This list is certainly incomplete because urea complexation[27,28] was used to remove the saturated fatty acids from the extract of methyl esters to simplify it for GCMS analysis. However, long-chain iso acids and those having a methyl group in the 2-position were also removed, either completely or partially. The ability of urea under more concentrated conditions to complex mono-branched fatty acid methyl esters was used to separate them from the multi-branched acids, and a repeat treatment of the non-complexing material after conversion to butyl esters was used to remove most of the remaining monomethyl components and approximately 20% of the dimethyl-substituted fatty acids. The branched-chain fatty acids of the original concentrate consisted of monomethyl acids 66·9%, dimethyl acids 26·9%, trimethyl acids 2·3% and monoethyl-substituted acids 3·9%. The nine tetramethyl acids were only present in trace amounts.

The substituted fatty acids ranged from nonanoic to stearic, with branches mainly on the even carbons (Table 8.4), although small

TABLE 8.4

Chain-length distribution of the most abundant mono-, di- and trimethyl-substituted fatty acids of lamb adipose tissue

Chain lengths	Substituents									
	4	6	8	10	12	4,8	4,10	4,12	4,6,8	4,8,12
10	0·61	0·26	0·06	—	—	0·25	—	—	—	—
11	0·31	0·12	0·06	—	—	0·75	0·8	—	—	—
12	1·64	0·02	0·33	0·02	—	2·11	0·51	—	—	—
13	0·77	0·04	0·29	0·05	—	1·52	0·66	—	0·11	—
14	6·55	3·13	3·62	2·22	1·38	1·99	1·13	0·77	0·10	0·42
15	2·22	1·78	1·78	1·14	0·68	1·50	0·84	0·65	0·04	0·68
16	4·96	2·22	2·39	2·56	3·98	0·91	0·58	0·88	—	0·15
17	0·45	0·34	0·37	0·07	0·57	—	0·22	0·28	—	—
18	0·17	0·18	0·18	—	—	—	—	—	—	—
Total	17·68	8·09	9·02	6·04	5·23	8·89	3·51	1·81	0·25	1·25

Results are wt % of fatty acids in concentrate of branched-chain fatty acids excluding iso and anteisoacids. Numbering from carboxylic end according to IUPAC.
Based on Smith *et al.*[26]

amounts of monomethyl substituents were also present on carbons 11, 13 and 15 numbered from the carboxylic ends as in the IUPAC system. The most common monomethyl branch occurred on carbon 4 with half as much on carbons 6 and 8 followed by 10 and 12. Extensive 4-methyl substitution of myristic and pentadecanoic acid was also reported by Miller *et al.*[29] Position 2 contained approximately 2%. The low substitution in the 2-position at all chain lengths suggests that it is not caused by loss of the longer-chain 2-substituted acids as urea complexes along with the straight-chain acids. It also indicates that the shorter-chain acids are synthesised directly rather than by chain shortening. Fatty acids with a 4-methyl group are more susceptible to termination during synthesis than normal, once they reach ten carbons in length, since the relative proportions of fatty acids with a 4-substituent are completely different from those of the normal-chain fatty acids in lamb fat. The low level of branching of the C18 acid, except for the 14- and 16-positions suggests that palmitic acid with methyl side chains nearer the carboxyl than C12 is a poor substrate for the elongation. The most common dimethyl-substituted acids, as might be expected, are those containing a 4-methyl group. There is clear discrimination against consecutive incorporation of residues from methylmalonyl-CoA so that the concentrations decrease in the sequence $4,8 > 4,10 > 4,12 > 4,6$ and the most common trimethyl-substituted acids are those with 4,8,12-substitution. As with the methyl substitution, the 4-position contained the greatest substitution with ethyl groups.[30]

Short-chain-length fatty acids are also present in adipose tissue and it is these, and in particular the branched-chain derivatives, which are thought to be the cause of the unacceptable flavour of lamb from animals fed diets high in cereals.[31,32] The concentrations are highest in parts of the subcutaneous fat, particularly that of the back[33] (Table 8.5). The major components so far identified are the four methyl derivatives of heptanoic and nonanoic acids. These fatty acids are present as glyceryl esters and as free acids, although the latter may have been produced by hydrolysis of the former under the conditions used for their extraction.

8.2.4 Triacylglycerol Composition of Lamb Fat

The triacylglycerol composition of most animal fats is extremely complex and no single procedure is yet adequate to give a complete analysis, although a combination of techniques can give good informa-

TABLE 8.5
Branched medium-chain-length fatty acids in sheep adipose tissue

Depot	Fatty acid			
	4-methyl heptanoate	6-methyl octanoate	4-methyl nonanoate	6-methyl nonanoate
Subcutaneous				
Brisket	$0.8 \pm 0.5(6)$	$1.3 \pm 0.4(6)$	$8.9 \pm 6.8(6)$	$0.5 \pm 0.3(6)$
Loin	$19.0 \pm 6.0(3)$	$12.3 \pm 6.0(3)$	$59.0 \pm 9.0(3)$	$8.4 \pm 2.2(3)$
Perinephric	$0.9 \pm 0.5(6)$	$0.7 \pm 0.3(6)$	$1.7 \pm 0.2(6)$	$0.4 \pm 0.2(6)$
Omental	$1.1 \pm 0.3(5)$	$0.4 \pm 0.2(5)$	$3.3 \pm 0.7(5)$	$0.5 \pm 0.1(5)$

Results are expressed as mg/kg, mean $\pm$ SD for the number of animals in parentheses.
Data from Johnson *et al.*[33]

tion about the proportions of the major components.[34] Techniques for the stereospecific analysis of the distribution of fatty acids on the three glycerol hydroxyls have been well established.[35] The triacylglycerols are treated with a Grignard reagent to give random 1,2- and 2,3-diacylglycerols. These are converted to phosphatidyl phenols and treated with phospholipase A. It is the stereospecificity of the enzymes for the 1,2-diacyl compounds which makes the analysis possible and only the fatty acids at the 2-position are released. The free fatty acids and lysophosphatidyl phenol are separated by thin-layer chromatography (TLC), recovered, converted to fatty acid methyl esters and analysed by GLC. The fatty acid composition of the 3-position is determined by the difference between the sum of positions 1 and 2 and the original triacylglycerol composition. Table 8.6 gives the positional distribution of the major fatty acids of internal and subcutaneous adipose tissue of sheep containing normal proportions of linoleic acid and those with high proportions of linoleic acid resulting from feeding protected sunflower seed. The distributions in the lambs and sheep fed the normal diet (A and C) are similar with saturated fatty acids mainly in the 1- and 3-positions and oleic and linoleic acids mainly in the 2-position. In contrast, *trans*-vaccenic acid is found mainly in the 3-position. There is a discrepancy in the proportions of linoleic acid in the 1- and 3-positions between the sheep (C) and lambs (A), with the lowest proportion in position 1 for the sheep and position 3 for the lambs. This does not appear to be an effect of the higher levels of

TABLE 8.6
Positional distribution of fatty acids in the triacylglycerols from sheep adipose tissue

Position		Internal				Subcutaneous			
	16:0	18:0	18:1 cis and trans		18:2	16:0	18:0	18:1 cis and trans	18:2
A Total	20·1	34·0	37·9		4·0	25·1	22·6	42·2	3·5
1	47·6ᵃ	42·4	20·9		25·2	49·7	48·0	16·8	22·7
2	26·8	16·6	47·5		65·2	21·3	16·5	44·4	66·3
3	25·7	41·0	31·6		9·3	28·8	35·4	38·6	11·0
B Total	23·7	26·2	26·3		23·7	19·2	21·5	34·7	20·2
1	52·7	54·0	24·2		12·2	49·9	58·6	55·9	11·3
2	22·3	12·9	35·4		56·2	20·1	12·1	38·4	56·4
3	25·0	33·2	40·3		31·5	30·0	29·2	41·0	32·3
C Total	21·4	34·7	27·5	4·5	2·4	23·2	25·8	33·3 4·2	2·2
1	48	43	7	37	11	61	45	7 34	12
2	28	14	59	11	75	16	15	59 11	68
3	24	43	34	52	14	23	40	33 55	20

A: Three lambs fed lucerne chaff plus rolled barley; B: five lambs fed as A but with added protected sunflower seed; C, four sheep fed hay and concentrates; A and B from Hawke et al.,[36] perirenal and inguinal; C from Christie and Moore,[37] intestinal and subscapular.
ᵃ Results are mean percentage distribution of fatty acids between the three ester positions of the triacylglycerols.

linoleic acid in the latter because when even more is present, in the lambs fed protected lipid (B), it goes into the 3-position, displacing stearic acid.

Methods are well established for the separation of triacylglycerols according to the number of carbon atoms in the fatty acids: by GLC with short packed or capillary columns, or high-performance liquid chromatography (HPLC). More recently, considerable progress has been made in obtaining separation based on the individual fatty acid composition using reverse-phase HPLC[38] and high-temperature GLC[39] although both systems suffer from non-linearity of the detection systems used; the laser mist detector and flame ionisation detector respectively. Even when these systems are fully worked out they will be unable to give the stereospecific distribution of the fatty acids, but HPLC has a clear advantage in that larger amounts can be separated and may be collected for subsequent analysis.

TABLE 8.7
Triacylglycerol composition of sheep fat according to carbon number and
degree of unsaturation

Carbon number	Triacylglycerol (moles %)					
	Saturated	Monoenoic	Dienoic	Trienoic	Tetraenoic	Total
44	1·7	0·3				
46	6·7	1·3	0·2	0·3	1·0	0·4
48	17·6	6·5	1·3	2·3	2·9	2·3
50	29·5	21·3	8·7	7·8	11·2	7·5
52	29·8	38·7	41·3	30·5	33·1	18·5
54	14·3	31·1	47·9	57·7	49·9	37·5
56	0·2	0·6	0·6	1·4	1·9	32·9
Total	31·3	45·8	19·4	2·6	0·8	100

Data from Kuksis.[40]

At present the newer techniques do not appear to have been applied
to sheep triacylglycerols. However, using a combination of TLC on
silver nitrate-impregnated plates and GLC, Kuksis[40] and his colleagues
determined the carbon number of sheep triacylglycerols according to
their degree of unsaturation (Table 8.7). Whereas the major saturated
triacylglycerols have carbon numbers of 50 and 52 and contain two and
one palmitic acids respectively, those triacylglycerols with two or more
double bonds have carbon numbers of 52 and 54; the latter being most
abundant.

8.2.5 Lipid Consistency

The consistency and related characteristics of fats may be determined by various standard procedures. Since fats consist of mixtures of
triacylglycerols containing different fatty acid combinations, they can
have no true melting point. The closed capillary tube melting point
method determines the temperature at which the highest melting
components dissolve in the liquid fat, hence one would expect it to be
affected most by the content of long-chain saturated triacylglycerols.
The open capillary tube melting (slip) point method (BS 684, 1.3)
determines the temperature at which the lipid becomes soft enough to
flow in the capillary under a fixed hydrostatic pressure. Hence values
determined by this method are usually 10–15°C lower than those

obtained by the closed tube method. The plastic behaviour of the fat may also be estimated by determining the solid fat index by dilatometry (BS 684, 1.12) and the content of solid fat by nuclear magnetic resonance (NMR). An overall index of the unsaturation of the fat and hence some indication of its likely consistency is given by determination of the iodine value or number (BS 684, 2.13). The latter was used more commonly before the advent of GLC made fatty acid compositions readily available. In a group of sheep with a two-fold difference in fatness, Callow[41] observed iodine numbers from 43·7 to 59·3 for the subcutaneous fat and from 41·7 to 58·5 for the intermuscular fat. This variation was animal-dependent and there were no significant differences between different subcutaneous sites as might be expected from the differences in fatty acid composition given in Table 8.1. Marchello and Cramer[42] obtained an iodine value of 38·5 for kidney fat, 45·6 for the loin subcutaneous and 47·3 for the leg subcutaneous fat. The melting points of these samples were 48·4°C, 43·1°C and 42·8°C respectively. In subcutaneous lamb fat[20] with 7·7% linoleic acid (Table 8.3) the iodine value reached 54·4 and the melting (slip) point was 36·6°C.

The latter study also presents evidence of the importance of the saturated fatty acids, particularly stearic acid, in determining the consistency of the fat. In a previous study,[43] the proportion of stearate in both the subcutaneous and perirenal fat was best correlated with their melting points, although the slopes of the regressions differed for the two depots. The steeper slope for the subcutaneous depot presumably stems from initial differences in the triacylglycerol composition. Perirenal lipid contains many di- and trisaturated glycerides and is hard whereas subcutaneous fat contains far fewer and is softer. As stearic acid rises in the latter there is a sharp increase in the content of these glycerides and the fat becomes much harder. Other workers have also observed high correlations between the stearic acid content of lamb fat and its melting point or firmness.[18,44] The decreased melting point of fat from lambs fed concentrates is probably caused by the low concentration of stearic acid present in the fat rather than the high concentrations of branched-chain fatty acids.[18,29,45–47]

8.2.6 Non-Fatty Acid-Containing Components

8.2.6.1 Colour
Mutton tallow is usually white but yellow fat has been reported to occur in sheep in Norway, Iceland, Ireland, New Zealand and

Australia. The occurrence of yellow fat has a genetic basis and is inherited as a single recessive gene.[48,49] The yellow pigments, which partition into the lipid on rendering, are xanthophylls. Two of the components are agreed on as lutein and flavoxanthin (lutein 5,8-epoxide) but there is disagreement as to whether the third pigment is lutein 5,6-epoxide[50] or auroxanthin (5,8,5′,8′-diepoxyzeaxanthin).[51,52] Total carotenoid concentrations in the yellow fat are in the range 200–300 μg/100 g fatty tissue which would give values of 240–360 μg/100 g tallow.[53]

8.2.6.2 Cholesterol

The content of cholesterol in sheep adipose tissue is reported to be 75 mg/100 g.[54] On the basis of a lipid content of 75% this would yield a concentration of approximately 0·1%. Trace amounts of steroid hormones would also be present but their quantities have not been determined.

8.3. BEEF LIPIDS

The fatty acid composition of the triacylglycerols of beef adipose tissue is similar to that of sheep but with somewhat less stearic acid so that it is slightly softer. This may result, in part from the slightly lower body temperature (*c.* 1°C) of cattle compared with sheep. In a comparison of Indian beef and mutton tallows the respective proportions of stearic acid were 17% and 30%.[55] The former is in the range of 6–40% quoted by others[56] (Table 8.8), although 40% is very high for unfractionated beef tallow in the light of reported compositions of fat from various sites within the beef animal (see below). As with sheep tallow the other major fatty acids are palmitic and oleic. The concentration of linoleic acid in Indian tallow (7·2%) exceeds the range quoted in Table 8.8, but in the absence of information on the type of diet given to the cattle it is difficult to account for this difference. The number of fatty acids reported was somewhat fewer than that of Marmer *et al.*[57] (Table 8.9) using highly polar GLC capillary columns, although most components would have been included in unresolved peaks on packed columns and the overall differences would not have affected the proportion of linoleic acid significantly. The comprehensive analysis of subcutaneous and perirenal fat from cattle (Table 8.9) reveals the presence of a range of branched-chain fatty acids: approximately half were identified as iso or anteiso acids which are believed to be synthesised by the rumen

TABLE 8.8
Fatty acid composition of edible beef tallow and lard

Fatty acid	Premier Jus and edible tallow	Lard and rendered pork fat
Shorter chain	<0·1	<0·5
14:0	1·4–6·3	0·5–2·5
14:1	0·5–1·5	<0·2
15:0	0·5–1·0	<0·1
iso 15:0	<1·5	<0·1
16:0	20–37	20–32
16:1	0·7–8·8	1·7–5·0
16:2	<1·0	
iso 16:0	<0·5	<0·1
17:0	0·5–2·0	<0·5
17:1	<1·0	<0·5
18:0	6·0–40	5·0–24
18:1	26–50	35–62
18:2	0·5–5·0	3·0–16
18:3	<2·5	<1·5
20:0	<0·5	<1·0
20:1	<0·5	<1·0
20:2		<1·0
20:4	<0·5	<1·0
22:0		<0·1

Taken from Spencer et al.[56]

micro-organisms or by the animal from iso and anteiso carbon residues derived from amino acid breakdown. Unlike sheep, cattle appear to be able to metabolise all their methylmalonyl-CoA to succinate and hence it is not used for fatty acid synthesis. The small quantities of multiple methyl-branched fatty acids which are found, such as phytanic acid (3,7,11,15-tetramethylhexadecanoic acid), are derived from chlorophyll. The only *trans* fatty acid reported, 18:1, was higher than the 1·7% in Indian beef tallow but agreed with other data on the *trans* fatty acid content of beef subcutaneous fat.[58] As expected, oleic acid was the major *cis*-monoenoic acid with a much smaller amount of palmitoleic.

Specific analysis of the monoenoic fatty acid composition of bovine perirenal fat has revealed a much wider spectrum of positional isomers of both the *cis* and *trans* fatty acids.[59] Monoenoic methyl esters were separated from other fatty acid methyl esters on an AgNo₃–silicic acid column and then the individual chain-length fractions were obtained

TABLE 8.9

Fatty acid composition of neutral lipid fractions of bovine adipose tissue as a function of tissue site and diet

Fatty acid	Site[a]		Diet[b]	
	Subcutaneous	Perirenal	Forage	Grain
14:0	3·28	3·68[c]	3·92	3·28[d]
9-14:1	0·86	0·31[c]	1·07	0·86
15:0	0·70	0·64[c]	0·66	0·70
iso 15:0	0·12	0·17[c]	0·39	0·12[d]
anteiso 15:0	0·13	0·23[c]	0·40	0·13[d]
16:0	25·99	26·66	27·07	25·99[d]
iso 16:0	0·11	0·16[c]	0·29	0·11[d]
9-16:1	3·30	1·34[c]	3·70	3·30[d]
17:0	1·99	2·09[c]	1·09	1·99[d]
anteiso 17:0	0·47	0·52	0·81	0·47[d]
9-17:1	1·46	0·66[c]	0·69	1·46[d]
18:0	12·50	24·71[c]	15·60	12·50[d]
trans 18:1	2·71	3·47[c]	3·93	2·71[d]
9-18:1	40·15	30·56[c]	33·32	40·15[d]
11-18:1	1·62	1·02[c]	1·14	1·62[d]
9,12-18:2	1·79	1·74	0·60	1·79[d]
9,12,15-18:3	0·14	0·16[c]	0·47	0·14[d]
anteiso 19:0	0·44	0·13[c]	0·19	0·44[d]
20:0	0·24	0·17[e]		0·24
11-20:1	0·18	0·11[c]	0·19	0·18[e]
8,11,14-20:3			0·16	
5,8,11,14,17-20:5			0·16	

Values are normalised % by wt of total fatty acids, for 10 crossbred steers per group.
[a] Steers fed on grain.
[b] Subcutaneous adipose tissue (backfat).
[c] Significantly different from subcutaneous, using Bonferroni *t*-test to compare sites ($P < 0.05$).
[d] Significantly different from forage.
[e] Not compared.
Taken from Marmer et al.[57]

by preparative GLC. Finally the individual *cis* and *trans* isomers were separated by AgNO$_3$ TLC, cleaved by permanganate–periodate oxidation and the resulting fatty acids identified and quantified by GLC.[60] Amongst the *cis* isomers the Δ^9 double bond predominated (Table 8.10), but it was less common in the *trans* isomers, particularly in the

TABLE 8.10
Distribution of double bonds in *cis* and *trans* monoenoic fatty acids in steer perinephric fat[a]

Position of double bond	*cis* isomers				*trans* isomers	
	14:1	16:1	17:1	18:1	16:1	18:1
5	3·2					
6	3·2	6·6	3·2		4·1	
7	12·2	28·0	6·3	0·5	3·3	1·0
8	5·4	Trace	34·0	1·1	9·2	2·1
9	76·0	58·8	56·5	85·4	43·2	5·0
10		Trace		2·2	6·7	11·9
11		1·2		8·6	9·5	46·9
12		5·3		2·2	11·9	6·0
13					9·5	6·6
14					1·0	7·4
15					1·6	5·5
16						7·6

[a] Values expressed in wt %.
Trace = less than 0·1%.
Data from Hay and Morrison.[59] Reproduced by permission of the American Oil Chemists Society.

octadecenoic acids in which *trans* vaccenic was most abundant, followed by the Δ^{10} isomer.

8.3.1 Depot Site and Fatty Acid Composition

Studies of the effect of body site on fatty acid composition revealed marked differences between subcutaneous and internal depots, but between different subcutaneous sites they were somewhat smaller.[61,62] Stearic acid varied most; from 6·0% to 11·5%, being highest in the precrural (groin) area and dorsal rib fat. Perirenal fat has been reported to contain stearic acid from 24% to 32%.[63–65] These higher proportions of stearic acid are accompanied by a decreased proportion of oleic acid, apparently the result of a lower synthesis from stearic acid by the desaturase of internal compared with subcutaneous depots. The proportion of palmitic acid varied from 23·9% to 33·6% in these studies with no consistent differences between internal and external fat, so the value of 37% for the top of the range for tallow,[56] like that for stearic acid, appears high.

8.3.2 Effect of Fatness, Age, Sex and Breed on Fatty Acid Composition

As with sheep, the main effect of age, sex and breed on cattle lipids depends upon carcass fatness, although most results are reported on an age basis. A specific age effect seems to be precluded since changes in composition occur at different ages in different studies and are affected by diet; smaller changes occur at younger ages in animals fed concentrates (grain) rather than forage. This suggests that the changes begin when the animal starts to fatten rapidly so that saturated fatty acids absorbed from the intestine are diluted more and more by unsaturated fatty acids synthesised within the tissue. Thus the subcutaneous fat becomes much softer with age as a result of a decrease in the proportion of stearic acid and an increase in oleic acid[64,66–69] (Table 8.11). Table 8.11 also reveals the rapid decrease in the shorter-chain milk-derived fatty acids during early growth. Wood[70] reported an extreme case of the subcutaneous lipid from an old fat steer which contained only 14% palmitic and 2·7% stearic acids, and which was completely liquid at 25°C. It is clear from these results that old fat animals cannot be the source of tallow with high concentrations of palmitic and stearic acid since the age-related changes in the internal fat depots resemble and may exceed those in the subcutaneous fat.[66]

The effects of sex and breed on the fatty acid composition of cattle are relatively small at normal slaughter weights and appear to depend mainly upon differences in fatness.[62–64,69,71–73] Steers and heifers are fatter than bulls and have lower proportions of stearic acid and more oleic acid. The proportion of palmitic acid may be higher, lower or the same. Differences between breeds are usually small. However, for a range of Aberdeen Angus crossbred steers of different sire breeds, Pyle *et al.*[69] reported age-corrected differences in the proportion of stearic acid in the brisket (chest) subcutaneous fat from 7·51% to 13·24%, and these were not lost when corrected for differences in fatness. Other studies have only revealed differences from 12·2% to 14·8% between breeds. Even the Zebu, *Bos indicus,* has similar fatty acid composition in its subcutaneous (hump) and perirenal fat to *Bos taurus.*[74]

8.3.3 Effects of Diet on Fatty Acid Composition

The effects of diet on the fatty acid composition of cattle are similar to those of sheep, except, as mentioned above, that feeding high

TABLE 8.11

Effect of age on the fatty acid composition of bovine subcutaneous lipid

Age (days)	56	112	168	224	280	336	392	448	Slaughter[a]
Average live wt (kg)	49	88	137	182	230	275	342	390	444
Fatty acid									
10:0	0.20^{bc}	0.07^{b}	0.07^{bc}	0.04^{b}	0.04^{b}	0.01^{c}	0.01^{c}	0.05^{bc}	0.00^{c}
12:0	0.64^{b}	0.17^{c}	0.16^{c}	0.06^{d}	0.12^{cd}	0.03^{d}	0.03^{d}	0.06^{d}	0.04^{d}
14:0	6.55^{b}	3.81^{c}	3.79^{d}	3.67^{c}	3.27^{cd}	3.22^{cd}	3.38^{cd}	3.57^{cd}	3.29^{d}
16:0	25.72^{bcd}	25.01^{bc}	24.10^{b}	24.23^{b}	26.98^{cd}	26.75^{d}	25.36^{bc}	26.21^{cd}	27.76^{d}
16:1	5.46^{b}	5.26^{b}	6.09^{bce}	6.59^{ce}	8.20^{d}	8.57^{d}	7.72^{de}	7.04^{e}	6.34^{de}
18:0	15.83^{b}	19.20^{c}	19.57^{c}	17.07^{b}	11.02^{d}	10.48^{d}	10.18^{d}	7.32^{d}	4.05^{e}
18:1	37.39^{b}	38.37^{b}	38.29^{bc}	40.93^{c}	44.12^{d}	44.72^{d}	46.55^{d}	48.03^{d}	49.61^{e}
18:2	4.55^{b}	4.54^{b}	3.92^{bd}	2.95^{c}	2.94^{cd}	2.73^{cd}	2.55^{c}	3.28^{cd}	3.42^{d}

Values are mean % of total fatty acids in tailfat lipid for 27 cattle of 3 breeds.
[a] Cattle were slaughtered on a constant finish basis between 420 and 602 days of age.
[b,c,d,e] Means in same line without a common superscript letter are significantly different ($P < 0.05$).
Taken from Hecker et al.[68]

concentrations of cereals does not cause deposition of increased quantities of branched-chain fatty acids. Most animals are raised either on forage or mixtures of forage and concentrates. The major differences between the body fats from animals fed on these diets are the higher proportions of saturated and *trans* unsaturated monoenoic acids found in forage-fed animals and lower proportions of linoleic acid (Table 8.9). In subcutaneous fat these were reported to be 15·6%, 4·06% and 0·6% for forage-fed animals and 12·5%, 2·7% and 1·8% respectively for stearic, *trans* monoenoic and linoleic acids in grain-fed steers.[57] However, the differences appear to be site-dependent, with palmitic rather than stearic acid being higher in the perirenal fat of hay-fed compared with grain-fed cattle.[64] This increased saturation was associated with a marked decrease in the proportion of oleic acid.[63] The extent of hydrogenation of polyunsaturated fatty acids is less in animals fed on concentrates so that the quantities of linoleic acid deposited increase. However, the effects are relatively small. Safflower oil added to the diet at 6% increased the deposition of linoleic acid by 30–50% but the final concentration was only 2·7–3·8%, depending on the depot.[75] However, in a more recent study in which cattle were fed concentrates in which maize and rapeseed supplied the fat, the mean concentrations of linoleic acid ranged from 4·4 to 5·6%.[76] Bypassing the bovine rumen with protected lipid or oilseed is as effective as in sheep in raising the linoleic acid content of tissue lipids.[77] Feeding protected safflower oil at 20% of the diet for 8 weeks increased the level of linoleic acid in the depots from a range of 3–5% to a range of 23–34%, with the highest concentration in the perirenal depot. The proportions of palmitic acid and 18:1 decreased, whereas stearic acid was unchanged.

8.3.4 Triacylglycerol Composition of Bovine Lipids

Stereospecific analysis of the fatty acid distribution in beef subcutaneous fat[34,78] shows that 18:1 is the main fatty acid in the 2-position followed by palmitic acid (Table 8.12). The latter is also the most abundant fatty acid on the 1-position, whereas 18:1 is the most abundant in the 3-position. The distribution of each fatty acid between the three positions (Table 8.12) is similar to that for lamb subcutaneous fat (Table 8.6) except for stearic acid, the highest proportion of which was present in position 3 in beef lipid rather than position 1, and linoleic acid which was spread widely over all three positions in beef fat, but concentrated in position 2 in lamb fat.

TABLE 8.12

Fatty acid composition and distribution in triacylglycerols from beef sub-cutaneous fat

	Fatty acids (%)					
	14:0	16:0	16:1	18:0	18:1	18:2
Fatty acid composition						
Total	4·0	26·7	6·0	16·7	32·7	4·7
At position 1	4	41	6	17	20	4
2	9	17	6	9	41	5
3	1	22	6	24	37	5
Distribution of each fatty acid						
At position 1	28·6	51·3	33·3	34·0	20·4	28·6
2	64·3	21·3	33·3	18·0	41·8	35·7
3	7·1	27·5	33·3	48·0	37·8	35·7

Data from Brockerhoff et al.[78]

However, stearic acid has been reported to be present in similar amounts in the 1- and 3-positions and linoleic to be mainly present in the 2- and 3-positions in equal quantities. The latter difference may stem from the much lower proportion of linoleic acid, 1·3% compared with 4·7%, present in the lipid.

The triacylglycerol compositions of tallow determined on the basis of carbon numbers by packed and capillary column GLC are shown in Table 8.13. Part of the difference between them stems from the

TABLE 8.13

Carbon number composition of tallow and lard triacylglycerols determined by GLC

Carbon number	Packed column[a] (mole %)		Capillary column[b] (weight %)	
	Tallow	Lard	Tallow	Lard
46	3		1·1	0·4
48	10	4	5·2	1·9
50	28	20	16·4	13·4
52	42	62	43·4	59·1
54	17	15	21·3	20·7
56			2·2	2·4

[a] Data from Karleskind et al.[79]
[b] Data from D'Alonzo et al.[193]

difference in units, mole % and weight % respectively, but direct comparison is not possible because of the absence of fatty acid composition data. However, they agree quite well. The packed column method has been standardised for IUPAC.[80] Reverse-phase HPLC analysis has been developed to the stage when almost complete resolution of triacylglycerols according to fatty acid composition is possible for the major components except for tripalmitin and stearoyldiolein.[38] However, isomers such as PPO and POP are not resolved. The major triacylglycerol in tallow was palmitoyl diolein (Table 8.14). Small quantities of glyceride in which the 1-ester is

TABLE 8.14
Triacylglycerol composition of tallow and lard determined by HPLC

Triacylglycerol	% by weight	
	Tallow	Lard
OLL[a]		0·5
OOLn		0·7
PLL		0·5
POLn	0·4	0·9
OOL	0·9	1·7
StLL		1·6
PoOO	1·1	
POL	2·7	6·5
MyOO/PPoO	3·8	3·2
MyOP	3·0	1·2
OOO	4·4	5·7
StOL	1·4	0·8
POO	23·0	26·4
PStL/PoStO	4·8	3·8
PPO	10·6	7·6
PPP/StOO	11·5	5·3
PStO	14·5	21·1
PPSt	3·3	2·4
StStO	4·0	1·9
PStSt	3·1	3·1
StStSt	1·6	0·5
Unknown	6·0	4·6

[a] Abbreviations for fatty acids: My, myristic; P, Palmitic; Po, palmitoleic; St, stearic; O, oleic; L, linoleic; Ln, linolenic. Positional isomers are not resolved.
Data from Perrin and Prévot.[38]

replaced by an ether or alkenylether are also present, the aldehyde from the latter amounting to 5 mg per kg of fat.[81]

8.3.5 Non-fatty Acid-Containing Components

The yellow colour of beef tallow, which varies from pale cream to deep yellow, results from the presence of β-carotene in the fat depots,[82,83] together with small quantities of α-carotene and xanthophyll. During rendering the pigments partition into the lipid fraction.[84] Factors affecting the development of yellow fat were reviewed by Morgan and Everitt.[85] The β-carotene is derived from the dietary pasture which is rich in carotenoid pigments. There is a marked genetic effect on the deposition of β-carotene, with the Jersey, Guernsey and South Devon breeds depositing much higher concentrations than others.[86,87] The depth of colour varies between fat depots in the carcass, and differences in carotene content have been reported. The acceptable level of β-carotene in beef adipose tissue has been reported to be less than 0·35 mg/100 g.[84] Other studies suggested that levels higher than about 0·2 mg/100 g were unacceptable,[88] but colour was not well correlated with values for carotene probably because the former included a contribution from haem pigments visible due to the poor development of the adipose tissue. Beef suet contains much lower levels, of the order of 70 μg/100 g.[89] Although cereal-based-diets and breed are important contributors to this low concentration, it is probable that some destruction of β-carotene, by oxidation, occurs during rendering. Retinol is reported to be present in beef suet at 52 μg/100 g.[89] Vitamin E is present in low levels in animal tissues although it is the major antioxidant present. The adipose tissue of cows contains 2 mg/100 g, but that of steers contains only 0·6 mg/100 g, while α-tocotrienol is present at 0·3 mg/100 g and 0·2 mg/100 g respectively.[90] The concentrations were highest in the autumn after the animals had been grazing. These workers reported the α-tocopherol level to be between 2 and 3 mg/100 g in suet derived from cows, and 1 mg/100 g in steer suet, similar to the 1·5 mg/100 g reported by Paul and Southgate.[89]

8.3.6 Steroids

Cholesterol is the major steroid present in animal tissues.[91–93] It is present in the membranous components of the cell and in the fat globule of fat cells.[94,95] A proportion of the cholesterol is esterified to long-chain fatty acids. Whereas Kritchevsky and Tepper[96] reported that 85% of cholesterol in minced beef was esterified, more recent studies suggest that 10% or less of beef muscle cholesterol is present as the ester.[97,98] In rat adipose tissue more than 75% of cholesterol was unesterified and in human adipose tissue this rose to more than 90%.[94] The early high values for cholesteryl esters may have resulted from the difficulty of precipitating free cholesterol with digitonin in the presence of large quantities of triacylglycerols.[94] Total cholesterol is usually determined after alkaline hydrolysis of a lipid extract, although direct saponification of the meat gives equivalent results.[99] If the lipid is to be extracted first it is important to use solvent mixtures which disrupt membrane structures such as chloroform:methanol or ether:ethanol[100] to obtain complete extraction. Various solvents including hexane, toluene, petroleum ether of boiling range 30–60°C, benzene and isopropyl ether have been used to extract cholesterol and other non-saponifiables from the alkaline hydrolysate. Subsequent determination of cholesterol may be by the Liebermann–Burchard reaction or by using cholesterol oxidase coupled to a chromogenic reaction as described in Chapter 7. However, both reactions are non-specific; the former assays unsaturated sterols and the latter sterols with a 3β-hydroxyl group. Hence they are unsuitable for the determination of cholesterol in compound fats containing vegetable oils. The AOAC method[101] overcomes this problem by converting the sterols to the trimethylsilyl derivatives, separating them by GLC and using 5 α-cholestane as an internal standard for quantification.

The reported concentrations of cholesterol in beef adipose tissue appear to have increased up to recent times. In a review of early studies Feeley *et al.*[102] (1972) gave an approximate value of 75 mg/100 g of adipose tissue. However, more recent studies have given higher values. Eichhorn *et al.*[103] found a significantly higher concentration in the subcutaneous compared with the perirenal adipose tissue of bulls and steers; 102 compared with 90 mg/100 g tissue. However, this difference was reversed in tissue from mature cows, 87 and 100 mg respectively.[104] Rhee *et al.*[105] reported 108 and 114 mg/100 g in intramuscular and subcutaneous adipose tissue respectively, and Hoelscher *et al.*[95] found 121 mg/100 g of subcutaneous adipose tissue. Allowing for the lipid content of adipose tissue, these values are in line with concentrations in tallow of 140 mg/100 g,[106,107] compared with the lower value of 70 mg/100 g given by Paul and Southgate.[89] However, Parodi[92] reported that cholesterol ranged from 114 to 203 mg/100 g in a single brand of unspecified animal fat. The reasons for the apparent change in cholesterol concentrations deter-mined over a period of years could be methodological, at least in part. Prolonged drying may have caused losses in some determinations.[108] Overestimates of cholesterol may arise if the saponification step is omitted or if antioxidants are not present during saponification when the colorimetric Liebermann–Burchard reaction is used.[109] It is probable that the leanness of contemporary cattle contributes to the higher concentrations. When lean mature cows were allowed to feed *ad libitum* and become fatter, the cholesterol concentrations in their adipose tissue fell by up to one-third.[104] However, any effect of leanness may be partly counteracted by the increase in the cholesterol content of adipose tissue as the animals age.[110]

In muscle tissue the reported concentrations of cholesterol have changed little over the years, compared with those for adipose tissue, providing further evidence against methodological reasons as the major cause of changes in the latter.[54,97,103–105] Values range generally from 50 to 70 mg/100 g tissue, but since the total lipid content of muscle is much less than in adipose tissue, the inclusion of muscle tissue during rendering will increase the concentration of cholesterol in the recovered lipid.

Small quantities of three other steroid groups occur in adipose tissue and would be expected to be present in the rendered fat. These consist of biosynthetic precursors of cholesterol such as lanosterol, steroid metabolites and steroids such as ergosterol, derived from dietary plant steroids and steroid hormones. Desmosterol and lanosterol are the most abundant followed by ergosterol but all are present at less than 0·5% of the cholesterol concentration.[91] Mordret *et al.*[93] reported a separation of all the non-saponifiables of fats using an OV17 GLC capillary column. They observed that 89·7% of the sterols in suet consisted of cholesterol, with 1·9% of campesterol and 3·8% of β-sitosterol and minor unidentified components making up the rest. Similar sterol composition occurs in the fat of Zebu cattle (*Bos indicus*).[74] Squalene, the precursor of the steroid nucleus, is also present in the non-saponifiable fraction of beef fat. The GLC chromatograms published by Tu *et al.*[91] suggested that it was a major component of the 400 mg/100 g of non-saponifiables in their beef fat, exceeding the cholesterol concentration. However, only 10 mg of squalene per 100 g of beef tallow was found by Fitelson,[111] and although his methodology was probably less sensitive it is unlikely to account for such a large difference. Squalene comprised 9·8% of the hydrocarbon fraction of tallow; the major components being phytane and pristane. The alcohol fraction contained phytol, and myristoyl, palmitoyl and stearoyl alcohols of which the latter was present in the highest proportion although absolute quantities were not reported.[112]

8.4. PORK LIPIDS

The major fat depot in the pig is the subcutaneous depot which forms a tube around the body and consists of 70–90% triacylglycerol. There is considerable intramuscular fat but the internal cavity fats are much less developed than in cattle or sheep. The major fatty acids present are oleic, palmitic and stearic, but there may may also be large quantities of linoleic acid, thus distinguishing lard from beef or mutton tallow in general, although there is some overlap at the lowest concentrations[56] (Table 8.8). Lard is much softer than tallow, not only because of the presence of more linoleic acid, but also because of its higher concentration of oleic acid and lower concentration of stearic acid. The ranges of fatty acids given by Spencer *et al.*[56] appear to be somewhat high for palmitic and stearic acids and low for linoleic acid based on recent reports in the literature of the composition of pig carcass fats. However, Hubbard and Pocklington[113] reported similar ranges for the fatty acid composition of various fat depots from European and Canadian pig carcasses. In view of the number of factors which affect the fatty acid composition of pigs it is not possible to ascribe specific causes for the composition of samples of unknown origin. Furthermore, since fat on meat is considered undesirable from a health point of view and is costly in terms of the feed required for its deposition, much research is aimed at producing leaner pigs and hence the reported compositions are often on pigs which are leaner than those in commercial production.

8.4.1 Depot Site and Fatty Acid Composition
Henriques and Hansen[114] demonstrated that the internal fat has a higher melting point and a lower iodine value than the subcutaneous fat, and that the inner subcutaneous fat layer is harder than the outer layer adjacent to the skin. Before the advent of GLC it was established laboriously that the latter difference arose from higher concentrations of palmitic and stearic acids in the inner layer and more oleic acid in the outer layer.[115,116] Across each subcutaneous layer the fatty acid composition is virtually constant,[117] but there is a distinct

TABLE 8.15

Fatty acid composition of pig adipose tissues at different sites

Depot site	Fatty acid (% by wt)					
	14:0	16:0	16:1	18:0	18:1	18:2
Subcutaneous: Loin outer	1·5	23·8	3·9	12·2	41·5	14·0
Loin inner	1·5	24·8	3·4	14·2	40·2	12·8
Belly	1·5	23·8	4·4	11·2	43·0	12·8
Internal: Perirenal	1·6	27·5	3·2	18·1	34·0	13·0
Intermuscular: Prescapular	1·5	23·9	4·1	11·9	43·1	12·6

Boars fed on a high energy diet to 90 kg live-weight.
Unpublished data of Wood, J. D., Buxton, P. J., Whittington, F. M. and Enser, M.

discontinuity at the connective tissue sheet separating the two layers, and there are changes adjacent to the skin and muscle.[118] The differences between the two layers are small; of the order 2% or less for each fatty acid involved (Table 8.15), most consistently for oleic and stearic acids. Linoleic acid is usually about 2% higher in the outer layer until its concentration reaches 30%, when the difference may be lost and even reversed at greater concentrations,[119–121] although Koch et al.[122] observed that the difference was retained even when the inner layer contained up to 38% linoleic acid. The factors which affect the variation in relative concentration of the other major fatty acids in the two layers are not clear. As the proportion of linoleic acid increases, the differences in stearic and palmitic acids between the two layers decreases[121,122] but in pigs which were virtually devoid of linoleic acid, the differences were quite small.[123] One would expect that the more stearic acid in the inner layer the greater would be the difference in concentration between the two layers, since stearic acid raises the melting point and the outer layer must remain semi-fluid at ambient temperature. However, such an effect only seems to happen when the stearic acid concentration of the inner layer exceeds 17%[119,124] which seldom occurs in pigs nowadays. In very fat pigs a third adipose layer becomes distinguishable in the back region and, although the trend towards higher proportions of saturated fatty acids continues as the distance from the skin increases, the differences are small and not significant.[125] Overall there are only minor differences between the outer backfat at the shoulder, loin and rump sites, and the belly fat.[76,126–129]

The perirenal and leaf fat of pigs contains higher proportions of saturated fatty acids and lower proportions of unsaturated fatty acids than the subcutaneous depots (Table 8.15). Stearic acid is 3–5% higher than in the inner subcutaneous fat whereas palmitic acid is raised by a smaller amount.[130] Oleic acid is approximately 2% lower as is linoleic acid,[122,126] although greater decreases of oleic acid have been reported,[127,129,131] whereas Wood *et al.*[129] found 21·9% linoleic acid in the perirenal fat compared with only 19·6% in the inner subcutaneous fat of the loin. The latter authors also found the intermuscular fat of the shoulder to closely resemble the outer loin subcutaneous adipose tissue in fatty acid composition. However, other workers have reported that intermuscular fat has a composition intermediate between that of subcutaneous and perirenal fat in respect of stearic acid.[132,133] Although the high proportion of linoleic acid (19%) in the fat of pigs from one study[129] may have contributed to these discrepant results, the relationship is clearly non-linear since the intermuscular fat was more like perirenal fat when the latter had 9% linoleic acid than when it had 3% linoleic acid.

8.4.2 Effect of Diet on Fatty Acid Composition

The potential for dietary variation of the fatty acid composition of pig depot fats is much greater than that for cattle and sheep because of the absence of a rumen in the digestive system. Hence unsaturated fatty acids, as well as saturated fatty acids, pass through the digestive system essentially unchanged and are deposited in the depots. The simple stomach of the pig also allows it to accept much higher concentrations of fat in the diet with the result that less common fatty acids present in some feeds can appear in the depot fat.

In the absence of significant amounts of fat in the diet palmitic, stearic and oleic acids make up more than 90% of the adipose tissue fatty acids[123,134] with smaller amounts of palmitoleic, myristic and arachidic acids. When fat is present in the diet it is deposited in the tissue together with the fatty acids which can be synthesised *de novo*. This is readily detected by the appearance in the fat depots of plant-derived linoleic and linolenic acids which cannot be synthesised by the pig. When linoleic acid is present in large quantities in the diet, its deposition in the fat lowers the melting point to the extent that the lipid becomes oily.[135] Linolenic acid[136,137] and the longer-chain poly-unsaturated fatty acids of marine oils[138,139] are also readily deposited,

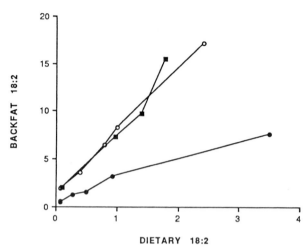

Fig. 8.1. Relationship between the linoleic acid content of the diet and the proportion of linoleic acid in pig backfat fatty acids. ○, from Babatunde *et al.*;[142] ■, from Dahl and Persson[141] both expressed as percentage by weight of linoleic acid in feed; ●, from Leat[140] expressed as percentage of dietary calories supplied by linoleic acid.

although their presence in the diet in significant amounts is usually avoided because of the oxidative instability they impart to the lard.

The concentration of linoleic acid in the adipose tissue lipids is proportional to the quantity in the diet or, more strictly, to the proportion of dietary calories which it supplies[140-145] (Fig. 8.1). The relationship remains linear until the lipids contain approximately 45% of linoleic acid, when dietary increments become less effective, although concentrations in backfat of over 50% have been reported[121] when pigs have consumed large quantities of unextracted sunflower seed.

Oleic acid is normally abundant in porcine depot lipids and changes in its deposition resulting from different dietary levels are often hard to detect. However, when diets consisting of more than 10% fat in which oleic acid is present at about 60% are fed, the subcutaneous fat may contain approximately 55% oleic acid.[76,146] Since such diets are designed to cause maximum deposition of oleic acid, it is difficult to envisage the source of the lard and rendered pork fat with 62% oleic acid reported by Spencer *et al.*[56] (Table 8.8). Unlike oleic, little eicosenoic (20:1) or erucic (22:1) acid occurs in pig fats normally, so that their deposition from the diet can be followed more easily. However, increasing chain length in the mono-unsaturated acids is associated with poorer deposition, and 2–3 times as much dietary erucic acid is required to give the same adipose tissue concentration as eicosenoic acid.[147,148]

Feeding large quantities of saturated long-chain fatty acids, unlike their unsaturated counterparts, fails to increase their concentration in adipose tissue.[134,146,148] Thus the upper range of these saturated fatty acids is unlikely to exceed that quoted by Spencer *et al.*,[56] as a result of vagaries in the supply of dietary saturated fatty acids. The biochemical mechanisms which regulate their concentration are not understood and poor digestion cannot account fully for the findings. However, studies with labelled saturated fatty acids clearly indicate that they are absorbed into the bloodstream and deposited in adipose tissue.[149]

Fatty acids with fewer than 14 carbon atoms are present in very small amount in the adipose tissue of mature pigs. Hence, their deposition after feeding coconut products or trilaurin is readily detectable.[127,130,150–154] However, whereas up to two-thirds of dietary linoleic acid may be deposited in the adipose tissue, only a quarter of the medium chain-length fatty acids (C8–C14) were deposited[154] under optimum conditions. Concentrations of lauric acid of 4–5% were observed in backfat with 10% of coconut oil in the feed, and myristic acid was between 9% and 16%.[130,152] The latter far exceeds the maximum of 2·5% suggested by Spencer *et al.*[56] for lard (Table 8.8). Less common fatty acids, such as those with a branched chain found in small amounts in ruminant fat, are also deposited by the pig when they are present in the diet.[155]

Whilst dietary fatty acids may have direct effects on depot fat composition in pigs, other aspects of diet are also important. If the caloric intake is decreased below the ad-libitum level there will be less substrate available for fatty acid synthesis and deposition. The net result is that dietary fatty acids form a higher proportion of the depot fat, and since these are usually of plant origin, there is an increase in the proportion of linoleic acid in the depots.[119,129,156] Pigs kept at low temperatures require more feed for temperature maintenance and hence have fewer calories available for fat synthesis, therefore softer dietary fatty acids increase in concentration in the adipose tissue. However, there is also an effect on the fatty acids synthesised *de novo* which increases the iodine value of the fat in pigs left in the cold.[114,117,157,158] Genetic selection for lean pigs may decrease appetite and hence fatty acid compositions resemble those of pigs fed a restricted diet.[159] However, since leaner pigs have less fat for trimming, the effect of these changes on lard fatty acid composition are likely to be small.

Copper, which is added to pig feed as copper sulphate to improve growth efficiency, changes the fatty acid composition of the depot fat and lowers its melting-point considerably.[160-166] It increases the desaturation of stearate to oleate so that the lipid is less saturated, and also the altered triacylglycerol composition may produce a larger change in melting point than would be expected from this 2–3% exchange of oleate for stearate. In the UK the amount of copper allowable in pig feed is being reduced so it is unlikely to have much effect on lard fatty acid composition.

8.4.3 Effects of Age, Sex and Breed on Fatty Acid Composition

The effects of age, sex and breed on the fatty acid composition of pig adipose tissue can be related to a large extent to their effects on animal fatness. Their action therefore is similar to that described above for changes in caloric intake. As pigs grow the proportion of energy available for fat deposition increases so that the rate of fatty acid synthesis *de novo* is raised. The main change is therefore a fall in the proportion of exogenous fatty acids in the depots revealed as a decrease in linoleic acid.[126,167,168] Other changes are less distinct but there is usually an increase in the proportions of stearic and palmitic acid;[120,126] oleic acid may be unaltered or decrease with age. Overall there is an increase in the saturation and hence the melting point with age. Because the age effect results from increased fatness, the change towards the production of leaner pigs will reduce the hardening.

At equal slaughter weights boars are leaner than gilts which in turn are leaner than castrated males (barrows). Hence the most fat trim and therefore lard comes from barrows. In the past all males, except those kept for breeding, were castrated. However, the pressure to produce lean meat has resulted in a decline in this practice, and the indications are that it will continue to drop. The fatty acid compositions are generally those expected from the fatness levels (Table 8.16) with proportions of linoleic acid decreasing in the order, boars > gilts > barrows.[119,122,125,129,169-175] The differences in linoleic acid concentration between boars and castrates occur in all depots, but to different degrees, and range from less than 1% to 10% depending on the study although the reasons for this are not always clear. The lower

TABLE 8.16
Effect of sex on the fatty acid composition of pig backfat

Fatty acid	Boars	Castrates	Gilts
14:0	1·6	1·6	1·5
16:0	26·8	27·8	26·8
16:1	3·5	3·3	3·2
18:0	15·0	15·8	15·4
18:1	42·3	41·8	43·5
18:2	10·2	8·6	9·0

Values are percentage by weight of total fatty acids from loin inner layer of pigs slaughtered at 68 kg live-weight.
Data from Enser, M. and Wood, J. D., unpublished.

linoleic acid in fat from barrows is usually associated with an increase in oleic, palmitic and/or stearic acids. However, if the comparison is made between boars and barrows of equal fatness the differences are reversed, and the fat of barrows contains significantly more linoleic acid.[175] But, since animals are slaughtered at fixed weights and hence different levels of fatness this observation will not affect commercial lard composition.

Whether breed produces differences in fatty acid composition which are independent of fatness or rate of fattening is not clear since the appropriate body composition and growth rate data are often not reported. In a comparison of boars and barrows of the Belgian Landrace and Pietrain breeds Bonneau *et al.*[174] observed that castration produced much wider differences in linoleic acid concentration in the Pietrain than in the Landrace. However, Wood[120] considered that differences between Pietrains and Large White pigs fed the same amounts of food were probably the result of greater fat deposition by Large Whites. Differences between breeds in linoleic acid content reported by Martin *et al.*[170] for several Lacombe lines and their crosses with Poland Chinas and Hampshires, and for Yorkshires were small and explicable in terms of fatness or rate of fattening. Overall, the effects of age, sex and breed are relatively small compared with the changes in fatty acid composition which can be produced by dietary means. However, within the limitations discussed above, it is clear that most long-chain fatty acids fed to pigs will be deposited in the adipose tissue and affect its composition. Once deposited the fatty acid has a good chance of being present when the animal is slaughtered since the turnover of the fatty acids is very slow, the half-life approaching the lifespan of bacon pigs in this country.[137,149]

8.4.4 Triacylglycerol Composition of Pig Lipids

Early studies demonstrated that lard was characterised by the presence of palmitic acid as the major component of the 2-position of the triacylglycerols.[176-178] This was confirmed when pancreatic lipase[179] was used to selectively remove the 1- and 3-fatty acids.[123,180-184] It was revealed that 81–96% of the palmitic acid could be present in the 2-position where it contributed 71–79% of the fatty acids. Using the data for the fatty acid composition at position 2, and assuming (i) that the fatty acid distributions are random at one position with respect to another, (ii) that positions 1 and 3 are occupied by identical proportions of saturated and unsaturated fatty acids, the tri-acylglycerol composition could be calculated in terms of saturated (S) and unsaturated (U) residues, including positional isomers.[185] Stereo-specific analysis[186,187] revealed that stearic acid occurs mainly in positions 1 and 3 with the former predominant, whereas oleic acid also occurs in these positions but with more in position 3[78,188] (Table 8.17). Linoleic acid is distributed similarly to oleic acid but with a greater preference for position 3. The similarity in fatty acid distribution between triacylglycerols from various sites in the carcass agrees with the analysis of Anderson et al.,[123] which showed that glycerides from different sites containing the same proportions of saturated and unsaturated fatty acids had similar fatty acid compositions.

The way in which fatty acid distribution changes in response to different overall fatty acid compositions was investigated by Christie and Moore.[189] This comprehensive study of 45 samples was based on triacylglycerols selected for their range of oleic acid concentrations. Despite this, some clear patterns emerged. For 14:0 the proportion of the total in positions 1 and 2 remained constant but since most was in position 2 this increased the most with an increase in total 14:0. For 16:0 the proportion of the total increased in positions 1 and 2 as the total amount increased but since most was in position 2, that showed the greatest increase. A constant proportion of 18:0 was present in position 1 whilst that in position 3 increased with concentration. However, since the former contained the highest proportion it increased most when total 18:0 increased. The proportion of 16:1 was

TABLE 8.17
Fatty acid composition in triacylglycerols from pig adipose tissues

Site	Position on glycerol	Fatty acid (%)					
		14:0	16:0	16:1	18:0	18:1	18:2
Backfat, outer	1	0·9	9·5	2·4	29·5	51·3	6·4
	2	4·1	72·3	4·8	2·1	13·4	3·3
	3	−0·2[a]	0·4	1·5	7·4	72·7	18·2
Backfat, inner	1	0·7	9·8	1·7	38·8	42·7	6·3
	2	3·5	72·1	3·7	3·8	14·0	2·9
	3	0·6	5·4	2·1	11·3	65·4	15·2
Perinephric	1	2·5	11·7	2·3	35·2	43·7	4·6
	2	4·8	75·4	4·1	2·8	10·2	2·7
	3	−1·6	5·0	2·3	12·1	68·5	13·7
Omental	1	1·5	15·1	2·4	37·9	38·8	4·3
	2	4·8	76·1	4·1	4·0	9·9	1·1
	3	0·6	12·0	1·6	19·0	58·4	8·4
Mesenteric	1	1·8	15·6	2·6	37·6	39·4	3·0
	2	4·5	78·4	3·9	3·0	8·9	1·3
	3	−0·3	2·3	1·6	19·1	66·6	10·7

[a] Negative values resulting from small losses in determinations.
After Christie and Moore.[188]

constant in all three positions over a range in total from 1% to 11%. As the total 18:1 in the lipid increased from 38% to 54%, the proportion in position 1 remained at 36% whereas that in position 2 increased and in position 3 fell. This resulted in the 18:1 forming a constant proportion of the position 3 fatty acids whilst it increased as a proportion of positions 1 and 2. Changes in 18:2 were similar to those of 18:1 but less marked with the proportion of total 18:2 in positions 1 and 2 increasing with the total amount in the lipid while the proportion in 3 fell. The decrease was small so that 18:2 as a proportion of the fatty acids in this position still increased from 10% to 20% as total triacylglycerol 18:2 rose from 4% to 11%. The generality of these results amongst different breeds, diet and growth regimes remains to be established. The concentration range of the fatty acids covered the normal levels for pig backfat, although the more saturated samples were obtained from the harder internal depots and some of the higher oleate samples from copper-fed pigs. However, the similarities between different depots and the absence of segregation of results from copper-fed pigs in the sample suggest they are quite robust. The marked decrease in melting point of fat from copper fed pigs,[190,191] in which there are relatively small changes in fatty acid composition, agrees with the expected changes in composition of the triacylglycerols.[162,184] The proportion of palmitic acid in position 2 falls steeply with a decrease in the total palmitic acid. It is replaced by 16:1, 18:1 and 18:2 which decreases the proportion of SSS and SSU triacylglycerols, those responsible for the firmness of the lard. As the second most abundant saturated fatty acid in pig triacylglycerols, stearic acid is a major component of these two solid glycerides and hence it is important in the control of the melting point of lard.[159] The softening effect of linoleic acid probably occurs through competition with stearate for positions 1 and 3, particularly the latter.

Pancreatic lipase hydrolysis, following separation of triacylglycerols according to their degree of unsaturation using silver nitrate TLC, made it possible to determine the triacylglycerol composition of lard—at least in terms of class of fatty acid, i.e. saturated, mono-unsaturated, di-unsaturated, etc.[162,192] The major glycerides, as expected from the total fatty acid composition and monoacylglycerol fatty acids were SSM and MSM followed by MSD, where S stands for saturated, M for mono-unsaturated and D for di-unsaturated, although the amounts of each varied between the two studies because of differences in the total fatty acid composition (Table 8.18). Christie & Moore[189] combined the the silver nitrate TLC with stereospecific analysis of the separated glyceride fractions (Table 8.19). In the lipid from the outer backfat, which is more unsaturated than that from the inner layer, oleic acid replaces stearic acid in position 3 of the SSM fraction. However, in the SMD fraction stearic acid concentration is maintained in position 3 in the outer fat, but in position 1 it falls, to be replaced by palmitic acid. The SMD fraction is noteworthy in that it shows differences in composition between the two layers despite the limited composition possible within the fraction. The total proportions of 18:1 and 18:2 in this fraction are virtually the same in the two layers yet their distribution reflects the changes expected from their higher total proportions in the outer layer. In position 1 oleic acid increases relative to linoleic and in position 3 this is reversed, compared with the inner layer (Table 8.19).

Karleskind *et al.*[79] reported the acylglycerol composition of lard using GLC on packed columns to separate the sample according to the number of fatty acid carbons (Table 8.13). Similar results were obtained during the collaborative study preceding the standardisation

TABLE 8.18
Triacylglycerol composition (moles %) of lard and pig backfat

Triacylglycerol	Lard[a]	Backfat[b]	
		Inner	Outer
SSS[c]	2·4	6·3	4·4
SSM	24·1	32·2	26·8
SMS	tr	0·7	0·5
SMM	3·5	5·2	5·0
MSM	29·8	25·5	28·1
MMM	8·5	5·0	6·6
SSD	4·5	6·5	5·9
SDS	0·1	0·2	0·3
SMD	0·1	1·2	1·6
MSD	14·5	7·7	9·1
SDM	0·3	0·9	1·1
Rest	12·2	8·6	10·6
Approximate fatty acid compositions (%):			
S	38·3	49·1	45·9
M	49·2	43·2	45·0
D	14·4	7·9	9·2

[a] Data taken from Blank *et al.*[192]
[b] Data taken from Christie and Moore.[162]
[c] S, saturated; M, mono-unsaturated; D, di-unsaturated fatty acids.

of the method by IUPAC.[80] This also reported that capillary columns could be used but they need to be short to prevent partial separation of isomers.

D'Alonzo *et al.*[193] reported the separation of lard on capillary columns to yield not only the triacylglycerol composition but also the diacylglycerol, monoacylglycerol and non-esterified fatty acids after derivatisation with [(N,O)-bis(trimethylsilyl)trifluoroacetamide]. Perrin and Prévot[38] used reverse phase HPLC to separate lard into most of its components, the remaining problem being the inability of the column to distinguish between positions 1 and 3.

8.4.5 Steroids

The cholesterol content of lard has been reported to be 75 and 95 mg/100 g and that of leaf lard to be 101 mg/100 g.[102,106] Reiser[54] quoted a value of 70 mg/100 g of pig adipose tissue which would produce lard with 86 mg of cholesterol/100 g. However, Pihl[194] and

TABLE 8.19

Stereospecific analyses of triacylglycerol species from pig inner and outer back fat

Fraction[a]	Position	Inner back fat							Outer back fat						
		14:0	16:0	16:1	18:0	18:1	18:2	18:3	14:0	16:0	16:1	18:0	18:1	18:2	18:3
SSS	1	—	14·5		85·5				1·2	12·6		86·2			
	2	6·2	80·1		13·7				5·9	90·9		3·2			
	3	1·6	44·9		53·5				1·6	39·9		58·5			
SSM	1	1·2	18·3	0·9	56·8	22·8			0·6	23·5	0·2	66·2	9·5		
	2	4·9	86·7	1·1	4·9	2·4			5·8	85·8	—	4·4	4·0		
	3	-0·7[b]	4·8	1·3	22·9	71·7			0·2	2·6	2·8	11·6	82·8		
SMM	1	0·8	3·7	3·1	13·6	78·8			—	5·8	3·0	10·4	80·8		
	2	3·8	74·6	4·5	3·0	14·1			5·1	75·6	4·2	2·6	12·5		
	3	-0·1	—	2·3	1·1	96·7			—	-0·4	2·7	1·4	96·3		
MMM + SSD	1	0·9	8·6	1·3	40·1	40·7	8·4		0·4	11·2	2·8	27·8	48·3	9·5	
	2	3·0	50·1	10·0	2·9	32·7	1·3		2·5	39·8	12·9	1·8	41·8	1·2	
	3	0·6	2·5	2·2	5·3	42·1	47·3		0·7	6·3	1·7	10·9	50·3	30·1	
SMD	1	—	5·5	2·1	15·9	49·3	27·2		0·3	10·8	2·8	9·7	54·3	22·1	
	2	4·3	67·7	3·6	3·4	10·8	10·2		4·7	67·8	5·3	3·1	10·9	8·2	
	3	0·2	0·3	1·2	2·9	34·7	60·7		-0·2	1·2	0·9	3·1	26·9	68·1	
Remainder	1	0·7	6·9	3·1	18·7	43·6	22·7	4·3	0·4	4·9	2·4	15·1	52·3	21·3	3·6
	2	2·7	40·0	7·8	3·3	20·4	20·8	5·0	3·0	44·7	8·2	1·9	21·9	17·9	2·4
	3	-0·7	7·1	1·4	2·0	36·8	44·4	9·0	-0·1	2·6	1·4	2·2	44·0	43·0	6·9

[a] Abbreviations for triacylglycerol species: SSS, trisaturated; SSM, disaturated-monoenoic; SMD, saturated-monoenoic-dienoic, etc.

[b] Negative values resulting from small losses in determinations.

Data taken from Christie and Moore.[188]

Bohac *et al.*[109] found pig subcutaneous adipose tissue to contain 99 and 107·4 ± 4·6 (SD) which could yield lard containing 124 and 134 mg/100 g assuming a tissue lipid content of 80% and complete recovery during rendering. There have been two reports of much higher cholesterol concentrations in pig adipose tissue with values from 358 to 537 mg/100 g.[195,196] The reasons for these high levels are not clear but may be methodological.[109]

Steroid hormones are also likely to be present in lard at low concentrations, but particular attention has been paid to a metabolic derivative, 5α-androst-16-ene-3-one, which is responsible for the unpleasant odour produced when some boar meat is cooked.[197] Its concentration in adipose tissue is low, 0·06–0·70 ppm, the highest levels occurring in old boars.[198,199] Although boars are being used increasingly for meat production, they are slaughtered before reaching sexual maturity to avoid the problem of boar taint so there should be no increase in tainted fat going for rendering.

8.4.6 Non-Fatty Acid-Containing Components

Total unsaponifiable matter in lard was reported to be 0·18% ± 0·02%[200] so that cholesterol accounts for over 50% of the total. However, Williams and Pearson[201] observed that only 17% was cholesterol but the rest appeared to have been oxidised. Their vitamin A plus squalene fraction at 19% seems high as other studies show squalene is present at very low levels; 3 mg/100 g[111] although it constitutes 89·6% of the hydrocarbon fraction of lard.[112] Somewhat surprisingly the latter authors also reported a phytol derivative in the alcohol fraction of lard, although the quantity of chlorophyll is low in pig feed. Palmitoyl alcohol and stearoyl alcohol were also present but none of the materials were quantified.

Paul & Southgate[89] reported only traces of vitamins in lard. Actual values determined by Piironen *et al.*[90] were: α-tocopherol 0·7 mg/100 g; α-tocotrienol 0·2 mg/100 g. Vitamin A is present at 40–200 U/100 g lard.[202]

8.5. FISH OILS

Fish oils are an important although relatively small component of worldwide oil and fat production amounting to approximately 2% of the total. They are produced in quantity from ten or so species or close

TABLE 8.20

Major commercial fish oils

Oil	Species	Locality caught
Anchovy/Anchoveta	*Engraulis capensis*	Off South and South West Africa
	Engraulis ringens	Pacific—off Chile and Peru
	Engraulis mordax	Pacific—off Southern California and Mexico
Capelin	*Mallotus villosus*	North Atlantic and Barents Sea
Herring	*Clupea harengus harengus*	North Atlantic, North Sea, Norwegian Sea
	Clupea harengus pallasi	Pacific
Horse mackerel	*Trachurus murphyi*	Pacific—off Chile
	Trachurus trachurus	Off South Africa
	Caranx trachurus	
Menhaden	*Brevoortia tyrannus*	Atlantic
	Brevoortia patronus	Gulf of Mexico
Mackerel	*Scomber scombrus*	North Atlantic, North Sea
	Scomber japonicus	Pacific
Norway pout	*Trisopterus esmarkii*	North Sea, North Atlantic, Norwegian Sea, Barents Sea
Sandeel	*Ammodytes tobianus*	North Sea
Sardine/Pilchard	*Sardinops sagax*	Pacific—off Chile and Peru, Atlantic—off USA and Canada
	Sardinops ocellata	Off South Africa
	Sardinops melanostica	Off Japan
Sprat	*Clupea sprattus*	North Sea

relatives (Table 8.20). However, the assignment of an oil to a particular species refers to the major species present although most catches will be mixed. In general, this section of the review will be limited to consideration of these oils although it will be necessary to consider information obtained by analysis of oil from individual fish or small samples which may or may not have been produced under commercial conditions.

8.5.1 Extraction Procedures

Fish oils may be produced from the offal remaining after filleting or from surplus table fish or from industrial fish caught specifically for the production of fish meal. There is little characteristic adipose tissue in fish. That which does occur is mainly associated with the skin, particularly in the lateral line area. The dark muscle, also present in the latter area, has a higher fat content than the white muscle, 18·3% compared with 7·6% in the mackerel.[203] However, within the muscle there may be considerable variation, with the ventral muscles, which enclose the body cavity, containing particularly high levels. The release of the fat from the muscle during processing requires cooking the fish up to approximately 90°C after which the liquor is strained off and the remaining oil forced out under pressure.[204] The oil and water phases are first separated by decanting followed by centrifugation. Whilst the bulk of the neutral lipids are removed from the flesh by this procedure, extraction of the phospholipid is only partial.

Although maximum lipid removal is desirable for the production of fish protein meal to avoid taint resulting from oxidative instability of its highly unsaturated fatty acid in the flesh of pigs and poultry, the presence of phospholipid and related arsenolipids in the oil is undesirable. These and other impurities are removed during refining of the oil[204] to give a product containing over 99% lipid. The major lipid component is triacylglycerol with some mono- and diacylglycerol and non-esterified fatty acids which should not exceed 4% by weight of oil. Non-saponifiable lipid is usually of the order of 2–3%.

8.5.2 Fatty Acid Composition

The fatty acid composition of fish lipids has been extensively reviewed by Ackman[205,206] who has contributed substantially to the analysis of the complex mixture in fish oils and partially hydrogenated fish oils by gas–liquid chromatography.[207,208] Although the techniques give accurate analyses, there have been questions about the relevance of particular studies, based on a small sample, to the composition of commercial oils.[209] Whilst this argument is correct it seems somewhat sterile. As a natural product, variation in composition of fish oil is to be expected, particularly since there is no control over the production of sea-caught fish. Amongst the fish themselves, features such as age, size and reproductive cycle affect the quantity and fatty acid composition. Environmental factors further complicate predictions of composition since they influence the feed supply which is a major source of variation and fish may alter their fatty acid composition according to water temperature. One can envisage that human intervention through level of fishing or changes in the relative distribution of species might well affect quantity and type of food available, and in relatively restricted waters such as the North sea, pollution might well have an affect. Because of the variation in fatty acid composition between oils produced in different years from the same species caught in the same area at the same time of the year general fatty acid composition data can only be used as a guide.[210] Clearly, the suitability of oil for processing into a particular product is best determined by analysis.

Fish oils contain a wider range of fatty acids, particularly unsaturated fatty acids, than the fats of land animals. This arises because fish behave like monogastric animals and absorb and deposit the wide range of fatty acids present in their food. Furthermore they convert other dietary components such as the alcohols of wax esters into fatty acids and deposit these also.[211] Fish also have the capacity to synthesise fatty acids *de novo* and to carry out desaturation and chain extension of existing fatty acids. Of the saturated straight-chain fatty acids from 14:0 to 24:0, palmitic acid is the most abundant followed by myristic and stearic acids. As in animal tissues 17:0 and 15:0 are the most abundant odd-chain-length fatty acids although usually present at less than 1%.

Monoenoic acids of the n-5, n-7, n-8, n-9, n-11 and n-13 series have been reported and the total monoenoic acids may range from 20 to 60% of the total fatty acids, depending on the oil. Those oils with low amounts usually contain mainly palmitoleic and oleic acids whereas those with higher concentrations contain mainly gadoleic and cetoleic acids although the oils from pacific herring with a high monoenoic acid content, around 55%, have mainly oleic acid followed by palmitoleic acid. Whilst many of the mono-unsaturated fatty acids could be produced endogenously through Δ^9 desaturation followed by chain extension of saturated fatty acids it is probable that the cetoleic and much of the gadoleic acid is produced by oxidation of the corresponding long-chain alcohols in the dietary copepod wax esters.[212]

Although linoleic acid is the major dienoic acid in fish oils it is present in amounts of only about 1% and its elongation products of the n-6 series are also present in small amounts. The n-3 fatty acids are of much greater significance in fish oils than in animal fats although α-linolenic acid, the precursor, is present in small amounts, as in animal fats. Eicosapentaenoic acid and docosahexaenoic acid are the major n-3 acids, reaching levels of 20% or more in some oils. Other polyunsaturated fatty acids of the n-1 and n-4 series have been reported although they are usually present in amounts less than 1%. They are derived from algae[213] whereas the n-3 polyunsaturates are present in both the phytoplankton[214] and copepods.[212]

Many minor components of fish oils were listed by Gunstone *et al.*[215] who investigated the occurrence of phytol-derived branched-chain fatty acids and also fatty acids containing a furan ring (Fig. 8.2). Whereas the phytol-derived fatty acids occur widely and data are available for herring and mackerel[215] and manhaden oils,[216] furan acids were found in mackerel but not herring or capelin. They occur mainly in triacylglycerols and cholesteryl esters at concentrations of approximately 0·2% in body oils but not in phospholipids. Isolation of the multi-branched and furan fatty acids was accomplished by urea complexation to remove the simpler saturated and unsaturated fatty acids, leaving the long-chain PUFA and multi-branched acids in solution. Further separation was by silver ion TLC.

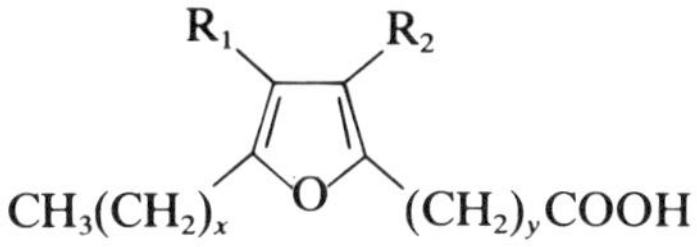

$$CH_3(CH_2)_x \underset{O}{\overset{R_1 \quad R_2}{\diagup\diagdown}} (CH_2)_y COOH$$

	R_1	R_2	x	y	Relative amounts
1.	H	CH_3	4	8	1·00
2.	CH_3	CH_3	2	10	0·55
3.	H	CH_3	4	10	0·44
4.	CH_3	CH_3	4	10	0·33
5.	CH_3	CH_3	4	8	0·11

Data from Gunstone *et al.*[215]

Fig. 8.2. Furan fatty acids of mackerel (*Scomber scombrus*) muscle oil.

Despite the earlier comments on the variability in the composition of fish oils certain characteristics are discernible in the fatty acid composition of fish oils and the author is indebted to Young[217] for his work in collecting together available literature data. As the iodine value increases across the species listed (Table 8.21) the proportion of palmitic acid tends to rise along with eicosapentaenoic acid whilst eicosenoic acid and docosenoic acid decrease. Alternatively one can divide them into two groups:[217] those with a high proportion of saturated and pentaenoic acids consisting of menhaden, sardine, horse mackerel and anchovy, species fished in the southern hemisphere or the southern part of the northern hemisphere, and the rest which have high proportions of monoenoic acids and lower proportions of saturated and pentaenoic fatty acids and are from fish in the north part of the northern hemisphere. Docosahexaenoic acid has a less distinctive distribution being slightly lower in capelin and herring and higher in Norway pout and particularly horse mackerel in which it appears to replace some of the eicosapentaenoic acid.

Tables 8.22–8.31 list the fatty acid composition of specific samples of the major fish oils and geographical and seasonal effects. However, care must be taken in attributing effects to geographical site when dates of the catch are not available. The large differences that can occur between different samples of an oil are demonstrated for anchovy/anchoveta in Table 8.22. The differences between the composition of the two Chilean samples is much greater than between samples from South Africa, Peru and Baja California. There is less than half as much 22:6 in the central Chilean oil compared with that from the north, the difference being made up by large increases in 20:1, 22:1 and 20:5. However, the high concentration of the latter merely brings it into line with all the other samples. The range of compositions to be found in menhaden oil and oils from different Sardinops species from different sites are shown in Tables 8.27 and 8.29. These also reveal differences of 100% or more in the content of some fatty acids such as 22:6. The specific effect of season on this variation is demonstrated in Table 8.23 for capelin, Table 8.24 for herring, Table 8.28 for Norway pout, Table 8.29 for Sardinops and Tables 8.30 and 8.31 for sandeel and sprat oils respectively. In general for the North Sea species there seems to be an increase in 20:5 and 22:6 in the summer followed by a winter decrease and their replacement by 20:1 and 22:1, although the timing may vary according to the particular species (Tables 8.23, 8.24, 8.28, 8.30). These changes

TABLE 8.21

Approximate fatty acid composition of fish oils

Fatty acid	Capelin	Herring	Sprat	Norway pout	Mackerel	Sandeel	Menhaden	Sardine/Pilchard	Horse mackerel	Anchovy
14:0	7	7	ND	6	8	7	9	8	6	9
16:0	10	16	16	13	14	15	20	17	24	19
16:1	10	6	7	5	7	8	12	9	7	9
18:1	14	13	16	14	13	9	11	12	13	13[a]
20:1	17	13	10	11	12	15	1	3	2	5[b]
22:1	14	20	14	12	15	16	0·2	3	2	2
20:5	8	5	6	8	7	9	14	17	11	17
22:6	6	6	9	13	8	9	8	9	16	9
Iodine value	100/140	110/140	125/150	140	135/160	140/175	150/175	140/200	180/195	170/200

[a] Contains element of 16:4 content.
[b] 20:1 and 18:4 combined.
ND, no data.
Source: Young.[217]

TABLE 8.22
Anchovy, anchoveta oil

Characteristics

Relative density	0·93 (21°C) (Young[217])
Refractive index, n_D^{60}	1·4668–1·4672 (Patterson[218])
Saponification value (mg KOH/g)	191–193·5 (Stansby[219])
Iodine value (Wijs)	180–198·5 (Stansby[219])
	min. 178, max. 184, av. 183 (Young[217])
Unsaponifiable matter (%)	3 max (Stansby[219])
	min. 0·3, max. 1·0, av. 0·7 (Young[217])

Fatty acid	Fatty acid composition (%)[c]				
Source:	S. Africa	Chile		Baja California	Peru
		North	Centre		
	NS	NS	NS	NS	NS
14:0	6·9	11·2	10·3	8·3	7·5
15:0	—	1·3	0·4	1·0	0·6
16:0	20·3	20·4	16·7	19·5	17·5
16:1	9·4	7·9	11·3	9·1	9·0
17:0	—	2·0	0·5	1·1	0·6
18:0	3·7	6·8	3·1	3·3	4·0
18:1	13·7[a]	12·2	9·0	16·9[a]	14·0[a]
18:2	1·0	3·3	1·3	0·9	1·9
18:3	—	0·8	0·3	0·6	1·3
20:1	3·5[b]	2·0[b]	7·8[b]	4·5[b]	4·8[b]
20:4	0·8	0·3	0·3	0·9	0·8
20:5	19·6	10·1	18·5	18·2	17·0
22:1	2·6	2·0	3·8	1·6	1·2
22:5	1·3	1·0	1·8	—	1·6
22:6	9·3	9·2	4·3	10·9	8·8
Other	7·9	9·5	10·6	3·2	9·4
Iodine value	—	—	163	185	181

[a] Combined 18:1 and 16:4 acids.
[b] Combined 20:1 and 18:4 acids.
[c] Ackman.[206]
NS = not stated.

TABLE 8.23
Capelin oil

Characteristics

Relative density	0·916–0·921 (Jangaard[220])
Refractive index, n_D^{50}	1·4620–1·4645, av. 1·4635 (Young[217])
Saponification value (mg KOH/g)	low 185, high 202, av. 189 (Young[217])
Iodine value (Wijs) winter:	low 94, high 143, av. 114 (Young[217])
summer:	low 134, high 164, av. 145 (Young[217])
Unsaponifiable matter (%) winter:	low 0·9, high 2·9, av. 1·5 (Young[217])
summer:	low 1·8, high 8·2, av. 4·8 (Young[217])

Fatty acid Source:	Fatty acid composition (%)						
	Norway (Ackman[206])			Norway (Young[217])		Canada Grand Bank (Jangaard[220])	Iceland (Young[217])
Date:	Jan. 1973	Mar. 1973	Aug. 1973	Winter	Summer	NS	NS
14:0	6·9	7·3	5·4	8·1	6·8	8·2	7·3
16:0	10·3	9·7	10·7	9·0	11·8	8·2	11·9
16:1	10·8	8·3	8·8	8·7	12·4	9·4	10·7
18:0	1·2	1·3	1·4	1·1	1·0	1·4	1·2
18:1	13·8	14·5	13·8	17·4	12·4	12·5	16·9
18:2	1·2	1·2	1·3	1·2	1·1	0·7	1·4
18:3	0·7	0·6	0·8	0·1	0·4	0·3	0·6
18:4	4·7	4·3	7·0	1·0	3·2	0·6	0·2
20:1	15·8	13·6	8·6	24·9	15·5	27·4	16·6
20:2	0·2	0·3	0·7	0·1	0·2	—	—
20:4	1·0	2·3	1·7	0·1	0·2	0·2	0·4
20:5	8·8	9·2	11·9	2·9	10·8	2·8	6·4
22:1	12·1	10·4	9·4	19·5	13·1	25·0	15·4
22:2	0·5	0·7	0·6	—	—	—	—
22:5	0·5	0·9	0·5	0·1	0·5	0·2	0·5
22:6	6·7	11·0	11·0	1·7	7·7	1·0	4·7
24:?	1·3	1·4	1·4	—	—	—	1·2
Other	3·0	3·0	3·0	4·1	2·9	2·1	4·4

NS not stated.

TABLE 8.24
Herring Oil

Characteristics

Relative density, 20°C	c. 0·9162 (Young[217])
Refractive index, 25°C	low 1·4730, high 1·4750, av. 1·4735 (Young[217])
	low 1·473, high 1·478 (Sonntag[221])
Refractive index, n_D^{60}	1·4591–1·4623 (Patterson[218])
Saponification value (mg KOH/g)	low 161, high 192, av. 183 (Young[217])
Iodine value (Wijs)	low 115, high 160, av. 140 (Sonntag[221])
	low 95, high 143, av. 110 (Young[217])
	winter 120–130, summer 140–160 (Stansby[219])
Unsaponifiable matter (%)	low 0·5, high 1·7 (Sonntag[221])
	low 1·2, high 1·7, av. 1·4 (Young[217])
	winter 0·9–1·3, summer 0·6–2·5 (Stansby[219])

Fatty acid composition (%)

Fatty acid	Norway (Ackman[206])			Pacific (Sonntag[221])		
Source:						
Date:	July 1974	Oct. 1974	Nov. 1975	— NS	Dorsal NS	Ventral NS
14:0	7·9	7·7	9·9	7·6	3·2	6·1
16:0	13·4	13·1	16·3	18·3	24·9	14·1
16:1	6·5	6·3	5·6	8·3	4·9	7·1
18:0	0·9	0·9	1·3	2·2	3·6	1·8
18:1	8·7	10·1	10·2	16·9	13·5	21·7
18:2	1·4	1·6	1·1	1·6	trace	1·7
18:3	1·9	2·0	0·6	0·6	a	a
18:4	4·6	3·6	0·8	2·8	—	—
20:1	13·6	11·8	15·2	9·4	19·2[a]	19·2[a]
20:2	0·5	0·6	0·7	—	—	—
20:4	0·4	0·5	0·3	0·4	b	b
20:5	7·2	6·4	1·7	8·6	4·6	4·6
22:1	18·6	19·9	27·0	11·6	19·3[b]	19·9[b]
22:2	1·0	0·4	0·6	—	—	—
22:5	0·5	0·6	1·3	1·3	—	—
22:6	8·5	9·8	1·1	7·6	3·8	3·8
24:?	1·4	1·7	1·3	—	—	—
Other	3·0	3·0	5·0	2·8	3·0	0

NS, not stated.
a Combined 18:3 and 20:1 acids.
b Combined 20:4 and 22:1 acids.

TABLE 8.25
Jurel, horse mackerel, maasbanker oil

Characteristics	
Relative density, 20°C	0·9227 (Young[217])
Refractive index	1·4741–1·4758 (Young[217])
Saponification value (mg KOH/g)	Maasbanker, S. Africa 193·7 (Young[217])
Iodine value (Wijs)	Chile, North 193 (Ackman[206])
	South 179 (Ackman[206])
	Maasbanker, S. Africa 158–170 (Stansby[219])
Unsaponifiable matter (%)	Maasbanker, S. Africa 3·6, 1·5, 1·3 (Young[217])

Fatty acid	Fatty acid composition (%)				
Source:	Chile (Ackman[206])		S. Africa, S. Atlantic (Young[217])		
Date:	North NS	South NS	June 1985	Nov. 1985	Mar. 1986
14:0	7·2	7·2	3·6	8·7	8·6
15:0	0·8	0·8	—	0·3	0·3
16:0	19·4	19·6	21·5	14·0	17·6
16:1	7·8	8·5	4·8	6·9	9·0
17:0	1·0	1·1	0·6	0·2	—
18:0	3·4	3·9	6·2	2·7	3·9
18:1	12·2	15·2	13·3	6·4	11·9
18:2	1·2	1·1	1·5	1·1	1·2
18:3	0·6	0·6	—	—	—
18:4	2·0	1·9	—	1·8	1·3
20:1	2·0	1·8	5·4	9·1	5·8
20:4	0·6	0·7	2·3	1·1	1·0
20:5	16·4	12·6	7·9	10·9	12·6
22:1	1·0	1·1	6·8	18·1	11·2
22:5	2·2	1·8	2·2	1·8	2·2
22:6	12·8	14·3	23·4	5·9	6·5
Other	9·4	7·8	0·5	11·0	6·9

NS, not stated.

probably relate to the use of copepods as the main dietary source after the winter period followed by an increase in the availability of phytoplankton fatty acids as the temperatures rise.

8.5.3 Triacylglycerol Structure

The distribution of fatty acids between the α- and β-positions of the triacylglycerols has been determined using pancreatic lipase[229] under conditions in which fatty acid selectivity did not occur. Commercial

TABLE 8.26
Mackerel oil

Characteristics

Relative density, Pacific, sp. gravity (15°C)	0·9301 (Tanikawa[223])
Refractive index, Pacific (20°C)	1·4811 (Tanikawa[223])
Saponification value (mg KOH/g)	c. 189 (Young[217])
	191·6 (Tanikawa[223])
Iodine value (Wijs)	low 136, high 157, av. 147 (Young[217])
	167 (Tanikawa[223])
Unsaponifiable matter (%)	low 0·4, high 1·3, av. 0·72 (Young[217])
	1·35 (Ackman[205])

Fatty acid	Fatty acid composition (%)	
Source:	Norway (Urdahl and Nygard[222])	NS (Ackman[205])
Date:	Autumn	NS
14:0	7·2	7·8
16:0	12·7	16·1
16:1	4·5	9·0
18:0	2·3	1·8
18:1	13·6	12·9
18:2	1·6	1·3
18:3	1·7	1·1
18:4	4·9	2·5
20:1	11·7	12·1
20:2	—	0·2
20:5	6·5	7·6
22:1	16·3	13·9
22:2	—	—
22:5	—	0·6
22:6	9·1	7·7
24	—	0·7
Other	7·9	4·7

NS, not stated.

herring and pilchard oils gave similar results. The general pattern[230] was that long-chain fatty acids occur in the α-position and polyunsaturated or short-chain fatty acids in the β-position (Table 8.32). However, for 20:5 similar total proportions of 14% in the two oils were distributed differently, with equal amounts in all three positions in pilchard oil triacylglycerols but 76% in the β-position of herring oil. In contrast, 22:6 had a similar distribution in both oils despite a

TABLE 8·27
Menhaden oil

Characteristics

Relative density (60°F)	0·912–0·930 (Young[217])
Refractive index, n_D^{65}	1·4590–1·4623, mean 1·4608 (Young[217])
Saponification value (mg KOH/g)	192–199, mean 196 (Young[217])
Iodine value (Wijs)	150–200 (Young[217])
Unsaponifiable matter (%)	0·6–1·6 (Stansby[219])

Fatty
acid

Fatty acid composition (%)[a]

Source:	Atlantic		East Gulf of Mexico		West Gulf of Mexico	
	Mean(23)	Range	Mean(39)	Range	Mean(65)	Range
14:0	8·9	6·6–12·3	9·0	7·8–10·9	9·0	8·2–11·1
14:1	0·3	0·2–0·4	0·3	0·2–0·4	0·2	0·1–0·3
15:0	0·6	0·5–1·1	0·7	0·5–0·8	0·6	0·4–0·8
16:0	18·2	14·3–20·8	20·0	16·9–22·2	20·1	18·1–22·8
16:1	10·8	7·7–15·1	12·2	10·5–14·8	12·0	10·6–12·9
16:2	1·4	0·9–2·0	1·9	1·3–2·6	1·8	1·4–2·3
16:3	1·6	0·9–3·0	2·2	1·5–2·7	2·4	2·0–3·1
16:4	1·2	0·5–2·1	1·0	0·4–2·1	1·2	0·5–2·1
17:0	0·9	0·6–1·3	0·9	0·6–1·1	0·8	0·3–1·0
18:0	3·4	2·5–4·0	3·5	2·7–4·2	3·4	2·9–3·9
18:1	9·8	6·5–12·0	10·2	8·1–12·6	12·2	6·6–15·6
18:2	1·3	1·0–1·6	1·2	0·8–1·9	0·9	0·6–1·5
18:3	1·4	0·7–2·4	1·3	0·8–1·6	1·1	0·6–1·6
18:4	3·2	1·5–4·6	2·2	1·7–2·8	2·1	1·5–2·6
20:1	0·9	0·5–1·4	0·9	0·6–1·3	1·4	0·5–1·9
20:4	2·3	1·4–4·3	2·2	1·6–2·6	2·3	1·3–4·1
20:5	14·7	12·3–18·1	13·5	11·4–16·8	13·3	11·5–17·7
21:5	0·7	0·6–0·8	0·6	0·5–0·9	0·6	0·6–0·8
22:1	0·3	0·2–0·4	0·3	0·2–0·4	0·4	0·2–0·4
22:5	2·3	2·0–3·2	2·6	2·0–3·5	2·5	1·7–3·7
22:6	9·4	4·5–14·5	7·7	5·7–10·6	6·4	4·2–8·6

[a] Joseph[210]; season 1982/3; no. of samples in parentheses.

two-fold difference in concentration whereas a lower 18:1 in pilchard
oil occurred though a decrease at the β-position.

8.5.4 Unsaponifiable Matter
The concentrations of unsaponifiable matter in different oils are shown
in Tables 8.22–8.31. The amounts vary according to the season and

TABLE 8.28
Norway pout oil

Characteristics

Relative density	
Refractive index	
Saponification value (mg KOH/g)	
Iodine value (Wijs)	141 (Young[217])
Unsaponifiable matter (%)	*c.* 5·5 (Young[217])

Fatty acid	Fatty acid composition (%)			
Source:	Norway (Ackman[206])			Denmark (Young[217])
Date:	Mar. 1973	Aug. 1973	NS, 1974	Mar. 1979
14:0	5·7	6·6	4·4	—
16:0	10·9	9·1	16·9	14·9
16:1	4·1	3·8	6·2	8·2
18:0	3·0	2·3	2·8	1·8
18:1	12·5	9·6	19·3	14·4
18:2	0·8	1·5	1·4	1·7
18:3	1·0	1·4	1·1	1·7
18:4	3·6	7·2	2·1	—
20:1	8·8	13·2	8·7	12·9
20:2	1·0	0·3	0·5	—
20:4	2·7	4·3	0·7	—
20:5	7·1	6·9	10·7	5·3
22:1	11·1	14·7	9·3	11·7
22:2	0·6	0·7	0·5	—
22:5	1·6	1·0	0·6	—
22:6	20·2	11·8	11·0	9·4
24:?	1·9	1·8	0·8	—
Other	3·0	3·0	3·0	18·0

NS not stated.

body composition of the fish. The seasonal effects are the result of high consumption of copepods which are rich in wax esters. Despite the conversion of much of their long-chain alcohols to fatty acids during passage through the intestinal wall, some are deposited intact within the tissues. Fatty alcohol concentrations are low in capelin in

TABLE 8.29
Sardine, pilchard oil

Characteristics

Relative density (s.g. 25/25°C)	0·914–0·921 (Sonntag[221])
Refractive index, n_D^{65}	1·4634–1·4648 (Sonntag[221])
n_D^{60}	1·4668–1·4672 (Paterson[218])
Saponification value (mg KOH/g)	188–199 (Sonntag[221])
Iodine value (Wijs)	159–192 (Sonntag[221])
	April 196. July 187, Oct. 203 (Spark[225])
	141, 158 (Young[217])
Unsaponifiable matter (%)	0·1–1·25 (Sonntag[221])
	April 1·6, July 2·6, Oct 6·4 (Spark[225])

Fatty acid composition (%)

Fatty acid	Ackman[206]	Stansby[219]	Stansby[219]	Urdahl and Nygard[222]		Spark[224]	Sonntag[221]	Young[217]
Source:	Pilchard (S. Africa)	Sardine (Peru)	Pilchard (*S. sagax*)	Pilchard (S. Africa)		Pilchard (S. Africa)	Sardine (Japan)	Pilchard (English Channel)
Date:	NS	NS	NS	Jan.	May	NS	NS	Jan.
14:0	7·8	8	7·6	7·7	12·0	4·6	6·6	—
15:0	0·4	trace	0·6	—	—	0·3	—	—
16:0	15·7	19	16·2	15·2	9·6	22·1	15·5	21·1
16:1	8·5	10	9·2	11·5	12·7	8·3	9·5	6·4
16:2	2·0	—	—	—	—	—	—	—
16:3	2·0	—	—	—	—	—	—	—
16:4	3·2	—	—	—	—	1·3	—	—
17:0	0·8	trace	0·7	—	—	1·1	—	—
18:0	3·7	3	3·5	2·4	1·6	5·2	3·7	3·6
18:1	9·3	14	11·4	11·3	7·5	14·1	17·3	15·9
18:2	1·5	1	1·3	0·9	0·7	1·6	2·5	2·1
18:3	1·1	trace	0·9	0·7	0·4	—	1·3	1·0
18:4	2·2	3	2·0	—	—	3·1	2·9	—
20:1	2·5	2	3·2	3·6	1·4	2·0	8·1	7·7
20:3	1·7	—	—	—	—	—	—	—
20:4	0·8	1	1·6	—	—	2·9	2·5	—
20:5	19·3	22	16·9	20·6	35·2	16·1	9·5	8·7
22:1	3·1	trace	3·8	2·3	0·9	0·6	7·8	6·1
22:5	2·4	2	2·5	—	—	0·8	2·8	0·9
22:6	6·5	4	12·9	13·3	4·4	11·4	8·5	12·4
Other	5·5	11	5·7	10·5	13·6	4·5	1·4	14·1
Iodine value	182	—	—	—	—	—	—	158

NS, not stated.

TABLE 8.30
Sandeel oil

Characteristics

Relative density (23°C)	0·9213 (Young[217])
Refractive index	
Saponification value (mg KOH/g)	180–190 (Young[217])
Iodine value (Wijs)	144–173 (Young[217])
Unsaponifiable matter (%)	2·4–6·3 (Bagge & Holmer[227])
	1·4–5·0 (Young[217])

Fatty acid	Fatty acid composition (%)				
Source:	Denmark (Holmer[228])	Denmark (Young[217])			
Date:	31.5.66	4.6.79	29.6.79	3.8.79	25.9.79
14:0	6·8	—	—	—	—
15:0	0·6	—	—	—	—
15:1	0·4	—	—	—	—
16:0	17·0	18·9	14·4	10·5	14·7
16:1	9·7	7·8	7·4	5·3	7·6
17:1 (16:2)	1·3	—	—	—	—
18:0	1·4	1·9	1·3	1·3	2·7
18:1	6·8	11·8	10·2	5·8	8·4
18:2	2·2	2·2	3·1	1·5	1·7
18:3	[a]	1·8	1·8	1·6	1·9
18:4	3·2	—	—	—	—
20:1	13·4[a]	12·7	15·0	20·9	11·7
20:2 (21:1)	0·3	—	—	—	—
20:4 ω3	0·5	—	—	—	—
20:5	10·8	10·0	9·6	7·3	9·3
22:1	17·0	11·2	17·1	21·7	12·1
22:5	0·8	—	0·6	0·7	0·9
22:6	7·1	8·1	7·0	6·6	13·9
Other	0·7	13·6	12·5	16·8	15·1
Iodine value	—	157	148	144	173

[a] Combined 18:3 and 20:1 acids.

TABLE 8.31
Sprat oil

Characteristics	
Relative density	
Refractive index	
Saponification value (mg KOH/g)	
Iodine value (Wijs)	125–147 (Sonntag[221])
Unsaponifiable matter (%)	1·0–1·8 (Sonntag[221])

Fatty acid	Fatty acid composition (%)		
Source:	North Sea (Young[217])		
Date:	Dec. 1978	Mar. 1979	Nov. 1979
16:0	16·4	15·9	17·2
16:1	5·9	7·9	6·6
18:0	1·8	2·4	2·3
18:1	15·1	17·2	14·7
18:2	2·0	2·0	1·9
18:3	1·7	1·7	1·8
20:1	10·8	10·0	10·0
20:5	6·3	5·5	7·2
22:1	16·2	13·5	12·6
22:4	0·9	0·5	0·6
22:5	0·7	—	0·8
22:6	8·6	7·0	10·6
Other	13·6	16·4	13·7

TABLE 8.32
Fatty acid composition in triacylglycerols of herring and pilchard oils

Fish	Position	14:0	16:0	16:1	18:1	20:1	22:1	20:5	22:6
Herring	Total	9	23	11	21	7	6	14	9
	α	8	20	12	25	13	14	4	2
	β	10	24	10	19	4	2	19	12
Pilchard	Total	12	20	10	12	4	3	14	17
	α	8	17	12	23	6	7	13	6
	β	14	22	10	6	3	1	14	22

After Brockerhoff and Hoyle.[229]

384

TABLE 8.33
Unsaponifiable matter of fish oils: composition (%)

Component	Spark[225] Maasbanker (S. Africa)	Spark[225] Pilchard (S. Africa)			Jangaard[220] Capelin					Bagge and Holmer[227] Sandeel (Denmark)			
		A	B	C	July	Sept.	Sept.	Oct.	Mar.				
Unsap. content in oil	—	—	—	—	3·20	3·10	7·99	4·47	1·36	2·6	2·7	5·6	6·3
Cholesterol	49·4	66·9	67·0	63·8	—	—	—	—	—	33/36[a]	22/27	20/22	17/18
Sterols	—	—	—	—	28	23	16	20	75	—	—	—	—
Glyceryl ethers	8·0	3·6	8·6	7·2	3	2	2	—	4	<10	<10	<10	<10
Hydrocarbons	—	—	—	—	6	7	6	15	16	<10	<10	<10	<10
saturated	12·4	18·9	6·1	5·1	—	—	—	—	—	—	—	—	—
squalene	2·3	2·1	2·8	1·7	63	68	76	65	5	—	—	—	—
Fatty alcohols	21·4	10·3	17·9	14·0	—	—	—	—	—	23/20[a]	40/51	35/40	43/41
Vitamin A	0·2	0·5	0·3	0·7	—	—	—	—	—	—	—	—	—

[a] Results of duplicate determinations.

TABLE 8.34
Fatty alcohols of commercial fish oils

Sample	Fatty alcohols (%)			
	16:0	18:1	20:1	22:1
Capelin, 1977	12·3	3·2	30·8	39·4
Herring, Atlantic 1973	9·5	4·5	29·1	43·9
Herring, Pacific 1973	18·4	35·3	5·0	5·6
Herring, Icelandic 1967	10·3	5·0	31·0	52·0
Mackerel, 1977	18·2	7·0	17·5	33·3

Adapted from Ackman.[205]

March when the total unsaponifiables are low, but have risen from 5% to 76% of the total by September when the unsaponifiables have reached 8% (Table 8.33). The pattern of deposited alcohols differs somewhat from that of the long-chain fatty acids derived from them. Whereas 22:1 alcohols are more abundant in fish than 20:1 alcohols (Table 8.34), their equivalent fatty acids are present in similar amounts in capelin, herring and mackerel oil. Whether this is a sampling effect, a less effective hydrolysis of the waxes with the longer alcohols during digestion or the provision of deposited 20:1 fatty acids from another source is not clear. Lean fish also have high concentrations of non-saponifiables suggesting that they may have particular functions in the fish or may be poorly utilised after deposition.

The other major component of the unsaponifiables are the sterols. Their concentrations in the oil appear to be relatively constant (Table 8.33) although their proportion in the unsaponifiables is affected by changes in other components. Based on the figures in Table 8.33, the cholesterol content of the quoted fish oils is much higher than that of animal fats with values of the order of 0·5–1·5 g/100 g. The glyceryl ethers consist of 1-alkyldiacylglycerols and 1-alkenyldiacylglycerols of which the former are more abundant. However, their concentrations in fish body lipids are much lower than in liver lipids, particularly of cartilagenous fish. Hydrocarbons consist of straight and branched chains of 14–33 carbons in length and it is thought that some may be derived from fatty acids by decarboxylation.[218] Squalene is an intermediate in cholesterol biosynthesis and is present in some liver oils in large amounts, but much less is present in the body lipids. This also applies to vitamins A and D. Tocopherol concentrations are also affected by season, changing from 25 μg/g for capelin oil in winter to 45 μg/g for oil from summer-caught fish. Samples of unspecified menhaden oil and anchovy oil are reported to contain 30 μg/g and more than 60 μg/g,[217] respectively.

ACKNOWLEDGEMENTS

The data presented in Tables 8.21–8.31 and 8.33 were collected by the International Association of Fish Meal Manufacturers for a technical bulletin on the composition of fish oils. The data in draft form were presented at the Leatherhead Food RA symposium on Fish Oils and Animal Fats in 1986. The published Technical Bulletin[217] is available free of charge from the International Association of Fish Meal Manufacturers, Hoval House, Orchard Parade, Mutton Lane, Potters Bar, Herts. EN6 3AR, UK.
The author thanks Mrs J. L. Roberts for checking the manuscript.

REFERENCES

1. Hagan, S. N., Murphy, E. W. and Shelley, L. M., *J. Am. Oil Chem. Soc.*, **50** (1967) 250–5.
2. Sahasrabudhe, M. R. and Smallbone, B. W., *J. Am. Oil Chem. Soc.*, **60** (1983) 801–5.
3. Folch, J., Lees, M. and Sloane Stanley, G. H., *J. Biol. Chem.*, **226** (1957) 497–509.
4. Bligh, E. G. and Dyer, W. J., *Can. J. Biochem. Physiol.*, **37** (1959) 911–17.
5. Prost, E. and Wrebiakowski, H., *Lebensmittel. Forsch.*, **149** (1972) 193–6.
6. Young, E. P., Kotula, A. W. and Twigg, G. G., *J. Anim. Sci.*, **42** (1976) 67–71.
7. Sheppard, A. J., *J. Am. Oil Chem. Soc.*, **40** (1963) 545–8.
8. Dole, V. P. and Meinertz, H., *J. Biol. Chem.*, **235** (1960) 2595–9.
9. Hara, A. and Radin, N. S., *Analyt. Biochem.*, **90** (1978) 420–6.
10. Maxwell, R. J., Marmer, W. N., Zubillaga, M. P. and Dalickas, A. G., *J. Ass. Off. Anal. Chem.*, **63** (1980) 600–3.
11. Chen, I. S., Shen, C.-S. J. and Sheppard, A. J., *J. Am. Oil Chem. Soc.*, **58** (1981) 599–601.
12. Maxwell, R. J., *J. Ass. Off. Anal. Chem.*, **70** (1987) 74–7.
13. Marmer, W. N. and Maxwell, R. J., *Lipids*, **16** (1981) 365–71.
14. Duncan, W. R. H. and Garton, G. A., *J. Sci. Fd Agric.*, **18** (1967) 99–102.

15. Kemp, J. D., Mahyuddin, M., Ely, D. G., Fox, J. D. and Moody, W. G., *J. Anim. Sci.*, **51** (1981) 321–30.
16. Garton, G. A. and Duncan, W. R. H., *J. Sci. Fd Agric.*, **20** (1969) 39–42.
17. Bensadoun, A. and Reid, J. T., *J. Nutr.*, **87** (1965) 239–44.
18. Busboom, J. R., Miller, G. J., Field, R. A., Crouse, J. D., Riley, M. L., Nelms, G. E. and Ferrell, C. L., *J. Anim. Sci.*, **52** (1981) 83–92.
19. Crouse, J. D., Kemp, J. D., Fox, J. D., Ely, D. G. and Moody, W. G., *J. Anim. Sci.*, **34** (1972) 388–92.
20. Gibney, M. J. and L'Estrange, J. L., *J. Agric. Sci., Camb.*, **84** (1975) 291–6.
21. Cook, L. J., Scott, T. W., Ferguson, K. A. and McDonald, I. W., *Nature*, **228** (1970) 178–9.
22. Mills, S. C., Searle, T. W. and Evans, R., *Aust. J. Biol. Sci.*, **32** (1979) 457–62.
23. Garton, G. A., Hovell, F. D. DeB. and Duncan, W. R. H., *Br. J. Nutr.*, **28** (1972) 409–16.
24. Smith, A. and Lough, A. K., *J. Chromatogr. Sci.*, **13** (1975) 486–90.
25. Abrahamsson, S., Stallberg-Stenhagen, S. and Stenhagen, E. In: *Progress in the chemistry of fats and other lipids*, Vol. 7, ed. R. T. Holman and T. Malkin, Pergamon Press, London, 1963, pp. 1–157.
26. Smith, A., Calder, A. F., Lough, A. K. and Duncan, W. R. H., *Lipids*, **14** (1979) 953–60.
27. Cason, J., Sumrell, G., Allen, C. F., Gillies, G. A. and Elberg, S., *J. Biol. Chem.*, **205** (1953) 435–47.
28. Schlenk, H. In: *Progress in the chemistry of fats and other lipids*, Vol. 2, ed. R. T. Holman, Pergamon Press, London, 1954, p. 243.
29. Miller, G. J., Kunsman, J. E. and Field, R. A., *J. Fd Sci.*, **45** (1980) 279–87.
30. Smith, A. and Calder, A. F., *Biomed. Mass. Spectrum*, **6** (1979) 347–9.
31. Wong, E., Johnson, C. B. and Nixon, L. N., *N.Z. J. Agric. Res.*, **18** (1975) 261–6.
32. Wong, E., Nixon, L. N. and Johnson, C. B., *J. Agric. Fd Chem.*, **23** (1975) 495–8.
33. Johnson, C. B., Wong, E. and Birch, E. J., *Lipids*, **12** (1977) 340–7.
34. Kukis, A., Marai, L. and Myher, J. J., *J. Chromatogr.*, **273** (1983) 43–66.
35. Brockerhoff, H., *Lipids*, **6** (1971) 942–56.
36. Hawke, J. C., Morrison, I. M. and Wood, P. R., *J. Sci. Fd Agric.*, **28** (1977) 293–300.
37. Christie, W. W. and Moore, J. H., *J. Sci. Fd Agric.*, **22** (1971) 120–4.
38. Perrin, J.-L. and Prévot, A., *Rev. Franc. Corps Gras*, **33** (1986) 437–45.
39. Termonia, M., Munari, F. and Sandra, P., *J. High Resol. Chromatogr.*, **10** (1987) 263–8.
40. Kukis, A. In: *Progress in the chemistry of fats and other lipids*, Vol. 12, ed. R. T. Holman, Pergamon Press, London, 1972, pp. 5–163.
41. Callow, E. H., *J. Agric. Sci., Camb.*, **51** (1958) 361–9.
42. Marchello, J. A. and Cramer, D. A., *J. Anim. Sci.*, **22** (1963) 380–3.

43. L'Estrange, J. L. and Mulvihill, T. A., *J. Agric. Sci. Camb.*, **84** (1975) 281–90.
44. Zygoyiannis, D., Stamataris, C. and Catsaounis, N., *J. Agric. Sci., Camb.*, **104** (1985) 361–5.
45. Tove, S. B. and Matrone, G., *J. Nutr.*, **76** (1962) 271–7.
46. Duncan, W. R. H., Ørskov, E. R. and Garton, G. A., *Br. J. Nutr.*, **31** (1972) 19A–20A.
47. Takahashi, T. and Oota, S., *Japn. J. Zootech. Sci.*, **56** (1985) 711–19.
48. Castle, W. E., *J. Hered.*, **25** (1934) 246–7.
49. Baker, R. L., Steine, T., Vabeno, A. W. and Breines, D., *Acta Agric. Scand.*, **35** (1985) 389–97.
50. Hill, F., *Irish J. Agric. Res.*, **1** (1962) 83–9.
51. Payne, E. and Twist, J. D., *Aust. J. Sci.*, **29** (1966) 140–1.
52. Crane, B. and Clare, N. T., *N.Z. J. Agric. Res.*, **18** (1975) 273–5.
53. Karijord, Ø., *Acta Agric. Scand.*, **28** (1978) 355–9.
54. Reiser, R., *J. Nutr.*, **105** (1975) 15–16.
55. Sreenivasan, B., *J. Am. Oil Chem. Soc.*, **45** (1968) 259–65.
56. Spencer, G. F., Herb, S. F. and Gormisky, P. J., *J. Am. Oil Chem. Soc.*, **53** (1976) 94–6.
57. Marmer, W. N., Maxwell, R. J. and Williams, J. E., *J. Anim. Sci.*, **59** (1984) 109–21.
58. Lin, K. C., Marchello, M. J. and Fischer, A. G., *J. Fd Sci.*, **49** (1984) 1521–4.
59. Hay, J. D. and Morrison, W. R., *Lipids*, **8** (1973) 94–5.
60. Hay, J. D. and Morrison, W. R., *Biochim. Biophys. Acta*, **202** (1970) 237–43.
61. Terrell, R. N., Lewis, R. W., Cassens, R. G. and Bray, R. W., *J. Fd Sci.*, **32** (1967) 516–20.
62. Terrell, R. N., Suess, G. G. and Bray, R. W., *J. Anim. Sci.*, **28** (1969) 449–53.
63. Roberts, W. K., *Can. J. Anim. Sci.*, **46** (1966) 181–90.
64. Leat, W. M. F., *J. Agric. Sci., Camb.*, **89** (1977) 575–82.
65. Garcia, P. T., Casal, J. J., Parodi, J. J. and Marangunich, L., *Meat Sci.*, **3** (1979) 169–77.
66. Pothoven, M. A., Beitz, D. C. and Zimmerli, A., *J. Nutr.*, **104** (1974) 430–3.
67. Leat, W. M. F., *J. Agric. Sci., Camb.*, **85** (1975) 551–8.
68. Hecker, A. L., Cramer, D. A. and Hougham, D. F., *J. Fd Sci.*, **40** (1975) 144–9.
69. Pyle, C. A., Bass, J. J., Duganzich, D. M. and Payne, E., *J. Agric. Sci., Camb.*, **89** (1977) 571–4.
70. Wood, J. D. In: *Fats in animal nutrition*, ed. J. Wiseman, Butterworths, London, 1984, pp. 407–35.
71. Sumida, D. M., Vogt, D. W., Cobb, E. H., Iwanaga, I. I. and Reimer, D., *J. Anim. Sci.*, **35** (1972) 1058–63.
72. Gillis, A. T., Eskin, N. A. M. and Cliplef, R. L., *J. Fd Sci.*, **38** (1973) 408–11.

73. Garcia, P. T., Casal, J. J. and Parodi, J. J., *Meat Sci.*, **17** (1986) 283–91.
74. Ramananarivo, R., Artaud, J., Estienne, J., Peiffer, G. and Gaydon, E. M., *J. Am. Oil Chem. Soc.*, **58** (1981) 1038–41.
75. Dryden, F. D. and Marchello, J. A., *J. Anim. Sci.*, **37** (1973) 33–9.
76. St John, L. C., Young, C. R., Knabe, D. A., Thompson, L. D., Schelling, G. T., Grundy, S. M. and Smith, S. B., *J. Anim. Sci.*, **64** (1987) 1441–7.
77. Cook, L. J., Scott, T. W., Faichney, G. J. and Lloyd Davies, H., *Lipids*, **7** (1972) 83–9.
78. Brockerhoff, H., Hoyle, R. J. and Wolmark, N., *Biochim. Biophys. Acta*, **116** (1966) 67–72.
79. Karleskind, A., Valmalle, G. and Wolff, J.-P., *Rev. Franc. Corps Gras*, **21** (1974) 617–22.
80. Pocklington, W. D. and Hautfenne, A., *Pure Appl. Chem.*, **57** (1985) 1515–22.
81. Schogt, J. C. M., Haverkamp Begemann, P. and Koster, J., *J. Lipid Res.*, **1** (1960) 446–9.
82. Palmer, E. J. and Eckles, C. H., *J. Biol. Chem.*, **17** (1914) 223–36.
83. Hill, F., *Inst. Meat Bull.*, **59** (1968) 6–16.
84. Morgan, J. H. L., Pickering, F. S. and Everitt, G. C., *Proc. N.Z. Soc. Anim. Prod.*, **29** (1969) 164–75.
85. Morgan, J. H. L. and Everitt, G. C., *N.Z. Agric. Sci.*, **4** (1969) 10–18.
86. Hammond, J., *Emp. J. Agric. Res.*, **3** (1935) 1–12.
87. Hirzel, R., Onderst. J. *et al.*, *Vet. Sci. Anim. Ind.*, **12** (1939) 449–53.
88. Forrest, R. J., *Can. J. Anim. Sci.*, **61** (1981) 575–80.
89. Paul, A. A. and Southgate, D. A. T., *McCance and Widdowson's The Composition of Foods*, 4th edn, HMSO, London, 1978.
90. Piironen, V., Syväoja, E. L., Varo, P., Salminen, K. and Koivistoinen, P., *J. Agric. Food Chem.*, **33** (1985), 1215–18.
91. Tu, C., Powrie, W. D. and Fennema, O., *Lipids*, **4** (1969) 369–79.
92. Parodi, P. W., *J. Am. Oil Chem. Soc.*, **52** (1975) 345–8.
93. Mordret, F., Prévot, A., Le Barbanchon, N. and Barboti, C., *Rev. Franc. Corps Gras*, **10** (1977) 467–75.
94. Farkas, J., Angel, A. and Avigan, M. I., *J. Lipid Res.*, **14** (1973) 344–56.
95. Hoelscher, L. M., Savell, J. W., Smith, S. B. and Cross, H. R., *J. Fd Sci.*, **53** (1988) 718–22.
96. Kritchevsky, D. and Tepper, S. A., *J. Nutr.*, **74** (1961) 441–4.
97. Tu, C., Powrie, W. D. and Fennema, O., *J. Fd Sci.*, **32** (1967) 30–4.
98. Hecker, A. L., Cramer, D. A., Beede, D. K. and Hamilton, R. W., *J. Fd Sci.*, **40** (1975) 140–3.
99. Adams, M. L., Sullivan, D. M., Smith, R. L. and Richter, E. F., *J. Ass. Off. Anal. Chem.*, **69** (1986) 844–6.
100. Rhee, K. S., Smith, G. C. and Dutson, T. R., *J. Fd Sci.*, **53** (1988) 969–70.
101. AOAC, *Official methods of analysis*, 14th edn, AOAC, Arlington, Virginia, 43.283, 1984.

102. Feeley, R. M., Criner, P. E. and Watt, B. K., *J. Am. Diet. Assoc.*, **61** (1972) 134–9.
103. Eichhorn, J. M., Wakayama, E. J., Blomquist, G. J. and Bailey, C. M., *Meat Sci.*, **16** (1986) 71–8.
104. Eichhorn, J. M., Coleman, L. J., Wakayama, E. J., Blomquist, G. J., Bailey, C. M. and Jenkins, T. G., *J. Anim. Sci.*, **63** (1986) 781–94.
105. Rhee, K. S., Dutson, T. R. and Smith, G. C., *J. Fd Sci.*, **47** (1982) 1638–42.
106. Punwar, J. K. and Derse, P. H., *J. Ass. Off. Anal. Chem.*, **61** (1978) 727–30.
107. Ryan, T. C. and Gray, J. I., *J. Fd Sci.*, **49** (1984) 1390–91.
108. Stromer, M. H., Goll, D. E. and Roberts, J. H., *J. Anim. Sci.*, **25** (1966) 1145–7.
109. Bohac, C. E., Rhee, K. S., Cross, H. R. and Ono, K., *J. Fd Sci.*, **53** (1988) 1642–4.
110. Angel, A. and Farkas, J., *J. Lipid Res.*, **15** (1974) 491–9.
111. Fitelson, J., *Ass. Off. Agric. Chem.*, **26** (1943) 506–11.
112. Maritano de Correche, M. and Oxley, R., *Gras Aceit.*, **36** (1985) 88–92.
113. Hubbard, A. W. and Pocklington, W. D., *J. Sci. Fd Agric.*, **19** (1968) 571–7.
114. Henriques, V. and Hansen, C., *Skand. Arch. Physiol.*, **11** (1901) 151–65.
115. Bhattacharya, R. and Hilditch, T. P., *Biochem. J.*, **25** (1931) 1954–64.
116. Banks, A. and Hilditch, T. P., *Biochem. J.*, **26** (1932) 298–308.
117. Dean, H. K. and Hilditch, T. P., *Biochem. J.*, **27** (1933) 1950–6.
118. Christie, W. W., Jenkinson, D. M. and Moore, J. H., *J. Sci. Fd Agric.*, **23** (1972) 1125–9.
119. Koch, D. E., Parr, A. F. and Merkel, R. A., *J. Fd Sci.*, **33** (1968) 176–80.
120. Wood, J. D., *Anim. Prod.*, **17** (1973) 281–5.
121. Marchello, M. J., Cook, N. K., Slanger, W. D., Johnson, V. K., Fischer, A. G. and Dinusson, W. E., *J. Fd Sci.*, **48** (1983) 1331–4.
122. Koch, D. E., Pearson, A. M., Magee, W. T., Hoefer, J. A. and Sweigert, B. S., *J. Anim. Sci.*, **27** (1968) 360–5.
123. Anderson, R. E., Bottino, N. R. and Reiser, R., *Lipids*, **5** (1970) 161–4.
124. Ingr, I., *Acta Vet. Brno*, **40** (1971) 163–70.
125. Villegas, F. J., Hedrick, H. B., Veum, T. L., McFate, K. L. and Bailey, M. E., *J. Anim. Sci.*, **36** (1973) 663–8.
126. Sink, J. D., Watkins, J. L., Ziegler, J. H. and Miller, R. C., *J. Anim. Sci.*, **23** (1964) 121–5.
127. Demarne, Y., Perazo-Castro, C. E., Henry, Y. and Flanzy, J., *Ann. Biol. Anim. Biochem. Biophys.*, **17** (1977) 137–46.
128. Jeremiah, L. E., *Meat Sci.*, **7** (1982) 1–7.
129. Wood, J. D., Buxton, P. J., Whittington, F. M. and Enser, M., *Livestock Prod. Sci.*, **15** (1986) 73–82.
130. Christensen, K. D., *Acta Agric. Scand.*, **13** (1963) 249–54.
131. Flanzy, J., Boudon, C., Leger, C. and Pihet, J., *J. Chromatogr. Sci.*, **14** (1976) 17–24.

132. Tsai, S. F., Witte, V. C. and Bailey, M. E., *J. Anim. Sci.*, **39** (1974) 317–24.
133. Vieites, C. M., Verges, J. B., Garcia, P. T., Marcelia, M., Luzzani, D., Ludden, L. B., Casal, J. J. and Basso, L. R., *Meat Sci.*, **23** (1988) 263–77.
134. Leat, W. M. F., Cuthbertson, A., Howard, A. N. and Gresham, G. A., *J. Agric. Sci., Camb.*, **63** (1964) 311–17.
135. Ellis, N. R. and Isbell, H. S., *J. Biol. Chem.*, **69** (1926) 219–38.
136. Beadle, B. W., Wilder, O. H. M. and Kraybill, H. R., *J. Biol. Chem.*, **175** (1948) 221–9.
137. Anderson, D. B., Kauffman, R. G. and Benevenga, N. J., *Lipids*, **7** (1972) 488–9.
138. Brown, J. B., *J. Biol. Chem.*, **90** (1931) 133–9.
139. Garton, G. A., Hilditch, T. P. and Meara, M. L., *Biochem. J.*, **50** (1952) 517–24.
140. Leat, W. M. F., *Br. J. Nutr.*, **16** (1962) 559–69.
141. Dahl, O. and Persson, K.-A., *J. Sci. Fd Agric.*, **16** (1965) 452–5.
142. Babatunde, G. M., Pond, W. G., Walker, E. F. and Chapman, P., *J. Anim. Sci.*, **27** (1968) 1290–5.
143. Nilsson, R., *Proc. 15th Eur. Meet. Meat Res. Work.*, Agricultural Research Council, Meat Research Institute, Langford, Bristol 1969, pp. 305–8.
144. Hanson, L. E., Allen, C. E., Meade, R. J., Rust, J. W. and Miller, K. P., *Feedstuffs*, **42** (1970) 16–18.
145. Brooks, C. C., *J. Anim. Sci.*, **33** (1971) 1224–31.
146. Roberts, J. L. and Enser, M., *Anim. Prod.*, **46** (1988) 502–3.
147. Walker, B. L., *Can. J. Anim. Sci.*, **52** (1972) 713–19.
148. McDonald, B. E. and Hamilton, R. M. G., *Can. J. Anim. Sci.*, **56** (1976) 677–80.
149. Cunningham, H. M., *J. Anim. Sci.*, **27** (1968) 424–30.
150. Chung, R. A. and Lin, C. C., *J. Fd Sci.*, **30** (1965) 860–4.
151. Flanzy, J., Francois, A. C. and Rerat, A., *Ann. Biol. Anim. Biochem. Biophys.*, **17** (1970) 137–46.
152. Creswell, D. C. and Brooks, C. C., *J. Anim. Sci.*, **33** (1971) 370–5.
153. Berschauer, F., Ehrensvärd, U. and Menke, K. H., *Arch. Tierernähr.*, **33** (1983) 826–42.
154. Berschauer, F., Ehrensvärd, U. and Menke, K. H., *Landwirtsch. Forsch.*, **37** (1984) 154–65.
155. Bastijns, L., *J. Sci. Fd Agric.*, **21** (1970) 576–8.
156. Hilditch, T. P., Lea, C. H. and Pedelty, W. H., *Biochem. J.*, **33** (1939) 493–504.
157. MacGrath, W. S., Vander Noot, G. W., Gilbreath, R. L. and Fisher, H., *J. Nutr.*, **96** (1968) 461–6.
158. Fuller, M. F., Duncan, W. R. H. and Boyne, A. W., *J. Sci. Fd Agric.*, **25** (1974) 205–10.
159. Wood, J. D., Enser, M. B., MacFie, H. J. H., Smith, W. C., Chadwick, J. P., Ellis, M. and Laird, R., *Meat Sci.*, **2** (1978) 289–300.
160. Taylor, M. and Thomke, S., *Nature*, **201** (1964) 1246.

161. Moore, J. H., Christie, W. W., Braude, R. and Mitchell, K. G., *Br. J. Nutr.*, **23** (1969) 281–7.
162. Christie, W. W. and Moore, J. H., *Lipids*, **4** (1969) 345–9.
163. Ho, S. K. and Elliot, J. I., *Can. J. Anim. Sci.*, **53** (1973) 537–45.
164. Ho, S. K. and Elliot, J. I., *Can. J. Anim. Sci.*, **54** (1974) 23–8.
165. Thompson, E. H., Allen, C. E. and Meade, R. J., *J. Anim. Sci.*, **36** (1973) 868–73.
166. Ho, S. K., Elliot, J. L. and Jones, G. M., *Can. J. Anim. Sci.*, **55** (1975) 587–94.
167. Elliot, J. I. and Bowland, J. P., *J. Anim. Sci.*, **30** (1970) 923–30.
168. Scott, R. A., Cornelius, S. G. and Mersmann, H. J., *J. Anim. Sci.*, **53** (1981) 977–81.
169. Brooks, C. C., *J. Anim. Sci.*, **26** (1967), 504–9.
170. Martin, A. H., Fredeen, H. T., Weiss, G. M. and Carson, R. B., *J. Anim. Sci.*, **35** (1972) 534–41.
171. Seerley, R. W., Emberson, J. W., McCampbell, H. C., Burdick, D. and Grimes, L. W., *J. Anim. Sci.*, **39** (1974) 1082–91.
172. Castell, A. G. and Mallard, T. M., *Can. J. Anim. Sci.*, **54** (1974) 443–54.
173. Bowland, J. P. and Newell, J. A., *Can. J. Anim. Sci.*, **54** (1974) 455–64.
174. Bonneau, M., Desmoulin, B. and Dumont, B. L., *Ann. Zootech.*, **28** (1979) 53–72.
175. Wood, J. D. and Enser, M., *Anim. Prod.*, **35** (1982) 65–74.
176. Hilditch, T. P. and Stainsby, W. J., *Biochem. J.*, **29** (1935) 91–9.
177. Meara, M. L., *J. Chem. Soc.* (1945) 23–45.
178. Quimby, O. T., Wille, F. L. and Lutton, E. S., *J. Am. Oil Chem. Soc.*, **30** (1953) 186–90.
179. Savary, P., Flanzy, J. and Desnuelle, P., *Biochim. Biophys. Acta*, **24** (1957) 414–23.
180. Mattson, F. H. and Lutton, E. S., *J. Biol. Chem.*, **233** (1958) 868–71.
181. Barford, R. A., Luddy, F. E., Herb, S. F., Magidman, P. and Riemenschneider, R. W., *J. Am. Oil Chem. Soc.*, **42** (1965) 446–8.
182. Stinson, C. G., deMan, J. M. and Bowland, J. P., *J. Am. Oil Chem. Soc.*, **44** (1967) 253–5.
183. Otake, Y., Nakazato, T., Sanada, T., Arai, T. and Namekawa, J., *Japn J. Zootech. Sci.*, **42** (1971) 551–8.
184. Amer, M. A. and Elliot, J. I., *Can. J. Anim. Sci.*, **53** (1973) 147–52.
185. Vander Wal, R. J., *J. Am. Oil Chem. Soc.*, **37** (1960) 18–20.
186. Brockerhoff, H., *J. Lipid Res.*, **6** (1965) 10–15.
187. Brockerhoff, H., *Arch. Biochem. Biophys.*, **110** (1965) 586–92.
188. Christie, W. W. and Moore, J. H., *Biochim. Biophys. Acta*, **210** (1970) 46–56.
189. Christie, W. W. and Moore, J. H., *Lipids*, **5** (1970) 921–8.
190. Moore, J. H., Christie, W. W., Braude, R. and Mitchell, K. G., *Proc. Nutr. Soc.*, **27** (1968) 45A.
191. Elliot, J. I. and Bowland, J. P., *Can. J. Anim. Sci.*, **49** (1969) 397–8.
192. Blank, M. L., Verdino, B. and Privett, O. S., *J. Am. Oil Chem. Soc.*, **42** (1965) 87–90.

193. D'Alonzo, R. P., Kozarek, W. J. and Wade, R. L., *J. Am. Oil Chem. Soc.*, **59** (1982) 292–5.
194. Pihl, A., *Scand. J. Clin. Lab. Invest.*, **4** (1952) 115–21.
195. Skelley, G. C., Borgman, R. F., Handlin, D. L., Acton, J. C., McConnell, J. C., Wardlow, F. B. and Evans, E. J., *J. Anim. Sci.*, **41** (1975) 1298–304.
196. Kellogg, T. F., Rogers, R. W. and Miller, H. W., *J. Anim. Sci.*, **44** (1977) 47–52.
197. Patterson, R. L. S., *J. Sci. Fd Agric.*, **19** (1968) 31–8.
198. Claus, R., Hoffman, B. and Karg, H., *J. Anim. Sci.*, **33** (1971) 1293–7.
199. De Brabander, H. F. and Verbeke, R., *J. Chromatogr.*, **363** (1986) 293–302.
200. Dahl, O., *Acta Agric. Scand. Suppl.* (1958), 5–67.
201. Williams, L. D. and Pearson, A. M., *J. Agric. Fd Chem.*, **13** (1965) 573–7.
202. Herb, S. F., Riemenschneider, R. W., Kaunitz, H. and Slanetz, C. A., *J. Nutr.*, **51** (1953) 393–402.
203. Ackman, R. G. and Eaton, C. A., *Can. Inst. Fd Sci. Technol. J.*, **4** (1971) 169.
204. Young, F. V. K., The production and use of fish oils. In: *Nutritional evaluation of long-chain fatty acids in fish oil*, ed. S. M. Barlow and M. E. Stansby, Academic Press, London, 1982.
205. Ackman, R. G., Fish lipids. In: *Advances in fish science and technology*, ed. J. J. Connell, Fishing News Books Ltd, Farnham, 1980.
206. Ackman, R. G., Fatty acid composition of fish oils. In: *Nutritional evaluation of long-chain fatty acids in fish oil*, ed. S. M. Barlow and M. E. Stansby, Academic Press, London, 1982.
207. Ackman, R. G. In: *Methods in enzymology, Vol. 14*, ed. J. M. Lowenstein, Academic Press, New York, 1969, pp. 329–81.
208. Ackman, R. G. In: *Progress in the chemistry of fats and other lipids, Vol. 12*, ed. R. T. Holman, Pergamon Press, Oxford, 1972, pp. 165–284.
209. Stansby, M. E., *J. Am. Oil Chem. Soc.*, **58** (1981) 13.
210. Joseph, J. D., *Marine Fish Rev.*, **47** (1985) 30.
211. Patton, J. S. and Benson, A. A., *Comp. Biochem. Physiol.*, **52B** (1975) 111.
212. Ackman, R. G., Linke, B. A. and Hingley, J., *J. Fish Res. Bd. Can.*, **31** (1974) 1812.
213. Ackman, R. G. and McLachlan, J., *Proc. N.Z. Inst. Sci.*, **28** (1977) 47.
214. Ackman, R. G., Tocher, C. S. and McLachlan, J., *J. Fish Res. Bd. Can.*, **25** (1968) 1603.
215. Gunstone, F. D., Wijesundera, R. A. and Scrimgeour, C. M., *J. Sci. Fd Agric.*, **29** (1978) 539.
216. Ratnayake, W. M. N., Olsson, B. and Ackman, R. G., *Lipids*, **24** (1989) 630.
217. Young, F. V. K., *The chemical and physical properties of crude fish oils for refiners and hydrogenators*, Fish Oil Bulletin No. 18, International Association of Fish Meal Manufacturers, Potters Bar, Herts, 1986.

218. Patterson, H. B. W., *Hydrogenation of fats and oils,* Applied Science, London, 1983.
219. Stansby, M. E., *Fish oils,* AVI Publishing Co. Inc., Westport, CN, 1967.
220. Jangaard, P. M., *The capelin,* Bulletin 186, The Fisheries Research Board of Canada, 1974, pp. 56–68.
221. Sonntag, N. O. V. In: *Baileys industrial oil & fat products,* Vol. 1, ed. D. Swern, Wiley Interscience, New York, 1979, pp. 442–53.
222. Urdahl, N. and Nygard, E., *Lipidforum Symposium Proceedings,* Oleochemicals, 1982, pp. 40–50.
223. Tanikawa, E., *Marine products in Japan,* Tokyo, 1971.
224. Spark, A. A., South African Association of Food Science and Technology, Symposium, 1972.
225. Spark, A. A., *South African fish oils,* Vol. 5/6, Namib und Meer, 1974/1975, pp. 37–48.
226. Mosher, W. A., Daniels, W. H., Celeste, J. R. and Kelley, W. H., *Commercial Fish Review,* **20**(11a) (1958) 1–6.
227. Bagge, A. and Holmer, G., *An investigation on the unsaponifiable matter of Danish fish oils from the period May–June 1977,* The Technical University of Denmark, 1977.
228. Holmer, G., *Studies over Danske Fiskoliers Sammensaetning,* Fish Ministry Research Laboratory, Copenhagen, 1967.
229. Brockerhoff, H. and Hoyle, R. J., *Arch. Biochem. Biophys.,* **102** (1963) 452.
230. Brockerhoff, H., *Comp. Biochem. Physiol.,* **19** (1966) 1.

9

Yellow Fats

R. A. Wilbey

Department of Food Science and Technology, University of Reading, White Knights, P.O. Box 226, Reading, RG6 2AP, UK

9.1. INTRODUCTION

The term 'Yellow Fats' is commonly used to cover the range of butters, margarines and other spreads sold in the retail market. The term may also include anhydrous milkfat (sometimes retailed as concentrated butter) and similar products such as ghee. This chapter is concerned primarily with milkfat, to complement the topics covered earlier in the book, bringing in margarine, reduced and low fat spreads as alternative products with overlapping characteristics.

Milkfat differs from the majority of fats available to the food industry in that, in its native form, it has organoleptic properties that are accepted (indeed usually desired) by the consumer, so that refining processes are not needed. By far the greatest quantity of the milkfat consumed worldwide is from cow's milk. Other sources include the buffalo, goat, camel and sheep.

9.2. NATURE AND PROPERTIES OF MILKFAT

Milkfat is secreted in the mammary gland principally in the form of globules enveloped by a milkfat globule membrane. The majority of globules (on a volume basis) are within the size range $1-10 \ \mu m$,[1] the average being typically less than $4 \ \mu m$. The milkfat globule membrane accounts for approx. 1.6% of the mass of the fat (i.e. $\approx 0.06\%$ in a typical bovine milk with 3.9% milkfat) and is composed principally of glycoproteins and polar lipids, particularly phospholipids[2]. A number of enzymes are associated with the membrane.

Triacylglycerols make up about 98% of the fat in the fat globules. Phospholipids, diacylglycerols and cholesterol plus smaller quantities of monoacylglycerols, free fatty acids, and cholesterol esters make up the remainder. The composition of milk lipids from a variety of species has been reviewed by Christie[3], examples being given in Table 9.1.

The fatty acids making up the triacyl glycerols may originate from the diet via the plasma lipids or may be synthesised in the mammary gland. In ruminants the dietary fats are usually extensively modified by the rumen microflora. This microbial activity brings about extensive hydrogenation of the dietary fatty acids and also leads to the production of short chain fatty acids, the presence of relatively high concentrations of butyric and hexanoic (caproic) acids being characteristic of ruminant milkfats. In bovine milkfat the level of butyric acid (3–4% on a mass basis) can account for up to 14% of the fatty acid molecules and be present in up to 40% of the triacylglycerols. Butyric and hexanoic acids are located at the sn-3 position, as is most of the octanoic (caprylic) acid. Myristic and palmitic acids tend to occupy the sn-2 while longer chain fatty acids similarly occupy the sn-1 position.

TABLE 9.1

Principal fatty acids in milk fats (wt. % of total) (typical values, adapted from Christie[3] plus data on camel milkfat[114])

Fatty acid	Man	Buffalo	Cow	Goat	Horse	Sheep	Camel
C4:0		3·6	3·3	2·6		4·0	0·1
C6:0	Tc	1·6	1·6	2·9	Tc	2·8	0·2
C8:0	Tc	1·1	1·3	2·7	1·8	2·7	0·2
C10:0	1·3	1·9	3·0	8·4	5·1	9·0	0·2
C12:0	3·1	2·0	3·1	3·3	6·2	5·4	0·9
C14:0	5·1	8·7	9·5	10·3	5·7	11·8	11·3
C16:0	20·2	30·4	26·3	24·6	23·8	25·4	26·6
C18:0	5·9	10·1	14·6	12·5	2·3	9·0	10·8
C16:1	5·7	3·4	2·3	2·2	7·8	3·4	11·7
C18:1	46·4	28·7	29·8	28·5	20·9	20·0	27·4
C18:2	13·0	2·5	2·5	2·2	14·9	2·1	3·1
C18:3	1·4	2·5	2·5		12·6	1·4	1·5
≥C20	Tc	Tc	Tc				0·6

Tc = Trace

Variations in feed and other factors will result in changes in the composition of the milkfat, often characteristic of the source and season, and reflecting different husbandry practices.[4] Feeding fresh grass will increase the proportion of C18 fatty acids at the expense of the C14 and C16. Substitution of C18:1 for C16 at sn-2 in triacylglycerols will reduce the melting point. Thus in spring and summer there will be a higher proportion of low melting triacylglycerols, giving rise to milkfats with less solid fat at temperatures between 0° and 20°C, and hence softer yellow fat products. The use of nuclear magnetic resonance (NMR) techniques is discussed in chapter 10. NMR is generally preferred to differential scanning calorimetry (DSC) as the data obtained is easier to interpret, particularly for routine monitoring of production.

Polymorphism may occur in milkfat. Rapid cooling can lead to crystals in the α-form with subsequent transition to more stable forms, normally forming mixtures of crystal forms.[5] In butter the β-form may predominate in the outer shell of fat globules while β' predominates in the interior and interglobular fat where lower melting, less saturated, fats occur. Milkfat is normally regarded as β' directing. However there is a tendency to form mixed crystals, especially with rapid cooling.

The unique flavour of milkfat has been the subject of extensive research, with one result that synthetic flavours have been developed for incorporation in table margarines. As yet researchers have not been able to match the organoleptic qualities of milkfat; one of the problems in this area may be that the melting properties of the milkfat affect the rate of release and hence the profile of the flavour as well as interaction between the mouthfeel and flavour. Concepts of an ideal flavour will also vary, depending on the experience of the taster.

Several reviews of flavour have been published e.g. Badings[6], Manning and Nursten.[7] The flavour in dairy based yellow fat products is primarily derived from compounds in the milkfat, though in emulsions there may also be a contribution from the aqueous phase.

In sweet cream butter and products derived directly from it, the flavour compounds are similar to those found in fresh cream, with some contribution from the aqueous phase and from the fat globule membrane material. Cultured-cream butters will also have the contribution of the fermentation products, particularly lactic and acetic acids plus diacetyl and dimethyl sulphide. The contribution of lactones to butter flavour is increased by heat treatment of milkfat, and is particularly important in baked goods.

9.2.1 Lipolysis

Milk contains a lipoprotein lipase, but with good manufacturing practices lipolysis should be minimal. Lipolysis may be induced by agitation and foaming; this is the result of mishandling the milk causing damage to the milk fat globule membrane (MFGM) so that the milkfat becomes accessible to the enzyme. Homogenisation can have a similar effect, increasing the surface to volume ratio of the milkfat and modifying the MFGM. Lipolysis has been reviewed by Deeth and Fitz-Gerald.[8]

The indigenous milk lipase is easily destroyed by pasteurisation, so lipolysis in dairy products is more likely to be due to microbial lipases, particularly the heat-resistant lipases produced by pseudomonas species.

Freshly secreted milk has ≈ 0.5 μmol free fatty acid (FFA)/ml (i.e., an acid value of about 0.7 on the fat) the result of incomplete synthesis of triacyl–glycerols rather than of any breakdown. This compares with a value of ≈ 2 μmol FFA/ml for milk with a rancid flavour. Short chain fatty acids are hydrolysed more readily from the milkfat and are the main contributor to the off-flavours, which may be described as 'unclean', 'bitter', 'butyric', 'rancid', etc. While low levels of butyric acid will contribute to the flavour of dairy products, higher levels lead to 'butyric' and other off-flavours. However the solubility of the butyric and hexanoic acids in water can lead to their removal during processing e.g. in the separation of buttermilk from buttergrains so that the decanoic (capric) and dodecanoic (lauric) acids are left as the major contributors to the off-taste. Dodecanoic acid is well known for its soapy taste.

In neutral products most of the free fatty acids will be as salts and have less effect on the flavour than at lower pH. The short chain fatty acids also have lower flavour thresholds in a lipid medium than an aqueous medium. (Table 9.2).

9.2.2 Oxidation

The oxidation of milkfat has been reviewed extensively by Richardson and Korycka-Dahl[11], but a more concise review of milk lipid deterioration is provided by Weirauch.[12] Oxidation of the unsaturated fatty acids, particularly the polyenoic fatty acids, in milkfat will give rise to a wide range of off-flavours. Unlike many fats however, very low levels of oxidation may enhance normal, acceptable, flavours, e.g. the enhancement of the creaminess of cream on whipping.

TABLE 9.2

Threshold values for added free fatty acids in good quality milk and sweet-cream butter to produce a rancid off-flavour. (Adapted from Deeth & Fitz-Gerald[8])

Free fatty acid		Threshold values ($mg\,kg^{-1}$)	
		Milk	Butter
4:0	butyric	46	11
6:0	caproic	30	52
8:0	caprylic	23	455
10:0	capric	28	162
12:0	lauric	30	128

Within the normal milk system there is a varying mixture of components with pro- and anti-oxidant activity. Riboflavine (when activated by light), metals especially copper (present in milk at ≈100 ppb) and metallo-enzymes such as xanthine oxidase are all potential pro-oxidants. Ascorbate and thiols may act as pro- or anti-oxidants, depending upon circumstances. Tocopherols, carotenoids, milk proteins and anti-oxidant enzymes (such as superoxide dismutase and catalase) provide anti-oxidants. Anti-oxidant activity may be augmented by thermal processing due to release of thiols from protein denaturation and the synthesis of Maillard reaction products, though denaturation of some enzymes may increase their pro-oxidant activity.

Many milk enzymes are associated with the MFGM, which also offers a binding site for metals. These pro-oxidant factors can outweigh the protective effect of tocopherols in the MFGM. Fermentation of milk or cream will lead to migration of copper from the milk serum to the MFGM, increasing its pro-oxidant activity. Thus the quantity and quality of MFGM associated with milkfat in a product can have a major effect on the stability of the milkfat in that product.

9.2.3 Methods of Analysis

In the past there has been a tendency for the dairy industry to remain independent of the other sectors of the oils and fats industry. The development of independent national standards has been a further complication,[13] though in recent years there has been increasing cooperation in the development and establishment of common analytical methodology, both nationally and internationally. For a number of years there has been cooperation between FAO/WHO committees[14] and the International Dairy Federation (IDF),[15] International Organisation for Standardisation (ISO)[16] and national organisations e.g. British Standards Institution (BSI[17]), the Association of Official Analytical Chemists (AOAC),[18] the American Oil Chemists Society (AOCS)[19] and the American Public Health Association (APHA).[20] An example of directly corresponding standards for butter is given in Table 9.3. A more comprehensive comparison of adopted methods was given by Tuinstra-Lauwaars et al.[21] IDF standards relating directly to milk fats and butter are listed in Table 9.4.

9.2.4 Anhydrous Milkfat
Production of anhydrous milkfat (AMF) starts with the raw milk. This must be of good quality and free of taints. The milk must be heated to at least 38°C[22] to ensure that all the fat is liquid; 50°C is a common separation temperature[23] though it has been suggested that the optimum separation efficiency may be achieved at about 63°C.[24] Cold separation of milk is seldom practiced as specially designed separators are needed to minimise damage to the fat globules. This damage can lead to premature destabilisation of the oil-in-water (o/w) emulsion of the cream, with higher fat loss in the skim milk and possibly plant blockage.

TABLE 9.3
Corresponding standards for analysis of dairy products published by the IDF, ISO and BSI

	IDF		ISO	BSI
6.A	1969	Acid Value	1740:1980	(BS 5086: Part 6: 1985)
7.A	1969	Refractive Index	1739:1975	
12.B	1988	Salt	1738:1980	(BS 5086: Part 4: 1985)
32	1965	Phytosteryl acetate test	3595:1976	
50.B	1985	Sampling techniques	707:1985	
54	1970	GLC of sterols	3594:1976	BS 5475: Part 2: 1977
74	1974	Peroxide Value	3976:1977	
80	1977	Water, MSNF & Fat	3727:1977	BS 5086: Part 2: 1984
94.A	1985	Yeasts & Moulds	6611:	
103.A	1986	Iron	6732:1985	
104.A	1984	pH	7238:1983	BS 5286: Part 7: 1985
112.A	1989	Water dispersion value	7586:1989	
122.A[a]	1988	Samples for micro. examn.	DIS 8261.2	
137.[a]	1986	Water content in butter	DIS 8850	(BS 5086: Part 3: 1984)

[a] Provisional standard; bracketed standards are equivalent but not identical.

TABLE 9.4

Principal IDF standards for analysis of milk, milkfats and butter (suffixes A and B indicate replacement of an earlier standard)

Number	Issued	Title
6.A[a]	1969	Determination of the acid value of fat from butter (reference method)
7.A[a]	1969	Determination of the refractive index of fat from butter (reference method)
8	1959	Determination of the iodine value of butterfat by the Wijs method
11.A[b]	1986	Determination of the milk solids not fat content (MSNF) of butter
12.B[a]	1988	Determination of the salt (sodium chloride) content of · butter (reference method)
23.A[b]	1988	Determination of the water content of butteroil by the Karl Fischer method
24	1964	Determination of the fat content of butteroil
30	1964	Count of contaminating organisms in butter
32[a]	1965	Detection of vegetable fat in milk fat by the phytosteryl acetate test
37	1966	Determination of soluble and insoluble volatile fatty acid values of milk fat
38	1966	Detection of vegetable fat in milk fat by thin layer chromatography of steryl acetates (confirmed 1983)
41	1966	Standard method for the count of lipolytic organisms
50.B[a]	1985	Milk and milk products—Guide to sampling techniques
54[a]	1970	Detection of vegetable fat in milk by gas-liquid chromatography of sterols
68.A	1977	Anhydrous milkfat, anhydrous butteroil or anhydrous butterfat, butteroil and ghee (compositional standards)
73.A	1975	Milk and milk products: count of coliform bacteria (reference and routine method).
74[a]	1974	Anhydrous milkfat: determination of the peroxide value.
75.B[b]	1983	Determination of the organochloride pesticide residues content of milk and milk products.
76.A	1980	Determination of the copper content of milk and milk products (photometric reference method).
80[a]	1977	Butter—Determination of water, solids-non-fat and fat contents on the same test portion.
83	1978	Standard method for the detection of thermonuclease produced by coagulase positive staphylococci in milk and milk products.
94.A[a]	1985	Milk and milk products—Detection and enumeration of yeasts and moulds.

TABLE 9.4—*contd.*

TABLE 9.4—contd.

Number	Issued	Title
99.A	1986	Sensory evaluation of dairy products (recommended general code—grading of butter—grading of milk powder).
100.A[b]	1987	Milk and milk products: Enumeration of microorganisms—colony count at 30°C.
103.A[a]	1986	Milk and milk products: Determination of the iron content (Photometric reference method.)
104.A[a]	1984	Butter—Determination of the pH of the serum.
112.A[a]	1989	Butter—Determination of water dispersion value.
122.A[a,b]	1984	Milk and milk products—Preparation of test samples and dilutions for microbiological examination.
137[a,b]	1986	Butter: Determination of water content (routine method)

[a] Corresponding to ISO/AOAC standards.
[b] Provisional standard.

The cream produced in the first stage separation will normally have a fat content of 35–40%. This warm cream should be pasteurised immediately, using a hold at 75°C for 15 s before cooling to 55–58°C. Where there is a risk of tainted milk affecting the cream quality, the cream may be vacuum treated to reduce the level of volatiles, as for buttermaking. The pasteurisation eliminates pathogens from the milk, reduces the number of potential spoilage organisms and will denature the lipase naturally present in raw milk. However the presence of high numbers of psychotrophic bacteria in the raw milk can result in the secretion of bacterial lipases which will withstand pasteurisation.

The pasteurised cream may be cooled and frozen for conversion to AMF at a later date, or may be fed to the second separation stage, to give a high fat cream of 70–75% milkfat. The separation is less efficient at this second stage and reseparation of the skimmed milk is often desirable; the recovered cream being fed back into the pasteurised cream.

The o/w emulsion in the high fat cream is then broken down by passing the cream through a high shear mixer, where the fat globules are damaged and fat released to form the dominant phase. In some smaller plants the higher shear may be applied by a serrated disc in the cream paring chamber of the second stage separator.

If butter is to be used as the source for the anhydrous milkfat then it should be melted and brought into the process at this stage. It may be necessary to add water to the melted butter as a washing stage. Washing will reduce the level of water-soluble components e.g. short chain fatty acids that may otherwise cause off-flavours. The overall process is outlined in Fig. 9.1.

The water-in-oil (w/o) dispersion is pumped to the third stage separator, where the aqueous portion is removed as a fraction rich in MFGM. If cream is used as the feedstock then this aqueous fraction may be used as a buttermilk equivalent, or recycled into the cream line. Where butter is the normal source or the plant throughput is high then two separators may be used in series at this stage. The design of the third stage separator differs from the cream separators in that the feedstock is distributed onto the outer surface of the disc stack rather than into distribution channels through the discs.

The milkfat produced by the tertiary separation will contain 0·2–0·4% water. This is reduced to less than 0·1% in the final polishing stage by heating the milkfat to 90°C and passing it through a vacuum drier. The milkfat should be cooled to ≈45°C for bulk storage,

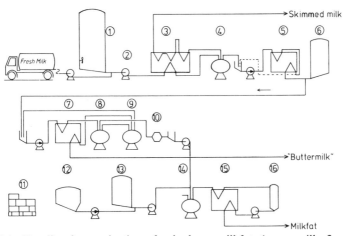

Fig. 9.1. Flowline for production of anhydrous milkfat. 1, raw milk; 2, pump; 3, milk preheater and skim pasteuriser; 4, milk separator; 5, cream pasteuriser; 6, cooled cream storage; 7, cream heater and buttermilk cooler; 8, second stage cream separator; 9, buttermilk separator; 10, emulsion breaker; 11, bulk butter supply; 12, butter melter; 13, melted butter tank; 14, milkfat separator; 15, milkfat heater and cooler; 16, vacuum drier.

packing or further processing. The bulk fat should be stored under nitrogen to minimise the risk of the development of oxidative rancidity and hence maximise the potential shelf life.[25]

For many uses, milkfat may be packed in polyethylene lined fibreboard boxes of 10–25 kg capacity. Handling problems may be minimised by supercooling the fat to ≤20°C and immediately filling into the container. Scraped surface heat exchangers may be used to cool the AMF to ≈10°C, extruding a semi-solid product into which nitrogen may be dispersed during the cooling to increase the shelf life. For international trade lined steel drums with a nitrogen flushed headspace are commonly used to ensure maximum product protection.

Trade in milkfat customarily uses the IDF standards, summarised in Table 9.5, and laboratory analysis will use the relevant IDF standard methods (Table 9.4). The major differences between the classes of milkfat result from the raw materials used and differentiation between products may be judged more critically on the basis of organoleptic than chemical analysis. Anhydrous milkfat (AMF), being produced from fresh cream or butter should have a characteristic clean, bland flavour and potentially have the longest shelf life. Anhydrous butter oil (ABO) is normally prepared from stored butter or cream and can be of similar quality to AMF but will often have a less bland flavour, reflecting the slight deterioration of the raw materials on storage. Should these flavours become pronounced then the product would need to be downgraded and sold as butteroil or butterfat.

Routine laboratory control in either production or purchase must include taste and odour at 20–25°C as the primary characteristic. Peroxide value (PV), free fatty acids and moisture should also be included in the routine checks. Where ABO is being produced from stored butter the PV and FFA values of the batches of raw materials should also be checked.

Microbiological tests should not be restricted to the final product. Both raw materials and process intermediates should be included in the daily schedule. Coliforms may be used as an indicator group for post processing contamination as, like the vegetative pathogens associated with raw milk, they will not survive the heat treatment inherent in the milkfat manufacturing process. The objective of the test schedule should be to identify both low grade ingredients which may indirectly compromise the quality of the product, and any breakdown in process hygiene.

The grade of milkfat required will vary with the end use. Essentially

TABLE 9.5
Summary of IDF Standard 68A: 1977
For anhydrous milkfat (AMF), anhydrous butteroil/anhydrous butterfat (ABO/ABF), butteroil or butterfat (BO/BF) and ghee

Parameter	Method (IDF)	AMF	ABO/ABF	BO/BF	Ghee
Ingredients		Prime quality—no neutralising substances	Variable age	Variable age	Milk, etc., from various species
Sampling	50B				
Milkfat %	24	≥99·8	≥99·8	≥99·3	≥99·6
Foreign fats	32, 37, 38, 54	Absent	Absent	Absent	Absent
Moisture %	23	≤0·1	≤0·1	≤0·5	≤0·3
FFA % as oleic	6A	≤0·3	≤0·03	≤0·03	≤0·03
Copper ppm	76A	≤0·05	≤0·05	≤0·05	≤0·05
Iron ppm	103A	≤0·2	≤0·2	≤0·2	≤0·2
Peroxide value mEq O₂/kg	74	≤0·2	≤0·3	≤0·3	≤1
Coliforms	[a]73A	Absent in 1g	Absent in 1g	Absent in 1g	Absent in 1g

TABLE 9.5—contd.

TABLE 9.5—contd.

Parameter	Method (IDF)	AMF	ABO/ABF	BO/BF	Ghee
Neutralising substances Antioxidants		Absent Any combination of propyl, octyl and dodecyl gallates with BHA or BHT up to a maximum of 200 ppm, but gallates not to exceed 100 ppm.	Traces only	Traces only	Traces only
Taste and odour		Clean, bland	No pronounced, unclean or other objectionable taste and odour.	Not too pronounced unclean or other objectionable taste and odour.	Not an objectionable taste and odour which does not meet edible requirements.
Structure		Smooth, fine grain	Smooth, fine grain	Smooth, fine grain	Mixture of crystals in semi-liquid

[a] Method 73A includes methodology for butter but does not specify a method for milkfat.

the blander the final food product then the better the quality demanded of the milkfat. Thus for use in a recombined milk or in an ice cream mix where fresh cream is customarily used only a first class milkfat will give acceptable results. However if the milkfat is to be used in a baked product then a milkfat of lesser quality may be used. This is because the baking process will bring about some oxidation of the fat as well as the breakdown of hydroxy acids to yield their corresponding lactones.[7] Also, some of the more volatile off-flavoured compounds will be driven off in the baking process. The situation is complicated by the concentration dependence of many of the flavour contributors; low levels may contribute to a 'creamy' flavour while higher levels give characteristic oxidised off-flavours.[6] The use of cultured butter, with its higher diacetyl level, as a raw material may be expected to produce a milkfat with a more pronounced flavour.

The very low level of non-fatty solids remaining in anhydrous milkfat results in a minimal retention of minerals. This is of particular interest where radioactive isotopes are of concern, since caesium and strontium are removed from milkfat while only a small proportion of the much shorter-lived iodine–131 remains[26] as shown in Table 9.6.

TABLE 9.6
Relative quantities of radioactive isotopes retained in dairy products
(After Lagoni *et al.*[26])

Radioisotope	Milk	Butter	AMF
Caesium-137	100	2·2	0
Iodine-131	100	3·5	2·1
Strontium-90	100	1·2	0

9.2.5 Fractionated Milkfat

Though milkfat has been used in a wide variety of products for many years, its melting properties are not optimal for every application. For the manufacture of biscuits and some types of pastry the melting range is rather low while for spreads there is too much solid fat for a plastic product at ≈10°C. There has consequently been interest in fractionating the milkfat so that the desired flavour characteristics are largely retained while the physical properties are modified.[27,28] Methods available are basically similar to those also used in the non-dairy fats industry, e.g. dry[28], solvent[29] and detergent[30] fractionation. Short-path distillation[31] has also been suggested as a fractionation method. Of these, dry fractionation is the commercially preferred process.

Dry fractionation involves cooling the molten milkfat at a controlled rate in a batch crystallising vessel. This is often a jacketed vessel with a large, slow moving paddle mixer. A flow diagram for dry fractionation is given in Fig. 9.2. Nuclei form both within the fat and on the cooling surface as the milkfat is cooled. Higher melting triacylglycerols will predominate in the crystals but the lower melting triacylglycerols may also be included to a greater or lesser extent, depending on the cooling conditions.

Milkfat crystals may be removed from the liquid fraction by filtration, on either a batch or continuous basis. At this stage control of process conditions, including temperature, is critical in order to maximise recovery of the solid fraction whilst draining off as much of the liquid fraction as possible. Retention of liquid fraction on the crystals will reduce the melting point of the high melting fraction. Recovery of high melting fraction (e.g. m pt 36°–40°C) will typically be less than 30%, with melting point dropping as yield increases due to incorporation of lower melting triacylglycerols. The low melting fraction would normally have a melting point of 23–26°C. (see Fig. 9.3).

The fractionation process may be repeated with the primary fractions to yield secondary fractions with more marked differences in melting characteristics. Small quantities of high (>40°C) and low (<20°C) melting second fractions may be prepared but their production is limited by the high cost of both the raw material and the additional processing. It is essential for the economics of fractionation

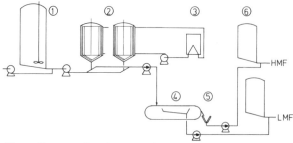

Fig. 9.2. Flow diagram for fractionation of milkfat. 1, bulk fat with pre-crystallisation option; 2, crystallisers; 3, attemperation system for programmed cooling; 4, batch or continuous filter; 5, remelt high melting fraction (HMF); 6, bulk storage of high and low melting fractions.

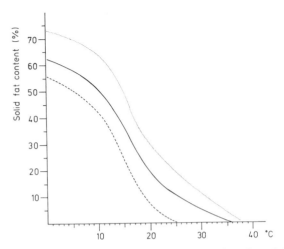

Fig. 9.3. Melting curves for UK milkfat with its high (· · · ·) and low (– – –) melting fractions. (Adapted from Rajah[32])

that a market be found for all of the fractions. The susceptibility of the AMF to oxidative rancidity increases with the degree of unsaturation of the fat, i.e. with decreasing melting point. Deterioration of the milkfat fractions may be minimised by carrying out the holding and cooling stages of the fractionation under nitrogen.

In addition to the tests outlined for milkfat, it is essential that the melting characteristics of any fractionated milkfat be checked and standardised. Slip and clear points are not sufficient for this and the supplier should check and standardise the solid fat content (SFC) of batches, preferably by using NMR as is current practice in most of the oils and fats industry and discussed in Chapter 10. Where fractions are being tailor-made for a particular product then a laboratory scale test for the 'functionality' of the fraction in that product may be required, e.g. test baking for biscuits and pastries, whipping for the creaming properties of shortenings.

On occasion it may be necessary to check milkfat samples for addition of fractions. In addition to unusual melting and chemical characteristics, the addition of high melting fractions to normal milkfat may be detected by crystallisation of the milkfat from hexane at 12·5°C.[33] High melting fractions give a 5–15% yield whereas normal milkfat should give a yield of less than 0·2%. This method cannot be applied to detection of the addition of soft fractions.

9.2.6   Chemical Modification

Interesterification may be used to modify the distribution of fatty acids both within and between the acylglycerol molecules. The treatment may be carried out under severe process conditions e.g. 100–200°C for 15 minutes to two hours in the presence of a sodium based catalyst, or by the use of enzymes.[34,35] In the latter case, however, lower temperatures and longer processing times are used.

A more saturated fat with higher melting point may be produced by hydrogenation.[36] However with both interesterification and hydrogenation the subsequent refining process (which is essential) will also remove many of the desired flavour components that give milkfat its advantage over other fat sources.

9.2.7   Extraction of Cholesterol from Milkfat

There has been much discussion of dietary cholesterol and, rightly or wrongly, criticism of milkfat on account of its relatively high cholesterol content. This has stimulated work on the extraction of cholesterol from milkfat. Some earlier workers used vacuum distillation[37] but recent interest has centred on supercritical extraction using carbon dioxide,[38] extraction with aqueous bile salts,[39] by precipitation of a cholesterol-saponin complex[40] or by absorption.[41]

The cholesterol extracted by these processes could be a valuable precursor for the manufacture of steroids and other products. Should reduction of cholesterol content become commercially exploited then the examination and identification of fat mixtures including milkfat (cf. IDF method 38) may become more difficult.

9.2.8   Ghee

Traditionally ghee has been prepared by heating cow's or buffalo's milk, boiling off the water then straining the fat off and cooling it slowly to give a coarse matrix of the higher melting triacylglycerides in the lower melting fraction.[42] The severe heat treatment, particularly in the presence of the other milk components, induces a pronounced cooked flavour in the milkfat. In addition to the lactones normally formed on heating milkfat, there will be Maillard products resulting from the interaction of the lactose and milk proteins. Oxidation of the milkfat will also occur. Flavour components may also be absorbed from the smoke, and the addition of spices will contribute as both a flavour and antioxidant.

On a larger scale, ghee may be prepared from cream or butter.[42,43] Either fresh cream, washed cream or cultured cream may be used.

Washed cream is diluted back with water then reseparated so that the milk-solids-not-fat (MSNF) level is reduced. The cream is boiled till most of the water is removed, i.e. at a temperature of <105°C, froth, scum and sediment being formed. Once most of the water has been removed (<1%) the temperature is allowed to rise to 110–115°C, not more than 120°C, and the ghee held till the desired colour and odour have been developed. The ghee is then strained to remove the sediment.

When ghee is prepared from butter, a milder initial process can be used, heating to ≈85°C and holding for 30 minutes to allow a proteinaceous film to form on the surface and the aqueous phase to separate below the fat. These layers may then be separated from the fat before further heating.

Ghee may also be prepared from anhydrous milkfat by subjecting the milkfat to an additional heat treatment to generate some of the cooked flavours, then warm filling the containers and allowing crystallisation to proceed slowly. Flavour compounds may also be added to create the desired flavour. Alternatively, fresh cream may be cultured with an aroma producing organism such as *Lactococcus lactis* biovar. *diacetylactis* to produce a sour cream before clarifying by heating at 115°C for 15 minutes, filtered and cooled.[44]

For international trade the testing procedures will normally follow the IDF standard 68A: 1977 (see Table 9.5). Standards for taste and odour will need further specification, particularly if a spice is to be added.

Vanaspati is a substitute for ghee, mainly produced from hydrogenated vegetable oils with melting points between 35° and 41°C, refined, flavoured and slowly cooled to produce a similar coarsely crystallised mass as for ghee. Addition of sesame oil at 5% to vanaspati eases detection of its addition to ghee as an adulterant.[42]

9.3. BUTTER

Most of the butter consumed in the UK is prepared from sweet cream and salted at a 2% level. This differs from normal European practice, where ripened butters are normally prepared with little or no salt addition.

Buttermaking started on a farmhouse scale with the batch churning of small quantities of cream separated from the milk by flotation. This process was often carried out overnight and the cream would normally be sour (or at least souring) by the time that it was made into butter. The introduction of mechanical separation enabled a rapid scale-up of cream separation capacity and efficiency on a factory scale. Butter-making was scaled up with larger batch churns, the traditional wooden drums being replaced by aluminium and stainless steel. This evolution was overtaken by the introduction of continuous buttermaking. The continuous manufacture by the 'Fritz' process[45] now accounts for most of butter production in the EEC.

The early continuous buttermakers had a capacity of ≈ 1 t/h but this has increased with time so that outputs of 5 t/h are the norm with outputs in excess of 10 t/h possible. This puts a major load on the handling of milk in the plant since each 10 t of butter produced is accompanied by production of about 10 t of buttermilk and over 180 t of skimmed milk. Thus for such a plant at maximum capacity (a 20 h processing day) the milk requirement will be in excess of 4000 tonnes and handling facilities for at least 3600 t/day of skim milk will be needed.

9.3.1 Sweet Cream Butter

Cream for buttermaking should be separated and standardised to the desired fat level for that particular machine, as the machines run most efficiently when supplied with a cream of constant fat content.[45] 40% fat is typical, but a range of 36–44% is common. The older batch churning process normally works better with lower fat contents e.g. $\approx 35\%$ milkfat. The cream should be pasteurised at 76°C for 15 s then cooled to <5°C. A flow diagram for production of sweet cream butter is given in Fig. 9.4.

Though good quality milk is desirable for production of good butter, the returns from buttermaking are often such that butter is a sink product for milk that is surplus or of unsuitable quality for production of more remunerative products. Cream rinsings or reject cream may also be salvaged as butter. Where off–flavours may be expected, whether due to handling problems or to 'natural' meadow or feed taints in the milk, volatiles may be removed by vacuum treatment. Mild problems can be treated in a Cream Treatment Unit (CTU) by heating the raw cream to 78°C then passing it into a chamber at low pressure equivalent to 62°C. The consequent evapora-tion of water and volatiles reduces the tainting. More rigorous removal of taints can be achieved by more severe vacuum treatment, e.g. by using a vacreator.[46]

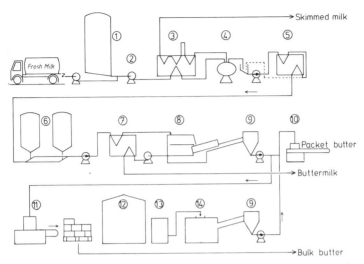

Fig. 9.4. Flow diagram for butter production. 1, raw milk, 2, pump, 3, milk preheater and skim pasteuriser, 4, milk separator; 5, cream pasteuriser; 6, ageing/ripening tanks, 7, cream preheater and buttermilk cooler; 8, continuous buttermaker; 9, butter silo; 10, butter portion packer; 11, bulk packer; 12, cold store; 13, attemperation; 14, reworker.

Cream for buttermaking should be held at <7°C for at least 4 h, preferably ≤5°C overnight, to age. This ageing process permits fat crystallisation to continue and rearrangement of crystal form to take place. The optimum temperature for fat crystallisation is lower than the optimum for buttermaking.

Cream must be fed to the continuous buttermaker at a constant rate, fat level, pH and temperature. The cream may be warmed by passing it through a heat exchanger immediately before entering the buttermaker. Ideally the temperature of the cream should be adjusted in the region 9–13°C to promote destabilisation of the emulsion and have ≈50% liquid fat in the fat globule. The Iodine Value (IV) has been used for assessing seasonal variation of the degree of unsaturation of the milkfat, though a direct measure of the melting properties by NMR would be preferable. The temperature of the cream should be held within ±0·25°C, with the heating water no more than 1°C higher than the cream exit temperature.

Most buttermakers use a two stage approach to the breakdown of the cream. Cream is fed into the top of the first cylinder of the buttermaker. This cylinder contains a beater (or dasher) rotating at high speed ($\approx$1000 rpm). Collision with the cream results in damage to the fat globules and partial aeration of the cream. Liquid fat released from the damaged globules aids aggregation of both damaged and undamaged globules to form small grains, dispersed in an aqueous phase containing particles of fat globule membrane, i.e. buttermilk. The mixture of fine buttergrains and buttermilk drops into a second cylinder where the grains are consolidated by a tumbling action at $\approx$30–35 rpm. Buttermilk drains from the cylinder through perforations, is collected and pumped away.

The buttermilk contains fragments from the milkfat globule membrane, finely dispersed fat particles, fat globules and some coarser fat particles. The larger fat particles may be recovered by flotation, by sieving and by centrifugation to reduce the fat content of the buttermilk to $\approx$0·5%. The relatively high ratio of membrane material in the buttermilk can cause problems in concentration and spray drying but confers valuable emulsifying properties on the product.

The buttergrains dropping into the working section contain a coarse dispersion of the aqueous phase. The working section consolidates the buttergrains during transfer via a pair of augers, squeezing surplus aqueous phase from the mass. Finer dispersion of the aqueous phase to form a water-in-oil emulsion is achieved by forcing the butter-mass through a combination of orifice plates and cutter blades.

Salt may also be added during the working operation. For a typical UK butter with 2% salt in a maximum 16% moisture, the salt addition rate is too high for addition as a saturated brine (26% salt) since the butter would need to be too dry before brine addition. Thus a 50% suspension of finely milled salt in water should be used, which requires constant stirring and brine circulation to prevent blockage of the equipment.

During the working process it is normal to include a vacuum treatment section so that the air content of the butter may be reduced and standardised. The final butter should have a moisture content $\leq$16% with the moisture and salt evenly dispersed in globules ideally less than 10 μm in size. The level of MSNF in the butter must not exceed 2%.[46] With good hygiene in the preparation of both the cream and the butter it is no longer necessary to wash the buttergrains (to reduce the microbial count and the levels of nutrients available to the microorganisms) and the residual MSNF will effectively substitute for milkfat in the product.

With economic pressures to maintain the moisture level as close as possible to 16%, close control of the moisture content is essential. Moisture content has been monitored continuously using the dielectric constant of the butter as it is extruded from the buttermaker. However with salted butter the presence of the salt alters the dielectric constant more than the water so accurate measurement is not possible. A later method incorporating gamma-backscatter as a second measurement does not appear to have lived up to its earlier promise.

9.3.2 Cultured Butters

Cultured, or lactic, butter is also nowadays prepared from pasteurised cream. Heat treatment conditions are often more severe than for sweet cream, e.g. up to 100–110°C with no hold. The cream may also be subjected to vacuum treatment. The heat treated cream is cooled to the fermentation temperature (typically 20–24°C) and the starter added at about 1%, depending on the activity of the starter preparation. A typical butter starter will contain strains of *Lactococcus lactis,* including the subspecies *cremoris* and the biovariant *diacetylactis* plus *Leuconostoc mesenteroides* subsp. *cremoris.* The latter two organisms will metabolise citrate in the cream to produce diacetyl and other flavour compounds in addition to the lactate produced by all these organisms.

Fermentation is allowed to proceed till the desired degree of acidity is achieved. This will vary with the taste preferences of the local market. In Sweden the normal acidity is achieved at a pH of 4·5–4·7, whereas the Netherlands set a maximum pH of 5·3. Biosynthesis of diacetyl is not significant above pH 5·2 so a less acid fermentation will result in a milder flavour. Cooling the cream to <10°C stops the fermentation and brings about further crystallisation of the milkfat. This cooling can be carried out within the fermentation, or ripening, vessel or by transfer through a heat exchanger. Fermentation continues during the cooling so the cooling process must be started before the desired pH is achieved so that the cooled mixed cream will be at the target pH.

Since the cooling follows a different pattern to that of sweet cream, the pattern of fat crystallisation is also different. Cultured creams normally give lactic butters with firmer textures than the corresponding sweet cream butter. There will also be seasonal effects. Slightly more spreadable butters with less seasonal variation in consistency may be prepared by modifying the time-temperature parameters in the cream handling. This is known as the Alnarp process,[47] and has been modified for sweet cream butters too.[48]

Apart from the cream ripening increasing the complexity of the process, the production of lactic butters also produces relatively large quantities of lactic buttermilk that are unsuitable for processing into powders. Most lactic buttermilk is thus used for animal feed where, in the absence of a subsidy, the return is low. Processes have been developed to overcome these problems, the best known being the NIZO process.[49]

In the NIZO process a sweet cream is used as the feedstock to the buttermaker, unripened buttergrains of 13–13·5% moisture being produced. During the compaction and working of these buttergrains a cultured milk blend and a starter concentrate are dosed into the working section of the buttermaker. The cultured milk blend is prepared from a milk fermented with starter including *Lactococcus lactis* biovar. *diacetylactis*, (aerated after culturing to increase the diacetyl level to ≈40 ppm) and a milk fermented with starter including *Leuconostoc mesenteroides* subsp. *cremoris*, this latter organism being capable of metabolising acetaldehyde produced by lactococci and hence avoiding the development of a 'green' off-flavour. The addition of cultured milk at ≈2% is not sufficient to reduce the pH of the butter to less than 5·3. The 'culture concentrate' is essentially a solution of lactic acid, prepared by incubating lactose-depleted whey with a culture of *Lactobacillus helveticus* then concentrating to at least 11% lactic acid. Addition of the 'culture concentrate' at ≈0·7% in addition to the cultured milk blend reduces the pH to less than 5·3 and gives a satisfactory match to the traditional product.

A further benefit claimed for the NIZO process is that the use of unripened cream minimises migration of copper from the serum to the fat globule. Copper is a powerful pro-oxidant with a significant effect at a 10 ppb level whereas milk contains ≈ 200 ppb. Oxidation is also promoted at lower pH values, an additional problem for ripened butters. An alternative method used to avoid this problem when using ripened cream is to wash the buttergrains then add a 'special salt'[50] containing 10% disodium hydrogen phosphate plus sodium carbonate to raise the pH to 6·5. The increased stability to oxidation should be balanced against other flavour problems, assuming that national regulations would permit this addition.

9.3.3 Butter Packing

Butter production seldom matches demand for the product so storage is needed, either for the cream or the butter (storage of the latter takes up less space and is more common unless production capacity is limited). Butter from the buttermaker may be immediately packed in consumer portions, though the greater part is bulk packed in 25 kg fibreboard cases lined with polyethylene or parchment. The bulk packed butter is then stored under deep freeze conditions till needed. Frozen storage for 3–6 months is common, but the flavour of butter stored for more than a year is likely to deteriorate.

Stored butter is very hard when defrosted and must be warmed (or tempered) and further blended (or reworked) to impart a degree of fluidity and plasticity to enable it to be packed. The tempered butter is normally comminuted (shived) before reworking. Sigma or 'Z' blenders are still used for small-scale reworking while continuous reworkers are now used for large scale packing operations. Batch blenders may also be used for the preparation of flavoured butters, e.g. garlic butter. The continuous reworkers are very similar to the working sections of continuous buttermakers. Additional water and salt may be added during the reworking to obtain a standard product with moisture as close as possible to the 16% maximum. As at the initial buttermaking, it is essential that any added water be finely dispersed within the continuous fat phase. Ideally the droplets should be less than 10 μm in diameter, evenly dispersed throughout the butter. Coarse moisture dispersion will lead to a shorter shelf life of the butter, and may be detected by placing a sheet of moisture indicator paper against the sample of butter. More severe problems with poor dispersion may be detected in the classic grading test, when moisture may be observed when the sample is taken. Where the moisture is poorly dispersed the droplets are often not discrete but form some continuous channels through the butter, with increased risk of microbial spoilage. This channeling may be detected by resistance measurement,[51,52] preferably using an alternating current to avoid polarisation effects.

The fat or lipid phase of the butter consists of a continuous phase of liquid fat in which crystals of the higher melting triacylglycerols are distributed. Temperature and the composition of the milkfat are major factors in the proportion of crystalline fat, the size of the crystals reflecting the rate of cooling, temperature cycling and shear conditions. A proportion of the milkfat will still be present within milk fat globules trapped in this matrix, and this will further modify the texture of the product. MFGM debris will be found in both the lipid and aqueous phases.

Most butter for retail use is packed in 250 g portions, using either parchment or aluminium foil. The packs are normally passed over a check weigher and through a metal detector before packing into fibreboard cases.

9.3.4 Quality Control and Analysis

In most dairies the routine chemical tests will be for fat, moisture, salt and/or pH. For routine examination in small dairies the moisture may be determined by loss of weight on heating a sample on a beaker over a bunsen flame. This method is covered by the British Standard routine method BS 5086 : Part 3 : 1984, requiring a balance accurate to ±5 mg. A similar method is routinely used in the USA.[53] The IDF provisional routine method (IDF 137: 1986) is slower, requiring a balance with 1 mg sensitivity and with drying completed in an oven at $102 \pm 2°C$ for 1 hour. More accurate but slower determinations to be used as a reference method use evaporation over a boiling waterbath then heating in an oven at $102 \pm 2°C$ to constant weight, e.g. IDF 80: 1977/BS 5086: Part 1: 1984.

Faster methods have been developed using infra-red moisture balances and microwave ovens to heat the sample. In larger dairies near infra-red (NIR) reflectance techniques may be employed for rapid moisture determination,[54] presenting the sample as a smooth surface in the open sample cup. With this technique, salt may only be estimated indirectly by difference.

Routine salt determination may use either an aqueous preparation of the butter (IDF 12B: 1988, BS 5086: Part 5: 1984) or by redissolving the residue from a fat determination.[53] The solution, or aliquots of the solution, may then be titrated with silver nitrate solution in the presence of a potassium chromate indicator solution. The BS reference method uses the digestion of the sample with excess silver nitrate in the presence of nitric acid, titrating the excess silver nitrate with potassium thiocyanate using an ammonium iron(III) sulphate indicator. The use of nitrobenzene was common in older methods of chloride determination[55] but has been discontinued on safety grounds. Chloride may also be estimated in the aqueous phase using an ion-specific electrode, or by X-ray fluorescence.[56]

The microbiology of butter has been reviewed by Murphy.[57] The microbiological quality of the butter will be reflected in the keeping quality of the product. Sampling must follow the appropriate standard, using sterile triers, knives and containers. The sample of the butter should be melted at max 40°C[53] or at 45 ± 1°C (Supplement No 1 (1970) to BS 4285: 1968). While these methods use a dilution of the whole sample, IDF 122A: 1988 uses 45°C but gives the option of either preparing a 10% dispersion in diluent, preferably peptone/saline solution, or preparing a 10% dilution of the aqueous phase only. Dispersion of the sample may be achieved using a peristaltic blender or stomacher with sterile bags, or by using a sterilised rotary blender. Dilutions may normally be used for colony counts on Yeastrel milk agar at 30 ± 1°C for 3 days to give a total mesophile count and at 22 ± 2°C for 5 days for a psychrotroph count. These counts would not be appropriate for cultured butters. Routine microbiological examination should also include colony counts for yeasts, moulds and coli-aerogened bacteria. Examination for caseolytic and lipolytic organisms may also be carried out, but enrichment techniques for pathogens should not be attempted in factory laboratories.

The total mesophile count will provide a guide to survivors of the pasteurisation of the cream plus post processing contaminants. The psychotroph count will reflect post processing contamination, a fruity odour associated with the plate indicating the presence of *Pseudomonas fragi* which will cause off-flavours in the butter. The coli-aerogenes, yeast and mould counts will also provide an indication of post processing contamination. Low levels of mould contamination may result from aerial contamination but higher levels of moulds or the presence of coli-aerogenes, yeasts and/or psychrotrophs indicates a breakdown in process hygiene. General microbiological standards for butter have been suggested by Davis[58] but standards suggested by Murphy for cream for buttermaking[57] should also be applicable to butter from a modern buttermaking plant e.g. a total mesophilic count of less than 1000 cfu ml^{-1}. Addition of herbs and other flavourings must be controlled to ensure that the additives are of good microbiological quality. The use of essential oils for the flavourings will minimise the microbiological risk; in fact the addition of some essential oils may extend the shelf life of the butter by inhibiting oxidation.[59]

The sensory aspects of butter quality have always been given a high profile in the testing of butter, since to the customer the sensory qualities are crucial. Most large scale producers adopted national grading schemes for their bulk butter; Davis[60] gives examples. In the UK grading standards are no longer set by MAFF but are included in the grading service operated by the Creamery Proprietors Association[61] and have since been included in the UK Intervention Board standards for bulk butter.

Bulk butter is normally graded at 10°C, several days after manufacture. Samples are removed from the block using a trier, the plug being smelt immediately to detect any off–odours. The surface of the core should then be examined for appearance and the presence of free moisture. A piece of butter is cut from the core using a knife or spatula and melted in the mouth to check the flavour. The grading system is summarised in Table 9.7. Scoring is normally conducted on the basis of an absence of off flavour etc, points being deducted for each defect.

Examination of packet butter will use a knife to cut across the pack so that surface defects will become apparent. A knife or spatula may then be used to remove a sample. Sensory work should be carried out in a well lit room free of odours and other distractions; ideally this room should be part of an air conditioned sensory suite.[62] The sensory evaluation of butter has been comprehensively covered by Bodyfelt *et*

TABLE 9.7
Summary of Creamery Proprietor's Association grading scheme for
UK butter.[61]

(a) Scoring

Attribute	Maximum points
Flavour and aroma	50
Body and texture	20
Colour appearance, finish and salt	20
Free moisture	10

(b) Grading

Grade	Total score	Flavour and aroma
Extra selected	≥93	& ≥47
Selected	85–92	& ≥44
Graded	75–84	& ≥40
No grade	≤74	or <40

al.[63] with particular reference to US practice. Lists of defects are commonly found in the literature but descriptors for the characteristics of a good quality butter are less often listed. Given a trained group of graders, a consensus of what is implied by good butter may be readily reached. However, if the sensory standard for butter is to be applied to competing spreads then a wider range is needed. Some common and suggested terms are listed in Table 9.8.

The sensation created by a product on the palate is time dependent, the resulting profile being an important characteristic of the product. Syarief *et al.* reported[64] sensory texture profile data for butter using principal component analysis, including definitions of the attributes, or character notes, used. The character notes were split between spreadability, first bite, mastication and afterswallow. The primary (independent) rheological parameters were found to be firmness and adhesiveness on first bite, then viscosity and cohesiveness on mastication.

Many methods have been developed for estimating the hardness of butter, the more popular being evaluated by an IDF expert group and reported in 1981.[65] The group favoured the cone penetrometer on grounds of ease, speed and cost. In the recommended method the butter should be held at 14°C for 10 days before testing at 14°C, the effect of consumer practice being imitated by a hold at 20°C for one

TABLE 9.8
Examples of descriptors that may be used for butter and other spreads

Attribute	Desirable	Undesirable/defect
Aroma and flavour	fresh, creamy lactic (if cultured) buttery	sour, cowy, musty, bitter, unclean, butyric, oxidised, cheesy, weedy, soapy, yeasty, too salty, cooked, malty, fruity, garlic/onions, metallic
Body and texture	smooth, waxy, closely knit, firm plastic	weak, fatty, sticky, crumbly, brittle, gritty, short
Colour and appearance	even coloured, natural, silky, creamy (if cultured)	mottled, streaky, wavy, dull, discoloured surface, unnatural colour
Absence of free moisture	smooth, dry	wet, leaky

day before keeping the sample at 14°C. A 200 g cone with a cone angle of 40° was preferred, with depth of penetration (p) being measured in tenths of a millimetre after 5 s. Under these conditions the apparent yield strength (AYS) was calculated as:

$$AYS = 470\,000/p^n \text{ kPa}$$

A value of $n = 2$ for p^n was chosen as an approximation, the true value of n being between 1·8 and 2·1. A spreadable butter was given as having an $AYS \leq 100$ kPa under these conditions. The test though simple requires a minimum sample size of 250 g and successive tests must be kept apart and away from the side of the smooth sample surface to avoid edge effects.

Comparison between instrumental tests, including the penetrometer, and sensory evaluation of spreadability and hardness of butter was carried out by Rohm and Ulberth.[66] With all four instruments (penetrometer AYS, referred to as AYV, extrusion force, penetration force and sectility) there was high correlation between the log of the value obtained and the temperature (T) over the range 3–15°C, i.e.

$$\log \text{Value} = C - KT$$

where C and K are constants. Sectility, the peak resistance to cutting, was measured with a 0·3 mm diameter stainless steel wire mounted in an Instron Universal Testing Machine running at 6 mm/min in accordance with DIN-Norm 10.331 (1979).

Sensory evaluation of the hardness and spreadability gave values that were linearly related to temperature for individual testers but there was considerable variation in the magnitude of the responses. For individuals there was good correlation between spreadability and hardness i.e.

$$\log \text{Hardness} = C + K \log \text{Spreadability}$$

The best agreement between sensory and instrumental methods was obtained with AYS and sectility, particularly between log spreadability and log AYS, log hardness and log sectility.

While the penetrometer may be adequate for simple QC tests, it oversimplifies the rheology so that one cannot distinguish between viscous and solid properties; nor can one be sure of the appropriateness of a single point measurement. More information can be obtained by non–destructive techniques such as visco-elastic rheometry.[67,68] Though visco-elastic rheometry is likely to remain a research tool rather than one for routine quality control, it may be used to check the appropriateness of single point measurements where less skill is needed.

Oiling off can be a problem resulting from the higher liquid fat levels associated with spreadable butters. A recommended method[65] for estimating oiling off is to take a plug of the butter, 15 mm diameter and 20–25 mm in height, place on a weighed filter paper and reweigh. The paper plus butter is then placed on an open grid and supported at the edge, holding for 4 h at 24°C followed by 30 min at 5°C. The plug of butter is then carefully scraped off and the paper is reweighed. The increase in weight of the paper is expressed as a percentage of the original butter weight. Duplicate results should not differ by more than 0·5% free oil.

9.4 MARGARINE

Margarine was invented as a substitute for butter, but over the last 120 years it has developed into a class of products in its own right, albeit competing in the same overall market for yellow fats. Margarine is generally defined as a water-in-oil emulsion containing a minimum of 80% fat and a maximum of 16% water. In some countries vitamin fortification may be required e.g. in the UK[69] fortification with vitamin A to 26·9–33·2 IU/g (equivalent to the vitamin A content of summer butter) and with vitamin D to 2·83–3·53 IU/g (approximately ten times that found in butter). In the USA inclusion of vitamin A is also required but addition of vitamin D is optional. β-carotene may play a dual role of providing both vitamin A activity and adding colour to the lipid phase (and hence the product). Other carotenoids, e.g. annatto, may also be used as colouring agents.

The inclusion of milkfat in margarine varies by country, in the UK this is limited to 10% of the fat content and has been used in the past to improve the flavour and perceived quality of the product. Higher usage rates have been used in other countries e.g. USA where butter-margarine blends have been sold and Sweden where the dairy industry has produced 'Bregott', containing a ratio of approximately 20% soybean oil: 80% milkfat in the lipid phase, by adapting buttermaking technology.[50,70] Bregott has been produced using both batch and continuous processes, carrying out the destabilisation of the cream in the presence of the added soybean oil. A more recent process avoids the buttermaking technology, taking a high fat cream from a second stage separator (cf preparation of anhydrous milkfat) and blending this with vegetable oils in a high shear mixer.[71] This process would retain fatty globules dispersed in the lipid phase, similar to that found in butter but not normally found in modern margarines.

Without the constraints of a single fat source, a wide range of products have been developed to meet specific market demands, e.g. culinary margarines and shortenings, packet margarines, soft margarines and all-vegetable fat products. The continuous lipid phase will determine the characteristics of the margarine.

The lipid phase of margarines will normally contain a blend of fats to give the desired consistency, both in terms of the melting (hence hardness) and the crystallisation characteristics.[72] On cooling the emulsion a rapid change from the α to the β' crystal form is desired. Further transformation to the β form is undesirable as it is associated with the formation of large crystals and the development of a coarse, almost sandy texture. Milkfat, partially hydrogenated versions of marine oils and of some liquid vegetable oils, tallow, palm and cottonseed oils form β' crystals while other commonly used oils and fats, typically with simpler composition and higher levels of C18 fatty acids, tend to form β crystals. This β tendency can be overcome by including β'-type oils in the fat blend and by blending partially hydrogenated β-type oils e.g. soybean oils for a soybean oil margarine.[73] Interesterification may also be used to modify either the crystallisation properties or melting characteristics of a fat or fat blend, particularly if a low level of *trans*-isomers is desired.

The crystallisation characteristics of the oil may also be modified by the type and quantity of the emulsifiers in the blend. However the main function of the emulsifiers is to aid and maintain the dispersion of the emulsion. This will not only reduce the energy needed to form the emulsion but, by reducing coalescence of droplets of the aqueous phase, will prevent leakage of poorly dispersed moisture on storage and prevent spattering when the margarine is used for frying. Both mono/diglycerides and lecithins have been used extensively in margarines. Mono–diglycerides can aid the stabilisation of soft margarines, where there are less crystals present. The degree of saturation of the mono–diglyceride may be reduced as the degree of unsaturation of the fat blend increases to give optimal emulsification.[74]

Oxidation is a potential problem, either during the storage of the fat ingredients, preparation of the lipid phase or in the product. This problem will increase with the degree of unsaturation of the fat blend, and may be minimised by holding the fats at the lowest practicable temperature for the shortest possible time. Contact with oxygen should be minimised, if necessary by purging the storage tank with nitrogen and holding the fat under a nitrogen blanket. Antioxidants may also be added to the fat.

The aqueous phase of margarine may be a simple salt solution but it is more common to include MSNF components to improve the organoleptic quality. These solids may be derived directly from milk, possibly fermented using a flavour producing culture as for butter ripening or, more commonly, using whey solids. Milk protein can contribute to the anti-oxidants by sequestering metals, e.g. copper.

9.4.1 Manufacture of Margarine

The manufacture of margarine involves the preparation of the two phases, emulsification, before or after heat treatment, cooling and crystallisation of the emulsion. Scraped surface cooling is the common feature of margarine production, normally carried out in 'Votator' type scraped surface heat exchangers. A flow diagram for margarine production is given in Fig. 9.5.

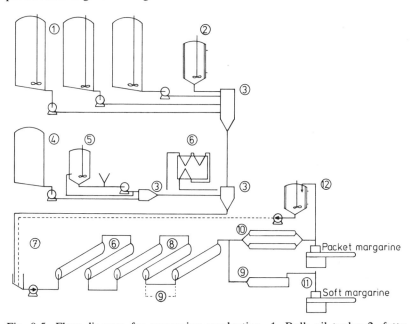

Fig. 9.5. Flow diagram for margarine production. 1, Bulk oil tanks; 2, fatty ingredients tank mix; 3, batch or continuous blending; 4, bulk water, skim or whey storage; 5, powder reconstitution; 6, heat treatment option; 7, high pressure pump; 8, scraped surface coolers; 9, worker unit; 10, resting tubes; 11, packer; 12, rework melting and return.

The two phases must be prepared separately. The lipid phase is prepared at a higher temperature than that of the melting points of the constituent fats. More saturated/higher melting emulsifiers may be handled by dissolving in a small portion of hot oil, then added to the bulk. Colouring and anti-oxidants may also be added at this stage. Polyunsaturated fats may need protection by inert gas blanketing during the preparation of the lipid phase. In any case the exposure of the fats to high temperatures and oxygen should be minimised.

A simple aqueous phase may be prepared using brine or acidified water, depending on local preferences as with butter production. The organoleptic quality will be improved by including some MSNF in the aqueous phase, for instance by dilution of a whey or skim milk concentrate. A 32% concentrate may be used for large scale manufacture and powdered sources for smaller scale operations. If a culture is to be used, the liquid should be pasteurised, cooled to the fermentation temperature (typically 20–25°C) and the starter culture added. The cultured aqueous phase should be cooled at the end of the fermentation to prevent further lactic acid production. Addition of salt to the aqueous phase may also inhibit further acid production. The starter microflora can inhibit the growth of other contaminant bacteria and may also have an anti-oxidant effect, reducing aldehydes to alcohols. If the aqueous phase is not to be cultured then salt and other ingredients, e.g. lecithin, citrate or other chelating agents, preservatives such as sorbate or benzoate (where permitted), should be added before pasteurisation.

The emulsion may be prepared either on a batch basis or as a continuous operation. For batch operations the lipid phase would be weighed into the tank then the aqueous phase added during vigorous mixing to ensure the formation of a coarse w/o emulsion. The temperatures of the ingredients should be adjusted to prevent fat crystallisation at this stage. The batch blending system may be automated using computer control linked to load cells on the blending tanks.

Continuous formation of the emulsion may be achieved by using metering pumps and an in-line mixing system. This approach may be taken a stage further to produce the lipid and aqueous phases from pre-mixes by multihead metering systems too.

Hygienic production of the margarine is essential to give a product that has a satisfactory shelf life and is free of pathogens and unwanted microflora. Pasteurisation of the margarine emulsion is desirable, preferably at the latest possible stage so that the risk of recontamination is minimised. In many margarine processes the pasteurisation is carried out during the preparation of the aqueous phase, wherein the greatest microbiological risk lies. The pasteurisation of the emulsion may be carried out using plate heat exchangers, shell and tube heat exchangers or scraped surface heat exchangers. The former are preferable in terms of capital and running costs, but must be capable of running at high pressures if a closed system is to be maintained. Turbulent flow conditions must be maintained to ensure that the emulsion is preserved.

Once the emulsion is cooled to the crystallisation point of the fatty components, scraped surface heat exchangers must be used, so that fat crystals forming on the heat exchange surface are rapidly removed and blended back into the bulk of the emulsion to form a seed, promoting the rapid formation of a fine crystalline matrix. A twin bladed shaft with typical rotational speeds of 100–500 rpm will give surface residence times of 60–30 ms with a blade speed of about $1 \cdot 5$–$8 \, \mathrm{m \, s^{-1}}$. Some shafts may be fitted with warm water circulation to prevent build-up of fat crystals on the shaft. The high shear forces developed as the product is scraped off the walls also promote the formation of a finer emulsion.

In all but the smallest pilot scale plants the cooling requirement is such that more than one scraped surface cooler is needed, plants with three units being common. The transfer of the margarine from one cooler to the next provides a short period of lower turbulence in which crystals may grow. In some plants this effect may be enhanced by inserting a worker or crystallisation unit before the final cooling stage to avoid excessive supercooling of the fat, rather than after the final cooler as is more common for the production of soft margarines. The worker unit consists of a tubular assembly with fixed pins and a rotating shaft bearing a second set of intermeshing pins. The shaft speed of the worker is normally variable to enable the texture of the product to be optimised.

For soft margarines the crystallisation and working should be sufficient so that the product will flow into the package during the filling operation. The shear forces in cooling and working should also ensure that the aqueous phase is finely dispersed in the fat. Ideally the droplets should be $3 \, \mu\mathrm{m}$ or less in size to ensure microbiological stability but droplets up to 20 μm may be found. There should not be excessive crystallisation after filling or a brittle product may result.[72,75] More than one worker unit may be used in the plant. Hard margarines would not normally require the use of a worker unit after the final cooler. Instead, the emulsion would be fed into one or preferably two resting tubes, which may contain baffles. Where there is a pair of tubes they would be filled alternately so the product is quiescent for a period of time, e.g. two min or more, so that maximum build-up of crystal structure is possible before extrusion. For whipped margarines nitrogen should be injected before crystallisation in the cooler(s) and a fine dispersion ensured by use of a high speed blender after cooling.

In most margarine production lines, particularly where there is a closed system feeding direct to the filler, an excess of margarine must be produced to ensure that the filler is not starved. The throughput of the margarine plant is adjusted so that the oversupply is minimised but when the filler stops the entire production will be surplus. Since scraped surface plants are not amenable to stopping and starting, it is normal practice to divert all surplus product and recycle it back to the emulsion balance tank. This rework may be melted in a tank, in line with a heat exchanger or by a combination of the two.

9.4.2 Laboratory Examination of Margarines

For routine analysis margarine may be examined in the same manner as butter, for instance as described in the APHA Standard Methods[53] and in the AOAC Official Methods of Analysis.[76] AOCS offer a routine method[77] for moisture and volatile matter, again for both butter and margarines, using a hot plate for heating the accurately weighed sample to not more than 130°C before cooling and reweighing. The IUPAC standard 2·801 describes two methods:

(i) A 'quick' method for determination of total fat in margarines by adsorbing a 1 g sample on a sodium sulphate column, eluting the fatty components with diethyl ether then evaporating off the solvent to recover and weigh the fatty material.

(ii) Estimation of the total fat and non-fatty components content by extraction of a 5 g sample with light petroleum, retaining the water and non-fatty components in a weighed pre-dried pumice. Fat content is again determined after solvent recovery, and the non-fatty solids estimated after drying the pumice at 103 ± 2°C.[78]

For routine proximate analysis the rapid analytical techniques used for butter would also apply.

With margarine the manufacturer has the flexibility to choose fats and other ingredients to give the specific properties desired of the product. This will in turn increase the range of tests that may be required. Examination of the lipid fraction should follow the methods described in earlier chapters.

The rheological properties of margarines cover a much wider range than for butter, from soft margarines that may be spread readily at 5–10°C to packet products designed to be stored at room temperature, or even at tropical temperatures. As with butter, the cone penetrometer (and its equivalents) has found wide acceptance for routine testing. The AOCS method[79] uses a closely specified cone with a 20° angle, describing the penetration in tenths of a millimetre and avoiding further calculation. Other cone angles are also used, e.g. 40° in Europe.[65] Haighton[80] suggested that differences in cone angle can be taken into account using the following relationship to calculate a yield value (YV):

$$YV = kWp^n$$

where k is a constant dependent on the cone angle, W is the weight of the cone, p is the penetration in tenths of a millimetre and n is a constant, the proposed value for n being 1·6. The YV is similar to the later AYS proposed by the IDF but the differing calculations (including the differing values for n) will give different values from the same experimental data—an illustration of the need to specify methods used.

The YV/AYS may be related empirically to the subjective evaluation of the hardness of the margarine. Since the hardness of the margarine will be related to the forces needed to disrupt the crystal network, there should also be a relationship between the solid fat content (SFC) and the firmness,[81] the firmness being proportional to the square of the SFC for margarines subjected to similar treatment. SFC may be estimated either from NMR or dilatometric examination of the blend, or calculated from the contribution of the component fats. This relationship will be subject to the modifying effects of processing and storage conditions on the product, since products that have undergone significant crystallisation after filling will have formed a network of larger interconnected crystals that will need more energy to disrupt than a system with small individual crystal clusters. Table

TABLE 9.9
Approximate relationship between yield value, solid fat content and equivalent temperature for corresponding firmness of butter.[32,80,81]

Subjective	Yield value (kPa)	Solid fat (%)	Butter (°C)
Very soft, not spreadable	<10	<12	>23
Soft, spreadable	10–20	12–17	20–23
Good, plastic and spreadable	20–80	17–34	15–20
Hard but still spreadable	80–100	35–39	14–15
At limit of spreadability	100–150	40–47	10–14
Too hard	>150	>47	<10

9.9 gives a summary of the relationship, including for comparison a similar estimation of the properties of a butter based on NMR data ignoring any modifying effects of production conditions and work softening.

In a butter or margarine the fat crystals will cover a range of sizes, though in a rapidly cooled product most crystals will be very small. For margarine the crystal sizes range from $0.1 \mu m$ to $20 \mu m$ with aggregates or spherulites up to $100 \mu m$,[82] though most crystals should be less than $1 \mu m$.[4-14] Where crystals have grown together a network of 'primary' bonds will become established. The large number of small individual crystals are so closely packed ($<0.01 \mu m$) that Van der Waals–London attraction leads to flocculation and the formation of a weaker network of 'secondary' bonds.

Disturbance of the structure by working will cause rupture of both types of bonds, though the secondary bonds will reform more rapidly over a period of hours. Primary bonds will only be reformed on storage as a result of further crystallisation, possibly attributable to temperature cycling. For margarines a balance is needed between the contributions of primary and secondary bonds, obtained by optimising the degree of working during manufacture.

The extent to which a product will soften on working will have an effect on the spreadability, the greater the worksoftening then the more spreadable the product will be for a given YV. The degree of worksoftening is also proportional to the amount of work done, so a standard worksoftening technique is essential. Haighton[83] suggested that margarine worksoftens to a greater extent than butter, and proposed a spreadability index (SI):

$$SI = C_u - 0.75(C_u - C_w)$$

where C_u is the yield value of unworked and C_w is the yield value of worked product. This index gave good correlation with assessment by panels of housewives.

Though the cone penetrometer has found wide use for routine control testing, the information obtained is incomplete, as discussed for butter. DeMan *et al.*[84] suggest that the cone penetrometer is unsuitable for evaluation of brittleness, which may however be evaluated by compression tests. In this work an Instron apparatus was used to compress 20 mm diameter cylinders from 20 mm high to 10 mm. Variations in the texture of the product with the cooling and tempering regime were also accompanied by variations in the SFC. The effect of the tempering regime on the SFC makes standardisation essential if SFC is to be used as an empirical guide to hardness but unless the particular production system can be reproduced on a laboratory scale it will be necessary to carry out further work to find the relationship between the production process and the standard tempering for that fat system. In more recent work on samples of shortenings and margarines, DeMan *et al.*[112,113] again found good correlation within the textural methods but not between texture and SFC values.

Other methods of examining the texture of margarines include creep analysis[85] and, for routine work, the sliding pin consistometer.[86,87]

Sensory evaluation of margarines is subject to the same needs as for butter but with the wider range of products any descriptive work is likely to produce a wider range of descriptors. Similarly, qualities that are abnormal for butter may be normal and possibly desirable in the margarine.

9.5 REDUCED AND LOW FAT SPREADS

For the purpose of this discussion, the term 'reduced fat spreads' includes those spreads of 41–79% fat content though some authorities may limit the term to those spreads with less than 60% fat. Reducing the fat content of the spread to less than 80% will bring the product outside most national standards for butter and margarine and give more scope for hybrid products.

Low fat spreads are generally defined as having less than half the fat of butter or margarine. Some national standards prescribe a narrow fat range, e.g. $40 \pm 1\%$ fat while others may use 40% as the maximum. The emulsion should be substantially water-in-oil (w/o).

As yet there are no statutory standards in the UK, but manufacturers have, in general, elected to add vitamins A and D to their products at the same levels as required for margarine. There have been a number of recent reviews of technical and marketing aspects[88-90] of this rapidly expanding sector of the yellow fats market.

9.5.1 Reduced Fat Spreads

Within the dairy industry, continuous buttermakers may be used to manufacture products with fat contents of 70–75%. In the UK the product 'Clover' produced by Dairy Crest Foods is an example with 75% fat, derived from milkfat and vegetable oils (including hydrogenated vegetable oil).

In salted spreads the level of salt will be determined, and limited, by the consumer's preference. Thus as the moisture content increases so the salt concentration in the aqueous phase will drop and its preservative effect will be less. The greater proportion of aqueous phase will also increase the probability of larger droplets of the aqueous phase being present. Both factors lead to a greater susceptibility to microbial spoilage.

Moisture dispersion in the product may be improved by inclusion of emulsifiers such as lecithin preparations or monoglycerides in the lipid phase. Stabilisation of the dispersion may also be achieved by increasing the protein content of the aqueous phase;[91] the protein will play a dual role as it has both emulsifying properties and reduces the mobility of the moisture droplets by increasing their viscosity. The presence of milk proteins will normally improve the palatability of spreads but may decrease their microbial stability. Where reduction of pH is used to improve microbial stability, and where a lactic flavour is desired, casein and whole milk protein preparations are not desirable as the solubility of casein is low in the pH range $\approx 4 \cdot 0 - 5 \cdot 3$, about the isoelectric point ($\approx \text{pH } 4 \cdot 6$). Whey proteins are far less susceptible to pH and are preferable for acidified products.

Conventional continuous buttermakers have been used for the production of reduced fat spreads, though these have usually been of 70–75% fat content. For hybrid products the non-milk fat may be added at most stages in the production, though addition at the buttermaker minimises the risk of accidental contamination of other products. One protein-enriched product with 55% milkfat has been produced via the butter churning route.

With decreasing fat content the need for more effective moisture dispersion favours the application of scraped-surface 'soft margarine' technology, and this approach has been used by margarine manufacturers to produce reduced fat products. Cooling and shear conditions should be optimised to balance sensory quality with microbial stability.

9.5.2 Low Fat Spreads

The early products were formed from dispersions of a salt solution in a blend of vegetable fats plus monoglyceride. Incorporation of milk proteins will improve the sensory qualities but reduces the stability of the emulsion, leading to phase inversion and processing problems as well as lower microbiological stability. For one group of products this problem was overcome by using much higher levels of casein enriched milk protein,[92] typically 13% in the aqueous phase, giving a significant viscous stabilising effect as well as altering the interface. This stabilising effect is not limited to milk proteins as the many claims made for low fat spreads include the use of gelatin and stabilisers in the aqueous phase.[93,94] Surface active agents may also be included in the aqueous phase.[95]

The method of production is similar to that for soft margarines, with production of the aqueous and lipid phases, blending then heat treatment and cooling of the emulsion. Control of the microbiological quality is critical for a satisfactory shelf-life and inclusion of pasteurisation as an integral part of the final heat treatment of the emulsion before packing is the safest option. Optimisation of shear conditions during the cooling and crystallisation stages is essential for product quality, and considerable problems may be encountered in scaling up processes.[96] Lipid phases may contain milkfat, vegetable fats or a blend. More saturated mono-di-glyceride emulsifiers may be used in the lipid phase of proteinaceous spreads[97] and monodiglycerides from fish oils may give better stability than those based on partially hydrogenated vegetable oil.[98] While batch formation of the emulsion is common, continuous dosing and blending systems may also be used. It is preferable to heat treat the emulsion rather than the discrete phases, to minimise the risks of recontamination. Though the easiest route for preparation of a low fat spread is via the formation of a w/o emulsion, claims have been made for the preparation of a fat-continuous product from an o/w emulsion. In this case shear forces invert the emulsion using churning or high pressure to give the w/o emulsion which will then be consolidated in the scraped surface heat exchanger.[99] The phase reversal route will give rise to some fat inclusions in the dispersed aqueous phase and thus improved organoleptic properties. Small quantities of fat may be included in the aqueous phase of some low fat spreads (as occurs naturally in butter) but the extent of this should be limited or the emulsion may become less stable as the proportion of continuous phase diminishes. Emulsions may also be dispersed in a continuous fat phase.[100]

With the establishment of low fat spreads in the market there has been a move further along the fat reduction route with claims for products containing less than 30% fat, 25% being typical. Examples of these claims are given in Table 9.10 below.

Stabilisation of low fat spreads by combinations of ingredients to give a high viscosity aqueous phase, and in some cases a gelled aqueous phase, can create processing problems and more needs to be known of the interrelationship of product and process.[89]

Whilst the low fat spreads should be substantially w/o emulsions, the structures can differ from simple dispersions and approach those of mixtures, i.e. a bicontinuous phase system. Simple cubic lattice or cubic inner-centered lattice models for the emulsion imply minimum fat contents of 43% and ≈34% fat respectively.[106] In theory emulsions with very high internal phases e.g. >90% can be produced using a range of droplet sizes and non-spherical droplets[107] though this has not yet been achieved commercially for fat-continuous emulsions. Commercial products with less than 10% fat are currently based on o/w emulsions and may be viewed as developments of cheese spreads.

TABLE 9.10
Examples of claims for very low fat spreads

Claim
Gellan gum blends in the aqueous phase of spreads with 20–30% fat.[101]
Low Dextrose Equivalent (DE) non-gelling starch hydrolysate with DE 4-25.[102]
Milk properties at ≥8% and modified starch at 0·1–1·2% in the aqueous phase.[103]
Proteinaceous aqueous phase containing gelatin and undissolved protein particles < 5 μm diameter.[104]
Shear churning an emulsion with aqueous phase containing a surfactant.[105]

These products, though structurally different, may be used as alternatives to the usual w/o based spreads. In this case however there will be no need for phase inversion on the palate and the microbiological stability would be expected to be lower than for the w/o based emulsions.

9.5.3 Routine Examination of Reduced and Low Fat Spreads

Fat-continuous systems may be readily distinguished from water-continuous systems by their lower conductivity, as discussed earlier in relation to butter.[51,52] Conductivity measurement has also been used for determining the emulsifying properties of protein preparations.[108] In low fat spreads the presence of more complex mixtures and possibly bicontinuous systems may account for a less clear discrimination between o/w and w/o systems. An alternative method[97] uses the behaviour of the emulsion in cold water, o/w emulsions dispersing while w/o emulsions remain intact; though a gelled o/w emulsion may take longer to disperse.

Texture studies have used a penetrometer[109] for evaluating hardness. More work is needed in this area since, with the much greater proportion of disperse phase, a more complex behaviour pattern may be expected than for margarines and butter. Microscopy is a valuable tool in structural studies of low fat spreads.

Routine compositional analysis may follow the standard methods for butter and margarine, with the caveat that checks should be made that the ingredients in a specific product do not interfere with that particular test method. This is particularly important when adopting rapid instrumental methods. Sorbic acid may be added to low fat spreads to improve the microbiological stability in some countries, e.g. at up to 2000 ppm in the UK.[110] Sorbic acid may be extracted from the sample by steam distillation from the acidified sample then determined by UV absorption or colorimetrically.[111] TLC and HPLC methods[109,115] are also available.

Microbiological evaluations of low fat spreads have used both APHA and IDF methods[109,111] As with butter and margarine, the finer the emulsion then the longer the potential shelf life.

REFERENCES

1. Walstra, P. and Jenness, R., *Dairy Chemistry and Physics*, Wiley Interscience, New York, 1984.

2. Mulder, H. and Walstra, P., *The Milk Fat Globule*, CAB/Pudoc, Farnham Royal, 1974.
3. Christie, W. W. In: *Developments in Dairy Chemistry, Vol. 2*, Fox, P. F. (ed.) Elsevier Applied Science, London, 1983.
4. Hawke, J. C. and Taylor, M. W. In: *Developments in Dairy Chemistry, Vol. 2*, Fox, P. F., (ed) Elsevier Applied Science, London, 1983.
5. Mortensen, B. K. In: *Developments in Dairy Chemistry, Vol. 2*, Fox, P. F. (ed.) Elsevier Applied Science, London, 1983.
6. Badings, H. T. In: *Dairy Chemistry and Physics*. Walstra, P. and Jenness, R. (eds) John Wiley, New York, 1984.
7. Manning, D. J. and Nursten, H. E. In: *Developments in Dairy Chemistry, Vol. 3*, Fox, P. F. (ed.) Elsevier Applied Science, London, 1985.
8. Deeth, H. C. and Fitz-Gerald, C. H. In: *Developments in Dairy Chemistry, Vol. 2*, Fox, P. F. (ed.) Elsevier Applied Science, London, 1983.
9. Scanlan, R. A., Sather, L. A. and Day, E. A., *J. Dairy Science*, **48** (1965) 1582.
10. McDaniel, M. R., Sather, L. A. and Lindsay, R. C., *J. Food Science*, **34** (1969) 251.
11. Richardson, T. and Korycka-Dahl, M. In: *Developments in Dairy Chemistry, Vol. 2*, ed. P. F. Fox, Elsevier Applied Science, London, 1983.
12. Weirauch, J. L. In: *Fundamentals of Dairy Chemistry*, 3rd edn, Wong, N. P. (ed.) Van Nostrand Reinhold, New York, 1988.
13. Kirk, R. S. In: *Challenges to Contemporary Dairy Analytical Techniques*. Royal Society of Chemistry, London, 1984, pp. 127–31.
14. FAO/WHO, World Health Organisation, Geneva, Switzerland.
15. International Dairy Federation, Square Vergot 41, 1040 Brussels, Belgium.
16. International Organisation for Standardisation, Rikilt, Bornstsesteeg 45, 6708 PD, Wageningen, The Netherlands.
17. British Standards Institution, 2 Park Street, London W1A 2BS.
18. Association of Official Analytical Chemists, Langhoven 12, 6721 SR Bennekom, Netherlands (European office).
19. American Oil Chemists' Society, 1608 Broadmoor Drive, Champaign, IL 61821, USA.
20. American Public Health Association, 1015 Eighteenth Street, NW Washington, DC 20036, USA.
21. Tuinstra-Lauwaars, M., Hopkin, E. and Boelsma, S. In: *IDF Bulletin 193*, International Dairy Federation, Brussels, 1985.
22. Rothwell, J. (ed.), *Cream Processing Manual*, 2nd edn, Society of Dairy Technology, Huntington, 1989.
23. Towler, C. In: *Modern Dairy Technology, Vol. 1*, Robinson, R. K. (ed.) Elsevier Applied Science, London, 1986.
24. Bird, J. In: *Milk Fat: Production, Technology and Utilization*, Burgess, K. J. and Rajah, K. K. (ed.), Society of Dairy Technology, Huntingdon, 1991.
25. Wilbey, R. A. In: *Milk Fat: Production, Technology and Utilization*,

eds K. J. Burgess and K. K. Rajah, Society of Dairy Technology, Huntingdon, 1991.
26. Lagoni, H., Paakola, O. and Peters, K. H., *Milchwissenschaft*, **18** (1963) 340–4.
27. Sherbon, J. W., *J. Am. Oil Chem. Soc.*, **51**(2) (1974) 22–5.
28. Antila, V., *Milk Industry*, **81**(8) (1979) 17–20.
29. Larsen, N. E. and Samuelson, E. G., *Milchwissenschaft*, **34**(11) (1979) 663–5.
30. α-Laval, *Dairy Handbook*, α-Laval, Lund, (1983?), pp. 203–5.
31. Arul, J., Boudreau, A., Makhalout, J., Tardif, R. and Bellavia, T., *J. Am. Oil Chem. Soc.* **65**(10) (1988) 1642–6.
32. Rajah, K. K., PhD thesis, University of Reading, 1988.
33. Muuse, B. G. and van der Kamp, H. J., *Neth. Milk and Dairy J.*, **39** (1985) 1–13.
34. Sonntag, N. O. V. In: *Bailey's Industrial Oil and Fat Products, Vol. I*, 4th edn, ed. D. Swern, John Wiley, New York, 1979.
35. Rattray, J. B. M., *J. Am. Oil Chem. Soc.*, **61** (1984) 61.
36. Smith, L. M. and Vasconcellos, A., *J. Am. Oil Chem. Soc.*, **51**(2) (1974) 26–30.
37. Bracco, UK Patent 1 559 064, 1980.
38. Anon., *Dairy Ind. Int.*, **55**(6) (1991) 37–8.
39. Montet, J. C., Lindheimer, M. H., Brun, B., Frankinet, J. and Molard, F., French patent application, FR 2 633 936 A1, 1990. Cited in *Dairy Sci. Abstr.*, **52**(8) (1990) 6186.
40. Riccomini, M., Wick, C., Peterson, A., Jimenez-Flores, R. and Richardson, T., *J. Dairy Science*, **73** (supplement 1) (1900) 107.
41. New Zealand Dairy Research Institute, European patent 318 326. Cited in *Light Food Products News*, **1**(1) (1990).
42. Srinivasan, M. R. and Anantakrishnan, C. P., *Milk Products of India*, Indian Council of Dairy Research, New Delhi, 1964, pp. 28–9.
43. van den Berg, J. C. T., *Dairy Technology in the Tropics and Subtropics*. Pucoc, Wageningen, 1988, pp. 187–93.
44. Yadav, J. S. and Srinivasan, R. A., *NZ J. Dairy Sci. Technol.*, **20** (1985) 29–34.
45. Wilbey, R. A. In: *Modern Dairy Technology*, Vol. 1, ed. R. R. Robinson, Elsevier Applied Science, London, 1986.
46. Butter Regulations 1966 (1966/1074), HMSO, London.
47. Samuelsson, E. G. and Mortenen, B. K., *XIX International Dairy Congress*, International Dairy Federation, Brussels, 1974, p. 662.
48. Dixon, B. D., *Aust. J. Dairy Technol.*, **25**(2) (1970), 82–4.
49. Netherlands Dairy Research Institute, British Patent 1 516 786, 1976.
50. Johansson, S. In: *Milkfat and its modification*, ed. R. Marcuse, Scandinavian Forum of Lipid Research, Göteborg, 1985.
51. Prentice, J. H., *J. Dairy Res.*, **20**(3) (1953) 327–32.
52. Prentice, J. H. In: *Humidity and moisture, Vol. 2*, ed. E. J. Admur, Reinhold, 1965.
53. *Standard Methods for the Examination of Dairy Products*, APHA,

Washington; 14th edn, ed. E. H. Marth, 1978; 15th edn Richardson, G. H. (ed.) 1985.
54. Weaver, R. W. V. In: *Challenges to Contemporary Dairy Analytical Techniques*, Royal Society of Chemistry, London, 1984, pp. 91–102.
55. Egan, H., Kirk, R. S. and Sawyer, R., *Pearson's chemical analysis of foods*, 8th edn, Churchill Livingstone, Edinburgh, 1981, p. 475.
56. Anon, *Food Processing*, (June 1988) 39.
57. Murphy, M. F., Microbiology of butter. In: *Dairy Microbiology Vol. II*, 2nd ed, ed. R. K. Robinson, Elsevier Applied Science, London, 1990.
58. Davis, J. G., In: *Quality Control in the Food Industry, Vol. 2*, 2nd edn, Herschdoerfer, S. M. (ed.), Academic Press, London, 1986.
59. Farag, R. S., Ali, M. N. and Taha, S. H., *J. Am. Oil Chem. Soc.*, **68**(3) (1990) 188–91.
60. Davis, J. G., *Dictionary of Dairying*: supplement to 2nd edn, Leonard Hill, London, 1965, pp. 1182–6.
61. *Rules of Grading Service*, Creamery Proprietors' Association, 19, Cornwall Terrace, London NW1 4QP, 1988.
62. Jellineck, G., *Sensory Evaluation of Food*, Ellis Horwood, Chichester, 1985.
63. Bodyfelt, F. W., Tobias, J. and Trout, G. M., *The Sensory Evaluation of Dairy Products*, Van Nostrad Reinhold, New York, 1988, pp. 376–417.
64. Syarief, H., Hamann, D. D., Giesbrecht, F. G., Young, C. T. and Monroe, R. J., *J. Text. Studies*, **16** (1985) 29–52.
65. IDF, *Evaluation of the Firmness of Butter*, Bulletin 135: 1981, International Dairy Federation, Brussels.
66. Rohm, H. and Ulberth, F., *J. Text. Studies*, **20** (1989) 409–18.
67. Bell, A. E., In: *Water and Food Quality*, Hardman, T. M. (ed.), Elsevier Science Publishers, London, 1989, pp. 251–7.
68. Mitchell, J. R., In: *Food Technology International Europe 1987*, ed. A. Turner, Stirling Publications, London, pp. 249–52.
69. Margarine Regulations 167/1967, HMSO, London.
70. Johansson, M. S. J., UK patent 1 582 806, 1980.
71. Larsson, M., *Food Manuf. Int.* (May/June 1990) 29.
72. Hoffmann, G., *The Chemistry and Technology of Edible Oils and Fats and their High Fat Products*, Academic Press, London, 1989.
73. Latondress, E. G., In: *Handbook of Soy Oil Processing and Utilization*, ed. D. R. Erickson, E. H. Pryde, O. L. Brekke, T. L. Mounts and R. A. Falb, American Soybean Association and AOCS, St Louis and Champaign, 1980.
74. Garti, M. and Remon, G. F., *J. Food Technol.*, **19** (1984) 711–17.
75. Chrysam, M. M., In: *Bailey's Industrial Oil and Fat Products, Vol. III*, 4th edn, ed. T. H. Applewhite, John Wiley, New York, 1985.
76. *Official Methods of Analysis*, 14th edn, AOAC, Arlington, 1984.
77. AOCS Method Ca 2b-38, *Official and Tentative Methods of the American Oil Chemists Society*, 3rd edn, Walker, R. C. (ed.), AOCS, Champaign, 1983.
78. IUPAC, *Standard Methods for the Analysis of Oils, Fats and Derivatives*, 7th edn, Blackwell, Oxford, 1987, pp. 259–63.

79. AOCS Method Cc 16–60, *Official and Tentative Methods of the American Oil Chemists Society*, 3rd edn, Walker, R. C. (ed.) AOCS, Champaign, 1983.
80. Haighton, A. J., *J. Am. Oil Chem. Soc.*, **36**(8) (1959) 345–8.
81. Haighton, A. J., *J. Am. Oil Chem. Soc.*, **53**(6) (1976) 397–9.
82. deMan, J. M. and Beers, A. M., *J. Food Text.*, **18** (1987) 303–18.
83. Haighton, A. J., *J. Am. Oil Chem. Soc.*, **42**(1) (1965) 27–30.
84. deMan, L., deMan, J. M. and Blackman, B., *J. Am. Oil Chem. Soc.*, **66**(1) (1989) 128–32.
85. de Man, J. M. and Gupta, S., *J. Am. Oil Chem. Soc.*, **62**(12) (1985) 1672–75.
86. Davey, K. R. and Jones, P., *J. Text. Studies*, **16** (1985) 75–84.
87. Davey, K. R. and Jones, P., *J. Text. Studies*, **18** (1988) 335–8.
88. Mann, E. J., *Dairy Industries International*, **54**(4) (1989) 9–10.
89. Moran, D. P. J., *Dairy Industries International*, **55**(5) (1990) 41–3.
90. Charteris, W. P. and Keogh, M. K., *J. Soc. Dairy Technol.*, **44**(1) (1991) 3–8.
91. Wilbey, R. A. In: *Low Calorie Products*, eds. G. G. Birch and M. G. Lindley, Elsevier Applied Science, London, 1988.
92. Strinning, O. B. S. and Thurell, K. E., UK Patent 1 455 146, 1973.
93. Wilton, I. E. M., Envall, L. O. G., Sundstroem, K. L. and Moran, D. P. J., US Patent 4 071 634, 1978.
94. Unilever Ltd, UK Patent, GB 1 564 800 (1980).
95. Konin. Brinkers Marg., UK Patent, GB 2 205 849 (1988).
96. Gerstenberg, G., *Dairy Ind. Int.*, **53**(11) (1988) 28–9.
97. Madsen, J., *Food Prod. Develop.* (April 1976) 72–80.
98. Madsen, J., *Res. Disclos.*, No. 249 (1985) 1.
99. Citation 2N11, *Light Food Prod. News*, **1**(2) (1990) 23.
100. Moran, D. P. J. and Sharp, D. G., US Patent, US 4 515 825, 1985.
101. Merck Co. Inc., *Light Food Prod. News*, **1**(2) (1990) 23.
102. Moorhouse, A. L. and Lewis, C. J., US Patent, US 4 536, 408, 1985.
103. Platt, B. L. and Gupta, B. B., UK Patent Application GB 2 193 221, 1988.
104. Bordor, J., van Heteren, J. and Verlhagen, L. A. M., German Patent DE 26 50 981 C2, 1985.
105. Moran, D. P. J. and Campbell, I. J., European Patent EP 0 098 663 31, 1987.
106. Glaeser, H., *Dairy Ind. Int.*, **55**(9) (1990) 29–31.
107. Moran, D. P. J., *Dairy Ind. Int.*, **55**(9) (1990) 31.
108. Kato, A., Fujishige, T., Matsudomi, N. and Kobayashi, K., *J. Food Sci.*, **50**(1) (1985) 56–8, 62.
109. Keogh, M. K., Quigley, T., Connolly, J. F. and Phelan, J. A., *Irish J. Food Sci. Technol.*, **12** (1988) 53–75.
110. *The Preservatives in Food Regulations 1979*, SI 1979 No. 752, HMSO, London.
111. Egan, H., Kirk, R. S. and Sawyer, R., *Pearson's Chemical Analysis of Foods*, 8th edn, Churchill Livingstone, Edinburgh, 1981, pp. 77–8.

112. deMan, L., deMan, J. M., and Blackman, B., *J. Am. Oil Chem. Soc.*, **68**(2) (1991) 63–9.
113. deMan, L., deMan, J. M., and Blackman, B., *J. Am. Oil Chem. Soc.*, **68**(2) (1991) 70–3.
114. Ahmed, M. M., PhD thesis, University of Reading, 1987.
115. Bullock, D. H. and Kenny, A. R., *J. Dairy Science*, **52** (1969) 625–8.

10

Analysis and Quality Control for Processing and Processed Fats

L. McGinley*

Formerly of BEOCO Ltd, PO Box 26, Regent Road, Bootle, Merseyside, L20 1EH, UK

10.1 INTRODUCTION

The large range of oils and fats available to the UK processor is spread over an edible oils market of just over 1 Mt. Table 10.1 gives estimates of the split between refined deodorised oils and hydrogenated oils, it relates to 1985. The disposition of these oils and fats, both unmodified and modified, in terms of utilisation in finished products is shown in Table 10.2, which is adapted from figures for 1984–85 published by Crawford.[1]

To ensure high quality final products, quality control (QC) has to start with raw materials, for as oils and fats are natural products so there are inherent variations in properties with which refiners have to contend. Hence to monitor raw materials, intermediate and final products QC should involve a system with a number of components:

(1) Specifications which may include
  (a) Physical properties such as colour, solid fat content (SFC), melting point, smoke point, refractive index (RI).
  (b) Chemical characteristics including free fatty acid content (FFA), peroxide value (PV), induction period (IP), anisidine value (AnV), iodine value (IV), trace elements.
  (c) Organoleptic quality: taste, odour, flavour stability.
(2) Sampling regime (Chapter 11, and Refs 2–5).
(3) Specifications for products at intermediate stages of processing.

* Present address: 21 Buttermere Avenue, Noctorum, Birkenhead, Merseyside L43 9RH

TABLE 10.1
UK Edible oils market

Oil type	Tonnes	
	Unhydrogenated edible (t)	Hydrogenated edible (t)
Rapeseed	171 000	46 000
Palm	152 000	23 000
Soyabean	63 000	50 000
Sunflower seed	41 000	3 000
Maize	14 000	1 000
Palm kernel	12 000	22 000
Coconut	6 000	5 000
Groundnut	4 500	500
Cottonseed	3 000	—
Animal	42 000	2 000
Marine	—	243 000
Others	100 000	2 000

TABLE 10.2
Disposition of edible oils for food use

Food product or use	(t)	Main oils, unmodified and modified
Margarine[a]	350 000	Hydrogenated marine oil, palm, soya, rapeseed, beef
Solid cooking fat		Hydrogenated marine oil, palm,
Retail	30 000	Soya, rapeseed, beef
Other	120 000	
Liquid cooking oils		Soya, rapeseed, sunflower, maize
Retail	70 000	
Catering	60 000	
Biscuits	130 000	Hydrogenated marine oils, palm, rapeseed, soya
Industrial frying	100 000	Soya, palm, palm olein, rapeseed, cottonseed, sunflower
Chocolate and sugar confectionery	50 000	Fractionated fats, palm kernel, coconut
Salad cream	10–20 000	Maize, soya, sunflower
Ice cream/filled milk	20–40 000	Palm kernel, palm

[a] Margarine is covered in Chapter 9.

(4) A system for feedback of analytical results to production for process regulation.
(5) Comprehensive record keeping and reporting.
(6) A complaints procedure.

The primary process in producing oil and fat products is the refining step and all quality aspects of oils and fats undergoing secondary processing depend upon the efficiency of the refining stage. While this chapter is not primarily detailing fat processing an outline of the appropriate unit process will be offered to put analysis and QC into context.

10.2. REFINING

Crude oils comprise from 94 to 98% triacylglycerols, depending on type and quality of the feedstock, while refined and deodorised oils are usually more than 99% triacylglycerols. The difference between the crude and refined oils represents the components of the crude oil which are removed during refining. The list below is not exhaustive but does illustrate the range of materials found in crude oils with which the refiner has to cope.

(a) Free fatty acids.
(b) Phosphatides, hydratable and non-hydratable.
(c) Pigments, such as carotenoids, chlorophylls, gossypols.
(d) Unsaponifiable matter; sterols, hydrocarbons, tocopherhols (the last are desirable).
(e) Partial acylglycerols.
(f) Free sugars and glycolipids.
(g) Oxidised lipids.
(h) Compounds containing trace elements, e.g. iron, copper, calcium, magnesium, sulphur.

The compounds found in group (h) might be present as copper and iron soaps, while the calcium and magnesium salts of phospholipids and phosphatidic acids are encountered as the greater part of the non-hydratable phosphatides which are described by Hvolby.[6] Sulphur compounds in rapeseed oil result from glucosinolates in the seed (see Chapter 5) and in fish oil from proteinaceous material.

All crude oils and fats contain free fatty acids but some compounds are specific to certain oils, e.g. gossypol in cottonseed oil; but while carotene is a significant minor component of palm oil carotenoids are also found in trace amounts in other oils. Hence the analyses carried out by the refiner are crucial both to quality and economical refining procedures.

10.2.1 Feedstock Quality

Pritchard[7] has covered the trade specification for the crude oils: soyabean, rapeseed, palm kernel, palm, coconut, and sunflower.

There are also British Standards (BS), from 1967, dealing with vegetable oils: crude coconut BS 628, crude groundnut BS 629, crude rapeseed (high erucic) BS 631, crude maize BS 651, crude palm kernel BS 652, crude soyabean BS 653, refined cottonseed BS 655. These specifications cover colour, density, RI, IV, saponification value, acidity and unsaponifiable matter.

Fatty acid compositions are covered by Codex Alimentarius specifications in Alinorm 79/17 Appendix XI, a revision of ranges for low erucic rapeseed, soyabean, arachis (groundnut), sunflower, maize, coconut, palm kernel, and palm oils is given in Alinorm 87/17 Appendix X published in 1987. They are also reviewed in Chapter 7.

However, a function of a wider range of analyses is to help the refiner choose the optimum refining conditions and decide the refining mode when alternative procedures are available, namely alkali versus physical (steam) refining.

In multi-feedstock refineries physical refining is becoming a common route, for suitable oils, to the edible product. The mode of refining chosen will depend upon the type and quality of oil; it is possible to classify oils and fats into two general groups. Typical members would be:

Group 1; Palm, palm kernel, coconut, tallow—low phosphatide oils.
Group 2; Soya, maize, sunflower, rapeseed—significant phosphatide levels.

Group 1 oils are readily amenable to physical refining if general quality criteria are satisfactory; the process is not as accommodating as alkali refining and poor quality oils will not satisfactorily refine this way. The Group 2 oils are usually alkali refined as the phosphatides, pigments and trace elements are more readily removed by this process. However, suitable pre-treatment to remove the gums, as discussed by Forster and Harper,[8] can render these oils physically refinable.

  Whichever process is used, physical or alkali refining, the difference between the initial quantity of crude oil and the final quantity of refined oil represents the loss. Analyses and methods to determine the loss, as well as oil quality, have to be employed. The methods will be outlined in the following sections.

Table 10.3 shows typical analytical characteristics of some crude oils. Chapter 7 presents some trading specifications which also reflect these criteria.

10.2.2 Alkali Refining

This outline of the process refers to continuous plants, more comprehensive discussions are given by Haraldsson,[9] Carr[10] and Braae.[11] The continuous process comprises three phases:

(1) Removal of gums (phosphatides) by either water alone for hydratable phosphatides, or mixed with acid, usually phosphoric, which reacts with both hydratable and non-hydratable phosphatides rendering them both separable. When only water is used for degumming, the separated phosphatides may be further purified and used for both food and non-food applications.[12]

(2) Free fatty acid neutralisation, by means of alkali, and centrifugal removal of soapstock. This step may incorporate the acid treatment and removal of non-hydratable gums if water only degumming was the first step. Due allowance has then to be made for extra alkali addition to neutralise the added acid.

(3) The neutralised oil is washed once or twice with softened water, calcium and magnesium ions should be minimal in order to ensure good washing of the oil and to obviate the fouling of centrifuge discs which can occur if the water has more than some 50 mg CaO/MgO per litre hardness. This is a QC test that should be done on the water.

10.2.2.1 Analysis for feedstock quality and refining yields

The analytical techniques for standard chemical tests will not be described in detail as many standard methods and texts are devoted to them.[2-5] However, where deemed appropriate an outline of methods will be given. The major components removed during refining are free fatty acids, phosphatides and impurities. Moisture and impurities (M & I) are included in the quality tests and specifications for crude oils and fats. Moisture may be determined by the method given in BS 684 Section 1:10 1982, and IUPAC Method 2.601; impurities are covered

TABLE 10.3
Typical analytical characteristics of some crude oils

	Palm kernel	Coconut	Palm	Tallow	Maize	Rapeseed	Soya[b]	Sunflower
% FFA	6	3	4	2	3	1	1·5	2
PV (meq/kg)	5	2	4	2	1	1	2·5	2
AnV	3	2	3	2	7	2	3	1
% M & I[a]	0·5	0·5	0·1	0·2	0·3	0·3	0·5	5
Phosphorus (mg/kg)	20	20	20	40	450	350	200	250
Carotene (mg/kg)	—	—	700	—	—	—	—	—
Iron (mg/kg)	5	5	3	2	2	2	2	2
Copper (mg/kg)	0·1	0·1	0·2	—	0·02	0·02	0·02	0·03

[a] Moisture and impurities (M & I).
[b] Following an initial water degumming.

by BS 684 Section 2:3, 1983, and IUPAC Method 2.604; as will be seen later M & I are incorporated into tests for yields.

10.2.2.2   Free fatty acids

The free fatty acid content is a factor which has a major effect on the final yield of refined oil, its determination is of prime importance. Rossell[2] reviews methods and the expression of results, which may be either as acid value, which is the number of milligrams of potassium hydroxide required to neutralise the acidity of 1 gram of fat, or as the percentage of free fatty acids present. Acid value determinations are covered by a number of standard methods, viz. AOCS method Cd3a-63, IUPAC Method 2.201 and BS 684 Section 2.10 1983; FFA and acid value are also covered by Cocks and van Rede.[3]

However, for refinery process control FFA as a percentage gives an instant indication of losses to be expected on refining an oil and an easily understood parameter, by process operators, to monitor the progress and efficiency of the operation. To ascertain the FFA percentage the 'conventional' molecular weight of the oil type has to be taken into account: Coconut and palm kernel oils are expressed as lauric acid (Mol. Wt 200), palm oil as palmitic acid (Mol. Wt 256) and all other oils as oleic acid (Mol. Wt 282). It would probably be more accurate to establish an average molecular weight from, say, gas chromatographic data but it is 'custom and practice' to use the predominant fatty acid. Classical methods for the determination of FFA and acid value are in contrast to a technique from Ekstrom[13] who describes an automated 'flow injection analysis' method which is now becoming commercially available and on-line analysis, with feedback, is the logical development of such automated systems.

10.2.2.3   Phosphatides

The final factor having a major effect on the yield and progress of refining is the level of phosphatides. A crude Group 2 oil, particularly sunflower and maize, when stored for some time may precipitate phosphatides in the so-called foots. Delvaux and Bertrand[14] review the composition and determination of foots which are assumed to comprise phospholipids, and their complexes, together with 'waxy' compounds. Their emphasis is on linseed oil but, as the authors point out, similar deposits are found in many vegetable oils.

A number of methods to determine foots are given: a simple volumetric test is to allow a fixed volume of oil, in a calibrated tube, to stand for 96 h at 15–20°C and then measure the sediment. This method for linseed oil is detailed in ISO Standard No. 150 (1980), Para. 11, and has been criticised by Delvaux and Bertrand for ignoring the moisture content of the oil which can profoundly influence the amount of sediment deposited. While the criticism is valid the method does offer a 'snapshot' of the likely degree of precipitation encountered at the time of sampling and thus indicates the degree of difficulty probably to be found when processing the oil. In practice in the author's laboratory we have found that good quality sunflower oil with levels of 0·5–0·8% phosphatide has generally given up to 4% by volume of foots; sometimes poor oil has been encountered well outside this, up to an astonishing 25%. At 0·16% moisture over half the phosphatides may precipitate out. Similar observations have been made on maize oil.

Chemical precipitation of foots, the phosphoric acid test (PAT), is the basis of IUPAC method 2.422, ISO Standard No. 150 (1980), Para. 12, for linseed oil. It can be applied to other oils and the final outcome is the direct weighing of acetone washed precipitated material, which comprises all the gums.

The ratio of non-hydratable to hydratable phosphatides may be ascertained by quantitatively removing the hydratable material after determining the total phosphorus level by any suitable method. Methods for phosphorus determination are given in IUPAC Method 2.421 which is based on ashing the oil and forming a phosphovanadomolybdic complex which is determined spectrometrically: AOCS Method, Ca12-55 (1973) determines phosphorus as molybdenum blue after ashing. Alternatively, if the equipment is available, a rapid analysis taking only a few minutes can be done by means of atomic absorption spectrometry using a graphite furnace technique as described by Slikkerveer et al.[15]

The quantitative removal of hydratable phosphatides can be done by heating 200 g of oil to 80°C in a beaker equipped with a stirrer, 4% of distilled water is added and agitation is continued for 30 min. The precipitated phosphatides are separated centrifugally at 3000–4000 rpm. The phosphorus content of the clear oil is determined and the hydratable phosphatide content is obtained by difference. The factor for converting phosphorus to phosphatide is discussed by Pardun[16] who proposes factors of 30–31·5 based on the composition of various lecithins. AOCS method Ca12–55 recommends a factor of 30.

The occurrence of significant amounts of non-hydratable phosphatides in soyabean oil has been attributed by Kock[17] to phospholipase activity during extraction. Some observations by Mounts[18] showed that apparently undamaged beans, left in the field, had both high levels of phosphorus and of non-hydratable phosphatide. Actual damaged beans had lower values:

Normal beans	1·8% phosphatide
Abnormal	2·4% phosphatide
Damaged	less than 1·5% phosphatide

The abnormal beans could not be water degummed to less than 0·5% phosphatides. Hence the presence of significant quantities of non-hydratable phosphatides usually predicts a poor quality oil, for in the case of soyabean oil from fresh good quality beans about 90% of the phosphatides are normally hydratable but when beans are damaged this can fall to about 50% over a period of time.[19] The value for rapeseed hydratable phosphatides is about 70–85%[12] and damage again causes an increase in non-hydratable phosphatide.

10.2.2.4 Refining factors and yields

The analyses so far considered make it apposite to introduce the consideration of monitoring refining loss, which is the difference between the yield of alkali refined oil and the quantity of crude oil processed. There are a number of standard tests to indicate loss: AOCS Methods Ca9a-52, 9b-52, 9c-52, 9d-52 and 9e-52: and ref. 3 pp. 329–32 offer standard laboratory procedures for determining yields of neutral from a particular crude. These methods require special refining equipment to accurately reproduce refinery conditions.

Modern refineries are equipped to measure, on line, instantaneous and average losses and to record final yields of oil. The efficiency of the process is given by the expression:

$$\text{Refining efficiency } (\%) = \frac{\text{Yield of neutral oil} \times 100}{\text{Neutral oil in crude oil}}$$

The refining factor (RF) by:

$$\text{RF} = \frac{\% \text{ Neutralising loss}}{\% \text{ FFA in crude oil}} = \frac{100}{\text{FFA recovered from soapstock } (\%)}$$

(assuming no saponification of neutral oil)

in subsequent phases of the process, e.g. bleaching, deodorising

$$RF = \frac{\% \text{ loss}}{\% \text{ FFA in crude oil}}$$

For a continuous plant the RF at the neutralisation stage is about 1·6–1·8.

The oil yield is governed by the levels of free fatty acid, impurities, phosphatides, the saponification losses, and neutral oil carried over in the soapstock. The neutral oil content of the crude oil is a key analysis for determining efficiencies and losses during refining and there are two widely used methods to determine this, the so-called 'Wesson Loss' and 'Neutral Oil and Loss (NOL)': these will be considered in turn.

10.2.5   Wesson loss

This classic determination is described by Jamieson[20] and is a test which follows the type of chemical treatment afforded an oil during alkali refining. However, while the principle of the method is simple the procedure has to be followed rigorously to obviate adventitious losses. An outline of the method is as follows. A weighed sample of oil is transferred, quantitatively, to a separating funnel by solution and washing with petroleum ether. A portion of 14% aqueous KOH solution is added and the mixture shaken, free fatty acids, phosphatides and other impurities react at this stage. Their removal is accomplished by the addition of 50% aqueous ethanol, with gentle shaking. A successful technique induces the separation of two phases.

Various washing stages have to be carried out carefully to avoid (a) saponification of neutral oil and (b) more likely, losses via emulsions.

The amount of neutral oil is obtained by direct weighing after solvent removal and the percentage calculated. Finally the alkaline soaps fraction is 'split' with hydrochloric acid and the percentage of FFA determined after their isolation. After these procedures it is possible to determine non-oil constituents by: $100 - (\% \text{ neutral oil} + \% \text{ FFA})$.

The final determination of FFA should correlate with the original chemical analysis thus giving a cross check on the technique. Jamieson indicates that with experience neutral oil replicates should not differ by more than 0·1%. This loss, as it is a chemically determined value, and insofar as the reactions are similar to the refining process, is a minimum value for the likely practical losses on the plant.

10.2.2.6 Neutral oil by chromatography

This is described in BS 684 Section 2:8 method 2, 1977, reconfirmed 1983, and is identical to the AOCS method Ca9f-57. The neutral oil is defined as being triacylglycerols plus unsaponifiable matter; the principle of the method is that the free fatty acids and miscellaneous polar materials are removed by passing them through a column of activated aluminium oxide. The adsorbent has to be activated to grade 4 on the 1–5 Brockman–Schodder scale. This grading is described by Lederer and Lederer,[21] the required degree of activation is obtained by the addition of 10% water to dried alumina and material of activity 4 should show a loss on heating of $11 \pm 1\%$. The chromatography apparatus is illustrated in Fig. 10.1.

Explicit instructions on the preparation of the chromatography column using a mixed solvent are given and have to be followed rigorously. A portion of 125 ml of mixed solvent is used to complete the chromatography. After chromatography the neutral oil is recovered by evaporating the solvent from the eluate on a water bath; then:

$$\% \text{ Neutral oil} = \frac{100 \times \text{Weight of eluate}}{\text{Weight of sample}}$$

$$\% \text{ Loss on column} = 100 - \text{Neutral oil}$$

An expression for refining efficiency has been given earlier but the refiner also needs to be able to predict losses; the following equations are standard expressions for alkali refining losses:

(a) Neutralising loss $= 0 \cdot 3 + 1 \cdot 25W$ when $W < 3\%$
(b) Neutralising loss $= 1 \cdot 35W$ when $W > 3\%$

W is defined as the 'Wesson Loss', as previously described, and is a minimum loss, the incorporation of the empirical factors make the 'neutralising loss' a practical value.

It is also feasible to define W from the chromatographic loss or even by the summation of analytical data (FFA + M & I + lecithin). This type of relationship was considered by King and Wharton[22] in a study of the Wesson Loss. However, they only considered the relationship of FFA plus M & I, nevertheless they concluded that the loss was

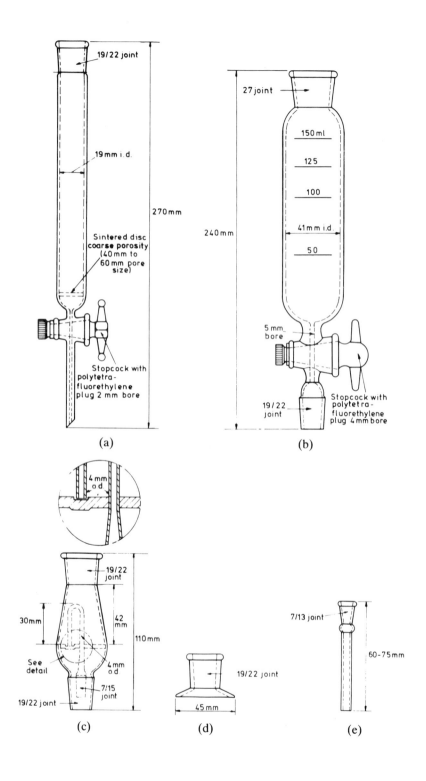

(a)

(b)

(c)

(d)

(e)

TABLE 10.4
Neutral oil and loss—crude oils (%)

Oil	FFA	M & I	Lecithin	Total	Chrom. loss	Wesson loss	Diff-erence
Rape (1)	1·08	0·12	1·21	2·41	2·42	—	0·01
Maize (1)	2·21	0·15	1·14	3·50	3·87	—	0·37
Rape (2)	0·95	0·12	1·36	2·43	—	1·92	−0·51
Maize (2)	1·9	0·25	1·1	3·25	—	2·88	−0·37

indicative of neutral oil content. Linteris and Handschumaker[23] carried out some of the early development work of the chromatographic method and they quote differences between losses determined by chromatography, chemically determined Wesson loss, and summation of analytical determinations. Their results showed Wesson and summation to be comparable while chromatography gave slightly higher results. In the author's laboratory chromatographic losses were also higher than summation, and Wesson loss was lower (Table 10.4).

10.2.2.7 Monitoring alkali refining

The refining processes are monitored usually by determination of free fatty acid reduction and dissolved soap in the oil. The FFA determination has been discussed; a comprehensive review of soap determination has been given by Rossell.[2] The usual method of analysis is by the Wolff titration, which is in BS 684 Section 2:5, 1986.

Typical values for soap contents during the process would be 1000 mg/kg in the neutralised degummed oil, which should be reduced to about 50 mg/kg after washing. The following bleaching step (see later) will normally remove traces of soap completely. The level of soap shows the operator whether or not the alkali addition is correct and also any faults in refining centrifuge settings, e.g. 'back pressures'.

In line refining loss monitors have been mentioned, but it is possible to determine losses accruing in the alkali washing stage by chemical means. The sodium balance method of Crauer and Sullivan[24] is such a

Fig. 10.1. Chromatography apparatus: (a) chromatographic column; (b) solvent reservoir; (c) flask with siphon arrangement; (d) weighing base; (e) extension tube. (Reproduced from BS 684 Section 2.8 by permission of BSI. Complete copies obtainable from BSI at Linford Wood, Milton Keynes, MK14 6LE.)

scheme. They base their calculations on the premise that all sodium in the treated feed appears in the soapstock and the refined oil. The formula derived is:

$$\% \text{ refining loss} = \frac{\% \text{ treat } (\% \text{ NaR} - \% \text{ NaS}) - 100 (\% \text{ NaO})}{\% \text{ NaS} - \% \text{ NaO}}$$

where % treat = percentage of reagent used on the crude oil.
  % NaR = percentage of sodium in the reagent.
  % NaS = percentage of sodium in the soapstock.
  % NaO = percentage of sodium in the refined oil.

The percentage treat can be determined directly by titration of a sample of the crude oil plus reagent. However, modern dosing pumps are usually sufficiently accurate for pump settings to be used to give '% treat'. The sodium in the alkali being used is determined by a standard acid/base titration. The total sodium in the soapstock is revealed by titrating an accurately weighed portion of soapstock (8–10 g), dispersed in 100 ml of distilled water, with 0·5 N sulphuric acid to pH 4. Either a pH meter or a methyl orange end point may be used. The pH 4 end point should be stable for 1 min before noting the titration. The sodium in the oil may be derived from the soap content, via the Wolff method, then:

$$\text{For Soapstock } \% \text{ NaS} = \text{titration} \times \text{N of acid}$$

$$\times \frac{2·3}{\text{Sample weight}}$$

$$\text{For refined oil } \% \text{ NaO} = \% \text{ soap} \times 0·075$$

$$\text{(for oleic soaps).}$$

10.2.3 Physical Refining

In the classification of oils into Groups 1 and 2 it is usual for the Group 2 oils to be alkali refined, but the possibility exists to physically refine them if a suitable 'wet' pretreatment is employed. It has to be borne in mind that the pretreatment has to substitute for the rigorous 'clean-up' afforded by alkali refining. The Group 1 oils normally respond to a 'dry' pretreatment. Both types of treatment are covered by Forster and Harper.[8]

Following a suitable pretreatment, the final step of which is bleaching with active earth, the labile material to be removed is mainly fatty acids and, particularly in the case of palm oil, carotenoids. This removal is accomplished in a high temperature (240–260°C) deodorisation step. However, before an oil can be considered suitable for deodorisation it should have, roughly, the following parameters:

Phosphorus	less than 5 mg/kg[6,25,26]
Chlorophyll*	less than 6 mg/kg[27]
Iron	less than 0·1 mg/kg
Copper	less than 0·05 mg/kg[28]

(*The level of chlorophyll in oils for hydrogenation should be lower, this will be seen later.)

Obviously these values will also apply to alkali refined oils prior to deodorisation.

10.2.3.1 Chlorophyll determination

The determination is a spectrometric one, most usually the AOCS-Cc13d-55. The absorbance of the sample is measured in the red portion of the visible spectrum at 630, 670 and 710 nm which is to cover the specific absorption maximum in the spectrum of chlorophyll. The method is only applicable to refined and bleached oils as further processing via hydrogenation or deodorisation removes absorbance at 670 nm.

Yuen and Kelly[29] discuss the method and conclude that there are two deficiencies for absolute determination. The first problem relates to the forms of 'chlorophyll' normally present, chlorophyll A and B and pheophytin A and B, which have absorbances at different wavelengths and of different intensities. Hence the measurement for chlorophyll may be better expressed as 'chlorophyll equivalent' after reading at the peak wavelength rather than at the wavelength of a particular component.

The second difficulty noted is the fact that calibration factors are for old instruments and it would seem to be advantageous to carry out the analysis using an instrument with a bandpass of 4 nm or less. If such a high performance instrument is available it could be used to analyse oils which may then be used to standardise other instruments, e.g. in 'round-robin' tests. As pyropheophytin may also be produced from chlorophylls during processing, the possibility of converting all pigments to this form and determining it spectrometrically is a consideration.

Using AOCS Cc13d-55 as the basis Tintometer Ltd, Waterloo Road, Salisbury, UK, have introduced an automatic Tintometer which is also equipped with suitable filters to give direct determination of Chlorophyll A and B.

10.2.3.2 Feedstock for physical refining

Generally, because alkali refining is better able to correct crude oil defects, oils for physical refining should be of low oxidation values, as indicated by totox (2PV + AnV), A232 conjugated dienes, low iron content and low, or labile, phosphatide content. Examples and suggested analyses are shown in Table 10.5.

10.2.3.3 Deterioration of bleachability index (DOBI)

This is a specific quality test for palm oil, which is now very often physically refined; a particularly detrimental type of oxidative deterioration involves the iron catalysed destruction of carotenoids and tocopherols. This, and other aspects of the chemistry of refining of palm oil, is discussed by Swoboda.[30] A feature of the work was the introduction of the DOBI test.[31] This involves the spectrometric assay of crude palm oil dissolved in, usually, pure iso-octane. As the method is concerned with relating the destruction of carotenoids with the concomitant increase in secondary products the analysis comprises the measurement of absorbances of 0·5–1% solutions of palm at 446 nm (carotene) and 269 nm (conjugated trienes).

TABLE 10.5
Characteristics of typical group 1 and 2 oils for physical refining (max. values)

Parameter	Group 1 (e.g. water-degummed soya)	Group 2 (e.g. crude palm)
% FFA	0·75	5·0
Carotene (mg/kg)	—	500
PV (meq/kg)	2·0	5·0
AnV	1·5	7·0
Totox value	6·5	17·0
A232	2·5	2·6
A268	0·3	0·7
Iron (mg/kg)	2·0	1·0
Phosphorus (mg/kg)	200	17·0

The DOBI is then given by:

$$\frac{A446}{A269}$$

The relation to quality is: DOBI 4–3 good, 3–2 average, 2–1 poor, less than 1 bad. Good and average oils will be suitable for physical refining. Poor oils will need maximum pretreatment with phosphoric acid and active bleaching earth, or alkali refining.

10.3. BLEACHING (ADSORBENT)

Norris[32] reviews the theory and practice of bleaching together with American colour standards. Not only does bleaching remove colour pigments but, in the physical refining process, traces of metals and phosphatides 'conditioned' by acid treatment with phosphoric or citric acid. When the unit process follows alkali refining then one function is to remove small amounts, say up to 50 mg/kg, of soap. However, if, for any reason, the alkali refining process becomes inefficient, with a concomitant rise in the level of soap present, while it may still be completely removed the bleaching earth will be partially deactivated, resulting in an increase in bleached oil colour. So the monitoring of refining streams for soap is important and may indicate the need for remedial actions to prevent repercussions downstream. In addition bleaching earths remove various polar compounds, and also hydroperoxides by catalytically decomposing them. This clean-up is essential for a following deodorisation step. If the oil is destined as the feedstock for hydrogenation the same considerations apply, as some of the compounds removed are catalyst poisons.

The production, structure and acid activation of bleaching earths is discussed by Morgan *et al.*[33] Zschau[34] points out that as well as being adsorbents, acid activated earths are catalysts for a number of reactions, for some adsorbed, or chemisorbed, colour bodies cannot be recovered from used earth by an extraction procedure and are assumed to have changed their structure due to the catalyic action of the earth. Because adsorptive capacity decreases with increasing temperature, and catalysis increases, Zschau infers there will be an optimum temperature range for bleaching.

Normal plant practice is to accomplish bleaching at 90–110°C with 0·25–2% earth. Vacuum bleaching is now widely used to inhibit oxidation during processing. The level of bleaching earth has to be optimised to reduce oil loss, for the earth absorbs about 25% of oil which cannot be recovered as prime quality. Hence close monitoring of bleached colour has to be carried out to prevent needless overdosing with earth.

10.3.1 Colour Measurement

Two useful reviews of colour measurement are by Rossell[2] and Naudet and Sambuc.[35] Probably the most widely used system of colour measurement is the Lovibond Tintometer. The manual version of the instrument uses a trichromatic system of coloured glass filters of varying depths of colour; they are red, yellow and blue: red and yellow filters are incremented from 0·1 units to 70 units and blue from 0·1 to 40. The units are additive, i.e. a 20 yellow slide has the same depth of colour as slides $10 + 9 + 1$, furthermore the field from three slides, red, yellow and blue of equal values is grey. There is also a set of neutral slides which are overlaid in the sample light path to compensate for 'brightness'. The oil, depending on its depth of colour, is read in cells having path lengths of $\frac{1}{4}$ in, $\frac{1}{2}$ in, 1 in, 2 in, $5\frac{1}{4}$ in or 6 in; 2·5 mm, 5 mm, 10 mm, 20 mm or 40 mm.

The colour of the oil under test is determined by illuminating it from a standard light source. A sample is contained in a cell of appropriate path length, which must be quoted when the result is declared. The light path goes through the oil and the colour filters by means of angled mirrors, the two colour fields are viewed side by side through a telescopic eye-piece. Matching of the two colour fields is done by manipulating the colour filter, using only the red and yellow if possible. The result would be declared for example as 5R 50Y in a $5\frac{1}{4}$ in cell. Green tinges from, say, chlorophyll residues will need a blue component in the result.

When the determination is carried out the sample must be clear, dry and bright. The oil or completely molten fat may be filtered through a Whatman No. 15 filter paper and sodium sulphate to ensure this. Optical faces of the cells should be clean and free from blemishes, the colour filters should also be inspected periodically to ensure no oil drips have penetrated onto them. After about 100 h of use the illuminating lamps should be changed to maintain a constant colour temperature. The British Standard for the determination of colour by this method is BS 684 Part 1, Section 1:14 1987 which supersedes the 1976 specification.

10.3.1.1 Spectrometric determination of colour

IUPAC method 2.103 is designed to give an objective measurement of oil colours. Crude oils and fats have characteristic transmission curves, while bleached oils all have similar curves which are without characteristic peaks. In order to obviate differences due to manual observations the method determines either the transmission curve between 400 and 700 nm, or transmission at definite wavelengths.

The transmission scales of the spectrometer have to be set to values specified in the method using a reference solution prepared from: hydrochloric acid, nickel sulphate and cobalt sulphate. The procedure includes tables to enable the transmittance to be calculated when applying Lambert's Law. Results are expressed either by means of a series of transmittances at various wavelengths or by plotting a graph of transmission against wavelength. While the method gives unambiguous results it is of interest only for special investigations.

An official AOCS method Cc13c-50, measures optical density at 460, 550, 620 and 670 nm in a 21·8 mm cell.

The photometric colour is given by:

$$\text{Photometric colour} = 1·29\,A460 + 69·7\,A550 + 41·2\,A620 - 56·4\,A670$$

This expression approximates to Lovibond red units of colour.

10.3.2 Bleaching Tests

These relate specifically to crude palm oil quality, which is a special case insofar as the final reduction in colour depends upon the thermal decomposition of carotene. One practical assessment of the bleachability of palm, which is quite widely appled in the refining industry, is the SCOPA bleachability test[36] which is now published as BS 684 2:27 1987. This simulates what happens in practice in physical refining as there is a pretreatment, an earth bleach, and a heat bleaching stage.

The principle of the method is to place 100 g of molten crude palm oil into a three-necked flask equipped with an agitator, thermometer and nitrogen blanketing inlet. The oil is heated to 90°C and 0·5% of a 20% solution of citric acid in distilled water is added. The mixture is agitated for 10 min under a blanket of nitrogen. After this time 2% standard bleaching earth (Tonsil FF is specified) is added and the temperature is raised to 105°C. Agitation is continued for 15 min then the oil is filtered through a Whatman No. 1 paper contained, for preference, in a Hartley funnel; if a Buchner funnel is used great care has to be taken to ensure no earth escapes around the paper periphery into the flask.

Next 90 g of filtered oil is transferred to a round bottomed flask connected to a vacuum pump (drawing 1–3 torr); the oil is heated to $260 \pm 5°C$ within 10 min. It should be held at this temperature for a further 20 min. Finally the oil is cooled and the colour read in a $5\frac{1}{4}$ in Lovibond cell and then in a 1 in cell. The whole test must be completed within $2\frac{1}{2}$ h. Satisfactory oils will probably produce colours of about 3·5 red or better in a $5\frac{1}{4}$ in cell.

There are variations of the SCOPA Test which are in use in individual laboratories. Morgan *et al.*[33] varied the method in their investigations into bleaching earths, insofar as they pretreated under nitrogen at 75°C using 0·15% of a 10% phosphoric acid solution for 15 min. The bleaching step, still under nitrogen, was at 95°C using 1% earth. Finally the vacuum bleach was done at 260°C for 30 min at 1–3 torr.

Shaw and Tribe[37] found that citric pretreatment was little different to phosphoric pretreatment. In the author's laboratory we also found this to be the case and use a modified SCOPA method in which 0·1% of 75% phosphoric acid is used for pretreatment before earth bleaching.

Tanz *et al.*[38] have shown the difference that the oxidation state of the crude oil can make to the final bleached colour. A palm oil of totox 16·2 had an earth bleached colour of 16·9 R ($5\frac{1}{4}$ in cell) and heat bleached colour of 4·5 R ($5\frac{1}{4}$ in) while palm of Totox 8·1 had corresponding values of 12·8 R and 3·2 R.

In contrast to the above bleachability tests a one step Direct Bleachability Test (DBT) has been proposed by Hoffman *et al.*[39] who claim that factory process simulation is no better at comparing bleachability than DBT. Their basic process is to agitate 200 g of oil, protected with a blanket of carbon dioxide, at $150°C \pm 0·5°C$ in the presence of 5% Tonsil Standard FF bleaching earth.

10.3.3 Quality Criteria and Autoxidation

Preventing the onset of autoxidation by the exclusion of air during critical high temperature phases of refining is standard practice, and there has to be a QC procedure to monitor the oil during these and earlier stages. Thomas[40] has outlined analytical characteristics relating to undesirable oxidative reactions during the refining of soya bean oil.

However, the points made are relevant to all oil types:

(1) Peroxide value, PV: measures content of hydroperoxides, it is useful in monitoring bleaching and deodorisation steps.

(2) Anisidine value, AnV: defines aldehydic secondary oxidation products, again involving bleaching and deodorisation steps.

(3) Absorbance at 232 nm, follows dienoic conjugation from formation of hydroperoxides, it is an indicator of crude oil quality.

(4) Absorbance at 268 nm, indicates trienoic conjugation, some naturally occurring but most from hydroperoxide decomposition, it is relevant to changes brought about by bleaching.

(5) FFA formed from lipolytic and hydrolytic damage, often parallels oxidative damage, and is indicative of crude oil quality.

(6) Phospholipid content, low non-hydratable phosphatide levels in seed oils generally indicates low degree of oxidation, again related to crude oil quality.

(7) Accelerated tests (see later) Rancimat, Swift, Schaal, are indicators of pro- and antioxidants and overall oxidation stability and thus deodorised oil quality.

Hydroperoxides should be reduced to zero at the deodorisation stage, which will be dealt with later. The determination of peroxide is standardised in BS 684, Section 2:14 1987. Autoxidation products from linoleate are four isomeric conjugated 9 and 13 hydroperoxyoctadecadienoates which have a UV absorbance at 232–233 nm. However, bleaching earths and deodorisation destroy hydroperoxides and may cause thermal isomerisation of polyunsaturated material. Hence the value of dienoic conjugation analysis is reduced on further processed oils.

Some conjugated trienes may arise after bleaching, usually by decomposition of hydroperoxides, because peroxide decomposition gives rise to aldehydes and some ketonic material, which may have carbonyl oxygen conjugated with diene. The type of cleavage possible from hydroperoxide can give rise to compounds such as decadienal whose 2 *trans*-4 *cis* isomer can be detected, as a 'fried flavour', at threshold concentration of one in 10^{10}. The aldehyde moiety is monitored via the anisidine value (AnV)—IUPAC Method 2.504. The chemistry of the oxidative processes are covered in some detail by Gunstone and Norris[41] and Hamilton.[42]

Swoboda[43] discusses these chemical species in relation to the spectrometric assay of palm oil quality. Considering the anisidine value, it is apparent that a double bond in the carbon chain will conjugate with the carbonyl double bond of the aldehyde group, which increases the absorbance at 350 nm, the wavelength for AnV, some four or five times. Hence 2-monoenals contribute a 'weighted' value to AnV, there is a larger enhancement from the 2,4-dienal structure, which is derived from linoleate.

The changes which occur to oxidised material during processing are illustrated in the results shown in Table 10.6. The absorbance values were carried out as specified in IUPAC Method 2.505. The Totox value drops steadily through the process; it should fall, by roughly 50%, at each stage, a failure to do so indicates a problem in the refining/deodorisation stream, for example air leaks; this, then, is of practical value for plant monitoring.

In Table 10.6 the method used for absorbances follows the IUPAC method. However, Swoboda[43] recommends that corrections should be made for other chromophores in the case of palm oil, i.e.

$$A_{232}^{1\%} \text{ (corrected)} = A_{232}^{1\%} - 0.06A_{446}^{1\%}$$

This is widely used in Malaysia to correct the value for oxidation products for the contribution due to carotene.

Similarly

$$A_{268}^{1\%} \text{ (corrected)} = A_{268}^{1\%} - 0.18A_{446}^{1\%}$$

TABLE 10.6
Oxidative quality assessment

	Crude	Bleached	Deodorised
Rapeseed oil			
PV (meq/kg)	1.0	0.4	Nil
AnV	1.4	0.9	0.5
Totox	3.4	1.7	0.5
A232	2.0	1.7	5.0
A268	0.5	0.5	0.4
Palm oil			
PV (meq/kg)	3.6	0.1	Nil
AnV	4.7	3.9	1.7
Totox	11.9	4.1	1.7
A232	2.2	1.7	2.0
A268	0.6	1.1	0.8

Actual carotene values may be obtained by determining $A_{446}^{1\%}$ and multiplying by the factor 383 in accordance with BS 684 Section 2.20 1977.

10.4. DEODORISATION

Deodorisation removes free fatty acids, various aldehydes and ketones, alcohols, hydrocarbons and miscellaneous compounds derived from the decomposition of peroxides and pigments, all of which contribute to taste and odour and whose removal results in a bland finished product. It is carried out at 230–260°C under a vacuum of 3–10 torr with steam sparging. At the relevant concentrations volatile flavour components have partial pressures approximating to those of the less volatile common fatty acids, hence flavour removal often parallels fatty acid removal. At an FFA content of 0·01–0·05% the oil is normally bland. The FFA level will never reach zero, for at about 0·005% the stripping steam is hydrolysing the oil at such a rate as to bring the FFA into equilibrium. The partial glycerides so produced are undesirable, so deodorisation times and temperatures have to be optimised.

Physical, or steam refining, is a deodorisation process but here up to 5% FFA may be removed; a temperature of 250–260°C is required to ensure adequate heat bleaching when palm is being processed. The deodorisation process is thoroughly reviewed by Brekke[44] and Norris.[45] Dudrow[46] has tabulated the relationship of temperature with time of deodorisation while Kochhar et al.[47] and Jawad et al.[48-50] have delineated the circumstances and the type of damage that can occur during deodorisation at extreme temperatures.

At the end of the deodorisation stage, when the oil is cooled to about 100°C, it is usual to dose it with 50–100 mg/kg of citric acid which chelates and thus deactivates trace element pro-oxidant catalysts, mainly copper and iron; the concentrations of non-chelated metals having some catalytic effect on oxidation is 0·005 mg/kg for copper and 0·03 mg/kg for iron.

10.4.1 Taste

All specifications for edible oils insist variously that they should be 'bland', 'neutral', have no 'off-flavours', or rancidity. Organoleptic evaluation of finished oils has to be carried out by a trained taste panel for the seemingly simple test may be fraught with problems arising from inconsistencies introduced by untrained, untested tasters. Sensory evaluation and taste panel standards are discussed by Mounts and Warner[51] and by Jackson.[52] There is also AOCS Recommended Practice for the Flavour Panel Evaluation of Vegetable Oils Cg2-83.

One way to select suitable taste panel members is the 'triangle test': three samples, two of which are identical, are submitted to a candidate to identify the odd one. A guess would have a one in three probability of being correct, but over several sets the accuracy of tasting can be ascertained. The requirements for suitable tasters are that they have the ability to detect known changes and, consistently, correctly identify a series of samples showing the variations normally encountered in processing edible oils.

Flavour intensity is usually based on an hedonic scale, often on a 1–10 ranking. The point of sample rejection will be set by the taste panel management. Both odour and taste should be assessed. Table 10.7 shows a typical scale.

The tasting should be carried out in a special area free from odours and interruptions. Ideally there should be individual booths to obviate panellists influencing each other. Each panel member will have a score sheet on which sample identification, score and comments are made. When a tasting session is completed the average score for each sample may be calculated. This average affords a check on individual performances and if a panellist is consistently different from the average then a re-test of ability may be called for.

TABLE 10.7
Taste scale

Score	Quality	Comments
10	Excellent	No detectable flavour
9	Very good	Suspicion of flavour
8	Good	Barely perceptible flavour
7	Fairly good	Very slight off-flavour
6	Fair	Slight off-flavour
5	Satisfactory	Noticeable off-flavour
4	Poor	More noticeable off-flavour
3	Very poor	Distinct off-flavour
2	Bad	Objectionable flavour
1	Very bad	Very objectionable flavour

10.4.2 Autoxidation and Rancidity

There are three phases in the autoxidation process:

(a) Initiation: free radical formation whose rate is governed to a large extent by the degree of unsaturation.
(b) Propagation: the level of free radical formation is now such that a chain reaction is sustained.
(c) Termination: the, by now, large number of free radicals start to combine, a great deal of damaging oxidation has now taken place.

Phase 'a' is the induction period (IP) and only very slight reversion flavours have been claimed for this stage. Phase 'b' is a rapidly accelerating rate of oxidation and marked flavour deterioration. The final phase has no effect on development of rancidity, according to Hudson.[53] The genesis of odoriferous materials has been touched upon but more complete descriptions of the flavour species are given by Hamilton,[42] Hudson,[53] Frankel[54] and Meara.[55]

10.4.2.1 Accelerated tests

As edible oils have a finite resistance to oxidation, for example, the generally accepted shelf-life for domestic bottled oils is of the order of 9–12 months, it is necessary to use accelerated methods to ensure that oil production is consistently resistant to oxidation. The basis of all accelerated methods is that the end of the IP is marked by a rapid change in the parameter used for measurement, e.g. PV, oxygen absorption, conductivity, odour.

10.4.2.2 Schaal oven test

This was originally developed by the biscuit industry, it is described by Joyner and McIntyre.[56] Briefly the method is to weigh 50 g of clean dry fat into a carefully cleaned 250 ml low form beaker. It cannot be over-emphasised that the vessel should be scrupulously clean to obviate adventitious contamination of the test material by, say, a film of partially oxidised material which would catalyse autoxidation. The beaker and contents are then placed in an incubator at $63° \pm 0.5°C$, normally every 24 h the odour is noted and a distinct rancid note is the end-point of the test.

The test temperature is nearer the practical storage temperatures than other accelerated tests (AOM, Rancimat, which will be discussed later) but even so the mechanism of oxidation at 63°C may be somewhat different from that at ambient temperatures. The Schaal test relates to storage at 21°C by a factor of 6–12 and to the AOM by about 0·04. Thus 12 h at 21°C corresponds to between 1 and 2 h at 63°C and 2–5 min at 98°C. The deployment of the Schaal test is probably something of an 'in-house' means of problem solving and trouble shooting as there are no recognised standards of performance.

As the oil is aged, under the test conditions, any chosen parameter indicating oxidation can be examined, viz. odour, taste, peroxide value or, an interesting possibility, refractive index, for according to Arya et al.[57] refractive indices show an increase of the order of 0·001 at the stage when a rancid odour is noticeable. The claim is made that the refractive index indicates more precisely than peroxide value, the end of the IP. Under the conditions of the Schaal test it is not really a good idea to withdraw too many samples from the 50 g test portion if the chosen analysis depletes the quantity significantly, for under these conditions the ratio of surface area to volume is important, and should be constant. Hence, it is advisable to keep a constant weight and set up multiple parallel tests for analysis.

10.4.2.3 Active oxygen method (AOM) (Swift stability test)

This is a widely quoted method for determining fat stability. There are standard methods available: AOCS method Cd12–57, IUPAC method 2.506; the BS method 684 Section 2:25 1979 has been withdrawn. The test measures the time, in hours, for a sample of oil held at $98° \pm 0°2·C$ and continuously aerated by bubbling air through the oil, to reach a specific peroxide value, quoted in millequivalents of oxygen per kilogram (meq/kg). The end points are arbitrary for, as Sonntag[58] points out, peroxide value and rancidity depend on the original material. Lard and hydrogenated shortenings, for example, show the onset of rancidity at peroxide values of 20 and 40, whereas for corn, sunflower and soyabean oils the peroxide values are between 125 and 150.

It has been found, at the Leatherhead Food RA, that the passage of air through the sample in the AOM or Rancimat (see later) tests distills substances volatile at the test temperature (Rossell, Private communication). The volatilised material may be pro- or antioxidant; hence if such compounds are known or suspected results have to be carefully interpreted, for reliability may be impaired.

The IUPAC method requires that two peroxide values, one hour apart, are done and must fall between 75 meq and 175 meq. The two values are plotted and the stability, in hours, is the value for the PV to reach 100.

10.4.2.4 Automated methods

Sylvester *et al.*[59] developed an apparatus to measure oxygen uptake of an oil in a closed vessel, the pressure drop being the indicator. An innovation, the FIRA/ASTELL apparatus, from the Leatherhead Food Research Association, is said[2] to be rather unreliable over very long periods.

An automated version of the AOM is now marketed by the Metrohm Company of Herisau Switzerland, as the Rancimat. Hadorn and Zurcher[60] proposed the method which, like the AOM, relies on bubbling air through the test sample at elevated temperature. The novel method of following the progress of the oxidation is that the volatile material eventually generated is swept from the oxidising fat by the air flow and is trapped in deionised water in a conductimetric cell. At the end of the IP there is a sharp rise in the conductivity of the water when the secondary oxidation products are evolved.

Six samples are accommodated in the equipment and are analysed simultaneously. Sample size is 2·5 g and the filtered, cleaned, dried air supply is 20 litres/h. The temperature of the analysis may be varied over a range, 100–140°C with an accuracy of ±0·1°C.

In an examination of temperature–IP relationship, using the FIRA/ASTELL apparatus, Meara and Rosie[61] and Kochhar and Rossell[62] have shown that with some products reaction rates had an Arrhenius relationship so that a connection often exists between determination at different temperatures in the 60–150°C range reported. It is desirable in QC testing to have a rapid result, but at temperatures of 100–140°C (in the case of the Rancimat) the temperature chosen has to give a meaningful result in a time commensurate with the production and despatch of the product under test. Table 10.8 shows a comparison of Rancimat values at 120°C and 140°C. The speed of analysis at 140°C is very useful and it may be that specifications for IPs at relatively high temperatures will evolve. The caveat is that reaction kinetics at very high temperature will depart from those at 100°C, the AOM level, and even more so from those at ambient temperature. This, and the results in Table 10.8, suggest that for QC work it will be prudent to collate a great deal of data before

TABLE 10.8
Rancimat induction periods

Oil		Hours at 120°C	Hours at 140°C	Hours at 120°/140°C
Rapeseed oil	IV 112	4·3	1·2	3·6
Hyd. rape	IV 70	28·2	6·5	4·3
Soyabean	IV 132	3·0	—	—
Hyd. soyabean	IV 105	5·1	1·2	4·3
Hyd. soyabean	IV 70	40·0	—	—
Palm oil	IV 53	10·3	2·6	4·0
Palm kernel	IV 19	11·0	2·2	5·5
Hyd. palm kernel	IV 2	66·0	14·1	4·7
Palm kernel +200 mg/kg tocopherols	IV 19	—	19·1	—
Hyd. palm kernel +200 mg/kg tocopherols	IV 2	—	24·5	—

calculating IPs at 'low' temperature from determinations at higher values. A typical graph is shown in Fig. 10.2, the induction times are determined by constructing tangents to the curves, as shown.

Temperature is not the only criterion governing the IP, the presence of pro- and antioxidants in a particular sample and their possible removal during testing, will affect the value as the addition of tocopherols to PK shows. Traces of oxidised material on glassware that is not scrupulously clean will also have a pro-oxidant effect, this aspect of cleanliness has to be a continuing theme for all stability tests. As fats can be stabilised by hydrogenation, which will be discussed later, it is pertinent to illustrate this here, and Table 10.8 includes some typical Rancimat IP results for hardened fats.

An Interlaboratory Test, between 11 laboratories, to ascertain the performance of the Rancimat, has been reported by Woestenburg and Zaalburg.[63] Their conclusions are that there is a satisfactory within-laboratory repeatability of 5%, and a between-laboratories reproducibility of 10%.

10.4.3 Colour Reversion in Edible Oils

Measurement of colour has been discussed earlier; the deodorisation process should almost always result in some heat bleaching, with a concomitant drop in the red contribution, then, if the oil has been improperly refined and bleached in the earlier process there may be a

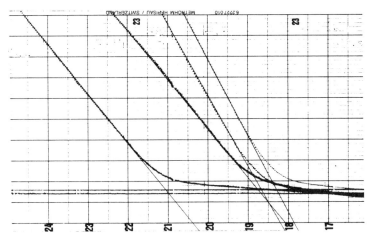

Fig. 10.2. Typical Rancimat graph.

rise in colour values, or a change in hue. For example residual chlorophyll masked by yellow or red in a bleached oil shows up as a green tinge when yellow/red chromophores are destroyed by deodorisation. Sometimes traces of nickel give rise to nickel soaps and green colouration, or nickel chlorophyll complexes are sometimes a problem (see later). Other impurities left by poor refining, traces of soap and/or phosphatides can give rise to poor finished oil colours, and possibly flavour problems.

Generally the fall in colour on deodorisation represents changes due to destruction of carotenoids; in addition there may be changes in tocopherols to dimerised products. This material can then give rise to colour darkening of the deodorised oil on storage. Lai *et al.*[64] suggest that enzymatic reactions, during soyabean flaking, convert γ-tocopherol to colour reversion precursors. Komoda *et al.*[65,66] have studied the phenomenon in soyabean oil in detail, and postulate the formation of tocored from dimerised γ-tocopherol. Tocored has a Chroman-5,6-quinone structure and causes a colour darkening, which reaches a peak and then falls during storage. In the author's laboratory we have observed the changes in both soyabean and maize oils. The degree of colour change, apart from the quality of the raw materials, seed or germ, is dependent on refining and deodorisation conditions, and exposure to light and air during storage. All these are factors that

TABLE 10.9
Characteristics of freshly deodorised oils

Oil	Lovibond Red	FFA (%)	AnV	P (mg/kg)	Fe (mg/kg)	Cu (mg/kg)
Coconut	0·7	0·01	0·5	0·9	0·4	0·003
Rapeseed	1·0	0·03	0·5	1·5	0·04	0·002
Soyabean	0·8	0·03	1·5	1·1	0·04	0·003
Sunflower	1·1	0·03	4·0	1·4	0·08	0·002
Hyd. rape	0·5	0·03	1·0	1·0	0·06	0·003

have to be taken into account by QC in order to minimise unwanted changes in a product.

Finally a useful summary of finished oil handling in Europe is given by Johansson,[67] Table 10.9 adapted from his work shows typical characteristics of freshly deodorised oils.

10.5. MODIFICATION PROCESSES

10.5.1. Hydrogenation

It may be defined as the direct addition of hydrogen to the pi bond of an unsaturated fatty acid in a triacylglycerol. The process is, actually, extremely complex because the reaction involves a great energy barrier and for the reaction to proceed a catalyst has to be used. This is usually specially prepared finely divided elemental nickel which, on a suitable support, can present as much as $100 \, m^2$ reaction surface per gram of nickel.

The reaction overall is the resultant of a number of concurrent reactions: as well as hydrogen addition to the ethylenic linkings there is migration of bonds along the hydrocarbon chain, and there is also a certain amount of geometrical isomerisation from *cis* to *trans* configuration. Hence, as the feedstock may comprise fatty acids containing up to three, or more, double bonds which may hydrogenate, or isomerise, both positionally and geometrically, all at different rates, the complexity of the reaction can be appreciated. There are a number of excellent texts dealing fully with the hydrogenation process: Patterson,[68] Ucciani,[69] Larsson,[70] Beckman[71] and Coenen.[72]

An important aspect of hydrogenation is selectivity which, as originally defined by Richardson et al.,[73] is the rate of conversion of a diene to a monoene compared with the rate of monoene to saturated. However, as all hydrogenation reactions are occurring simultaneously, it is the relative rates that impart selectivity. There are a number of types of selectivity which are discussed by Patterson,[68] Gray and Russel[74] and Coenen.[72] A simplified summary for the selectivities is:

Selectivity 1 (S1):	the rate of hydrogenation of linoleic to oleic acid compared with rate of oleic to stearic acid.
Specific isomerisation (Si):	the number of *trans* double bonds formed per double bond hydrogenated.
Triacylglycerol selectivity (ST):	measures different reactivities of individual fatty acid chains in the molecule. A high ST minimises the formation of 'tri-stearins' by ensuring random distribution of fully saturated acids.

Patterson[68] and the other texts referred to all describe the process variables that affect selectivity. Coenen[72] also summarises the relationship of selectivity to the melting behaviour of end products. He gives the following examples:

Confectionery fats which have high solids at ambient temperatures and zero or very low contents at body temperature. High values for S1, Si, and ST promote the formation of tri-elaidin, stearo-di-elaidin, and di-stearo-elaidin all of which melt between ambient and body temperature. Tri-stearins impart a 'waxy' sensation and a high ST inhibits their formation. A flatter solid fat content (SFC) curve (see later) is required for dough fats which encompass some solids at 40°C. For these, lower selectivities are needed.

To increase the stability of oils for frying a 'brush' hydrogenation is performed, e.g. lowering the IV of soyabean oil by 20–25 units. This is done as selectively as possible, but minimising the formation of *trans* material in order to try and maintain a liquid end product. Now, because high polyunsaturation promotes selectivity there is a tendency for the hydrogenation of linolenic acid at 130–140°C to be selective, but as the linolenic level falls then these low temperatures are not conducive to selectivity and some *trans* isomers and stearins will form. It is then necessary to carry out a subsequent winterisation (see later) to remove the high melting components and produce a clear liquid oil.

The promotion of selectivity can be monitored (a) by the formation of *trans* acids (which is high during selective hydrogenation); standard methods for *trans* acids are, IUPAC method 2.207, an infra-red method, and 2.208 a combination of thin layer chromatography and capillary gas–liquid chromatography (GLC); (b) by changes in the ratios of stearic, oleic and linoleic acids by GLC. Fatty acid composition by GLC of methyl esters is covered by IUPAC methods 2.301 and 2.302, BS Section 2:34 and 2:35 1986 and AOCS methods Ce1-62 and Ce2-66 1973. Chapter 7 covers the determination of *trans* acids. The use of fatty acid composition to calculate selectivities is covered by Allen.[75]

The subject of GLC, both packed and capillary column, is described by Hammond[76] and Ackman.[77] In hydrogenated vegetable oils the use of capillary column GC alone to give a good approximation to the *trans* acid figures arrived at by the special methods above is possible, but this has not been confirmed for hydrogenated fish oils.

In all the decisions regarding hydrogenation conditions the laboratory has to play a crucial part in specifying the conditions as well as final products.

10.5.1.1 Hydrogenation feedstock quality

The starting oil must be degummed, neutralised and bleached, it should be dry, as any moisture present during the reaction could cause hydrolysis and thus an increase in FFA. Bleaching has to ensure low chlorophyll levels, for work by Abraham and deMan[78] has shown it to be a catalyst poison affecting rates of reaction and production of *trans* isomers. Moreover, chlorophyll and pheophytin may form nickel complexes that impart an undesirable green colour to hydrogenated oils. The level of chlorophyll in oils destined for hydrogenation should be very low, even compared with ordinary edible oils.

Patterson,[68] quoting from some work by Ottesson, gives the dose of nickel inactivated by 1 mg/kg of catalyst poison: for phosphorus (as lecithin) it was 0·0008%; nitrogen (as amino acids) 0·0016%; sulphur 0·004%. In general a specification (max. values) for oils to be presented to the hydrogenation plant would be FFA 0·1%, soap content 1 mg/kg (for any level of soap can cause filtration problems), moisture 0·05%, phosphorus 5 mg/kg, sulphur 3 mg/kg and chlorophyll 0·1 mg/kg.

Analysis of feedstock for hydrogenation—sulphur. All analyses except sulphur have been referred to earlier. Sulphur is the most potent catalyst poison but is, fortunately, limited to rapeseed and fish oils in significant amounts. Rapeseed oil has had most attention as in fish oils the sulphur is present, mainly, in amino acids. According to Devinat *et al.*[79] rapeseed sulphur occurs in a number of forms: volatile, thermolabile and non-volatile inorganic forms. The various sulphur compounds have differing poisoning effects on catalysts, for example Abraham and deMan[80] found allyl isothiocyanate to be more potent than either heptyl isothiocyanate or 2-phenylethylisothiocyanate. Earlier Daun and Hougen[81] identified four of seven volatile sulphur compounds they isolated from rapeseed oil: 5-vinyl-2-oxazolidinethione, butenyl, pentenyl, and phenylethylisothiocyanate.

There is not, as yet, a standard method for determination of sulphur or the sulphur containing compounds in rapeseed oil. However, a number of methods are applicable and these are outlined below under four headings, it should be remembered that some of the methods have been adopted from the petrochemical industry.

(1) Combustion methods; the oxygen (Schoniger) flask method has been reviewed comprehensively by Macdonald.[82]
(2) The Parr oxygen bomb method.
(3) Raney nickel reduction and desulphurisation.
(4) Volatile sulphur compounds.

Combustion methods. The principle is that a sample of oil is burned in a stoppered glass vessel filled with oxygen. The resulting sulphur oxides are trapped and determined. Belisle *et al.*[83] modified the Schoniger flask combustion method by designing an oxygen flow-through system. An oxygen flow apparatus devised by Bladh is shown by Persmark[84] and a description of the method given by Halvarson and Hoffman,[85] an outline of which is: 10 g of oil is dissolved in dimethyl carbonate and is combusted on injection into a flowing oxygen atmosphere. The sulphur oxides formed are passed into hydrogen peroxide solution to form sulphuric acid. The solution is titrated with barium ions using Thorin as indicator and a spectrometer to detect the end point at 525 nm. The detection limit is said to be about 1 mg/kg sulphur in oil.

Oxygen bomb methods. Unlike the oxygen flask method, which is carried out at atmospheric pressure the oxygen bomb uses combustion at high oxygen pressure, typically 30 atm. The Parr bomb is almost synonymous with the technique and is supplied by Parr Instrument

Co., Moline, Illinois. A bibliography of analytical methods to disclose the products of combustion is given in Parr manual 207M supplied with the bomb. Abraham and deMan[86] used the technique in conjunction with ion chromatography to determine sulphur in canola oil (which is from low erucic low glucosinolate rapeseed cultivars). After collecting washings from the bomb in distilled water, they were passed through a strong anion exchange column using a multi-component solvent as eluant. The peaks from the ion chromatograph were compared with a calibration curve from standard sulphate solution. Ion chromatography is a specific form of HPLC using phases suitable for the separation of different ionic species. The detector used by Abraham and deMan measured the difference in conductivities of the eluting sample ions and eluant ions.

Raney nickel determinations. The principle here is that the sample is treated with hydrogen and Raney nickel catalyst, any sulphur derived from the reduction and disruption of sulphur compounds is revealed by acid treatment of the nickel and determination of evolved hydrogen sulphide. Baltes[87] has reported a simple and quick method which has been used for marine oils and rapeseed oils. Granatelli[88] devised a Raney nickel procedure for determining down to 0·1 mg/kg of sulphur in non-olefinic hydrocarbon solutions. Daun and Hougen[89] using a modified version of the apparatus have applied the technique to sulphur compounds in rapeseed oil.

In the Granatelli method organically bound sulphur is reduced to nickel sulphide, hydrogen sulphide is liberated by the addition of acid and absorbed in a caustic–acetone solution. The absorbed hydrogen sulphide is then titrated with mercuric acetate with dithizone as indicator. Granatelli states that the presence of olefins introduced an appreciable error. Therefore, in contrast to the Baltes method which stipulates mild hydrogenation conditions, fully hydrogenating any rapeseed oil being analysed should give the best results. This is the principle adopted in the Nordic method for sulphur determination[90] but using commercial nickel catalyst rather than Raney nickel.

Volatile sulphur compounds. Gas chromatography of volatile sulphur compounds, as described by Daun and Hougen[81] depended upon the use of a flame photometer for peak detection. The same type of detector was also used by Abraham and deMan.[91] The general technique to isolate volatiles is to strip them from the sample by

TABLE 10.10
Sulphur determined in rapeseed oils (mg/kg)

	Analytical method				
	Volatiles (GLC)	Raney nickel (Baltes)	Raney nickel (Granatelli) (Daun and Hougen) (modification)	Flask (Oxid'n)	Bomb (Oxid'n)
Results from	—	2·7	6·5	12·1	—
Ref. 85	—	0·2	2·0	5·0	—
	—	6·9	11·3	18·8	—
	0·291	—	1·3	—	17·5
Results from	0·128	—	0·36	—	16·5
Ref. 86	0·322	—	1·2	—	15·7
	Trace	—	0·2	—	13·6

sparging and trap the effluent in liquid nitrogen. The utility of flame photometry for the detection of sulphur is that the system has a sensitivity of 10 000 to 1 for sulphur relative to carbon.

Sulphur summary. The methods considered: oxidation, desulphurisation and volatiles determination are responding to the different properties of the sulphur compounds. This can be seen from data compiled from Refs 85 and 86 and displayed in Table 10.10. In the light of such results Abraham and deMan[86] suggest the use of the following terms for sulphur analysis in Canola oil: volatile sulphur; Raney nickel sulphur; and total sulphur. The level of sulphur acceptable for commercial processing therefore depends on the method of analysis used, and often this is not specified. A total sulphur level of between 5 and 10 ppm is, however, satisfactory for most purposes. There is, obviously, still a fruitful field of work to be exploited in the formalisation of methods of analysis for sulphur in the oils and fats area.

10.5.1.2 Hydrogenation process control—melting point

Hydrogenated fats often incorporate melting point in their specification; but fats have a melting range, the end of which is the melting point, and the value of which depends upon the pretreatment and thus the degree of crystallisation. There are a number of definitions and methods for the melting points of fats, Rossell[2] and deMan *et al.*[92] have published good reviews and comparisons of methods. The slip, or open tube (OT), melting point is covered by BS 684 Section 1:3 1986 and AOCS Cc3-25 standard procedures. Incipient fusion (IF) may be determined at the same time as the slip melting point, it is defined as the temperature at which a meniscus may be discerned on the fat column in the melting point tube. The capillary melting point, or complete fusion (CF), is also an AOCS standard procedure, Cc1-25 as is the Wiley melting point, Cc2-38. The flow and drop point (DP) are sometimes quoted and this is described in BS 684 Section 1:4 1983.

Melting point results are affected by the amount of solid fat present at the chosen end point in the melting range. Table 10.11 shows some data from the author's laboratory relating SFC to various melting point determinations for shortenings, lard, palm and palm kernel oils. The fact that there are appreciable levels of solids at the complete fusion end point, which is taken as when no visible fat can be seen in the column, implies that complete fusion is not the true complete melting point as determined by solid fat content. The 'rule of thumb' that at the slip melting point (OT) there is about 5% solids is borne out.

TABLE 10.11
Solid fat content at determined melting point

Material	Average melting point °C (three analyses)					% solids at the melting point				
	OT	IF	CF	DP	Wiley	OT	IF	CF	DP	Wiley
Short. 1	42·8	29·7	44·8	42·5	44·6	5	16	2	5	3
Lard 1	42·4	33·5	49·0	43·7	46·7	8	13	2	7	5
Short. 2	42·1	28·3	43·2	43·4	44·7	6	16	5	5	3
Short. 3	39·0	28·2	41·1	40·6	43·0	6	13	4	5	3
Short. 4	37·2	27·6	38·0	38·2	38·3	5	12	5	4	4
Lard 2	35·8	29·6	41·7	32·9	39·6	4	9	2	6	3
Palm Rearr.	35·4	28·1	41·0	36·3	40·6	7	10	3	6	4
Lard	29·9	26·3	34·5	30·5	33·3	5	8	4	5	
P.K.	28·1	26·5	28·3	28·1	29·4	6	17	5	6	4
					Average	5·8	12·7	3·6	5·4	3
										3·6

Slip-melting points. For process control the slip melting point is a simple procedure but both AOCS and BS methods specify 16-h stabilisation periods, this is too long for inclusion in process control. We have carried out some work on the aspect of stabilisation procedures. In all cases the fats were completely melted, filtered and dried and then drawn by capillarity into a BS specified tube, four replicate tubes were analysed per test. Heating rate was as described in BS684 Section 1:3. The stabilisation techniques tried were:

(1) Capillaries chilled at −10°C for 5 min.
(2) Fat solidified in capillary at −10°C then held at 15°C for 30 min.
(3) Capillary held at 15°C for 30 min, no 'shock' chill.
(4) Capillaries chilled at 0°C and held at 0°C for 3 h.
(5) As per BS 684 Section 1:3.

Six different fats were analysed for each treatment.

Statistical analysis of the data showed the sum of variances ranked the methods 5, 3, 4, 2, 1, i.e. the BS method had the least spread and the 'quick' method the most. On average the 'quick' method was about 1°C lower than BS. However, the 'quick' method was found to be satisfactory for plant quality control, when melting points are required.

Berger *et al.*[93] have examined tempering techniques for slip points of palm oil products, they conclude that the most important factor is the tempering (stabilisation) temperature, but it is dependent on the nature of the sample.

Rate of solidification. Huizenga[94] comments that the rate of solidification of hardened fats is not necessarily a function of the overall solids profile, but is affected by triacylglycerols with the highest melting point initiating solidification. He describes an ingenious method for rapidly indicating the rate; the apparatus comprises a thick stainless steel plate whose base contains a thermostatted water chamber and the top face machined grooves running downwards. The plate may be tilted between 30°, 45° and 60°. The assembly is completed by a transparent cover. Temperature and tilt are fixed at values depending on oil type; 600 μl of fat at 50°C ± 1°C is applied to the top of a groove. The length, in centimetres, taken for the fat to solidify is noted.

The classical way to assess the speed of crystallisation is by means of a cooling curve. Rossell[2] has reviewed the two main procedures, Shukoff and Jensen. The Shukoff method is standardised by the IUPAC[4], Method 2.132 and the Jensen by the method in BS 684 Section 1:13 1983. The methods are empirical and strict adherence to the specification is necessary.

Wilton and Wode[95] show examples of the Shukoff flask method in factory control and illustrate cooling curves for a variety of materials.

Determination of solids profile. Until about 1968 dilatometry was the almost universal method for determining the relative proportions of solids to liquids in melting fats. An excellent review of the topic is given by Hannevijk *et al.*[96] Standard European methods are specified by IUPAC method 2.141 and BS 684 Section 1:12 1983, American practice is covered by AOCS Cd10-57. In a review of dilatometry Rossell[2] discusses these three procedures.

True solids can now be measured by wide line nuclear magnetic resonance (NMR). There are two systems in use, pulsed and continuous wave, which are discussed by Sleeter[97] and Waddington.[98,99] IUPAC method 2.150 deals with SFC determination by NMR, both pulsed and continuous wave methods are covered.

As in the case of melting points the solids content determined depends on the thermal pretreatment of the sample. The IUPAC standard describes three methods. Treatment 1 comprises melting fat at 80°C, holding at 60°C for 5 min, then holding at 0°C for 60 min and, finally, at the temperature of measurement for 30 min. Method 2 involves melting fat as for 1 but tempering takes some 43 h before holding at measuring temperature for 60 min. Finally Method 3 is a complex tempering cycling procedure taking 30 h before sequential reading of the sample at chosen temperatures.

Treatment 1: This is applicable to fats which do not need a stabilisation of their crystalline form, such as fats used in margarine shortenings, and some confectionery applications.

Treatment 2: For fats showing a high degree of polymorphism and which may crystallise in an unstable crystalline form without suitable pretreatment.

Treatment 3: This is for investigative and development work, it ensures an absolutely stable crystalline form in equilibrium with the lipid phase.

Waddington[98] observes that the relationship between dilatation and NMR is linear below solids contents of 20% but above this a quadratic relationship is appropriate. Conversion tables from dilatation to NMR solids have been published by van den Enden *et al.*[100,101] However, once NMR equipment is installed it is the author's experience that all internal specifications are in terms of NMR solids only. Tabular relationships with dilatometric values may occasionally be of some use in external transactions, but nowadays, certainly in the UK, NMR seems to be universal.

Refractive index and hydrogenation control. The standard methods for refractive index (RI) are from IUPAC Method 2.102, BS 684 Section 1:2 1984 and AOCS method Cc7-25. The present discussion is with regard to RI as a very rapid process control method for hydrogenation, throughout which process the RI falls proportionally to the drop in IV, only the Abbé refractometer is referred to.

For the determination of hydrogenated fats the refractometer prism should be at $60 \pm 1°C$, as all fats, except some fully hydrogenated ones, are liquid at this temperature. There is a relationship between temperature and RI which falls by 0·00036 for each °C rise. Hence, unless the temperature is accurately controlled variations in readings equivalent to IV changes of 1 or 2 units are possible.

The RI is influenced by the structure and composition of fatty acids and triacylglycerols. It decreases with increasing chain length but increases with the number of double bonds, and, significantly in hydrogenation, conjugation. As hydrogenation of an oil proceeds the number of double bonds falls and thus the RI. In addition to this effect the transformation of *cis* to *trans* double bonds also results in a fall of RI, viz. at 40°C triolein RI is 1·4622 and trielaidin 1·4612 (Litchfield[102]). Therefore, there are a number of competing effects and for a successful control procedure a sample of feedstock and a series of samples of product during the reaction are analysed for RI and IV and a plot of RI versus IV constructed. This procedure has to be carried out for each type of feedstock processed and for any given set of hydrogenation parameters; temperature, pressure, catalyst type and concentration, the finished product from any one feedstock will have a typical IV, which can be determined during the reaction from the RI. Latondress[103] has put forward some methods for bleaching and hydrogenation control procedures, but he tends to limit the effectiveness of RI alone to IVs above 95, when dealing with soyabean oil. Below IV 95 he advocates a parallel analysis such as congeal point, AOCS method Cc14-59, to indicate *trans* formation. Probably a 'quick' NMR profile would be a better indicator. The philosophy is, of course, that, depending on hydrogenation conditions, widely differing solids profiles can obtain for a similar IV.

Analysis for nickel. At the completion of the reaction the hydrogenated oil is filtered free of catalyst, the oil should now be clear and bright. Nevertheless there may be nickel soaps present and a simple qualitative test with ammonium sulphide will reveal 5–10 mg/kg. If on the addition of 1 ml of 20% ammonium sulphide to 50 ml of oil, there is a 'grey cast' after heating at 65°C for 5 min then nickel is present.

The normal requirement for finished hydrogenated oils is for less than 1 mg/kg of nickel, post-treatment either with citric or phosphoric acid, and bleaching earth or alkali refining should result in levels of nickel of 0–0·2 mg/kg. Analytical methods have to quantify this level of the element.

Simple, fairly accurate spectrometric methods for trace elements in oils are given by Newlove.[104] The principle is that the oil is shaken with a solution of a suitable colour producing reagent and a solvent to dissolve the oil, the colour complex, and excess reagent. The absorbance is read at an appropriate wavelength. Methods for copper, iron and nickel are given.

Nowadays, more rapid accurate and sensitive analyses are available via atomic absorption (flame and furnace) and plasma emission (inductively coupled and direct current) spectrometry. Sleeter's review[97] covers instrumental analysis of trace elements, which includes nickel.

Olejko[105] gives a flameless AA method for trace elements, including nickel; it is sensitive to 525 pg and will easily determine 0·1 mg/kg. Dijkstra and Meert[106] using DC plasma emission spectrometry for vegetable oils found a detection limit of 0·07 mg/kg for nickel. AOCS method Ca15-75 is a flame AA method in which the oil is directly aspirated into the flame, in MIBK solvent the detection level is 2 mg/kg. As hard oils may foul the burner a graphite furnace (flameless) method Ca18-79, still using MIBK solvent, but with the addition of a few drops of concentrated nitric acid may be used; this will determine nickel down to 0·1 mg/kg. It should be noted that any suitable solvent free of trace metal may be used and the MIBK solvent recommended in some AOCS methods is not always of sufficient purity in some parts of the world.

The drawback to these procedures is that the instruments are all expensive, with plasma emission types costing more than AA. However, with the adverse consequences to QC and QA (quality assurance) of incorrect results, the speed, precision and accuracy of such instruments means that their costs are soon recouped.

10.5.2 Interesterification

This is a 'blanket' term covering a number of interchange reactions of fats with other components. The various types of reaction are:

(1) Alcoholysis: (a) with monohydric alcohols to give, for example, methyl esters; (b) with polyhydric alcohols to form, say, monoacylglycerols.

(2) Acidolysis: reactions with acids to give acid interchange.

(3) Transesterification: (a) rearrangement of single fats; (b) rearrangement of blends of oils and fats.

Sonntag[107] has reviewed the processes in detail: in this chapter only the widely used reactions 1b and 3a and 3b will be considered.

10.5.4.2.1 *Monoacylglycerol preparation and analysis*

In essence the process is to agitate the fat or blend with about 20% by weight of glycerol under vacuum at 180–190°C in the presence of a catalyst, usually 1% sodium hydroxide based on the glycerol content. The reaction is an equilibrium one, and to ensure reaction progresses, water has to be removed while retaining glycerol. After 90–100 min reaction the catalyst may be neutralised with a stoichiometric amount of acid, often phosphoric acid. Excess glycerol is stripped off carefully in order to avoid reversion.

The feedstock has to be good quality neutralised and bleached oil, low in iron content (less than 1 mg/kg) which can cause colour problems. It has to be dry, this, of course, can be done *in situ*. A typical specification for a plastic monoacylglycerol would be: Lovibond colour 4–6 Red ($5\frac{1}{4}$ in), 'mono' content 40% min., free glycerol 1·5%

max., soap 0·3% max., FFA (olcic) 1·5% max. and moisture 2·0% max. Of these only the 1-monoacylglycerol and glycerol determination have not been mentioned, reactions and methods for these are discussed by Rossell.[2] A method due to Kruty et al.[108] has been used for many years in the author's laboratories without major drawbacks. Free glycerol and 1-monoacylglycerols are determined in two separate stages after oxidation by excess periodic acid, the Malaprade reaction, which is the basis of chemical determinations. Standard methods are given in IUPAC Method 2.322 and AOCS method 12 Cd11-57 which confirms two methods, one for monoacylglycerol levels up to 15% and the other for levels over this.

10.5.2.2 Transesterification (rearrangement)

The discussion here will be limited to interactions between separate triacylglycerols, i.e. ester–ester interchange. Such reactions can have diverse applications, for example they can be used to modify crystal habits by, say, the reduction of StPO glycerols in lards; zero *trans* acid feedstock for high polyunsaturated fatty acid blends (for margarine or shortening) can be produced by transesterifying fully hydrogenated fat with liquid oils.

The process is a catalytic one; the feedstock, a dry, neutralised, bleached fat, or blend, is heated to reaction temperature, normally 90–120°C in a reactor which is equipped with an agitator and is also capable of being evacuated, for ease of processing. The catalyst used is 0·1–0·2% of an alkali metal or 0·2–0·3% of sodium methoxide.

Shortly after the addition of the catalyst the reactants become orange-brown in colour; this is the first control check that has to be made. Once the colour develops the reaction will be completed within about 30 min, at this stage there is an equilibrium mixture where there is random distribution of fatty acids on the glycerol molecules. Problems that may be encountered in starting the reaction due to catalyst inactivation were set out by Boothby (private communication); 0·02% water will inactivate 30% of the added catalyst, 0·1% lauric acid 13%, oleic acid 10%, and 1·0 meq of peroxide 5%. On the basis of these figures a specification for feedstock would include: moisture less than 0·01%, FFA (as oleic) less than 0·05%, peroxide less than 1 meq/kg.

Changes due to transesterification. When fats containing significant numbers of fatty acids are randomly rearranged the potential number of triacylglycerols is extremely large. From probability considerations, knowing the molar percentages of component fatty acids, the molar percentage of any component of a randomised mixture may be calculated. If A, B and C are molar percentages of fatty acids then percentage of triacylglycerols containing (a) one component fatty acid, (b) two component fatty acids and (c) three component fatty acids are:

(a)
$$\% \, GA_3 = \frac{A^3}{10\,000}$$

(b)
$$\% \, GA_2B = \frac{3A^2B}{10\,000}$$

(c)
$$\% \, GABC = \frac{6ABC}{10\,000}$$

expression (b) is derived for isomers (AAB), (ABA) and (BAA); (c) for (ABC), (ACB), (CBA), etc.

The possibility of triacylglycerol formation from n different fatty acids was formulated by Daubert:[109]

(1) For all possible isomers differentiated there will be n^3 triacylglycerols.

(2) Positional but not optical isomers, differentiated then $\dfrac{n^3 + n^2}{2}$ triacylglycerols possible.

(3) No isomers differentiated then $n^3 + 3n^2 + 2n$ triacylglycerols possible.

10.5.2.3 *Control analyses*

To define the end products of the reaction the most widely used methods to correlate the randomisation of the acyl groups and the changes in physical properties this brings about are:

(1) Melting point.
(2) Solids profile (by NMR, dilatometry).
(3) Cooling curves.
(4) Triacylglycerol composition (by TLC, GLC).

The data in Table 10.12 were provided by Boothby (Private communication).

TABLE 10.12
SFC and slip mpt before and after interesterification

Material		% solids						Slip mpt	
		20°C		30°C		35°C		(°C)	(°C)
		Before	After	Before	After	Before	After	Before	After
Palm/PK	70:30	14·0	28·0	5·0	9·5	1·5	2·5	—	—
	50:50	15·0	26·5	3·5	5·5	Tr.	0·5	—	—
	30:70	23·0	27·0	2·0	1·5	Tr.	0·5	—	—
Palm/soya	80:20	35·0	41·0	14·5	21·0	5·5	9·5	—	—
	70:30	28·0	30·0	11·0	14·5	5·0	7·0	—	—
	60:40	23·5	29·0	9·5	11·5	4·0	4·0	—	—
Palm oil		—	—	—	—	—	—	37·0	43·0
Hydrog. fish		—	—	—	—	—	—	37·2	34·7
Tallow/rapeseed		—	—	—	—	—	—	43·6	27·5
Hydrog. tallow/sunflower		—	—	—	—	—	—	60·0	43·0

Melting point changes. Table 10.12 includes change in slip melting pt for single oils and blends. In general single naturally occurring oils show an increase and blends and hydrogenated oils a decrease in melting point.

Solids profile. Table 10.12 also shows solids changes before and after interesterification, which were determined using pulsed NMR with stabilisation treatment 1.

Cooling curves. For monitoring interesterification reactions a differential cooling curve proposed by Jacobsen et al.[110] has been used successfully in the author's laboratories for some years. The technique is to measure the temperature difference, by means of matched thermocouples, between a liquid oil and the product under test. The temperature changes are plotted automatically using a pen recorder as the two oils are cooled rapidly from 100°C under controlled conditions. Solidification of the test sample occurs after 6–7 min. The heat of crystallisation keeps the temperature of the test sample above that of the liquid oil over certain ranges and thus produces peaks in the curve being plotted. Rearranged and partially rearranged material can be distinguished. Figure 10.3 illustrates this.

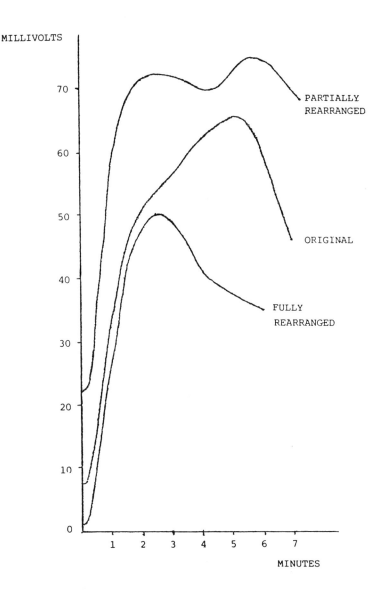

Fig. 10.3. Cooling curves during rearrangement of lard.

TABLE 10.13
TLC of interesterified hydrogenated lard/sunflower oil 40:60

Triacylglycerol	Before inter. (%)	After inter. (%)	Theoretical random (%)
S3	37·7	6·7	6·9
S2U	0·5	30·1	29·7
SU2	16·6	39·6	42·8
U3	45·0	23·6	20·6

S, saturated acyl; U, unsaturated acyl.

Triacylglycerol composition. Thin layer chromatography (TLC) is comprehensively reviewed by Hamilton.[111] The technique, using argentation TLC, was used to monitor the rearrangement of an hydrogenated lard–sunflower blend by Chobanov and Chobanova.[112] Table 10.13 was adapted by Boothby (Private communication) from their results. That randomisation produces large changes can be seen without calculating the theoretical composition. The technique is useful as an investigative technique but is not of much utility for process control.

The use of GLC for the analysis of triacylglycerol composition has been reviewed by Litchfield,[102] pp. 104–38. Rossell[113] examined the interesterification of palm kernel oils via their GLC analysis; he used a 60-cm column with 3% OV1 on a silica support which was temperature programmed from 250 to 350°C. The analysis may be done in 20–30 min. The technique is applicable to materials having significant chain-length differences.

10.5.3 Winterisation and Dewaxing

Thomas[114] reviews the process. Winterisation, which is a special case of fractional crystallisation, was, historically, a naturally occurring process in wintertime when the higher melting triacylglycerols of cottonseed oil were found to separate, in the Southern US, at 4–6°C. Nowadays the process is carried out in refrigerated lagged tanks provided with a means of gentle agitation to prevent temperature gradients and yet permit the growth of large, easily filterable, crystals.

Partially hardened soyabean oil, of IV 105–110, is a good candidate for winterisation to produce stable frying and salad oils. List and

Mounts[115] give a very full account of the preparation of salad oil by this route.

10.5.3.1 Dewaxing

Sunflower, maize, rice bran and safflower oils often cloud on cooling, despite their high levels of unsaturation. This effect is due to the presence of waxes, esters of long-chain alcohols and fatty acids, which are present in seed coats and pass into the oil when undecorticated seeds are expelled and extracted.

Haraldsson[9] shows how the dewaxing process is incorporated into a modern refinery. He notes that wax contents can vary from a few hundred to over 2000 mg/kg but only below 10 mg/kg is the oil stable to refrigeration. The classic dewaxing process is not dissimilar to winterisation but it is usual to use a filter aid to help the filtration process.

Phospholipids, in particular lecithin, inhibit crystallisation and separation of wax, as explained by Rivarola *et al.*[116] Methods for the determination of wax are discussed in Chapter 7.

10.5.3.2 Finished winterised and dewaxed oils

The so-called salad oils will remain limpid at refrigeration temperatures. The finished oils have to conform to the normal quality standards for edible oils but in addition oils which are expected to be refrigerated must pass the 'cold test' set out in AOCS specification Cc11-53; the oil has to remain clear after $5\frac{1}{2}$ h held at 0°C under standard conditions. In practice a satisfactory oil for bottling for the domestic market would remain clear for at least 24 h. Indeed cold tests in excess of 100 h are not uncommon; this, though, would not be a realistic QC requirement simply because 4 days is too long for a decision.

10.5.4 Fractionation

'Dry' and detergent fractionation processes are used to separate 'solid' higher melting triacylglycerols from the liquid oil components. The 'dry' and detergent processes relating to palm oil are discussed by Deffense[117] and with respect to lauric oils by Rossell.[118] Thomas[114] reviews all types of fractionation processes.

10.5.4.1 Palm oil fractions

Quality standards and identity characteristics for palm and palm fractions have been published by The Standards and Industrial Research Institute of Malaysia (SIRIM) in MS, 815 and 816 (1983), the Palm Oil Refiners Association of Malaysia (PORAM) Technical BROCHURE (1984/1985), and the Palm Oil Research Institute of Malaysia (PORIM) Technology, 4 (May 1981).

The IVs of solid fractions, via a 'dry' process, range from about 22 to 50 and the liquid fraction 56–61. It may be obvious, but care should be exercised when using palm fractions in blends not to merely reconstitute palm oil. Berger[119] gives a wide ranging review of palm oil products and their uses.

Solvent fractionation of palm oil. The use of solvents produces relatively clearly defined fractions which, from palm oil, may be used as, or in, a cocoa butter substitute (CBS). The term CBS is one of a number of definitions relating to confectionery fats which is covered by the general term cocoa butter replacer (CBR) for partial or whole replacement of cocoa butter. A specific case is cocoa butter equivalent (CBE) which is completely compatible with cocoa butter. Definitions and properties of commercial CBRs have been given by Keel.[120]

The characteristics of a CBR depend upon the fact that in a typical cocoa butter the mol. % concentrations of POP, POSt and StOSt are 12, 35 and 25; while in a typical palm they are 24, 7 and 0·5. Suitable treatment with a polar solvent, e.g. acetone, will isolate the palmitic 2-oleo glycerols which produce a good CBS. For a CBE a mixture of palm fractions with, say, shea or illipe (borneo tallow) fractions is used as the last two are rich in StOSt and POSt.

Typical fractionation processes and production of CBS's and CBE's are disclosed in a number of patents. A selection is summarised as follows: the use of acetone in the fractionation of palm and of borneo tallow,[121] the fractionation of partially hydrogenated palm oil,[122] the use of shea butter and palm fractions,[123] 'fractioned extraction' procedures,[124,125] the use of nitropropane and/or acetone for palm crystallisation[126] and the partial hydrogenation of palm fractions.[127]

The construction of isodilatation curves, which is a useful technique for assessing the compatibility of cocoa butter with CBRs is described by Rossell.[128] The phase behaviour of CBRs and cocoa butter is developed by Paulicka[129] to show the concept of compatibility in physically significant terms.

10.5.4.2 Lauric oil fractions

Rossell[118] and Timms[130] show that the oleins are obtained in greater yield than stearins, but are less commercially desirable. Hydrogenating them is the usual route to increase their utility, for example for use in biscuit creams. Both authors list properties and uses for lauric oil fractions and their hydrogenated derivatives.

10.6. APPLICATIONS OF PROCESSED OILS AND FATS

A very good summation of fats in food fat systems is given by Bessler and Orthoefer.[131] On the basis of how lubricity will effect properties, including mouthfeel, they give a classification, suggested by Cochran.[132]

Group A   This comprises the polyunsaturated triacylglycerols with melting points from −13°C to 1°C and whose functions are suggested as being both nutritional and the provision of lubricity at 5°C.
Examples: (LLL), (PLL), (OLL), (StLL).

Group B   Unsaturated triacylglycerols with melting points from 6°C to 23°C, provide lubricity at 25°C.
Examples: (OOO), (StOL), (OOP), (StOO).

Group C   Melting from 27°C to 41°C they can stabilise aeration, act as a moisture barrier, and provide lubricity at 37°C.
Examples: (PPL), (StPL), (PPO), (StStO).

Group D   High melting material from 56°C to 65°C which provides structure at 25°C and lubricity at cooking temperatures.
Examples: (PPP), (StPP), (StStP), (StStSt).

St = stearic, P = palmitic, O = oleic, L = linoleic.

In natural and processed fats there is, obviously, a great deal of overlapping of the above groups. Hence, it is the trend to pre-dominance of any one group that determines behaviour and, so, end use. As well as lubricity, resistance and stability to oxidation and heat follows through the scheme.

10.6.1 Salad oils

These have been introduced in considering dewaxing and winteris-ing. Suitable salad oils are those that contain significant amounts of Group A triacylglycerols, i.e. dewaxed corn and sunflower oils, rapeseed and soyabean oils. The latter oils may be partially ('brush') hydrogenated and winterised to lower the linolenate level to about 3% or less in order to improve flavour and oxidative stability.

As has been noted the main criterion for resistance to refrigeration is the cold test. Weiss[133] gives a quick method (due to Pohle) for assessing resistance to crystallisation which involves chilling the sample at −60°C for 15 min then transferring it to 10°C, if crystals persist after 30 min at this temperature then the oil is unsatisfactory.

10.6.2 Cooking oils

In this context it is frying oils and fats that are being considered and whose function is to act as a heat transfer medium. They must be nutritionally sound for as cooking proceeds water is usually displaced, fat is absorbed and so becomes a food ingredient. The processor provides cooking oils for three main areas of use; domestic, catering and industrial.

10.6.3 Domestic oils

Criteria are similar to salad oils with regard to appearance and maintenance of limpidity at refrigeration temperature because bottled oils are widely used for deep and shallow frying as well as salad uses, e.g. mayonnaise preparation. Some QC tests involving domestic scale frying should also be done to ensure there is no untoward flavour development. Typical properties of a bottled oil would be that the FFA be 0·05% or less, M & I nil, smoke point 210–230°C, closed cup flash point 330°C, and cold test 24 h.

10.6.4 Catering and Industrial Frying

A rugged oil is required for these applications and could include more of group C and D material.[131] Mouthfeel would dictate a limit to Group 4 material, while at the other end of the scale more than 1–2% linolenic acid is detrimental, in this type of application, where the fat is subjected to almost continuous heating. McGill[134] reviews the chemistry of this type of frying; Chang et al.[135] also discusses the reactions encountered during frying and Stevenson et al.[136] consider QC in the use of deep frying oils. Berger[137] discusses 'The Practice of Frying' in depth.

A simple performance test which indicates the ruggedness of a heavy duty frying oil is to heat 400 g of the oil in a stainless steel beaker 10 cm in diameter for 48 h at 190°C; it is important to ensure the surface area to volume is constant. After this time the Lovibond colour in a 1 in cell should have changed from 1R to no more than 2R, the viscosity at 100°C should be 11 cS max., the smoke point 195°C min. and the FFA (oleic) 0·2% max.

The foaming propensity of oils increases as they 'age' during frying, and a method to measure foaming in fresh and used oils is described by Bracco *et al.*[138]

10.6.5 Shortenings

The use of fats as shortenings for bakery products is discussed by Thomas.[139] Prepared shortenings are plastic fats containing a crystal mass trapping liquid oil and plasticity is conferred because the crystals are individual, and can move in relation to each other. The ratio of the two phases must be correct to impart the correct degree of mobility.

Shortenings usually comprise one or more partially hardened base stocks, for hard fats will extend the plastic range. However, solids contribution from different ingredients may have different effects on the final shortening. High melting fats will impart greater firmness and 'body' than the same proportion of solids from lower melting ingredients. Typical shortening formulations would, in the temperature range 15–35°C, have solids ranging from 30 to 10%.

Apart from actual solids content the solid moiety should crystallise in a suitable form. Triacylglycerols are often polymorphic, i.e. they can have multiple melting points depending on thermal treatment. The main polymorphs are: alpha (α), beta prime (β') and beta (β). The α forms are waxy, β' fine grained and the β modifications coarse grained. To prepare plastic shortenings the desirable form is β'; in a blend of fats it is likely that the highest melting fat will 'seed' and set the crystal pattern for the blend.

Of the three main forms α is least stable and lowest melting, β is most stable and highest melting. This type of polymorphism is monotropic (one form stable throughout its existence) therefore transformations occur only in one direction: $\alpha \rightarrow \beta' \rightarrow \beta$. Most fats possess an α form, though often unstable, some have both β' and β forms while others have stable β' or stable β; Table 10.14 shows the crystal modification of hydrogenated fats in their stable state. In some cases, e.g. palm kernel oil, the β' form is sufficiently stable to persist for several weeks, or even months, and may be regarded as having pseudo-stability.

TABLE 10.14
Stable polymorphs of fats

Beta prime (β')	Beta (β)
Cottonseed[a]	Soyabean[a]
Palm	Sunflower[a]
Interesterified lard	Lard
Tallow	Groundnut[a]
High erucic rapeseed[a]	Low erucic rapeseed[a]
Fish oil[a]	Maize[a]
(Palm kernel)[b]	Palm kernel
(Coconut[b])	Coconut
	Cocoa butter

[a] In the hydrogenated form.
[b] Pseudo stable.

Timms[140] has published a comprehensive review of the phase behaviour of fats. The ideal modification for shortening blends is β and blending and formulation has to accomplish this. When formulating and monitoring for quality there is a sound base for ensuring that blends of oils and fats containing fatty acids with as wide a range of chain lengths as possible is used. Moreover, any hardstock likely to influence crystallisation should come from the β grouping.

Lauric acid fats should be omitted from any formulations that could conceivably be used for frying as admixtures of more than 5–10% of these fats cause excessive foaming. Shortenings may contain added emulsifiers, such as monoacylglycerol, to produce high ratio shortenings suitable for high sugar/liquid bakery recipes. Once blends are formulated they are processed through a scraped surface heat exchanger—Weiss[133] outlines this process.

10.6.5.1 Quality testing

Tests for quality: taste, FFA, PV, IP, solids profile, monoacylglycerol analysis (in high ratio shortening) have been mentioned.

Consistency testing of shortenings is most simply measured by the use of a penetrometer; methods are given in BS 684 Section 1:11, 1983 and AOCS method Cc16-60. deMan[141] in a review of fat consistency and measuring methods shows relationships between penetrometer readings and yield and hardness values.

The fatty acid composition by GLC will need ascertaining for shortenings blended to conform to the nutritional goals set out by the

National Advisory Committee on Nutritional Education (NACNE). These are that the polyunsaturated to saturated fatty acid ratio, P/S, is increased, to increase polyunsaturated intake by 25% with a concomitant saturated acid reduction of 15%. In general nutritional guidelines from NACNE, the Committee on Medical Aspects of Health (COMA), the US Senate Committee on Nutrition and the World Health Organisation (WHO) recommend that the aim should be to have no more than 30% of total energy from fat of which 10% or less is saturated.

REFERENCES

1. Crawford, R. V., *The role of fats in human nutrition*, ed. F. P. Padley and J. Podmore, in collaboration with J. R. Brun, R. Burt and B. W. Nichols, Published for Society of Chemical Industry by Ellis Horwood, Chichester, 1985, pp. 182–99.
2. Rossell, J. B., *Analysis of oils and fats*, ed. R. J. Hamilton and J. B. Rossell, Applied Science Publishers, London, 1986, pp. 1–90.
3. Cocks, L. V. and van Rede, C. (eds), *Laboratory handbook for oils and fat analysis*, Academic Press, London, 1966, pp. 43–57.
4. International Union of Pure and Applied Chemistry, *Standard methods for the analysis of oils, fats and derivatives*, 7th revised and enlarged edn, ed. C. Paquot and A. Hautfenne, Blackwell, Oxford, 2.001, 1987, pp. 16–17.
5. *Official methods and recommended practices of the American Oil Chemists Society*, 3rd edn, including additions and revisions, American Oil Chemists Society, Champaign, Illinois, 1987, c. 1–47.
6. Hvolby, H., *J. Am. Oil Chem. Soc.*, **48** (1971) 503–9.
7. Pritchard, J. L. R., *J. Am. Oil. Chem. Soc.*, **60** (1983) 322–32.
8. Forster, A. and Harper, A. J., *J. Am. Oil Chem. Soc.*, **60** (1983) 265–74.
9. Haraldsson, G., *J. Am. Oil Chem. Soc.*, **60** (1983) 251–6.
10. Carr, R. A., *J. Am. Oil Chem. Soc.*, **53** (1976) 347–52.
11. Braee, B., *J. Am. Oil Chem. Soc.*, **53** (1976) 353–7.
12. Szuhaj, B. F. and List, G. R. (eds.), *Lecithins*, American Oil Chemists Society Monograph 12, 1985.
13. Ekstrom, L.-G., *J. Am. Oil Chem. Soc.*, **58** (1981) 935–8.
14. Delvaux, E. and Bertrand, J. E., *Analysis and characterisation of oils, fats and fat products*, Vol. 1, (ed. H. A. Boekenogen, John Wiley Interscience Publishers, London, 1964, pp. 281–307.
15. Slikkerveer, F. J., Braad, A. A. and Hendrikse, P. W., *Atom. Spectrosc.*, **1** (1980) 30–2.
16. Pardun, H., *Fette, Seifen. Anstrichm.*, **83** (1981) 240–2.
17. Kock, M., *J. Am. Oil Chem. Soc.*, **60** (1983) 198–202.

18. Mounts, T. L., *Recent developments in vegetable oil technology*, American Soyabean Association, Brussels, 1979, pp. 37–59.
19. Mounts, T. L., List, G. R. and Heakin, A. J., *J. Am. Oil Chem. Soc.*, **56** (1979) 883–5.
20. Jamieson, G. S., *Vegetable fats and oils*, 2nd edn, Reinhold Publishing Corporation, New York, 1943, pp. 454–6.
21. Lederer, E. and Lederer, M., *Chromatography a review of principles and applications*, 2nd edn, Elsevier Publishing Co., London, 1957, pp. 25–8.
22. King, R. R. and Wharton, F. W., *J. Am. Oil Chem. Soc.*, **25** (1948) 66–8.
23. Linteris, L. L. and Handschumaker, E., *J. Am. Oil Chem. Soc.*, **27** (1950) 260–3.
24. Crauer, L. S. and Sullivan, F. E., *J. Am. Oil Chem. Soc.*, **38** (1961) 172–4.
25. Grothues, B., *Proceedings of the Second American Soyabean Association Symposium on Soyabean Processing*, 1981, Brussels.
26. Evans, C. D., List, G. R., Beal, R. E. and Black, L. T., *J. Am. Oil Chem. Soc.*, **51** (1974) 444–8.
27. Roden, A. and Ullyot, G., *J. Am. Oil Chem. Soc.*, **61** (1984) 1109–11.
28. Willems, M. G. A. and Padley, F. B., *J. Am. Oil Chem. Soc.*, **62** (1985) 454–9.
29. Yuen, W. and Kelly, P., *Analytical chemistry of rapeseed and its products*, ed. J. K. Daun, D. I. McGregor and E. E. McGregor, The Canola Council of Canada, Manitoba, 1980, pp. 139–42.
30. Swoboda, P. A. T., *J. Am. Oil Chem. Soc.*, **62** (1985) 287–91.
31. Swoboda, P. A. T., *PORIM Bulletin*, **5** (1982) 28.
32. Norris, F. A., *Bailey's industrial oil and fat products*, Vol. 2, 4th Edn, ed. D. Swern, John Wiley, Chichester, 1982, pp. 292–310.
33. Morgan, D. A., Shaw, D. B., Sidebottom, M. J., Soon, T. C. and Taylor, R. S., *J. Am. Oil Chem. Soc.*, **62** (1985) 292–9.
34. Zschau, W., *Palm oil product technology in the eighties (a workshop)*, Palm Oil Research Institute of Malaysia, Paper TP3, June 1983.
35. Naudet, M. and Sambuc, E., *Analysis and characterisation of oils, fats and fat products*, Vol. 2, ed. H. A. Boekenogen, John Wiley Interscience Publishers, London, 1968, pp. 45–98.
36. Palm oil bleachability test. Leaflet by Seed Crushers and Oil Producers Association, 6, Catherine St, London 3rd revision.
37. Shaw, D. B. and Tribe, G. K., *Palm oil product technology in the eighties (a workshop)*, Palm Oil Research Institute of Malaysia, Paper TP5, June 1983.
38. Tanz, T. S., Teoh, K. T. and Lee, Y. Y., *Palm oil product technology in the eighties (a workshop)*, Palm Oil Research Institute of Malaysia, Paper T11, June 1983.
39. Hoffmann, G., Nijzink, T. and Recourt, J. H., *Rev. Franc. Corps Gras*, **22** (1975) 511–15.
40. Thomas, A., Analytical control during soyabean oil refining, *Proceedings of Second ASA Symposium on Soyabean Processing*, American Soyabean Association, Brussels, 1981.

41. Gunstone, F. D. and Norris, F. A., *Lipids in foods, chemistry, biochemistry and technology*, Pergamon Press, Oxford, 1983, pp. 58–69.
42. Hamilton, R. J., *Rancidity in foods*, ed. J. C. Allen and R. J. Hamilton, Applied Science Publishers, London, 1983, pp. 1–20.
43. Swoboda, P. A. T., *Palm oil product technology in the eighties (a workshop)*, Palm Oil Institute of Malaysia, Paper T23, June 1983.
44. Brekke, O. L., *Handbook of soy oil processing and utilisation*, ed. D. R. Erickson, E. H. Pryde, O. L. Brekke, T. L. Mounts and R. A. Falb, American Soya Bean Association and American Oil Chemists Society, 1980, pp. 155–91.
45. Norris, F. A., *Bailey's industrial oil and fat products, Vol. 3*, ed. T. H. Applewhite, John Wiley, Chichester, 1985, pp. 127–65.
46. Dudrow, F. A., *J. Am. Oil Chem. Soc.*, **60** (1983) 272–4.
47. Kochhar, S. P., Jawad, I. M. and Rossell, J. B., Leatherhead Food Research Association Research Report No. 385, 1982, (available to members only).
48. Jawad, I. M., Kochhar, S. P. and Hudson, B. J. F., *J. Fd Technol.*, **18** (1983) 353–60.
49. Jawad, I. M., Kochhar, S. P. and Hudson, B. J. F., *Lebensm.-Wiss. U-Technol.*, **16** (1983) 289–93.
50. Jawad, I. M., Kochhar, S. P. and Hudson, B. J. F., *Lebensm.-Wiss. U-Technol.*, **17** (1984) 155–9.
51. Mounts, T. L. and Warner, K., *Handbook of soy oil processing and utilisation*, ed. D. R. Erickson, E. H. Pryde, O. L. Brekke, T. L. Mounts and R. A. Falb, American Soya Bean Association and American Oil Chemists Society, 1980, pp. 245–66.
52. Jackson, H. W., *Bailey's industrial oil and fat products, Vol. 3*, ed. T. H. Applewhite, John Wiley, Chichester, 1985, pp. 243–72.
53. Hudson, B. J. F., *Rancidity in foods*, ed. J. C. Allen and R. J. Hamilton, Applied Science Publishers, London, 1983, pp. 47–57.
54. Frankel, E. N., *Handbook of soy oil processing and utilisation*, ed. D. R. Erickson, E. H. Pryde, O. L. Brekke, T. L. Mounts and R. A. Falb, American Soya Bean Association and American Oil Chemists Society, 1980, pp. 229–44.
55. Meara, M. L., *Fats and oils: chemistry and technology*, ed. R. J. Hamilton and A. Bhati, Applied Science Publishers, London, 1980, pp. 193–213.
56. Joyner, N. T. and McIntyre, J. E., *Oil and Soap*, **15** (1938) 184–6.
57. Arya, S. S., Ramanujam, S. and Vijayaraghavan, P. K., *J. Am. Oil Chem. Soc.*, **46** (1969) 28–30.
58. Sonntag, N. O. V., *Bailey's industrial oil and fat products, Vol. 2*, 4th edn, ed. D. Swern, John Wiley, Chichester, 1982, pp. 407–525.
59. Sylvester, N. D., Lampitt, L. H. and Ainsworth, A. N., *J. Soc. Chem. Ind.*, **61** (1942) 165.
60. Hadorn, H. and Zurcher, K., *Deut. Lebensm. Rundschau*, **70** (1974) 57–65.
61. Meara, M. L. and Rosie, D. A., Leatherhead Food Research Association Technical, Circular No. 487, 1971, (available to members only).

62. Kochhar, S. P. and Rossell, J. B., Leatherhead Food Research Association Research Report No. 412, 1983 (available to members only).
63. Woestenburg, W. J. and Zaalburg, J., *Fette Seif. Anstrchm.*, **88** (1986) 53–5.
64. Lai, M.-T., Lin, W.-M., Chu, Y.-H., Chen, S.-L. Y., Kong, K.-S. and Chen, C.-W., *J. Am. Oil Chem. Soc.*, **66** (1989) 565–71.
65. Komoda, M., Onuki, N. and Harada, I., *Agric. Biol. Chem.*, **31** (1967) 461–9.
66. Komoda, M. and Harada, I., *J. Am. Oil Chem. Soc.*, **46** (1969) 18–22.
67. Johansson, G. M. R., *J. Am. Oil Chem. Soc.*, **53** (1976) 410–13.
68. Patterson, H. W. B., *Hydrogenation of fats and oils*, Applied Science Publishers, London, 1983.
69. Ucciani, E., *Proceedings, Third ASA Symposium of Soybean Processing, Hydrogenation of Soy Oil*, Antwerp, American Soybean Association, Brussels, 1983, pp. 7–13.
70. Larsson, R., *J. Am. Oil Chem. Soc.*, **60** (1983) 275–80.
71. Beckman, H. J., *J. Am. Oil Chem. Soc.*, **60** (1983) 282–90.
72. Coenen, J. W. E., *J. Am. Oil Chem. Soc.*, **53** (1976) 382–9.
73. Richardson, A. S., Knuth, C. A. and Milligan, C. H., *Ind. Engng Chem.*, **16** (1924) 519–22.
74. Gray, J. I. and Russell, L. F., *J. Am. Oil. Chem. Soc.*, **56** (1979) 36–44.
75. Allen, R. R., *Bailey's industrial oil and fat products, Vol. 2*, 4th edn, ed. D. Swern, John Wiley, Chichester, 1982, pp. 1–95.
76. Hammond, E. W., *Analysis of oils and fats*, ed. R. J. Hamilton and J. B. Rossell, Applied Science Publishers, London, 1986, pp. 137–206.
77. Ackman, R. J., *Analysis of oils and fats*, ed. R. J. Hamilton and J. B. Rossell, Applied Science Publishers, London, 1986, pp. 137–206.
78. Abraham, V. and deMan, J. M., *J. Am. Oil Chem. Soc.*, **63** (1986) 1185–8.
79. Devinat, G. S., Biasini, G. S. and Naudet, M., *Rev. Franc. Corps Gras.*, **27** (1980) 229–36.
80. Abraham, V. and deMan, J. M., *J. Am. Oil Chem. Soc.*, **64** (1987) 855–8.
81. Daun, J. K. and Hougen, F. W., *J. Am. Oil Chem. Soc.*, **54** (1977) 351–3.
82. Macdonald, A. M. G., *The Analyst*, **86** (1961) 3–12.
83. Belisle, J., Green, C. D. and Winter, L. D., *Anal. Chem.*, **40** (1968) 1006–7.
84. Persmark, U., *Rapeseed*, ed. L-A. Appelqvist and R. Ohlson, Elsevier Publishing Co., Amsterdam, 1972, pp. 174–94.
85. Halvarson, H. and Hoffman, I., *Analytical chemistry of rapeseed and its products*, ed. J. K. Daun, D. I. McGregor and E. E. McGregor, The Canola Council of Canada, Manitoba, 1980, pp. 136–8.
86. Abraham, V. and deMan, J. M., *J. Am. Oil Chem. Soc.*, **64** (1987) 384–7.
87. Baltes, J., *Fette Seifen. Anstrichm.*, **69** (1967) 512–14.
88. Granatelli, L., *Anal. Chem.*, **31** (1959) 434–6.

89. Daun, J. K. and Hougen, F. W., *J. Am. Oil Chem. Soc.*, **53** (1976) 169–71.
90. IAFMM, International Association of Fish Meal Manufacturers, Hoval House, Orchard Parade, Potters Bar, Hertfordshire, UK, Fish Oil Bulletin No. 16, 1983.
91. Abraham, V. and deMan, J. M., *J. Am. Oil Chem. Soc.*, **62** (1985) 1025–8.
92. deMan, J. M., deMan, L. and Blackman, B., *J. Am. Oil Chem. Soc.*, **60** (1983) 91–4.
93. Berger, K. G., Siew, W. L. and Oh, F. C. H., *J. Am. Oil Chem. Soc.*, **59** (1982) 244–9.
94. Huizenga, T., *Proceedings, Third ASA Symposium of Soybean Processing, Hydrogenation of Soy Oil,* Antwerp, American Soybean Association, Brussels, 1983, pp. 14–21.
95. Wilton, I. and Wode, G., *J. Am. Oil Chem. Soc.*, **40** (1963) 707–11.
96. Hannevijk, J., Haighton, A. J. and Hendrikse, P. W., *Analysis and characterisation of oils, fats and fat products, Vol. 1,* ed. H. A. Boekenogen, John Wiley Interscience Publishers, London, 1964, pp. 119–82.
97. Sleeter, R. T., *Bailey's industrial oil and fat products, Vol. 3,* ed. T. H. Applewhite, John Wiley, Chichester, 1985, pp. 167–242.
98. Waddington, D., *Analysis of oils and fats,* ed. R. J. Hamilton and J. B. Rossell, Applied Science Publishers, London, 1986, pp. 341–99.
99. Waddington, D., *Fats and oils: chemistry and technology,* ed. R. J. Hamilton and A. Bhati, Applied Science Publishers, London, 1980, pp. 25–45.
100. van den Enden, J. C., Rossell, J. B., Vermaas, L. F. and Waddington, D., *J. Am. Oil Chem. Soc.*, **59** (1982) 433–8.
101. van den Enden, J. C., Haighton, A. J., van Putte, K., Vermaas, L. F. and Waddington, D., *Fette Seif. Anstrchm.*, **80** (1978) 180–6.
102. Litchfield, C., *Analysis of triglycerides,* Academic Press, London, 1972, pp. 230–2.
103. Latondress, E. G., *J. Am. Oil Chem. Soc.*, **44** (1967) 154A, 156A, 192A.
104. Newlove, T. H., *Laboratory handbook for oils and fat analysis,* eds. L. V. Locks and C. van Rede, Academic Press, London, 1966, pp. 346–51.
105. Olejko, J. T., *J. Am. Oil Chem. Soc.*, **53** (1976) 480–4.
106. Dijkstra, A. J. and Meert, D., *J. Am. Oil Chem. Soc.*, **59** (1982) 199–204.
107. Sonntag, N. O. V., *Bailey's industrial oil and fat products, Vol. 2,* 4th edn, ed. D. Swern, John Wiley, Chichester, 1982, pp. 97–173.
108. Kruty, M., Segur, J. B. and Miner, C. S., *J. Am. Oil Chem. Soc.*, **31** (1954) 466–9.
109. Daubert, B. F., *J. Am. Oil Chem. Soc.*, **26** (1949) 556–8.
110. Jacobsen, G. A., Tiemstra, P. J. and Pohle, W. D., *J. Am. Oil Chem. Soc.*, **38** (1961) 399–402.
111. Hamilton, R. J., *Analysis of oils and fats,* ed. R. J. Hamilton and J. B. Rossell, Applied Science Publishers, London, 1986, pp. 243–311.

112. Chobanov, D. and Chobanova, R., *J. Am. Oil Chem. Soc.*, **54** (1977) 47–50.
113. Rossell, J. B., *J. Am. Oil Chem. Soc.*, **52** (1975) 505–11.
114. Thomas, A. E., *Bailey's industrial oil and fat products, Vol. 3*, ed. T. H. Applewhite, John Wiley, Chichester, 1985, pp. 1–39.
115. List, G. R. and Mounts, T. L., *Handbook of soy oil processing and utilisation*, ed. D. R. Erickson, E. H. Pryde, O. L. Brekke, T. L. Mounts and R. A. Falb, American Soya Bean Association and American Oil Chemists Society, 1980, pp. 193–214.
116. Rivarolo, G., Anon, M. C. and Calvelo, A., *J. Am. Oil Chem. Soc.*, **65** (1988) 1771.
117. Deffense, E., *J. Am. Oil Chem. Soc.*, **62** (1985) 376–85.
118. Rossell, J. B., *J. Am. Oil Chem. Soc.*, **62** (1985) 385–90.
119. Berger, K. G., *Fette Seif. Anstrichm.*, **88** (1986) 250–8.
120. Keel, J. P., Leatherhead Food Research Association, Scientific and Technical Survey No. 100, 1977 (available to members only).
121. B.P. 827, 172, Best, R. L., Crossley, A., Paul, S., Pardun, H. and Seeters, C. J., Unilever Ltd.
122. B.P. 861, 016, Crossley, H. and Paul, S., Unilever Ltd.
123. B.P. 925, 806, Best, R. L., Davies, A. C., Paul, S. and Soeters, C. J., Unilever Ltd.
124. B.P. 953, 451, N.V. Twincon Ltd.
125. B.P. 953, 452, N.V. Twincon Ltd.
126. B.P. 1, 230, 317, Kao Soap Co.
127. B.P. 1, 349, 846, Beresford, M. W. and Rossell, J. B., Unilever Ltd.
128. Rossell, J. B., *Chemy Ind.* (1973) (17) 832–5.
129. Paulicka, F. R., *Chemy Ind.* (1973) (17) 835–9.
130. Timms, R. E., *Fette Seif. Anstrichm.*, **88** (1986) 294–300.
131. Bessler, T. R. and Orthoefer, F. T., *J. Am. Oil Chem. Soc.*, **60** (1983) 1765–8.
132. Cochran, W. M., A.E. Bailey Award Lecture, North Central AOCS Meeting March 19th, 1975 (unpublished).
133. Weiss, T. J., *Food oils and their uses*, 2nd edn, Ellis Horwood, Chichester, 1983, p. 18.
134. McGill, E. A., *Bakers Digest*, **54** (1980) 38–42.
135. Chang, S. S., Paterson, R. J. and Ho, C., *J. Am. Oil Chem. Soc.*, **55** (1978) 718–22.
136. Stevenson, S. G., Vaisey-Genser, M. and Eskin, M. A. N., *J. Am. Oil Chem. Soc.*, **61** (1984) 1102–8.
137. Berger, K. G., *PORIM Technology No. 9* (May 1984), Palm Oil Research Institute of Malaysia.
138. Bracco, U., Dieffenbacher, A. and Kalarovic, L., *J. Am. Oil Chem. Soc.*, **58** (1981) 6–12.
139. Thomas, A. E., *J. Am. Oil Chem. Soc.*, **55** (1978) 830–33.
140. Timms, R. E., *Progr. Lipid Res.*, **23** (1984) 1–38.
141. deMan, J. M., *J. Am. Oil Chem. Soc.*, **60** (1983) 82–87.

11

Sampling for Analysis

The Late A. Thomas
Unimills, Hamburg, FRG

11.1 GENERAL INTRODUCTION

Sampling is a most important aspect of quality assessment. It is therefore appropriate to consider it in some depth, an aspect that some books on analytical methodology fail to do. A detailed discussion of sampling involves a thorough knowledge of statistics which not only makes for difficult reading, but also may be of little practical benefit to the practising technologist. Instead this chapter reviews the subject from a pragmatic point of view ranging from the taking of sample increments from a parcel of goods, through to the test portion on which the analysis is performed.

It is a prime requirement that both the sampler and more particularly the analyst are aware of the requirements and in many cases the confidence limits of both operations, i.e. sampling and analysis. All too often the analyst is prepared to work to very fine limits which are negated by the sampling procedure used for his starting material.

Arguably there is also a responsibility for the analyst to disclose deficiences in the sample to his client as regards packaging, labelling and size before commencing the analysis. An appreciation of these requirements, amongst others, will be found in the following text.

11.1.1 Scope

The estimate of the value of a characteristic is dependent not only on the method of analysis but also on the type of sampling. Sampling is a random process. Selecting a truly random sample from a large bulk of material is not as simple a procedure as might appear; it must be based on statistical theories. If reliance is placed on human judgement the result is likely to be biased. Representative sampling has to take into account a number of basic considerations, namely:

—purpose of sampling and subsequent analysis, e.g. process control or checking on contract/agreed specifications;
—physical state of product (liquid, semi-solid, solid, powder, lumps, etc.);
— form of presentation (bulk, pre-packed);
— characteristic to be analysed, i.e.
    —commodity defects,
    —compositional characteristics,
    —net content,
    —health-related properties;
—the type of examination for which the samples are required (e.g. physical, chemical, sensory, bacteriological);
—effect of sampling procedure on characteristic to be analysed (e.g. microbiology, oxidation characteristics);
—homogeneous or heterogeneous distribution of relevant attributes within the material to be sampled.

11.1.2 Vocabulary of Sampling Terms

Whereas the emphasis in the context of routine commercial and technical purposes would normally be on experience and speed of execution the sampling procedures for, for example, EC subsidy purposes have to strictly follow formal/legal sampling directives.

Wherever the emphasis is being placed, there is a need for standard terminology to avoid misunderstandings, especially so since traditional usage in the field of agricultural food products can differ from internationally standardised terminology. The vocabulary of sampling terms used here is in line with ISO and Codex recommendations, and also explains in more detail the general theory of sampling.

11.1.2.1 Acceptable quality level (AQL)

For a given sampling plan the quality of a lot expressed as the percentage of defective items in the lot, or as the mean value of the inspected characteristic, both associated with a low probability of rejection (usually in the region of 5%). *Acceptance number* is the maximum number of defective items found in the sample, below which the lot is acceptable.

11.1.2.2 Analysis sample
The sample which is analysed. It is sometimes also referred to as test sample and normally prepared from the laboratory sample.

11.1.2.3 Attribute
See characteristic.

11.1.2.4 Batch
See lot.

11.1.2.5 Bulk sample/blended bulk sample
A combined aggregation of the increments, e.g. the quantity of oil seeds obtained by combining and mixing the primary samples.

11.1.2.6 Characteristic
A property which helps to differentiate between items. The differentiation may be either quantitative (by variables) or qualitative (by attributes).

11.1.2.7 Composite sample
A sample consisting of portions from each item, taken *in proportion* to the quantity of product in each item selected.

11.1.2.8 Consignment
A quantity of some commodity (e.g. oil seed or oil) dispatched or received at one time and covered by one particular set of documents (e.g. shipping document or contract). A consignment may consist of one or more lots. Sometimes a consignment is also referred to as 'parcel' or 'lot'.[1]

11.1.2.9 Consumer risk
For a given sampling plan, the probability of acceptance (usually in the region of 10%) of a lot having the limiting or rejectable quality level.

11.1.2.10 Contract sample
See laboratory sample.

11.1.2.11 Defective item
An item containing one or more defects, i.e. failure to meet one or more of the specified requirements.

11.1.2.12   Gross sample
See bulk sample.

11.1.2.13   Increment
A quantity of material taken at one time from a larger body of material, e.g. a small quantity of oil seeds taken from a single position in the lot.[2] Increment is sometimes also referred to as *primary sample*.[1]

11.1.2.14   Inspection
The process of examining (including looking over), measuring, testing, gauging or otherwise comparing an item with one or more requirements of a standard or specification. The relative amount of sampling done within a sampling plan is often referred to as the *inspection level*, e.g. numbers ranging from 1 to 5.[3]

11.1.2.15   Item
A defined quantity of material on which a set of observations may be made.
Note: The terms 'individual' and 'unit' are sometimes used as synonyms of 'item'.

11.1.2.16   Laboratory sample
A sample, usually derived from the bulk sample, prepared for sending to the laboratory and intended for inspection or testing. Laboratory sample[2] is sometimes also referred to as contract sample.[1] In short: anything that is sent to a laboratory.

11.1.2.17   Lot
A stated quantity of some commodity (e.g. oil seed or oil), manufactured under conditions that are presumed uniform or presumed to be of uniform characteristics, usually being a proportion of a consignment. The term lot (or batch) normally means inspection lot in sampling, i.e. a quantity of material from which a sample is drawn and inspected.

11.1.2.18   Lot size
The number of items or quantity of material constituting the lot.

11.1.2.19 Mean value

For a series of n observations $X_1, X_2, \ldots, X_n$, the arithmetic mean of these observations:

$$\bar{x} = \frac{1}{n} \sum_{i=1}^{n} x_i$$

Note: The mean value, $\bar{x}$, calculated for a sample of n items, taken at random from a population, is an estimation of the *true mean* μ which would result from measuring the characteristic in question on the totality of the items in this population.

11.1.2. Primary sample

See increment.

11.1.2.21 Package/unit

That part of a consignment that is packed separately.

11.1.2.22 Parcel

See consignment.

11.1.2.23 Rejectable quality level

The quality of a lot expressed as the percentage of defective items in the lot which is considered to be unacceptable.

11.1.2.24 Sample (general term)

One or more items taken at random from a population (or lot) and intended to provide information representative of the population. A sample is considered to be representative if it has been taken according to a standard method. Terms such as average sample, sub-sample, part sample and quality sample (see e.g. IASC/standard procedure for sampling oils and fats)[1] can cause confusion when cited out of context and should preferably be referred to as primary, bulk, contract sample, etc.

11.1.2.25 Sample size

The number of items or quantity of material constituting the sample.

11.1.2.26 Sampling

The procedure used to draw and constitute a sample.

11.1.2.27 Sampling error [4]

If a sample of n separate items is drawn at random and a mean value, $\bar{x}$, of some property is measured, this is distributed approximately normally about the true mean, μ, with standard deviation $\delta/\sqrt{n}$, where δ is the standard deviation of the items in the consignment. These items may be jars, cans, etc., or increments taken from the bulk. This result is sufficiently true as long as the sample size is small compared with the consignment size, which is normally the case. If the sample forms more than 10% of the consignment a correct expression for the standard error of the mean is

$$\delta \sqrt{\frac{N - n}{n(N - 1)}}$$

where N represents the total number of items in the material sampled. This correction to δ becomes progressively smaller as n approaches N and disappears when $n = N$ (i.e. when the whole of the consignment is tested). Provided that the sample size is relatively small the sampling standard error is independent of the size of the consignment. The justification for the common practice of taking a number of items related to the size of the consignment (e.g. fixed proportion or square root) is simply that with a large consignment the consequence of error is more serious and, hence, the accuracy of sampling needs to be greater.

11.1.2.28 Sampling plan

The rules stating the sample size to be taken from a lot and the acceptance/rejection criteria to serve as the basis for a decision as to the acceptance or rejection of the lot.

11.1.2.29 Standard deviation

The square root of the variance. The standard deviation calculated for a sample of n random items is an estimation of the true standard deviation.

11.1.2.30 Test portion

The measured amount of the test sample on which the analysis is performed.

11.1.2.31 Variance

The variance (v) is a measure of dispersion based on the mean squared deviation from the arithmetic mean.

$$v = \frac{1}{n-1} \sum_{i=1}^{n} (x_i - \bar{x})^2$$

11.1.2.32 Variable

Essentially synonymous with characteristic.

11.1.3 Values of Characteristics

An agreed value of a characteristic can be subject to different assumptions. For example, a specified maximum level of 5 ppm iron in a crude oil may be interpreted in different ways. It could mean that no single item in a lot may exceed 5 ppm iron or it could mean that the average of a number of items may not have an iron content greater than 5 ppm. Generally, in the case of oils, fats and fatty foods the second definition applies. In addition, the average figure in the second case could be obtained either by analysis of a blended bulk sample or by analysing each of the sampled items separately and then calculating the average. Again, in the case of oils, fats and fatty foods it is preferred practice to analyse a blended bulk sample, unless—and this would be the exception rather than the rule—information on the distribution of the characteristic is required.

Consideration must of course also be given to the nature of the items forming the sample. Where the product is pre-packed each package will constitute an item for the purpose of sampling. If the product is supplied in bulk it will be necessary to take an increment and each increment will constitute a sample item.

11.1.4 General Principles for Selecting Sampling Procedures

It is only possible to give general guidelines for sampling plans and procedures. For commodity defects, such as discoloration and presence of extraneous matter, and compositional characteristics a 'variables' sampling scheme is normally more appropriate than classical attribute sampling. Compositional characteristics can encompass homogeneously distributed variables such as FFA in oils and heterogeneous characteristics such as moisture in oils and hulls in oilseeds. In the case of a uniform variable one basically applies a sampling plan according to which individual items are taken at random from the whole consignment and blended into one bulk sample from which contract and analysis samples are then prepared. In the case of heterogeneously distributed characteristics additional samples have to be taken from zones or part of the consignment known or suspected to be deviating from the bulk.

There appears to be no consensus on a specific sampling plan for net content in the case of pre-packed goods. The concept of average net contents is generally accepted, whereby the average content of the items of the lot shall be equal to or greater than the declared contents, with additional tolerances for the individual items, and be based on quality control programs agreed between buyer and seller.

Health-related properties, for example microbiological status, microbial by-products, environmental contaminants, require specific sampling plans and procedures appropriate to each individual situation. Sampling for microbiological assessment, for example, must take into account the possibility of heterogeneous distribution of microorganisms and must ensure that no extraneous microorganisms are introduced during sampling. Sampling for assessment of residues of bleaching earth in an oil must take into account that these types of residues tend to settle on the bottom. Quantitative analysis for aflatoxins in a consignment of groundnuts requires special sampling plans in view of this toxin normally not being distributed uniformly, tending to be concentrated in damaged nuts.

'Health-risk' associated characteristics attract a low value AQL (e.g. $0 \cdot 1$–1%) whereas those for compositional characteristics such as moisture normally attract a higher value AQL (e.g. 5 or 10%). For a given AQL the lower the inspection level number the greater is the risk of passing poor quality lots. Level 4 within a scale of 1–5 (see also Section 11.1.2) is probably a normal level for sampling lots. Where health risks are of major concern inspection level 5 may be more appropriate.

11.1.5   Method of Sampling and Sampling Equipment

The appropriate method of sampling will also depend on the type of container from which the samples have to be taken, e.g.

—horizontal or vertical land tanks,
—silos,
—ship, coaster, barge tanks,
—rail or road tank cars,

—pipelines,
—consignment in packages such as barrels, drums, tins, cases, bags, bottles.

Sampling equipment or apparatus will depend on the type of container, the physical state of the product and also on special requirements such as sampling for assessing the bacteriological status. The equipment or apparatus should conform to national and international standards.

11.1.6 Labelling and Handling of Samples

The sample container itself must be in line with the property/attribute to be determined and also lend itself to transport storage and mailing prior to analysis. Attention should be paid in particular to the following general requirements:

—cleanliness of the container;
—inertness of all parts of the container to the sample;
—robustness to withstand transport hazards;
—suitability for preserving the sample unchanged (e.g. by preventing undesirable access of light, heat and air into the sample and the passage of moisture and other volatiles out of the sample).

The sample itself must not be accessible without breaking the seal. The information on the label of the sample container must be indelible and contain sufficient information to identify the samples, and the corresponding consignment, without ambiguity.

11.1.7 Sample Report

The sample report must ideally be based on the following check list:

—designation of the material (name, grade, specification);
—name of producer, seller, buyer, etc.;
—date and place of production;
—ship or vehicle carrying the consignment;
—from/to/date of arrival;
—quantity of the consignment;
—lot or batch number;
—bulk or packages;
—number and date of bill of lading, type, member and date of contract;

—sampler's name;
—date of sampling;
—method of sampling and sample container;
—place and point of sampling;
—general reference to the condition of the consignment;
—any unusual circumstances that may have influenced sampling
   (e.g. relative humidity, temperature, etc.).

11.1.8 National and International Bodies

Various bodies have concerned themselves with sampling procedures and standard analyses:

Association Française de Normalisation (AFNOR), American Oil
Chemists' Society (AOCS), Association of Official Agricultural
Chemists of the USA (AOAC), American Society for Testing and
Materials (ASTM), British Standards Institution (BSI), Board of
Grain Commissioners for Canada, Deutsche Gesellschaft für Fett-
wissenschaft (DGF), Deutsche Industrienormen (DIN), European
Economic Community, International Association of Seed Crushers
(IASC), International Union of Pure and Applied Chemistry
(IUPAC), International Standards Organisation (ISO), Interna-
tional Olive Oil Council (IOOC), Netherlands Normalisatie Insti-
tuut (NNI), Federation of Oils, Seeds, and Fats Associations Ltd
(FOSFA International), Codex Alimentarius Committee on Meth-
ods of Analysis and Sampling, National Cottonseed Products
Association (NCPA), National Institute of Oilseed Products
(NIOP), National Soybean Processors Association (NSPA), Oil-
seeds Control Board of South Africa, US Department, of Agricul-
ture, Federal Grain Inspection Service (FGIS), Palm Oil Refiners
Association of Malaysia (PORAM), Palm Oil Research Institute of
Malaysia (PORIM), Netherlands Oils, Fats and Oilseeds Trade
Association (NOFOTA).

The recommendations of these various bodies for specific products are
referenced under the respective product headings. The reader should
refer to the full standard where appropriate.

11.2. OILSEEDS

11.2.1 Type of Analyses

Oilseeds are generally traded on the basis of moisture and admixture with foreign and damaged seed although these are only indirect

parameters of the essential bulk criteria, namely oil content and protein content. On the other hand, the moisture content can in many instances be an indication of ease of processing and oil quality. The degree of seed damage will similarly enable a prediction of crude oil quality contained in the seed. Although there is a trend to generally specify oil and protein content, this will in the foreseeable future not obviate the need for specifications in terms of damaged, immature seed, etc. Trace contaminants generally do not feature in trading contracts although they are in part covered by legislation, e.g. aflatoxins in groundnuts. For EC-grown rapeseed and sunflower seed, moisture, oil content and foreign matter, and in the case of rapeseed erucic acid and glucosinolate contents, must meet certain limits for the seed to qualify for subsidy and intervention rules.

In international trade about 80% of all oilseeds, oils and fats are handled on FOSFA contracts.[5] Quality attributes and methods of analysis are defined and the expertise of FOSFA approved analysts monitored by a Quality and Laboratory Committee. Other major trade associations are the NSPA, NIOP and NOFOTA. A comprehensive list with addresses is given in a publication by IASC.[1]

Some examples of typical oilseed specifications and gradings have been reviewed by Pritehard.[5]

When sampling oilseeds it must be borne in mind that these do not normally represent a perfectly homogeneous lot. There may be 'moisture pockets'. Sand and other inorganic impurities will tend to settle to the bottom. Hulls will tend to 'migrate' to the top. Special cases of a heterogeneous distribution can, for example, be 'fines' in copra and aflatoxins in groundnuts. The following is largely based on ISO standards, IASC recommendations, and FOSFA contract conditions.

11.2.2 Transport/Storage Conditions

The handling of oleaginous seeds and fruits during transport and storage has a decisive influence on the quality of the crude oils. Damage and high moisture contents generally lead to hydrolytic and oxidative impairment of the oil. A comparison of intact and damaged palm kernels, for example, demonstrated higher initial FFA levels and an increased rate of enzymatic lipolysis on storage in the case of the damaged kernels.[6]

A high moisture content accelerates the uptake of oxygen and the corresponding release of carbon dioxide. This leads to generation of heat which in turn promotes lipolysis, growth of microorganisms (which in turn can lead to the formation of mycotoxins such as aflatoxins), formation of undesirable colour and odour, and, ultimately, to clumping of seed in silos and even spontaneous combustion.

Oilseeds should be dried before storage to minimise these effects. The critical moisture content correlates roughly with the hygroscopic equilibrium at 75% relative humidity and varies between 6 and 13%, depending on the protein and carbohydrate content. In the case of copra, rapeseed and sunflower seed the critical moisture content is *c.* 7%, in the case of soyabeans *c.* 13%.

11.2.3 Sampling Devices[1,2]

For the reasons outlined earlier samples should preferably be taken from the product stream during loading and unloading. This is facilitated by the fact that transport and storage in bulk has almost entirely superseded that in bags. However, special grades of oilseeds intended for direct consumption—after cleaning and roasting—are still largely traded in bags. For these reasons a full array of sampling devices covering both sampling from bulk and from bags must be available. Apart from this, standardised equipment is required for reducing bulk samples to contract samples and contract samples to analysis samples. These are depicted (in Figs 11.1–11.9) (taken from ISO 542–1980[2] which is in line with the recommendations of the IASC.[1] Needless to say, all sampling apparatus must be clean, dry and free from foreign odours. In essence the following types of apparatus are relevant for oilseeds:

11.2.3.1 Sampling from bags
Sack-type spears or triers, cylindrical and conical samplers, and hand scoops (Figs 11.1–11.5).

11.2.3.2 Sampling from bulk
Shovels, hand scoops, cylindrical samplers, conical samplers, mechanical samplers, and other apparatus for drawing small periodical samples from a flow of oilseeds (Figs 11.1–11.6).

11.2.3.3 Mixing and dividing
Shovels, quartering irons, riffles and other dividing apparatus (Figs 11.1, 11.7–11.9).

Dimensions in millimetres

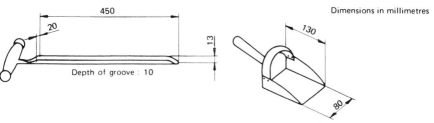

Fig. 11.1. Sampling spear (open trier).   Fig. 11.2. Hand-scoop.

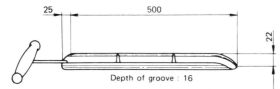

Fig. 11.3. Divided sampling spear (open trier).

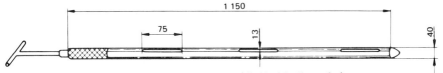

Fig. 11.4. Cylindrical sampler (divided bulk probe).

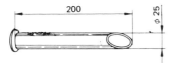

Fig. 11.5. Running iron (sack-type trier).

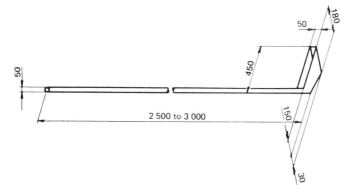

Fig. 11.6. Falling stream sampler (Pelican type).

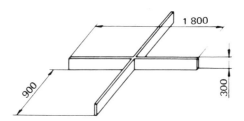

Fig. 11.7. Quartering irons.

11.2.4   Limitation of the Size of Lot[1,2]

Unless stated otherwise, a consignment of oilseeds is normally examined in lots of up to 500 t for large seeds (e.g. copra, palm kernels) and medium-sized seeds (e.g. groundnuts, soyabeans, sunflower seed) and of up to 100 t for small seeds (e.g. rapeseed, linseed). If, for example, an ocean-going vessel has in one particular hold a consignment just exceeding 500 t then it would obviously be nonsensical to base subsequent sampling and analysis on two 'formal' lots.

Most of the oilseeds are received from ocean-going vessels or from river transport for unloading into silos or warehouses. In both cases sampling normally takes place at the point of transfer from the vessel. In the case of transfer from vessel to vessel or to railcars, sampling is very similar. Cars should be sampled individually.

11.2.5   Method of Drawing Samples[1,2]

11.2.5.1   General

Sampling should be done by superintendents appointed by buyers and sellers, or at least by experienced staff if it involves quality control outside contract obligations. It is the responsibility of each sampler to obtain representative samples and to prepare detailed reports and labels.

Obviously damaged seed, e.g. by sea water, or showing other defects, e.g. foreign odour or discoloration, should be assessed separately.

It is not unusual for example, for, large stones, ball-bearings or insects to be readily apparent in the lot to be sampled but not in the sample. For this reason samplers must record all unusual conditions and other pertinent information to avoid, for example, a lot being

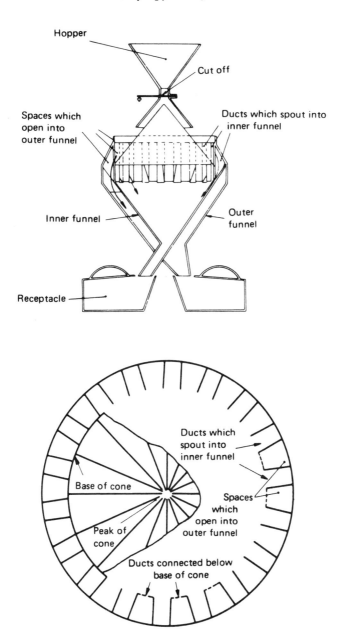

Fig. 11.8. Conical divider.

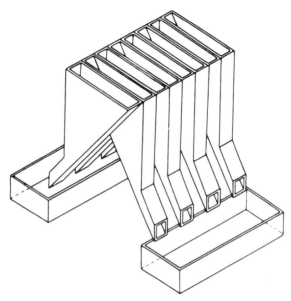

Fig. 11.9. Multiple-slot divider.

inadvertently misgraded. Corresponding auxiliary samples must not be mixed with the representative bulk or contract sample.

11.2.5.2 Drawing of primary samples

The samples must be drawn from products in bulk or in bags using the sampling apparatus mentioned previously.

Bags. Primary samples (increments) should be drawn from at least 2% of the bags forming the lot. Hand scoops or cylindrical samplers are appropriate in the case of opened bags. If the bags are closed spears or triers may be used (Figs 11.1–11.5).

Bulk. Sampling should preferably be done from the flow of the product. Care must be taken to ensure that the primary samples are drawn through the whole section of the seed and at time intervals depending on the rate of flow. The Ellis Cup is a manual sampling device designed to draw a sample from seed moving on a conveyor belt. The pelican (Fig. 11.6), a pouch attacked to a long pole, is designed to collect seed from a falling stream.

Of all approved sampling devices the diverter type mechanical sampler draws the most representative samples. Diverters vary in design, are powered electrically or pneumatically, and all operate on the same principle. Installed at the end of a conveyor belt or within a spout they draw their samples by periodically moving a pelican-like device through the seed stream. The frequency of these 'cuts' is regulated by timer controls. After the seed enters the primary sampler, it flows through a tube into a divider to reduce the sample size.

For sampling from cars, weigh scales, etc., the primary samples should be drawn at three levels from an adequate number of points:

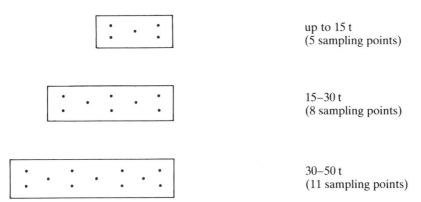

up to 15 t
(5 sampling points)

15–30 t
(8 sampling points)

30–50 t
(11 sampling points)

Especially the FGIS has issued elaborate patterns for sampling from cars.

One of the most common errors in sampling is the failure to obtain all required samples from an obviously non-uniform lot. Three samples must be drawn whenever a lot is obviously not uniform, namely a sample of the entire lot, of the inferior portion and of the remainder of the lot.

11.2.5.3 Contract/laboratory sample

The bulk sample is mixed and divided into the required number of contract samples, usually three, by the use of apparatus mentioned previously.

For some seeds, e.g. copra and unshelled groundnuts it is standard practice[1,2] to screen out fines and coarse admixture and to add back in proportion to the bulk sample.

11.2.5.4   Reduction of contract sample to test sample [7]

Prior to analysis the contract sample must be reduced in size by suitable dividing apparatus such as quartering irons, conical or multiple slot dividers which will ensure uniform distribution of the components of the contract sample in the analysis sample. Depending on the type of contract, or for example, EC regulation, large-sized impurities may have to be removed, and of course weighed, from the contract sample prior to preparing the analysis sample. This is not to be confused with the determination of fines, oleaginous and non-oleaginous impurities in oilseeds by sieving the analysis sample in accordance with for example, ISO 658–1980.[8] The actual method of contract sample reduction into test portions for extraction has been described in detail for palm kernels.[6]

11.2.5.5   Size of sample

Sample sizes given as guideline values in Table 11.1 are usually adequate. Quite often contract samples for large and medium-sized seeds do not exceed 4 kg (for lots of up to 500 t). According to general FOSFA recommendations copra samples sent to laboratories should not be less than 10 kg. Larger laboratory samples may be required especially when if the characteristic to be examined is distributed very heterogeneously, for example, as in the case of aflatoxins.

For EC-grown rapeseed and sunflower seed subject to intervention and subsidy rules the sample sizes in Table 11.2 apply.

Copra is especially difficult to sample representatively because of its size, its inhomogeneity, and handling and storage having a relatively

TABLE 11.1
Recommended sample sizes

Seed	Lot (t)	Primary sample (kg)	Bulk sample (kg)	Contract or laboratory sample (kg)	Test sample (c. min. cf. ISO664)[7] (g)
Large seeds	Up to 500	1	Up to 200	6	600 (1000 for copra)
Medium-sized seeds	Up to 500	0·5	Up to 100	5	500
Small seeds	Up to 500	0·1	Up to 20	2	200

TABLE 11.2
Recommended sample sizes for EC-grown rapeseed and sunflower seed subject to intervention and subsidy rules

	Lot	Test sample
Sunflower seed	Up to 1000 t	500 g
Rapeseed	Up to 500 t	200 g

large influence on subsequent analyses. For this reason FOSFA[9] have issued special instructions which are in line with the above general recommendations but draw attention to special sample storage conditions, etc.

11.2.6 Packaging and Labelling of Samples

When a seed contract sample is to be analysed for admixture, oil content and FFA in oil it is traditionally packed in a strong cotton or linen bag and sealed, or alternatively in a woven polypropylene bag within a sealed cotton or linen bag. IASC[1] and FOSFA[10] have issued standard procedures for the packaging and labelling of contract samples. In the case of process samples or samples not having a specific contract significance one can also use strong paper, paperboard containers, tins, glass jars and plastic containers.

Samples for the determination of moisture content, or for other analyses in which it is important to avoid loss of volatile matter (for example examination for evidence of chemical treatment), should be packed in moisture-tight containers. Glass jars are usually not recommended because of the risk of damage from breakage. Polythene or PVC jars with screw cap lids of the same material are eminently suitable. Such plastic containers also have an advantage over cotton or linen bags for samples in which the oil content is to be determined since this obviates duplicate determination of the moisture content which should always be simultaneously quoted together with the oil content, unless the oil content is quoted on dry matter. Apart from being used as a simple mathematical correction factor for the oil content in samples subject to longer periods of storage in moisture-transmitting containers the moisture content can itself have a direct influence on the amount of oil extracted in a standard test. Generally, the amount of extractable lipids decreases with decreasing moisture content; this is mainly due to phospholipids being more strongly retained in drier seed.

The points to be observed in connection with labels for samples, despatch of samples and sampling report have been dealt with under the general introduction (Section 11.1.6).

11.2.7 Special Cases of Heterogeneous Distribution

The distribution pattern of a mycotoxin within a commodity is governed by the nature of the mycotoxin,[11] the level of contamination, and the type of commodity. One of the problems associated with the sampling of agricultural products for aflatoxins is that contamination is generally restricted to a very small percentage of the sample. Generally, contamination is concentrated in seeds that have been damaged by insects and subsequently been stored at relatively high moisture levels.

It has, for example, been reported[12] that 12 kernels within a 2-kg sample of groundnuts contained aflatoxin at levels between 300 and 1 100 000 ppb, demonstrating that only *c.* 0·25% of the kernels were contaminated and that the level of contamination varied widely. It has been suggested that the highly localised contamination of groundnuts by aflatoxin can be represented by the negative binomial distribution.[12] Velasco and Whitaker[13] showed that in the case of cottonseed the aflatoxin content of 20 laboratory samples taken from one bulk sample varied between 0 and 400 ppb. Because of this extreme distribution representative sampling is difficult, variation among replicates tends to be great, and the true aflatoxin concentration in a given lot cannot be determined by working on one 2–5 kg bulk sample as provided for in the general recommendations for sampling and analysing oil seeds.

According to Horwitz[14] sampling contributes at least two thirds of the total variability of the aflatoxin content, subsampling of the ground sample about 20% and the actual analysis the remaining 13%. TLC is still the most commonly used procedure in routine mycotoxin analysis although HPLC may become of greater interest in view of its potential for automation. This topic is reviewed in greater depth by P. M. Scott in Chapter 4.

A practicable remedy of this sampling problem consists of taking not only more but also larger bulk samples. An alternative is the sampling of meal produced from the oilseed. Through the inevitable mixing and homogenisation during pre-expelling and extraction the concentration of mycotoxin in the meal is less heterogeneous than in the oilseed. Even so, distribution in the meal is still not homogeneous and special provision is made for this in ISO 5500[15] (see Section 11.3; oilseed residues). However, not all oilseed is converted into meal and there is now also EC-legislation[16] on maximum permissible aflatoxin concentrations for imported commodities such as cottonseed, babassu nuts, copra, groundnuts and palm kernels.

According to the USDA sampling plan[12] for bagged kernels a bulk sample of at least 66 kg is taken from each lot, irrespective of the lot size, by sampling every fourth bag. The 66 kg sample is then subdivided into 3×22 kg contract samples from which a 1100 g analysis sample is prepared by suitable subdivision. According to the so-called TPI plan for raw groundnut kernels, aimed at detecting those batches which contain 30 ppb aflatoxin, the bulk sample, obtained from 25% of the bags in a lot, is 10·5 kg which is then successively subdivided into $3 \times 3·5$ kg and test samples of 350 g and 100 g. For mycotoxins other than aflatoxins a similar procedure can be used.

It is important to inspect the laboratory sample visually before actually subdividing—preferably by grinding in a suitable hammer, disc or blade mill, followed by subsampling—and then analysing it. As stated earlier, mycotoxin-infested kernels are usually insect-damaged and hence holed and discoloured. This is the basis for screening kernels to be used, after roasting, for direct consumption as such or as peanut butter. High-speed sorting machines identify and separate discoloured kernels in a stream of kernels. It must also be pointed out, however, that kernels to be used in or as foodstuff are normally bought under special contracts and are of premium quality.

According to a provisional code of practice under the Dutch Food and Drugs Act as from Sept. 1988 four samples of 7·5 kg each are taken from a parcel of 20 t of hand-picked, selected groundnut kernels, each sample being taken from at least 5% of the number of bags. Each sample is analysed for the aflatoxin B1 content. The parcel is considered acceptable for human consumption when each result shows aflatoxin B1 present in a concentration of 3 μg/kg or less. The parcel must be rejected if one or more of the samples show an aflatoxin B1 content of more than 3 μg/kg. This sampling plan is based on the assumption that only one kernel out of 15 000 kernels is likely to be highly infested with aflatoxin B1.

11.3. OILSEED RESIDUES[15]

The methods of sampling oilseed residues are analogous to those for oilseeds. Generally, oilseed residues are more homogeneous than the corresponding oilseeds. For bagged products ISO 5500,[15] recommends the procedure in Table 11.3.

Bulk products should preferably be sampled whilst in motion. If automatic instruments are used, they should have a slot opening which is at least three times the size of the largest particles. Sampling of laden vehicles should be done from as many points as possible.

Table 11.4 gives an indication of the required sizes of the various sample types. Laboratory samples should be packed in bags of closely woven cloth, in polyethylene bags, or in tins. Samples for the determination of moisture or other volatile matter should be packed in airtight and watertight containers.

The test samples, usually c. 100 g, are prepared from the laboratory samples by grinding to c. 1 mm particle size and mixing.[17]

In the case of mycotoxins or poisonous seed residues suspected to be non-uniformly distributed, separate bulk samples have to be prepared for each lot and separate laboratory samples prepared from each bulk sample. ISO 5500/Appendix C[15] recommends the number of bulk samples given in Table 11.5.

TABLE 11.3
Recommended sample sizes for oilseed residues in bags

No. of bags	
In the lot	To be sampled
Up to 10	Each bag
11–100	10, sampled at random
More than 100	Approx. square root of the total No., sampled at random

TABLE 11.4
Recommended sample sizes for oilseed residues in bulk

Lot	Increment	Bulk sample	Laboratory sample
Up to 500 t	min. 100 g	10–50 kg	1–2 kg (in bags or tins)

TABLE 11.5
Recommended sampling scheme for analysis of mycotoxins or poisonous seed
residues in non–homogeneous products

No. of bags in lot	No. of bulk samples from bags Lot weight (tonnes)	Minimum no. of separate bulk samples
1–20	Up to and incl. 1	1
21–200	Over 1 upto and incl. 10	2
201–800	Over 10 upto and incl. 40	3
More than 800	Over 40 upto and incl. 500	4

11.4. OILS AND FATS

11.4.1 Type of Analyses

Oilseeds are generally processed by a combination of expelling and solvent extraction. With oleaginous fruits such as a palm fruit and olives mere heating or boiling with water suffices to liberate the oil. Animal fats are obtained by wet or dry rendering of fatty tissue.

Crude oils and fats contain minor components which are undesirable from the point of view of taste, appearance, further processing, etc., seed particles, inorganic dirt, phophatides, carbohydrates, fatty acids, trace metals, pigments, waxes, volatile and non-volatile oxidation products of fatty acid groups, in some instances trace components such as polycyclic aromatic hydrocarbons, gossypol, sulphur compounds, pesticide residues, and solvent—normally technical hexane—residues.

The aim of refining oils and fats is to remove such undesirable components without affecting the concentration of desirable constituents such as vitamins and polyunsaturated fatty acids and without significant loss of the major glyceride components.

Specifications for crude, semi-refined and fully refined oils and fats normally encompass the following analyses: FFA, moisture, particulate impurities, phosphatides, peroxide value, iodine value, colour, melting point, solid fat content, fatty acid composition; while specifications for fully refined products also include taste/odour, peroxide value, trace contaminants and oxidation tests.

If the oil or fat is completely liquid at the time of sampling most of these attributes are distributed homogeneously. Possible exceptions are moisture, particulate impurities and phosphatides. These tend to

settle to the bottom of unstirred containers in the case of crude
oils/fats ultimately leading to the formation of bottom sludges with
higher FFA levels. Caution must be exercised when determining
peroxide values since these can easily be higher near the surface of a
tank where the oil is saturated with atmospheric oxygen. The
determination of trace contaminants, such as polycyclic aromatic
hydrocarbons, pesticide residues and suspected residues from previous
cargoes of chemicals, entails the use of specially clean sample containers, preferably not made of plastic.

For oils and fats ready to be consumed the Codex Committee on
Oils and Fats has suggested that for *bulk* material a number of items
be individually procured and then combined to form *one* blended bulk
sample for analysis. This Codex Committee has also suggested that for
small 'retail' items, again in the light of experience in the processing
procedures of the oils and fats industry, it may be assumed that there
are no significant differences between items produced as part of one
lot. Consequently, for both additives and contaminants, as well as for
conventional commodity characteristics it normally suffices to analyse
one sample from bulked and homogenised items taken from a lot. The
Codex Committee has recommended to use the sampling procedures
given in ISO 5555–1983.[18]

These ISO procedures take into account that crude and semi-refined
oils and fats do not necessarily have the same degree of homogeneity
as refined oils and fats and hence may need special sampling
procedures. The ISO procedures are practically identical with the
corresponding recommendations issued by IASC, aimed at drawing
sufficiently representative samples from a consignment under practical
conditions.

Typical specifications for various oils and fats have been reviewed by
J. R. Pritchard.[5]

11.4.2 Transport/Storage Conditions

Since bulk oils and fats can deteriorate after the sample is taken,
adequate steps should be taken to preserve the initial quality.
Furthermore, sampling and inspection should study conditions for
storage. Samples taken and analyses of them should, among other
things, give guidance on whether the commodity is in a suitable
condition for long-term storage or transport. In view of this, it is
appropriate to review the influence of transport and storage conditions
on quality attributes.

If crude oils and fats are to be stored for a longer period of time or to be transported over larger distances it is advisable to remove insoluble impurities by sedimentation, filtration or centrifuging, to dry the oil to at least 0·2% moisture, and to avoid excessive heating.

Seed particles and cell fragments contain fat-splitting enzymes which in the presence of traces of moisture can lead to gradual hydrolysis of glycerides and a corresponding increase in free fatty acids, while protein itself is a nutrient for microorganisms. Phosphatides gradually become hydrated, even in the case of only traces of moisture being present, and the resulting wet sludges settle to the bottom of the tank, in turn accelerating the above mentioned hydrolysis reaction. Hydrolysis is also promoted by higher temperatures. Tanks should therefore always be clean and dry before use. Since free fatty acids themselves promote hydrolysis by increasing the solubility of water in the fatty phase the increase in acidity during transport and storage tends to increase with starting concentration of free fatty acids.

Quality deterioration can also be caused by oxidation. In the case of crude oils this generally means higher processing costs and mostly also an inferior quality of the corresponding raffinate. Oxidation is promoted by light, temperature, catalysts and concentration of oxygen in the oil. Oxidation can hence be minimised by storage without heating, or in the case of solid fats to not more than 10–15°C over the melting point, by avoiding frequent pumping, since this tends to saturate the oil with air, and by avoiding contact with rust. The addition of antioxidant and synergists is primarily effective with animal fats, most vegetable oils already containing sufficient concentrations of natural antioxidants. Most crude vegetable oils can be stored for some months without significant loss in quality if the above precautionary measures are observed. Animal fats can only be stored for relatively short periods of time without loss in quality. Continuous monitoring via acid value, peroxide value, and other data characterising the state of oxidation during storage is recommended.

Since natural antioxidants, such as tocopherols and phenols, are partially removed during the refining process semi- and fully-refined oils often show a lower oxidation stability than the corresponding crude oils. Bulk raffinates are best stored in closed containers with a relatively small ratio of surface area to volume in order to minimise access and infusion of air. Heating—if at all—should be done uniformly without local overheating, the temperature ideally not being higher than *c.* 10°C above the melting point. Aeration can be further reduced by saturating the oil with nitrogen immediately after deodorisation and by using a bottom-loading pipe.

The storage containers are preferably made of aluminium or stainless steel but mild steel can also be used if access of air can be widely reduced and if cooling is adequate. Raffinates can be kept at rest in large storage tanks for several weeks without the peroxide value beginning to rise; the peroxide value primarily tends to increase each time the oil is pumped. Raffinates are transported in bulk in coasters and trucks, in drums, tins, and bottles; solid fats can also be transported in block or powder form if the melting point is sufficiently high.

The Codex Alimentarius Committee for Oils and Fats has in 1987 issued a draft international code of practice for the storage and transport of edible oils and fats in bulk. This code is largely based on publications by Berger.[19] The most important features of this advisory code are outlined below, Other publications on the storage, handling and application of fats have also been considered.[20,21]

11.4.2.1 Storage installations and transport

Mode of transport. By far the greatest quantities of oils and fats are transported with *sea-going vessels, coasters and tank barges.* The vessel normally has separate compartments, with a capacity ranging from 200 to 1000 t, which should preferably be serviced each with their own pump so as to avoid cross-contamination.

The average size *road tankcar* carries up to 25 t of fat or oil in a tank of cylindrical or oval shape mounted on a chassis. The tank slopes towards the outlet and is normally lagged with mineral wool, the whole being covered with metal sheathing. Although the temperature will not normally change by more than 1°C/h during transport, the vehicles are normally fitted with steam pipes at the bottom of the tank in order to be able to deal with higher melting fats after a long journey.

Fat transported by *rail tank car* will usually arrive in a solid state and the steam lines in the tank will have to be connected to a steam supply at the point of reception.

A decreasing quantity of fat is being handled in *drums* because of the additional labour involved. If the contents must be melted before discharge an immersion heater or, preferably, an electric heating jacket can be used. Fat may also be supplied in plastic-lined *cardboard boxes* containing up to 25 kg each.

Tanks. The most suitable shape for land tanks is the vertical cylindrical tank with a convex roof. Where possible, relatively tall, narrow tanks are preferred to minimise exposed surface areas. Tanks should always be insulated. Tank bottoms should be conical or sloped to be self-draining. The size must be related to the expected storage period.

Tanks of mild steel should preferably be coated. Most mild steel tanks in modern ships are coated to prevent corrosion of the tank and contamination of the cargo. Damage to coatings can be caused by abrasion or by unsuitable cleaning methods leading to local corrosion. Tanks should always be inspected before use.

Copper and its alloys should not be used for any part of the installation. Gauges containing mercury should not be used. Glass equipment should be avoided in situations where breakage might lead to product contamination.

All tanks for solid or semi-solid products should be installed with heating facilities in order to obtain homogeneous products when they are unloaded. Heating by hot water (about 80°C) circulated through coils is a good procedure because it is least likely to cause local overheating. When heating with steam at a pressure of, for example, 1.5 kg/cm^2 gauge, corresponding to *c.* 127°C, a maximum heating rate of 5°C/24 h should be maintained if no provision exists for mixing or stirring.

Especially in temperate and cold climates tanks and tankcars should be lagged. This not only saves energy and helps to preserve oil quality but also reduces the risk of moisture condensation and hence corrosion.

Pipelines. Pipelines should reach the bottom of the tank; there should be a proper drain-out pipeline at the base of each tank so that it can be completely drained. Mild steel is acceptable for all crude and semi-refined oils and fats; 316 stainless steel should preferably be used for refined products. A pipeline pigging system should be provided.

All flexible hoses used to connect pipelines during loading and unloading must be of inert material and be suitably reinforced. Especially in cold climates pipelines should be lagged and be provided with heating, either by steam tracing lines or electrical heating tape.

TABLE 11.6
Recommended loading/unloading temperatures for edible oils (see also ISO 5555, annex A)

Products	Minimum °C	Maximum °C
Palm oil	50	55
Palm stearin	55	65
Palm olein	30	35
Palm kernel/coconut oil	40/35	45
Tallow	55	60
Greases	45	55
Lard	50	55
Fish oils	25	35
Liquid vegetable oils (e.g. soyabean, rapeseed)	15	30
Hardened oils	5 above slip melting point	15 above slip melting point

Loading and unloading. Solid and semi-solid products in tanks should be heated up slowly until they are liquid and completely homogeneous before transfer. The temperatures in Table 11.6 are guideline values for both crude and processed fats/oils and are designed to minimise damage to the oil or fat.

To prevent excessive crystallisation during short-term storage and shipping, oil in bulk tanks should be maintained within the temperature ranges given in Table 11.7.

Where a number of products have to be unloaded through a common pipeline, the system must be cleared completely between

TABLE 11.7
Recommended temperature during storage and transit of edible oils

Products	Minimum °C	Maximum °C
Palm oil	32	40
Palm olein	25	30
Palm stearin	40	45
Tallow	44	49
Fish oil	20	25
Palm kernel and coconut oil	27	32
Liquid vegetable oils	15	20
Hardened oils	Ambient	5 above slip point

different products or grades. The order of loading or discharge should be carefully chosen to minimise the chance of damage:

- Fully refined oils before partly refined.
- Partly refined oils before crude oils.
- Edible oils before technical grades.
- Fatty acids or acid oils last.

Contamination. Although the transport of non-edible products prior to edible oils and fats in the same tank or container is not strictly desirable it cannot yet be entirely avoided because of transport economics. It is virtually impossible in most cases to devise suitable analytical techniques to detect contamination with residues of previous cargoes if a record of the ship's or tankcar's log is not available. Such contamination can range from, for example, crude fish oil in palm oil to chemicals in an edible oil or fat. Whereas in the first case serious product defects in terms of keepability of the fully refined palm oil cannot be excluded, the second case can be far more serious in view of the potential health hazards involved. Even very small residues of some chemical which is highly reactive towards an oil or fat may cause considerable damage to the oil without this necessarily being detected by routine tests. For this reason the previous cargoes carried in a ship's tank should in time be restricted to toxicologically safe compounds and in any case be declared to the charterer of the tank and the records made available to all parties involved. Where tanks have been used for non-edible materials, the greatest care must be taken by cleaning and inspection (cleanliness should be certified!) that all residues have been totally removed.

Regular maintenance checks by suitably qualified staff should include condition of tank coating, hoses and other ancillary equipment.

Where doubts exist about cleanliness of pipes and manifolds, even after pigging or steam flushing, first runnings from the delivery line into a separate container should be drawn for inspection.[21] Tank sediments should also be kept separate from the bulk.

11.4.3 Sampling Devices[18]

Many forms and types of sampling instruments exist and only the most common ones are mentioned here. Sampling instruments must be inert to oils and fats. Copper or alloys containing copper should not be used. Other basic requirements are: ease of cleaning, practical size, robustness.

A *sample bottle or can* is suitable for sampling large vessels and tanks carrying liquid oil. It consists of a bottle or metal container attached to a handle long enough to reach the lowest part of the bulk to be sampled. It has a removable stopper with an attached chain. This

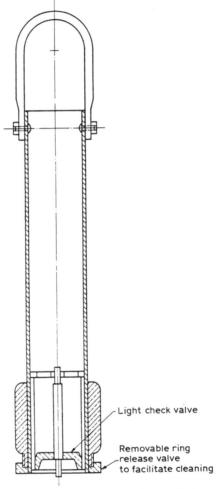

Fig. 11.10. Valve sampling cylinder.

device is lowered to the desired depth, where the stopper is pulled out and the container allowed to fill.

A *sampling tipping dipper* consists of a cylinder, normally *c.* 15 cm long and 5 cm diameter, carrying an extension with a hole at its closed end and a wire handle at the open end; the handle has a metal catch and a rope. The cylinder is inverted by inserting the catch into the hole, and then lowered into the tank. At the required depth, the rope is pulled to release the catch, whereupon the cylinder rights and fills itself.

A *valve sampling cylinder* (see Fig. 11.10) consists of an open-headed cylinder with a bottom valve which remains open owing to the pressure of the oil on the valve whilst the instrument is being lowered through the oil, ensuring that an even flow of oil passes into the cylinder. When lowering is stopped, the valve closes and a sample of the liquid is taken from the depth reached by the instrument.

A *bottom or zone sampler* is suitable for withdrawing samples at any level from tanks of liquid oil. Of the various designs one is shown in Fig. 11.11. It consists of a container with a valve on a central spindle. To withdraw a bottom sample, the apparatus is attached to a cord or chain and lowered empty to the bottom of the tank, where the valve

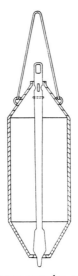

Fig. 11.11. Bottom sampler or zone sampler

automatically opens and the container fills from the bottom. On withdrawing the sampler, the valve automatically closes again. To withdraw a sample at a particular level, the apparatus is lowered empty to the required level and then, by means of an additional cord attached to the top of the central valve spindle, the valve is opened and the container allowed to fill. When the sampler is full, the valve is allowed to close and the container is withdrawn.

A *continuous average sampler* can be used to obtain an average sample in one operation. It consists of a tube with a central spindle attached at the bottom to a piston which has holes in its periphery. According to the position of the piston, these holes may be blocked or open to allow liquid to enter the tube. The sampler is lowered at a constant rate through the liquid with the piston in the 'open' position. The holes are automatically closed, and the piston is locked, when the instrument reaches the bottom of the tank; the tube is then withdrawn by means of a cord. The instrument can also be fitted with a device for remote control of the piston, enabling it to be used for bottom or zone sampling.

The *sampling tube* shown in Fig. 11.12 consists of two concentric tubes closely fitted into each other throughout their entire length, so that one tube can be rotated within the other. Longitudinal openings are cut into each tube. In one position the tube is open and admits the oil; by turning the inner tube it becomes a sealed container. The inner tube may be approximately 2–4 cm in diameter and may not be divided in its length. The two tubes are provided with V-shaped ports at their lower ends, so placed that oil contained in the instrument can be drained through them when the longitudinal openings are closed.

A *valve sampling tube* (see Fig. 11.13) consists of a metal tube, with a valve at the base connected by a central rod to a screwed handle at the top. When the handle is screwed down, the valve is kept closed. The tube is inserted into the oil with the valve open, allowing the oil to enter while the displaced air passes through an air hole at the top of the tube. When the base of the tube touches the bottom of the container, the valve automatically closes. The handle is then screwed tight, keeping the valve shut, and the tube containing the sample is withdrawn. Markings on the exterior of the valve sampling tube at intervals of length show the depth of liquid being sampled. Sampling tubes of various lengths are used, sufficiently long to reach the bottom and convenient for sampling tanks, wagons or cars.

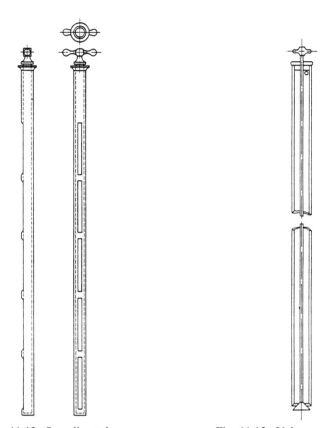

Fig. 11.12. Sampling tubes.   Fig. 11.13. Valve sampling tube.

A *compartmented valve sampling cylinder* (see Fig. 11.14) consists of a cylinder divided into a number of separate compartments by specially designed pistons fixed to a central spindle which can be rotated, raised or lowered in relation to the cylinder. The pistons are disposed at calculated intervals. Orifices, located in the cylinder wall of each compartment, can be opened or closed by lowering or raising the central spindle. The compartmented valve sampling cylinder is completely immersed in the tank with the orifices closed. These are then opened by rotating the central spindle and lowering it in relation to the cylinder. The orifices are then closed by the reverse procedure,

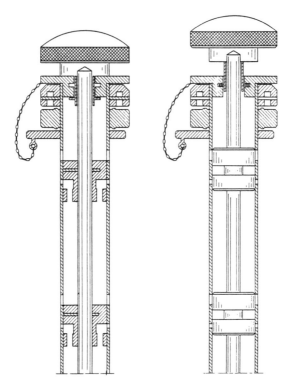

Fig. 11.14. Compartmented valve sampling cylinder.

and the sampling cylinder withdrawn. The apparatus is so designed that the volume of the sample taken in each compartment is equivalent to the volume of liquid in the tank at the level of the orifices, i.e. the proportion of the total sample taken into one compartment is equal to the proportion of the total volume of the tank occupied by the oil at the level of the orifices of that compartment. The accuracy of sampling is not affected if the tank is elliptical or if the apparatus is tilted inside the tank.

A *sampling trier* for liquid oils is divided into compartments along its length. It is of D-shaped cross-section and is opened and closed by means of a shutter which moves up and down throughout the entire length. It may be from 25 to 50 mm in diameter. The instrument is inserted closed, the shutter pulled out to admit oil, and the tube is then closed and withdrawn. Open triers are use for hard fats. They are of semi-circular or C-shaped cross-section, and drill out a core of fat from the fat.

For sampling oils during transfer a tap or so-called drip-cock is recommended. This method only works satisfactorily if the product is entirely liquid and contains no components which could cause blockage. The tap or drip-cock should be fed from a nozzle suitably large in diameter, fixed in the centre of the main discharge pipeline, and facing the flow of liquid. Taps let flush into the side or bottom of the pipeline are far less acceptable. The tap or drip-cock should be introduced into a horizontal section of the pipeline on the pressure side of the pump. Samples from very large bulk quantities can often be conveniently taken during transfer by means of frequent removals of material from the flow at regular intervals when the tank is being emptied. Alternatively, sampling may be carried out by means of a side stream tapped from the main stream. Here sampling must be done sufficiently long so that any initial surges of water and extraneous matter do not impair the representativeness of the bulked sample.

11.4.4 Methods of Sampling[18]

Sampling procedures depend on the types of containers, i.e. land tanks, ships tanks, rail and road tankcars, weigh scales, containers, pipelines, consignments in packages such as drums, tins, bottles.

11.4.4.1 Sampling from weigh scales, containers and vertical land tanks

The preliminary operations include a check as to whether there is a bottom layer of sediment or moisture. As far as possible, any water should be removed and quantified before sampling is begun. The whole product should be sampled as nearly liquid as possible. This may require heating the tank to the temperatures indicated in Section 11.4.2.1. If there is still doubt as to homogeneity the sampling procedure, for example using a zone sampler (Fig. 11.11) or a valve sampling cylinder (Fig. 11.10), must be adapted accordingly. ISO 5555 in such cases recommends taking increments at depths of every 30 cm, from top to bottom, until the deviating zone is reached. In this zone increments would have to be taken more frequently. Then three samples would have to be prepared as follows:

(a) a sample of the clear oil,
(b) a sample of the separated layer,

(c) a bulked composite sample in proportion to the respective sizes of the two layers.

In practice it is more common, certainly less open to controversy, to homogenise the tank contents before sampling. If the contents of the tank are relatively homogeneous, the same sampling instruments as above are used but in this case it usually suffices to take three increments only, namely one at $c.\ \frac{1}{10}$ depth of the contents level, one from the middle and one from $\frac{9}{10}$ depth. The bulk sample is prepared by blending one part each of the top and bottom increments with at least three parts from the middle.

A more accurate method is probably one which entails taking samples foot by foot, from surface to tank bottom, using a valve sampling tube. A representative composite sample is obtained by mixing equal quantities of the sub-samples.

11.4.4.2 Sampling from ships' tanks

The sampling procedure depends on whether the cargo is homogeneous or not. Usually, sampling is done during transfer (see Section 11.4.4.4). If this is not possible, the procedure outlined under Section 11.4.4.1 applies. Each tank must be sampled separately. The number of bulk samples to be prepared as a function of the size of the consignment has been advised in ISO 5555[18] as in Table 11.8.

11.4.4.3 Sampling from road and rail tankcars, horizontal, cylindrical and oval tanks

The procedure to be adopted depends on whether the oil is homogeneous or not. Corrective measures must be taken to ensure

TABLE 11.8
Recommended number of bulk samples to be taken from each ship or land tank

Mass of tank contents (t)	Number of bulk samples to be taken
Up to and including 500	1
Over 500 up to and including 1 000	2
Over 1 000	1 for every 500 t or part thereof

that the volume of increments taken is truly proportionate to the cross-sectional area of the tank from which the sample is taken.

If the material is homogeneous increments from the zones 1, 2, 3, 4 and 5 in the case of a cylindrical tank or from zones 1, 2, 3 and 4 in the case of an oval tank as shown in Fig. 11.15 are taken with, for example, a zone sampler (Fig. 11.11) or a valve sampling cylinder (Fig. 11.10) and blended into a bulk sample according to the comparative area in the column at the right of the diagrams.

If the material is heterogeneous this procedure has to be modified in the same manner as that described under Section 11.4.4.1. In addition to the increments from the zones 1–5 or 1–4 extra increments are taken from the deviating zone and blended into a representative average sample for this zone.

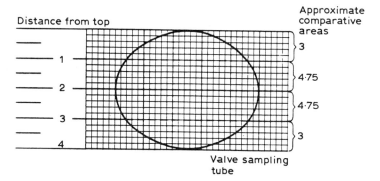

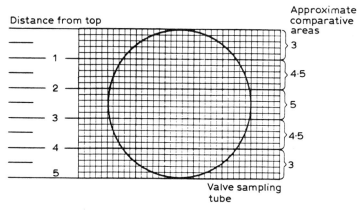

Fig. 11.15. Cross-sections of typical tanks. (a) oval cross–section horizontal tank; (b) cylinder cross–section horizontal tank.

11.4.4.4 Sampling during transfer

This method obviously only applies if the material is pumpable and contains no large particulate components. For taking a pipeline sample a special tap or drip cock is used. The oil should be collected in a clean container and the time recorded for a particular increment so that the correct proportional quantities from the increments can be taken to compose the bulk sample. A separate increment should be taken at the beginning of the sampling operation; the quantity should be determined by weighing, measuring or gauging and the proportion required for the bulk sample estimated (Table 11.9).

Sampling in the context of process control, e.g. from a neutraliser/bleacher vessel, can be done by pumping a sample into a closed loop between the vessel and the sampling point. Such loops can be equipped with tracing and air flushing and are in principle also capable of providing samples under vacuum conditions.

11.4.4.5 Sampling from packaged consignments

If a consignment consists of a relatively large number of separate units, for example drums, cases, tins, bottles or paper bags it will in most cases not be possible to sample each separate unit. A number of units are therefore chosen at random from the consignment on the assumption that the consignment is reasonably homogeneous. The corresponding bulk sample is made by blending the various increments. Recommendations for the number of packages to be selected for sampling are given in Table 11.10.[18] These recommendation are of an order of magnitude corresponding to inspection levels of 3–4 out of a maximum of 5 (see ISO 3951, 1981, Table 3).[3]

To obtain increments from solid fats an open trier may be used. It is inserted through the opening of the drum, probing the whole depth of the contents in as many directions as possible. Soft pastes and

TABLE 11.9
Recommended minimum size of bulk samples taken during transfer

Minimum size of bulk sample	
Mass of tank contents (t)	Minimum size of bulk samples (1)
Up to and including 20	1
Over 20 up to and including 50	5
Over 50 up to and including 500	10

TABLE 11.10
Number of items to be sampled in consignments packaged in separate units

Size of packages	Number of packages in the consignment	Number of packages to be sampled
Over 20 kg up to 5 t	1–5	All
maximum	6–50	6
	51–75	8
	76–100	10
Over 5 kg up to and	1–20	All
including 20 kg	21–200	20
	201–800	25
	801–1 600	35
	1 601–3 200	45
	3 201–8 000	60
	8 001–16 000	72
	16 001–24 000	84
	24 001–32 000	96
	More than 32 000	106
Up to and including	1–20	All
5 kg	21–1 500	20
	1 501–5 000	25
	5 001–15 000	35
	15 001–35 000	45
	35 001–60 000	60
	60 001–90 000	72
	90 001–130 000	84
	130 001–170 000	96
	More than 170 000	108

semi-liquid products are similarly sampled using a chamber trier or a valve sampling tube (see Fig. 11.13).

For larger packages containing liquid fats and oils increments are taken with sampling tubes, valve sampling tubes or, for example, chamber triers. In the case of smaller bottles and tins it is obviously more convenient to mix the whole contents of all selected packages and to take a bulk sample from this.

If fats, for example, flaked or powdered hydrogenated fats, are in a loose lumpy condition a representative sample may be obtained by collecting, from different parts, sufficient amounts of different sized lumps, crushing these and reducing the sample by quartering.

11.4.4.6   Preparation of laboratory (contract) samples

According to Table 11.8 one bulk sample (*c.* 1–2 kg) should be available for every 500 t or part thereof of a particular consignment. Laboratory (contract) samples can be prepared either from the combined bulk samples or from each of the bulk samples. It is common practice to adopt the first approach. For contract purposes 3–4 laboratory samples of 250–500 g each are prepared (one for buyer, one for seller, two in reserve for possible arbitration).

11.4.5   Packaging and Labelling of Oil Samples

Contract samples should be packed in clean, dry, airtight containers, preferably made of glass and filled to the top. If the product has to be heated for filling purposes the recommended temperature limits referred to under Section 11.4.2.1 should not be exceeded. If oxidation criteria such as peroxide value are to be determined it is advisable to use brown glass. Glass jars or wide-necked bottles should be closed with screw caps; care should be taken to avoid plastic lining if the sample is to be examined for trace contaminants such as plasticisers, plastic monomers, and other compounds possibly present from previous cargoes.

Whatever the method of closure, for contract analyses the container should be sealed in such a way that the container, the closure, and the label form one unit to prevent any tampering. All sample containers must carry a label. This label should at least feature the following information:

—place and date of sampling,
—name of oil and quantity of parcel,
—type of sample (i.e. tank, pipeline, etc.),
—adequate identification of the vehicle in which the consignment was transported and of the tank in which it was stored,
—whether the consignment was in packages; total number of packages; percentage/number of packages selected for bulk sample.

For arbitration or similar purposes the samples should be kept for at least 3 months protected against direct daylight and in a cool place (*c.* 20°C).

11.5. LECITHINS

11.5.1 Applications/Required Analyses

Phospholipids are essential constituents of the protoplasm of animal and plant cells. Oilseeds are one of the richest phospholipid sources.

During prerefining of crude vegetable oils, especially soyabean and rapeseed oil, the major part of the phosphatides is removed as sludge by hydration with water followed by settling or centrifuging. Drying of this sludge gives 'lecithin' which is used, possibly after further modification, in the edible sector as emulsifier, anti-spattering agent dispersant, mould release agent, viscosity reducing agent, etc. It also finds application in pharmaceuticals, toilet preparations, animal feedstuffs, and special technical products.[22]

A typical composition of soyalecithin is as follows:

Phosphatidyl choline (lecithin)	14%
Phosphatidyl ethanolamine (cephalin)	13%
Phosphatidyl inositol	10%
Phosphatidic acid	8%
Other phospholipids	7%
Polar, non-phosphatidic lipids	6%
Sugars	7%
Neutral oil	35%

The following analytical data normally cover the various lecithin specifications:

—oil content (i.e. 100% minus acetone-insoluble part),
—toluene-insoluble matter (a measure of the degree of contamination with particulate dirt),
—acid value (a measure of the degree of intended or unintended hydrolysis),
—moisture (normally max. 1%),
—peroxide value,
—colour,
—viscosity,
—emulsifying capacity (a variety of tests is available),
—bacteriology.

11.5.2 Transport/Storage

Under dry conditions and at ambient temperatures lecithin can be stored for several months without loss in quality or change in analytical data.

Lecithin swells in the presence of water, leading to flocculation of highly viscous particles, and ultimately to phase separation.

The viscosity of lecithin decreases markedly with temperature. After storage in, for example, drums at ambient temperature for a longer period of time it is advisable to gently heat up to c. 40°C, hence lowering the viscosity to enable representative sampling and adequate emptying of the containers.

Lecithin is normally transported in bulk at c. 40–50°C. For unloading and loading it is advisable to use a positive displacement pump of adequate capacity. Storage temperature should preferably be maintained at 50–60°C; significantly higher temperatures over prolonged periods of time can lead to discoloration.

The storage tank, preferably of stainless steel or at least suitably coated, must be absolutely dry, a vertical cylindrical type with a conical bottom generally being recommended. The tank should be insulated to prevent condensation of water which would lead to phase separation (see above) and in turn promote bacterial growth. The tank contents should be agitated intermittently to prevent thixotropically induced viscosity increases and to safeguard homogeneity. Sufficient agitation is normally achieved by a so-called 'recirculation system'; a frame type agitator is a good alternative.

Contamination with other cargoes must be strictly avoided since the possibilities of rectifying any flavour or odour defects by refining are very limited in view of lecithin beginning to decompose at temperatures above c. 100°C and reacting with acids and bases.

11.5.3 Methods of Sampling

These are analogous to those described under Section 11.4.4. Obviously, greater care than with oils and fats must be taken to establish whether phase separation has occurred and to take this into account if the contents cannot be adequately homogenised before sampling. If free water is visible it is likely that the bacteriological quality has suffered.

If a consignment consists of a relatively large number of separate, smaller packages the previous recommendations (ISO 5555;[18] see also Section 11.4.4.5) as to the number of packages to be sampled apply. It is common practice to sample a number of drums equivalent to the square root of the total number of drums in a consignment of up to 50 t.

11.5.4 Samples
The packaging, labelling and general handling is analogous to that for oils and fats.

11.6. PREPACKED GOODS

Examples of prepacked goods in this context can be oil in bottles or tins, spreads and margarine in packs or tubs, peanut butter in jars, chocolate bars in wrappers, etc. For the purpose of sampling each package constitutes an item or increment. The determination of compositional characteristics will in most cases be the purpose of sampling and subsequent analysis. As already outlined in the general

TABLE 11.11

Number of items to be inspected from lots of different sizes and at different inspection levels for variables acceptance sampling plans with unknown standard deviation

Size of lot (number of items)		Number of items to be inspected				
		Inspection level				
		1	2	3	4	5
Up to	15	3	3	3	3	5
	16–25	3	3	3	4	7
	26–50	3	3	4	5	10
	51–90	3	3	5	7	15
	91–150	3	4	7	10	20
	151–280	3	5	10	15	25
	281–400	4	7	15	20	35
	401–500	4	7	15	25	35
	501–1 200	5	10	20	35	50
	1 201–3 200	7	15	25	50	75
	3 201–10 000	10	20	35	75	100
	10 001–35 000	15	25	50	100	150
Over	35 000	20	35	75	150	200

Selected from ISO 3951: 1981.[3]

introduction (see especially Sections 11.1.3 and 11.1.4) one can assume that the variable in question will be distributed uniformly throughout the consignment. For these types of foods this normally also applies to microorganisms. Consequently, individual items taken at random according to a 'variables' sampling plan can be blended into one bulk sample and need not be analysed separately. As a general guide to the number of packages to be sampled one can use tables from ISO 3951:1981.[3]

ACKNOWLEDGEMENTS

The author wishes to thank the International Organisation for Standardisation for permission to reproduce the figures. These have been taken from ISO Standards 542 (1920) and 5555 (1983). The complete standards may be obtained from the International Organisation for Standardisation, Case postale 56 CH-1211, Geneve 20 Switzerland, and from many National Standardisation Organisations.

Thanks are also due to the Management of Unimills International for their encouragement and support.

REFERENCES

1. *Oilseeds, oils and fats*, 3rd edn, IASC Sub-Committee, International Association of Seed Crushers, Salisbury Square House, 8 Salisbury Square, London EC4P 4AN, April 1980.
2. International Organisation for Standardisation, *Oilseeds sampling*, ISO 542: 1980 (see also revision in ISO/DIS 542 submitted on 24 November 1988).
3. International Organisation for Standardisation, *Sampling procedures and charts for inspection by variables for percent defective*, ISO 3951:1981.
4. Steiner, E. H., Statistical methods in quality control, In: *Quality control in the food industry*, Vol. 1, ed. S. M. Herschdoerfer, Academic Press, London, 1984, pp. 222–3.
5. Pritchard, J. R., *J. Am. Oil Chem. Soc.*, **60** (1983) 322–32.
6. Kershaw, S. J. and Hardwick, J. F., *Oléagineux*, **40** (1985) 397–406.
7. International Organisation for Standardisation, *Oilseeds—reduction of contract samples to analysis samples*, ISO 664: 1977.
8. International Organisation for Standardisation, *Oilseeds—determination of impurities content*, ISO 658: 1980.
9. FOSFA, Official Method, Instructions for Sampling at Port of Discharge and for Analysis of Samples, 1982.

10. FOSFA, Official Method, Packaging of Oil Seed Samples, 1982.
11. Bender, A. E., Health problems in food: chemical aspects, In: *Quality control in the food industry,* Vol. 1, ed. S. M. Herschdoerfer, Academic Press, London, 1984, pp. 38–9.
12. Coker, R. D., High performance liquid chromatography and other chemical quantification methods used in the analysis of mycotoxins in foods, In: *Analysis of food contaminants,* ed. J. Gilbert, Elsevier Applied Science Publishers, London, New York, 1984, pp. 215–23.
13. Velasco, J. and Whitaker, T. B., *J. Am. Oil Chem. Soc.,* **52** (1975) 191–4.
14. Horwitz, W., Sampling and preparation of sample for chemical examination, paper presented at the AOAC Symposium, Scottsdale, Arizona, Sept. 17, 1986.
15. International Organisation for Standardisation, *Oilseed residues—sampling,* ISO 5500: 1986.
16. EC directive 86/354/EEC of 21 June 1986.
17. International Organisation for Standardisation, *Oilseed residues—preparation of test samples,* ISO 5502: 1983.
18. International Organisation for Standardisation, *Animal and vegetable fats and oils—sampling,* ISO 5555: 1983.
19. Berger, K. G., *J. Am. Oil Chem. Soc.,* **62** (1985) 438–42.
20. Atkinson, R. E., *Storage, handling and application of fats,* National Renderers Assocation, London, 1986.
21. Subramaniam, S., *J. Am. Oil Chem. Soc.,* **62** (1985) 443–8.
22. Szuhaj, M. and List, G., In *Lecithins,* American Oil Chemists Society monograph, 1985.

Index